Novel Aspects of Lipoprotein Metabolism with Focus on Systemic Inflammation

Novel Aspects of Lipoprotein Metabolism with Focus on Systemic Inflammation

Special Issue Editor
Robin P.F. Dullaart

MDPI • Basel • Beijing • Wuhan • Barcelona • Belgrade

Special Issue Editor
Robin P.F. Dullaart
Department of Endocrinology,
University of Groningen and
University Medical Center
Groningen
The Netherlands

Editorial Office
MDPI
St. Alban-Anlage 66
4052 Basel, Switzerland

This is a reprint of articles from the Special Issue published online in the open access journal *Journal of Clinical Medicine* (ISSN 2077-0383) from 2019 to 2020 (available at: https://www.mdpi.com/journal/jcm/special_issues/lipoprotein_metabolism_inflammation).

For citation purposes, cite each article independently as indicated on the article page online and as indicated below:

LastName, A.A.; LastName, B.B.; LastName, C.C. Article Title. *Journal Name* **Year**, *Article Number*, Page Range.

ISBN 978-3-03928-214-2 (Pbk)
ISBN 978-3-03928-215-9 (PDF)

© 2020 by the authors. Articles in this book are Open Access and distributed under the Creative Commons Attribution (CC BY) license, which allows users to download, copy and build upon published articles, as long as the author and publisher are properly credited, which ensures maximum dissemination and a wider impact of our publications.

The book as a whole is distributed by MDPI under the terms and conditions of the Creative Commons license CC BY-NC-ND.

Contents

About the Special Issue Editor vii

Preface to "Novel Aspects of Lipoprotein
Metabolism with Focus on
Systemic Inflammation" ix

Nicolas Vuilleumier, Sabrina Pagano, Christophe Combescure, Baris Gencer, Julien Virzi, Lorenz Räber, David Carballo, Sebastian Carballo, David Nanchen, Nicolas Rodondi, Stephan Windecker, Stanley L. Hazen, Zeneng Wang, Xinmin S. Li, Arnold von Eckardstein, Christian M. Matter, Thomas F. Lüscher, Roland Klingenberg and Francois Mach
Non-Linear Relationship between Anti-Apolipoprotein A-1 IgGs and Cardiovascular Outcomes in Patients with Acute Coronary Syndromes
Reprinted from: *J. Clin. Med.* **2019**, *8*, 1002, doi:10.3390/jcm8071002 1

Nicolas Vuilleumier, Sabrina Pagano, Fabrizio Montecucco, Alessandra Quercioli, Thomas H. Schindler, François Mach, Eleonora Cipollari, Nicoletta Ronda and Elda Favari
Relationship between HDL Cholesterol Efflux Capacity, Calcium Coronary Artery Content, and Antibodies against ApolipoproteinA-1 in Obese and Healthy Subjects
Reprinted from: *J. Clin. Med.* **2019**, *8*, 1225, doi:10.3390/jcm8081225 15

Sabrina Pagano, Alessandra Magenta, Marco D'Agostino, Francesco Martino, Francesco Barillà, Nathalie Satta, Miguel A. Frias, Annalisa Ronca, François Mach, Baris Gencer, Elda Favari and Nicolas Vuilleumier
Anti-ApoA-1 IgGs in Familial Hypercholesterolemia Display Paradoxical Associations with Lipid Profile and Promote Foam Cell Formation
Reprinted from: *J. Clin. Med.* **2019**, *8*, 2035, doi:10.3390/jcm8122035 31

Josephine L.C. Anderson, Sabrina Pagano, Julien Virzi, Robin P.F. Dullaart, Wijtske Annema, Folkert Kuipers, Stephan J.L. Bakker, Nicolas Vuilleumier and Uwe J.F. Tietge
Autoantibodies to Apolipoprotein A-1 as Independent Predictors of Cardiovascular Mortality in Renal Transplant Recipients
Reprinted from: *J. Clin. Med.* **2019**, *8*, 948, doi:10.3390/jcm8070948 51

Sigurd E. Delanghe, Sander De Bruyne, Linde De Baene, Wim Van Biesen, Marijn M. Speeckaert and Joris R. Delanghe
Estimating the Level of Carbamoylated Plasma Non-High-Density Lipoproteins Using Infrared Spectroscopy
Reprinted from: *J. Clin. Med.* **2019**, *8*, 774, doi:10.3390/jcm8060774 62

Lídia Cedó, Srinivasa T. Reddy, Eugènia Mato, Francisco Blanco-Vaca and Joan Carles Escolà-Gil
HDL and LDL: Potential New Players in Breast Cancer Development
Reprinted from: *J. Clin. Med.* **2019**, *8*, 853, doi:10.3390/jcm8060853 71

Laura G.M. Janssen, Matti Jauhiainen, Vesa M. Olkkonen, P.A. Nidhina Haridas, Kimberly J. Nahon, Patrick C.N. Rensen and Mariëtte R. Boon
Short-Term Cooling Increases Plasma ANGPTL3 and ANGPTL8 in Young Healthy Lean Men but Not in Middle-Aged Men with Overweight and Prediabetes
Reprinted from: *J. Clin. Med.* **2019**, *8*, 1214, doi:10.3390/jcm8081214 92

James P. Corsetti, Charles E. Sparks, Richard W. James, Stephan J. L. Bakker and Robin P. F. Dullaart
Low Serum Paraoxonase-1 Activity Associates with Incident Cardiovascular Disease Risk in Subjects with Concurrently High Levels of High-Density Lipoprotein Cholesterol and C-Reactive Protein
Reprinted from: *J. Clin. Med.* **2019**, *8*, 1357, doi:10.3390/jcm8091357 103

Luca Liberale, Federico Carbone, Giovanni G. Camici and Fabrizio Montecucco
IL-1β and Statin Treatment in Patients with Myocardial Infarction and Diabetic Cardiomyopathy
Reprinted from: *J. Clin. Med.* **2019**, *8*, 1764, doi:10.3390/jcm8111764 115

Hanna Wessel, Ali Saeed, Janette Heegsma, Margery A. Connelly, Klaas Nico Faber and Robin P. F. Dullaart
Plasma Levels of Retinol Binding Protein 4 Relate to Large VLDL and Small LDL Particles in Subjects with and without Type 2 Diabetes
Reprinted from: *J. Clin. Med.* **2019**, *8*, 1792, doi:10.3390/jcm8111792 133

Erwin Garcia, Maryse C. J. Osté, Dennis W. Bennett, Elias J. Jeyarajah, Irina Shalaurova, Eke G. Gruppen, Stanley L. Hazen, James D. Otvos, Stephan J. L. Bakker, Robin P.F. Dullaart and Margery A. Connelly
High Betaine, a Trimethylamine N-Oxide Related Metabolite, Is Prospectively Associated with Low Future Risk of Type 2 Diabetes Mellitus in the PREVEND Study
Reprinted from: *J. Clin. Med.* **2019**, *8*, 1813, doi:10.3390/jcm8111813 149

Gemma Chiva-Blanch and Lina Badimon
Cross-Talk between Lipoproteins and Inflammation: The Role of Microvesicles
Reprinted from: *J. Clin. Med.* **2019**, *8*, 2059, doi:10.3390/jcm8122059 163

Kayla A. Riggs, Parag H. Joshi, Amit Khera, Kavisha Singh, Oludamilola Akinmolayemi, Colby R. Ayers and Anand Rohatgi
Impaired HDL Metabolism Links GlycA, A Novel Inflammatory Marker, with Incident Cardiovascular Events
Reprinted from: *J. Clin. Med.* **2019**, *8*, 2137, doi:10.3390/jcm8122137 175

Eline H. van den Berg, Eke G. Gruppen, Hans Blokzijl, Stephan J.L. Bakker and Robin P.F. Dullaart
Higher Sodium Intake Assessed by 24 Hour Urinary Sodium Excretion Is Associated with Non-Alcoholic Fatty Liver Disease: The PREVEND Cohort Study
Reprinted from: *J. Clin. Med.* **2019**, *8*, 2157, doi:10.3390/jcm8122157 184

Yijing Yu, Fitore Raka and Khosrow Adeli
The Role of the Gut Microbiota in Lipid and Lipoprotein Metabolism
Reprinted from: *J. Clin. Med.* **2019**, *8*, 2227, doi:10.3390/jcm8122227 200

Peter R. van Dijk, Amaal Eman Abdulle, Marian L.C. Bulthuis, Frank G. Perton, Margery A. Connelly, Harry van Goor and Robin P.F. Dullaart
The Systemic Redox Status Is Maintained in Non-Smoking Type 2 Diabetic Subjects Without Cardiovascular Disease: Association with Elevated Triglycerides and Large VLDL
Reprinted from: *J. Clin. Med.* **2020**, *9*, 49, doi:10.3390/jcm9010049 220

About the Special Issue Editor

Robin P.F. Dullaart, MD, PhD, is Associate Professor of Endocrinology at the University of Groningen, The Netherlands. R.P.F. Dullaart has held the position of Chair of Clinical Education and Training Program of Clinical Endocrinology and Diabetology. He has served as Course Director of the International Bachelor of Medicine Groningen Program. He was a member of the board of the European Society of Clinical Investigation for more than 20 years, most recently serving as President of the European Society of Clinical Investigation Trust Foundation.

The research of R.P.F. Dullaart is focused on lipid metabolism and the role of lipoprotein subfractions therein, low grade inflammation including recently developed nuclear magnetic spectroscopy glycoprotein inflammation markers, HDL functionality, and on branched chain amino acids as predictors of diabetes, NAFLD, and hypertension. R.P.F. Dullaart has authored and co-authored 425 publications. His work was been cited over 8400 times and his h-index is 46.

Preface to "Novel Aspects of Lipoprotein Metabolism with Focus on Systemic Inflammation"

Twenty-three papers were submitted to this Special Issue of JCM. Two were considered not suitable for the topic of this Special Issue, one paper was withdrawn by the authors, and four papers were rejected. As a result, this Special Issue contains 16 outstanding papers with contributors from Europe and North America.

Four research papers studied the role of autoantibodies against apolipoprotein A-I (apo A-I) in cardiovascular disease. Anti apoA-I autoantibodies i) were shown to predict a new adverse atherosclerotic event in patients who had experienced a coronary event, ii) were observed to predict cardiovascular mortality in renal transplant patients, iii) were positively related to macrophage cholesterol content but were paradoxically also associated with ABCA1-mediated cholesterol efflux capacity in vitro, and iv) were found to induce foam cell formation, but were unexpectedly associated with lower LDL cholesterol in children with familial hypercholesterolemia.

The other original research papers dealt with a variety of topics. At a glance: A North European population-based study on incident cardiovascular disease reported that low serum activity of the antioxidative enzyme paraoxonase-1 predicts incident cardiovascular events in subjects with concurrently high HDL cholesterol and hsCRP levels. Systemic oxidative stress, as inferred from higher plasma levels of free thiols, was found to be unaltered in well-controlled non-smoking type 2 diabetic patients; higher free thiol levels were paradoxically associated with higher triglycerides and greater numbers of large VLDL particles. In a multiethnic US cohort, higher plasma levels of GlycA—a nuclear magnetic resonance-based glycoprotein marker of systemic inflammation—was found to predict incident cardiovascular disease which was, in part, attributable to impaired HDL-mediated cholesterol efflux capacity.

The topics of the four review papers dealt with i) the emerging role of oxidatively modified lipoproteins and HDL in breast cancer development, ii) the importance of microvesicles, in concert with oxidized LDL, on the onset and progression of atherothrombosis by promoting vascular inflammation, iii) the dual role of statin treatment to ameliorate low-grade inflammation in addition to its LDL cholesterol lowering effects, and the potential benefit of combining statin therapy with canakinumab, an interleukin-1β inhibitor, and iv) the influence of the gut microbiome on bile acid metabolism, short chain fatty acid signaling, and endocrine cell function via GLP-1/GLP-2, thereby affecting an important pathway for postprandial lipid metabolism.

We hope that the items covered in this Special Issue of JCM are relevant for your future research. We hope that you thoroughly enjoy reading this Special Issue!

Robin P.F. Dullaart
Special Issue Editor

Article

Non-Linear Relationship between Anti-Apolipoprotein A-1 IgGs and Cardiovascular Outcomes in Patients with Acute Coronary Syndromes

Nicolas Vuilleumier [1,2], Sabrina Pagano [1,2,*], Christophe Combescure [3,4], Baris Gencer [5], Julien Virzi [1,2], Lorenz Räber [6], David Carballo [5], Sebastian Carballo [5], David Nanchen [7], Nicolas Rodondi [8,9], Stephan Windecker [6], Stanley L. Hazen [10,11], Zeneng Wang [10], Xinmin S. Li [10], Arnold von Eckardstein [12], Christian M. Matter [13,14], Thomas F. Lüscher [14,15], Roland Klingenberg [13,14,†] and Francois Mach [5,†]

1. Division of Laboratory Medicine, Diagnostic Department, Geneva University Hospital, 1205 Geneva, Switzerland
2. Department of Internal Medicine specialities, Medical Faculty, Geneva University, 1205 Geneva, Switzerland
3. Division of Clinical Epidemiology, University Hospital of Geneva, 1205 Geneva, Switzerland
4. Center of Clinical Research, University of Geneva, 1205 Geneva, Switzerland
5. Division of Cardiology, Cardiology Center, Geneva University Hospital, 1205 Geneva Switzerland
6. Department of Cardiology, Bern University Hospital, University of Bern, 3010 Bern, Switzerland
7. Center for Primary Care and Public Health (Unisanté), University of Lausanne, 1015 Lausanne, Switzerland
8. Institute of Primary Health Care (BIHAM), University of Bern, 3010 Bern, Switzerland
9. Department of General Internal Medicine, Inselspital, Bern University Hospital, University of Bern, 3010 Bern, Switzerland
10. Department of Cardiovascular and Metabolic Sciences, Lerner Research Institute, Cleveland Clinic, Cleveland, OH 44195, USA
11. Department of Cardiovascular Medicine, Heart and Vascular Institute, Cleveland Clinic, Cleveland, OH 44195, USA
12. Institute of Clinical Chemistry, University Hospital of Zurich, 8091 Zürich, Switzerland
13. Department of Cardiology, University Heart Center, University Hospital of Zurich, 8091 Zürich, Switzerland
14. Center for Molecular Cardiology, University Heart Center, University Zurich, 8091 Zürich, Switzerland
15. Heart Division, Royal Brompton and Harefield Hospital and Imperial College, Uxbridge UB9 6JH, London, UK
* Correspondence: Sabrina.Pagano@hcuge.ch
† These authors contributed equally to this work.

Received: 14 May 2019; Accepted: 8 July 2019; Published: 9 July 2019

Abstract: Autoantibodies against apolipoprotein A-I (anti-apoA-I IgGs) are prevalent in atherosclerosis-related conditions. It remains elusive whether they improve the prognostic accuracy of the Global Registry of Acute Coronary Events (GRACE) score 2.0 (GS) in acute coronary syndromes (ACS). In this prospective multicenter registry, 1713 ACS patients were included and followed for 1 year. The primary endpoint (major adverse cardiovascular events (MACE)) was defined as the composite of myocardial infarction, stroke (including transient ischemic attack), or cardiovascular (CV) death with individual events independently adjudicated. Plasma levels of anti-apoA-I IgGs upon study inclusion were assessed using ELISA. The association between anti-apoA-I IgGs and incident MACE was assessed using Cox models with splines and C-statistics. One-year MACE incidence was 8.4% (144/1713). Anti-apoA-I IgG levels were associated with MACE with a non-linear relationship ($p = 0.01$), which remained unchanged after adjusting for the GS ($p = 0.04$). The hazard increased progressively across the two first anti-apoA-I IgG quartiles before decreasing thereafter. Anti-apoA-I IgGs marginally improved the prognostic accuracy of the GS (c-statistics increased from 0.68 to 0.70). In this multicenter study, anti-apoA-I IgGs were predictive of incident MACE in ACS independently of the GS but in a nonlinear manner. The practical implications of these findings remain to be defined.

Keywords: acute coronary syndrome; biomarkers; anti-apolipoprotein A-I autoantibodies; GRACE score; C-statistics

1. Introduction

Over the last decades, the insight into the pathogenesis of atherosclerosis evolved from a lipid-centered disease to a predominantly T-helper (Th)-1 driven immune response against various antigens and auto-antigens within atherosclerotic plaques [1–3]. Among the latter, both cellular and humoral autoimmune responses against different membrane proteins, native or modified lipoproteins have been shown to modulate the course of atherogenesis either in a pro- or anti-atherogenic manner [4,5]. Reflecting this intrinsic biological complexity, several autoantibodies have been detected in sera of individuals and associated either with an increased or decreased cardiovascular (CV) risk [6–10].

Among autoantibodies targeting lipoproteins potentially associated with increased CV risk, those directed against apolipoprotein A-I (apoA-I), the major protein of high-density lipoprotein (HDL), appear as appealing candidates. Indeed, most observational studies performed so far indicate that antibodies against anti-apoA-I (anti-apoA-I IgGs) were an independent CV risk factor predictive of poor prognosis, both in the general population and in different CV risk settings, associated with a systemic pro-inflammatory profile and atherosclerotic plaque vulnerability [11–22]. Furthermore, in vitro and in vivo studies indicate that these autoantibodies can promote the loss of HDL anti-oxidative function [15], induce sterile inflammation, atherogenesis, myocardial necrosis, and death in experimental mouse models through activation of a specific innate immune receptor complex consisting of Toll-like receptor (TLR)2, 4, and CD14 [23–25].

Nevertheless, it is unknown whether anti-apoA-I IgGs provide incremental prognostic value over the Global Registry of Acute Coronary Events (GRACE) risk score 2.0 [26] after acute coronary syndromes (ACS), and whether such an association could follow a nonlinear relationship with the hazards, as recently shown for HDL-cholesterol regarding incident overall mortality [27]. Finally, we explored the associations between anti-apoA-I IgGs and other clinical and biological variables available for this cohort [28,29], such as lipid profile, high sensitivity C-reactive protein (hs-CRP), N-terminal pro-brain natriuretic peptide (NT-proBNP), high-sensitivity troponin T (hs-cTnT), trimethyl-amine-N-oxide (TMAO), red blood cells, white blood cells, and platelets counts.

2. Experimental Section

2.1. Study Population

These findings are derived from the prospective multicenter Swiss ACS Cohort [28,30,31], as part of the Special Program University Medicine (SPUM), which enrolled patients referred for coronary angiography with a primary diagnosis of ACS at one of the participating University Hospitals (Zurich, Bern, Lausanne, Geneva; Clinical Trials Registration number: NCT01000701; www.spum-acs.ch) between December 2009 and October 2012. Of 2168 included Biomarker Cohort 1 patients, 1713 ACS patients had available anti-apoA-I IgG measurements in addition to the previously described clinical and laboratory data and were included in this study.

Briefly, inclusion criteria were: (1) patients older than 18 years admitted within 5 days (preferably within 72 h) after pain onset with the main diagnosis of S–T segment (ST)-elevation myocardial infarction (STEMI), non-ST-elevation myocardial infarction (NSTEMI), or unstable angina (UA); (2) persistent ST-segment elevation or depression, T-inversion or dynamic Electrocardiogram (ECG) changes, or new left bundle branch block (LBBB); (3) evidence of positive troponin by local laboratory reference values with a rise and/or fall in serial troponin levels; (4) known coronary artery disease, specified as status after myocardial infarction (MI), coronary artery bypass graft (CABG), percutaneous coronary intervention, or newly documented ≥50% stenosis of an epicardial coronary artery during the initial catheterization. Exclusion criteria comprised of (1) severe physical disability; (2) inability to comprehend study; (3) less than 1 year of life expectancy for non-cardiac reasons. Patients were followed-up for 1 year. All subjects gave written informed consent according to the declaration of Helsinki, and the study was approved by the respective institutional review boards.

2.2. Definitions of Study Outcomes

The primary outcome of this study consisted of major adverse CV events (MACE), defined as a composite of myocardial infarction (MI), stroke (including transient ischemic attack), or CV death, as described previously [28,30,31]. Participants were first followed up at 1-year post-ACS by telephone by a trained study nurse to attend a clinical visit. If patients were unable to attend the clinic visit, follow-up was performed in the following order: (1) phone call, (2) postal mail or email, (3) through family members, or (4) via primary care physician or cardiologist. Reviews of medical records and clinical visits were conducted for the 1-year outcomes. Clinical endpoints were adjudicated by a panel of three independent certified senior cardiologists blinded to anti-apoA-1 IgG results using standardized pre-specified adjudication forms.

2.3. Biochemical Analyses

Blood was drawn from the arterial sheath at coronary angiography and centrifuged at 2700 × g for 10 min at room temperature to obtain serum, and then frozen and stored in aliquots at −80 °C. Routine laboratory tests (including lipid profile, hs-CRP, NT-proBNP, hs-cTnT, TMAO, red blood cells, white blood cells, and platelets count) were performed according to laboratories of each Institution, and the corresponding detailed analytical aspects have been reported earlier [28,29]. Estimated glomerular filtration rate (eGFR; in mL/min per 1.73 m^2) was calculated for each cohort using the modification of Diet in Renal Disease study (MDRD) equation.

Anti-Apolipoprotein A-I IgG Levels

Anti-apoA-I IgGs were measured as previously described [11–13,16,20–22], using frozen EDTA plasma aliquots, stored at −80 °C until analyses. Maxisorp plates (Nunc^{TM}, Roskilde, Denmark) were coated with purified, human-derived delipidated apolipoprotein A-I (20 µg/mL; 50 µL/well) for 1 h at 37 °C. After being washed, all wells were blocked for 1 h with 2% bovine serum albumin (BSA) in a phosphate buffer solution (PBS) at 37 °C. Participants' samples were also added to a non-coated well to assess individual non-specific binding. After six washing cycles, a 50 µL/well of signal antibody (alkaline phosphatase-conjugated anti-human IgG; Sigma-Aldrich, St Louis, MO, USA), diluted 1:1000 in a PBS/BSA 2% solution, was added and incubated for 1 h at 37 °C. After washing six more times, phosphatase substrate p-nitrophenyl-phosphate-disodium (Sigma-Aldrich, St Louis, MO, USA) dissolved in a diethanolamine buffer (pH 9.8) was added and incubated for 20 min at 37 °C (Molecular Devices^{TM} Filter Max F3, Molecular Devices, San Jose, CA, USA). Optical density (OD) was determined at 405 nm, and each sample was tested in duplicate. Corresponding non-specific binding was subtracted from mean OD for each sample. The specificity of detection against lipid-free and unmodified apoA-I has been determined previously by conventional saturation tests, Western blot, and LC-MS analyses [12]. At an intermediate value of 0.6 OD, the interassay coefficient of variation was 9% (n = 5), and the intra-assay CV was 5% (n = 5).

2.4. Statistical Methods

The data were expressed as medians ± interquartile ranges for continuous variables and as numbers and percentages for categorical variables. The correlation between anti-apoA-I IgG and other biomarkers was evaluated by nonparametric approach (Spearman rank correlation). Given the fact that anti-apoA-I IgG seropositivity cut-off has not been defined nor validated on EDTA plasma, continuous anti-apoA-I IgG levels in OD were categorized according to quartiles and/or used as a continuous variable to assess the association with clinical outcomes. Time-to-first event or composite events were analyzed censoring patients at 365 days or last valid contact date. Univariate and adjusted (for GRACE score 2.0) associations of anti-apoA-I IgG categories with study endpoint was evaluated using Cox proportional hazards models and expressed with hazard ratios (HRs) and 95% CI. The predictive value of anti-apoA-I IgG over and above a reference model was assessed by Harrell's C-statistics calculated from a Cox proportional hazards regression model based on a logistic model using the GRACE risk score as a reference [26]. To identify a possible nonlinear relationship between MACE risk and anti-ApoA-I IgG continuous values, we used a Cox regression model with splines (with a degree of freedom 2) [32]. A two-sided p-value <0.05 was considered statistically significant. Statistical analyses were performed using STATA software® (Version 15, STATA Corp., College Station, TX, USA).

3. Results

3.1. Baseline Demographic and Biological Characteristics

Baseline characteristics of patients and pertinent data are summarized in Table 1. During the 1-year follow-up, 144 patients (8.4%) met the composite endpoint of MACE. Among them, 57 had a myocardial infarction, 26 had a stroke, and 68 died from CV causes.

3.2. Associations with Major Adverse Cardiovascular Events (MACE) at 1-Year of Follow-Up

Traditional CV risk factors, as well as the GRACE risk score 2.0, hs-CRP, hs-cTnT, NT-proBNP, TMAO levels, and anti-apoA-I IgG, upon study inclusion, were found to be higher in individuals with MACE at 1-year post-ACS than in those without (Table 1). As shown in Tables 2 and 3, most of the traditional CV risk factors, history of MI, renal function, and conventional biomarkers were significantly associated with MACE at 1 year in univariate analyses. Of note, HDL cholesterol was not associated with the study endpoint. As previously reported in the SPUM-ACS cohort [33], lower low-density lipoprotein (LDL) and total cholesterol levels were significantly associated with a decreased risk of MACE.

Anti-apoA-I IgG levels were associated with MACE (Table 3). Cumulative incidence curves are represented in Figure 1 by anti-apoA-I IgGs quartiles. The adjustment for the GRACE score did not modify the association (Table 4). The relationship between the incidence of MACE and anti-apoA-I IgG levels was not linear since the cumulative incidence and the hazard ratio did not change progressively with the anti-apoA-I IgG level (Figure 1, Table 3). The non-adjusted relationship between the one-year risk of MACE and anti-apoA-I IgGs is depicted in more detail in Figure 2a. An inverted U-shaped reminiscent association was apparent for MACE, with a progressively increasing risk up to an anti-apoA-I IgG level of 1.0 OD, before an apparent decrease, difficult to interpret due to the broadness of the 95% confidence interval. In an additional Cox model accounting for this nonlinearity and adjusted for the GRACE score, the relation between the hazard and MACE remained identical (Figure 2b). With this model, the non-linearity of the relationship ($p = 0.0110$) and the association between MACE and anti-apoA-I IgG level ($p = 0.0068$) were statistically significant. Of note, anti-apoA-I IgG values spanning from 0 to 1.0 OD represented 77% of all anti-apoA-1 IgGs values, whereas values above 1.0 represented 23% of all anti-apoA-I IgG values on this cohort. Secondary analyses revealed that the main driver of anti-apoA-I IgG-related MACE risk in the composite first endpoint was the recurrence of MI as anti-apoA-I IgG levels were not predictive of stroke or CV deaths (Figure S1, Supplementary Material).

Table 1. Baseline characteristics.

Patient Characteristics	All Patients (n = 1713)	MACE * (n = 144)	No MACE * (n = 1569)
Age, years			
Median (IQR)	64 (54–74)	76 (64–81)	63 (53–73)
BMI, kg/m^2			
Median (IQR)	26.5 (24.2–29.4)	26.0 (24.1–30.4)	26.5 (24.2–29.4)
Missing data	23	9	14
Female, n (%)	363/1713 (21.2%)	35/144 (24.3%)	328/1569 (20.9%)
Current smoker, n (%)	690/1680 (41.1%)	44/138 (31.9%)	646/1542 (41.9%)
Diabetes history, n (%)	310/1713 (18.1%)	47/144 (32.6%)	263/1569 (16.8%)
MI history, n (%)	246/1713 (14.4%)	31/144 (21.5%)	215/1569 (13.7%)
Hypertension history, n (%)	993/1713 (58.0%)	100/144 (69.4%)	893/1569 (56.9%)
Dyslipidemia history, n (%)	1031/1711 (60.3%)	82/143 (57.3%)	949/1568 (60.5%)
Valvular disease history, n (%)	33/1713 (1.9%)	10/144 (6.9%)	23/1569 (1.5%)
Statin use, n (%)	503/1702 (29.6%)	50/141 (35.5%)	453/1561 (29.0%)
Beta-blocker use, n (%)	410/1699 (24.1%)	54/141 (59.3%)	356/1558 (22.8%)
GRACE score 2.0			
Median (IQR)	123 (106–142)	144 (120–165)	121 (104–140)
Hs-CRP, mg/l			
Median (IQR)	3.0 (1.2–8.0)	5.8 (2.3–18.9)	2.7 (1.1–7.6)
Missing data	203	11	192
Hs-cTnT, pg/L			
Median (IQR)	0.22 (0.07 0.75)	0.51 (0.13–1.92)	0.21 (0.06–0.70)
Missing data	194	10	184
TMAO, µmol/L			
Median (IQR)	2.93 (1.99–4.97)	4.25 (2.12–7.35)	2.86 (1.98–4.85)
Missing data	360	38	322
Anti-ApoA-I IgGs, OD			
Median (IQR)	0.66 (0.44–0.97)	0.76 (0.49–0.98)	0.65 (0.44–0.97)
HDL, mmol/L			
Median (IQR)	1.13 (0.94–1.39)	1.15 (0.94–1.46)	1.12 (0.94–1.38)
missing data	65	8	57
LDL, mmol/L			
Ledian (IQR)	3.08 (2.34–3.84)	2.72 (2.04–3.45)	3.11 (2.39–3.86)
Missing data	68	8	60
Total cholesterol, mmol/L			
Median (IQR)	4.86 (4.10–5.70)	4.5 (3.7–5.2)	4.9 (4.1–5.8)
Missing data	49	7	42
Triglyceride, mmol/L			
Median (IQR)	1.02 (0.69–1.60)	0.95 (0.65–1.33)	1.04 (0.70–1.60)
Missing data	58	7	51
Creatinine			
Median (IQR)	76.0 (65.0–91.0)	89.0 (71.8–114.3)	75.0 (65.0–90.0)
eGFR			
Median (IQR)	90.9 (73.5–108.8)	74.6 (55.8–92.3)	92.1 (75.4–109.6)
Missing data	5	0	5
Systolic blood pressure			
Median (IQR)	129 (114–145)	129 (110–142)	129 (115–145)
Diastolic blood pressure			
Median (IQR)	75.0 (65.0–84.0)	72.0 (61.0–80.3)	75.0 (65.0–84.0)
Missing data	17	0	17
Renal failure (eGFR < 60), n (%)	214/1708 (12.5%)	43/144 (29.8%)	171/1564 (9.0%)

* one-year follow-up; Abbreviations: BMI—body mass index; SBP—Systolic blood pressure; DBP—diastolic blood pressure; MACE—major adverse cardiovascular events; IQR—interquartile range; MI—myocardial infarction; GRACE—Global Registry of Acute Coronary Events; Hs-CRP—high sensitivity C-reactive protein; Hs-cTnT—high-sensitivity troponin T; TMAO—trimethyl-amine-N-oxide; Anti-ApoA-I IgGs—autoantibodies against apolipoprotein A-I; OD—optical density; HDL—high-density lipoprotein; LDL—low-density lipoprotein; eGFR—estimated glomerular filtration rate.

Table 2. One-year cumulative major adverse cardiovascular events (MACE) incidence according to univariate analyses.

Predictors	Levels	N MACE/N Total	One-Year Cumulative Incidence of MACE (95% CI)	p-Value *
Age	26–54 y	17/450	3.8 (2.0–5.6)	<0.0001
	55–64 y	22/443	5.0 (3.0–7.0)	
	65–74 y	32/418	7.7 (5.1–10.3)	
	≥75 y	73/402	18.4 (14.5–22.1)	
BMI, kg/m²	20–24.9	49/554	9.0 (6.5–11.3)	0.1569
	25–29.9	50/767	6.6 (4.8–8.4)	
	30–34.9	26/283	9.3 (5.8–12.6)	
	≥35	10/86	12.0 (4.7–18.8)	
GRACE score	≤99	9/318	2.9 (1.0–4.7)	<0.0001
	100–119	25/453	5.6 (3.4–7.7)	
	120–139	29/456	6.4 (4.1–8.7)	
	140–159	37/303	12.3 (8.5–16.0)	
	≥160	44/183	24.3 (17.8–30.3)	
Gender	female	35/363	9.8 (6.6–12.8)	0.3095
	male	109/1350	8.2 (6.7–9.6)	
Diabetes history	no	97/1403	7.0 (5.6–8.3)	<0.0001
	yes	47/310	15.5 (11.3–19.4)	
MI history	No	113/1467	7.8 (6.4–9.2)	0.0124
	yes	31/246	12.9 (8.5–17.0)	
Hypertension history	no	44/720	6.2 (4.4–7.9)	0.0041
	yes	100/993	10.2 (8.3–12.1)	
Dyslipidemia history	no	61/680	9.1 (6.9–7.9)	0.4057
	yes	82/1031	8.0 (6.4–12.1)	
Valvular history	no	134/1680	8.1 (6.8–9.4)	<0.0001
	yes	10/33	30.3 (12.7–44.3)	
Current smoker	no	96/990	9.6 (7.7–11.4)	0.0251
	yes	44/690	6.5 (4.6–8.3)	
Hs-CRP, mg/L	0–0.99	12/299	4.1 (1.8–6.3)	<0.0001
	1–1.99	18/280	6.5 (3.5–9.3)	
	2–9.99	55/614	9.1 (6.8–11.4)	
	≥10	48/317	15.4 (11.3–19.4)	
Hs-cTnT, ng/L	0–0.14	40/647	6.3 (4.4–8.1)	0.0001
	0.15–0.52	29/379	7.8 (5.0–10.5)	
	>0.52	65/493	13.3 (10.3–16.3)	
TMAO, µmol/L	<2	25/340	7.4 (4.6–10.2)	0.0001
	2–2.99	12/350	3.5 (1.5–5.4)	
	3–3.99	13/206	6.4 (3.0–9.7)	
	4–6.99	27/257	10.6 (6.7–14.3)	
	≥7	29/200	14.7 (9.6–19.4)	

* Log-rank test. Abbreviations: BMI—body mass index; GRACE—Global Registry of Acute Coronary Events; MI—myocardial infarction; Hs-CRP—high sensitivity C-reactive protein; Hs-cTnT—high-sensitivity troponin T; TMAO—trimethyl-amine-N-oxide.

Table 3. One-year cumulative major adverse cardiovascular events (MACE) incidence according to univariate analyses.

Predictors	Levels	N MACE/N Total	One-Year Cumulative Incidence of MACE (95% CI)	p-Value *
Anti-ApoA-I IgGs, OD	≤0.45	31/442	7.1 (4.7–9.5)	0.0323
	0.46–0.65	24/396	6.1 (3.7–8.5)	
	0.66–0.95	48/425	11.5 (8.4–14.5)	
	>0.95	41/450	9.2 (6.5–11.9)	
HDL, mmol/L	<1	41/511	8.2 (5.7–10.5)	0.6210
	1–1.3	48/626	7.8 (5.6–9.8)	
	>1.3	47/511	9.3 (6.7–11.8)	
LDL, mmol/L	<2.5	59/492	12.3 (9.3–15.1)	0.0005
	2.5–3.5	46/580	8.0 (5.8–10.2)	
	>3.5	31/573	5.5 (3.6–7.3)	
Total cholesterol, mmol/L	<4.5	98/606	11.4 (8.8–13.9)	0.0027
	4.5–5.5	41/573	7.2 (5.1–9.3)	
	>5.5	28/485	5.9 (3.7–7.9)	
Triglyceride, mmol/L	<0.80	46/527	8.8 (6.3–11.2)	0.0653
	0.80–1.30	56/564	10.1 (7.6–12.6)	
	>1.30	35/564	6.3 (4.2–8.3)	
Creatinine, μmol/L	<70	31/606	5.2 (3.4–7.0)	<0.0001
	70–85	35/550	6.5 (4.4–8.5)	
	>85	78/557	14.2 (11.2–17.1)	
NT-proBNP, ng/L	<200	18/511	3.6 (1.9–5.2)	<0.0001
	200–1000	36/514	7.1 (4.8–9.3)	
	>1000	80/493	16.5 (13.1–19.7)	
SBP, mmHg	<120	47/561	8.4 (6.1–10.7)	0.5490
	120–140	60/649	9.4 (7.1–11.6)	
	>140	37/503	7.5 (5.2–9.8)	
DBP, mmHg	<70	59/602	9.9 (7.5–12.3)	0.2120
	70–80	49/565	8.8 (6.4–11.1)	
	>80	36/529	6.9 (4.7–9.0)	
eGFR <60 mL/min	no	101/1595	6.9 (5.6–8.1)	<0.0001
	yes	43/257	20.3 (14.7–25.6)	

* Log-rank test. Abbreviations: MACE—major adverse cardiovascular events; Anti-ApoA-I IgGs—autoantibodies against apolipoprotein A-I; OD—optical density; HDL—high-density lipoprotein; LDL—low-density lipoprotein; NT-proBNP—N-terminal pro-brain natriuretic peptide; SBP—Systolic blood pressure; DBP—diastolic blood pressure; eGFR—estimated glomerular filtration rate.

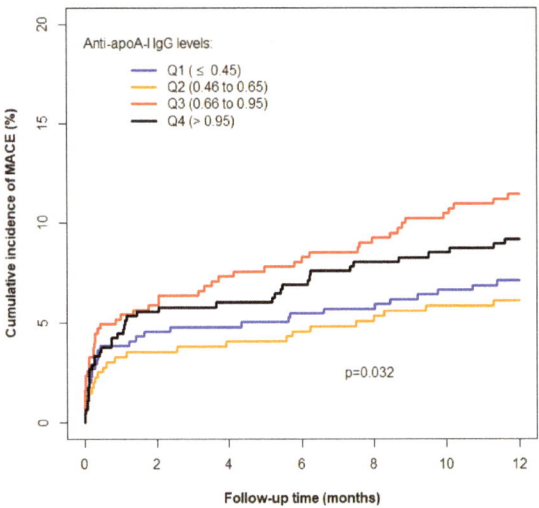

Figure 1. Cumulative incidence of major adverse cardiovascular events (MACE) (myocardial infarction, stroke, TIA, or cardiovascular (CV) death) according to autoantibodies against apolipoprotein A-I (anti-apoA-I IgG) optical density (OD) quartiles.

Table 4. Autoantibodies against apolipoprotein A-I (Anti-apoA-I IgGs) and major adverse cardiovascular events (MACE) (myocardial infarction, stroke, or cardiovascular (CV) death) risk according to Cox-regression analyses.

Predictors	Unadjusted HR (95% CI)	p-Value	Adjusted HR (95% CI) **	p-Value
Anti-ApoA-I IgGs OD				
≤0.45, Q1	Reference group	0.0338 *	Reference group	0.0496 *
0.46–0.65, Q2	0.85 (0.50–1.46)	0.5625	0.82 (0.48–1.40)	0.4682
0.66–0.95, Q3	1.64 (1.04–2.57)	0.0322	1.56 (0.99–2.44)	0.0558
>0.95, Q4	1.30 (0.82–2.08)	0.2657	1.23 (0.77–1.97)	0.3808
GRACE score 2.0				
≤99	Reference group	<0.0001 *	Reference group	<0.0001 *
100–119	1.96 (0.92–4.20)	0.0931	1.99 (0.93–4.26)	0.0772
120–139	2.27 (1.07–4.79)	0.0321	2.25 (1.07–4.76)	0.0332
140–159	4.44 (2.14–9.19)	<0.0001	4.44 (2.14–9.20)	<0.0001
≥160	9.44 (4.61–19.35)	<0.0001	9.36 (4.57–19.18)	<0.0001

* p-Value for the overall association between the factor (all levels) and risk of MACE; ** Adjusted for the GRACE score 2.0. Abbreviations: HR—hazard ratio; OD—optical density; GRACE—Global Registry of Acute Coronary Events.

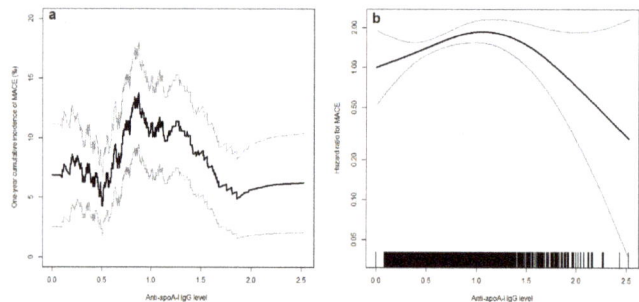

Figure 2. (**a**) Cumulative incidence of major adverse cardiovascular events (MACE) at one year according to autoantibodies against apolipoprotein A-I (anti-apoA-I IgG) levels (in optical density (OD)) in univariate analyses (black line). (**b**) Hazard ratio for MACE according to anti-apoA-I IgG levels (black line) adjusted for the categories of the Global Registry of Acute Coronary Events (GRACE) score 2.0. The vertical black lines above the x-axis represent the distribution of anti-apoA-I IgG levels in the study population: each vertical line represents the value of one participant. In both panels, the grey lines represent the 95% confidence interval. The distribution of anti-apoA-I IgG values was as followed: from 0 to 0.5 OD: n = 544 (31.8%); >0.5 and ≤1.0: n = 774 (45.2%); >1.0 and ≤1.5: n = 292 (17.0%); >1.5 and ≤2.0: n = 89 (5.2%); >2.0: n = 14 (0.8%).

C-statistics confirmed that anti-apoA-I IgG levels were a modest but significant predictor for MACE. When added to the GRACE risk score for MACE prediction at one year, anti-apoA-I IgG marginally improved the prognostic accuracy of the model: C-statistics changed from 0.680 for Cox model with the categories of a GRACE risk score to 0.698 for Cox model combining the categories of GRACE risk score and anti-apoA-I IgG levels in a continuous way.

3.3. Anti-apoA-I IgG Associations in Acute Coronary Syndrome (ACS)

To further explore the associations between anti-apoA-I IgGs and patients' clinical and biological characteristics in ACS, we analyzed the distribution of the latter according to anti-apoA-I IgG quartiles. As shown in Tables 5 and 6, an inverse association was observed with the prevalence of diabetes that decreased among the increasing anti-apoA-I IgG categories, and a trend toward increased representation of valvular disease along anti-apoA-I IgG categories was noted. Furthermore, Spearman correlation coefficients showed modest but statistically significant positive associations between anti-apoA-I IgG levels and NT-proBNP and hemoglobin levels, while negative associations were retrieved between anti-apoA-I IgG and neutrophil, eosinophil, and platelet counts (Table S1, Supplementary Material).

Close to significant associations ($p \leq 0.1$) were noted with age, hs-cTnT, hemoglobin, hematocrit, leucocytes, creatinine, total cholesterol, and triglyceride levels. No other significant association was retrieved, notably with TMAO levels (Table S1). None of these anti-apoA-I IgG associations would have remained significant after the application of Bonferroni correction (significance threshold: 0.05/24: $p = 0.002$).

Table 5. Exploring the factors associated with autoantibodies against apolipoprotein A-I (anti-apoA-I IgG) levels.

	Anti-apoA-I IgG Levels					
	All Patients	≤0.45 Q1	>0.45 and ≤0.65 Q2	>0.65 and ≤0.95 Q3	>0.95 Q4	P *
Age, years						
median (IQR)	64 (54–74)	63 (53–74)	63 (54–74)	64 (54–72)	65 (55–75)	0.1787
26–54 y	450 (26.3%)	122 (27.6%)	111 (28.0%)	109 (25.6%)	108 (24.0%)	
55–64 y	443 (25.9%)	121 (27.4%)	102 (25.8%)	114 (26.8%)	106 (23.6%)	
65–74 y	418 (24.4%)	99 (22.4%)	92 (23.2%)	110 (25.9%)	117 (26.0%)	
≥75 y	402 (23.5%)	100 (22.6%)	91 (23.0%)	92 (21.6%)	119 (26.4%)	
BMI, kg/m^2						
median (IQR)	26.5 (24.2–29.4)	26.2 (24.2–29.4)	26.4 (24.2–29.3)	26.8 (24.3–29.7)	26.6 (24.2–29.4)	0.6434
20–24.9	554 (32.8%)	154 (35.2%)	128 (32.4%)	134 (32.2%)	138 (31.3%)	
25–29.9	767 (45.4%)	189 (43.2%)	187 (47.3%)	182 (43.8%)	209 (47.4%)	
30–34.9	283 (16.7%)	75 (17.1%)	62 (15.7%)	73 (17.5%)	73 (16.6%)	
≥35	86 (5.1%)	20 (4.6%)	18 (4.6%)	27 (6.5%)	21 (4.8%)	
missing data	23	4	1	9	9	
Female, n (%)	363/1713 (21.2%)	102/442 (23.1%)	87/396 (22.0%)	93/425 (21.9%)	81/450 (18.0%)	0.2669
Current smoker, n (%)	690/1680 (41.1%)	174/436 (39.9%)	147/390 (37.7%)	183/419 (43.7%)	186/435 (42.8%)	0.2875
Diabetes history, n (%)	310/1713 (18.1%)	97/442 (21.9%)	66/396 (16.7%)	83/425 (19.5%)	64/450 (14.2%)	0.0177
Hypertension history, n (%)	720/1713 (42.0%)	248/442 (56.1%)	234/396 (59.1%)	247/425 (58.1%)	264/450 (58.7%)	0.8191
MI history, n (%)	246/1713 (14.4)	53/442 (12.0%)	52/396 (13.1%)	73/425 (17.2%)	68/450 (15.1%)	0.1416
Cholesterolemia history, n (%)	1031/1711 (60.3%)	276/441 (62.6%)	247/395 (62.5%)	257/425 (60.5%)	251/450 (55.8%)	0.1311
Valvular history, n (%)	33/1713 (1.9%)	4/442 (0.9%)	6/396 (1.5%)	8/425 (1.9%)	15/450 (3.3%)	0.0572
GRACE score						
median (IQR)	123 (106–142)	123 (106–142)	123 (106–140)	123 (104–144)	123 (106–143)	0.7916
≤99	318 (18.6%)	83 (18.8%)	74 (18.7%)	84 (19.8%)	77 (17.1%)	
100–119	453 (26.4%)	121 (27.4%)	110 (27.8%)	100 (23.5%)	122 (27.1%)	
120–139	456 (26.6%)	115 (26.0%)	103 (26.0%)	115 (27.1%)	123 (27.3%)	
140–159	303 (17.7%)	84 (19.0%)	65 (16.4%)	76 (17.9%)	78 (17.3%)	
≥160	183 (10.7%)	39 (8.8%)	44 (11.1%)	50 (11.8%)	50 (11.1%)	
CRP, mg/L						
median (IQR)	3.0 (1.2–8.0)	2.60 (1.10–7.70)	2.70 (1.00–6.40)	3.30 (1.40–8.25)	3.10 (1.30–9.20)	0.1704
0–0.99	299 (19.8%)	77 (20.2%)	80 (22.7%)	63 (16.6%)	79 (19.9%)	
1–1.99	280 (18.5%)	76 (19.9%)	67 (19.0%)	71 (18.7%)	66 (16.6%)	
2–9.99	614 (40.7%)	149 (39.1%)	141 (39.9%)	166 (43.8%)	158 (39.8%)	
≥10	317 (21.0%)	79 (20.7%)	65 (18.4%)	79 (20.8%)	94 (23.7%)	
missing data	203	61	43	46	53	
Hs-cTnT, ng/L						
median (IQR)	0.22 (0.07–0.75)	0.20 (0.06–0.67)	0.21 (0.06–0.66)	0.24 (0.07–0.87)	0.26 (0.07–0.83)	0.1947
0–0.14	647 (42.6%)	172 (45.0%)	155 (43.7%)	158 (41.1%)	162 (40.7%)	
0.15–0.52	379 (25.0%)	98 (25.7%)	86 (24.2%)	93 (24.2%)	102 (25.6%)	
>0.52	493 (32.5%)	112 (29.3%)	114 (32.1%)	133 (34.6%)	134 (33.7%)	
missing data	194	60	41	41	52	
TMAO						
median (IQR)	2.93 (1.99–4.97)	2.85 (1.89–4.97)	2.96 (2.01–4.98)	2.93 (2.00–4.46)	3.05 (2.07–5.43)	0.4744
<2	340 (25.1%)	98 (28.3%)	74 (23.9%)	83 (25.1%)	85 (23.2%)	
2–2.99	350 (25.9%)	87 (25.1%)	84 (27.2%)	86 (26.0%)	93 (25.3%)	
3–3.99	206 (15.2%)	42 (12.1%)	51 (16.5%)	60 (18.1%)	53 (14.4%)	
4–6.99	257 (19.0%)	66 (19.1%)	50 (16.2%)	63 (19.0%)	78 (21.3%)	
≥7	200 (14.8%)	53 (15.3%)	50 (16.2%)	39 (11.8%)	58 (15.8%)	
missing data	360	96	87	94	83	

* Kruskal-Wallis or Chi-2 tests (Fisher exact test for liver history). Abbreviations: IQR—interquartile range; BMI—body mass index; MI—myocardial infarction; GRACE—Global Registry of Acute Coronary Events; CRP—C-reactive protein; hs-cTnT—high-sensitivity troponin T; TMAO—trimethyl-amine-N-oxide.

Table 6. Exploring the factors associated with autoantibodies against apolipoprotein A-I (anti-apoA-I IgG) levels.

	Anti-apoA-I IgG Levels					
	All Patients	≤0.45 Q1	>0.45 and ≤0.65 Q2	>0.65 and ≤0.95 Q3	>0.95 Q4	P *
SBP, mmHg Median (IQR) Missing data	129 (114–145) 0	129 (115–145) 0	129 (115–144) 0	128 (112–145) 0	130 (114–146) 0	0.9714
DBP, mmHg Median (IQR) Missing data	75 (65–84) 17	75 (66–84) 7	75 (65–84) 2	73 (65–84) 4	74 (63–84) 4	0.6476
Hemoglobin, g/L Median (IQR) Missing data	138 (127–149) 87	137 (124–149) 30	138 (126–148) 19	139 (129–148) 19	139 (128–150) 19	0.3715
Hematocrit, % Median (IQR) Missing data	41 (38–43.3) 76	41 (37–43) 26	40 (37–43) 14	41 (38–44) 18	41 (38–44) 18	0.2161
Leucocytes, G/L Median (IQR) Missing data	9.6 (7.4–12.1) 73	9.5 (7.3–12.0) 25	9.2 (7.1–12.0) 14	9.8 (7.9–12.1) 17	9.8 (7.6–12.3) 17	0.0340
Erythrocytes, T/L Median (IQR) Missing data	4.5 (4.1–4.8) 84	4.5 (4.1–4.9) 28	4.4 (4.1–4.8) 17	4.5 (4.2–4.8) 19	4.5 (4.1–4.8) 20	0.3240
Lymphocytes, % Median (IQR) Missing data	1.62 (1.16–2.35) 395	1.67 (1.20–2.55) 118	1.54 (1.09–2.31) 95	1.69 (1.19–2.34) 87	1.58 (1.11–2.22) 95	0.0526
Neutrophils, G/L Median (IQR) Missing data	7.3 (5.3 to 10.0) 412	7.0 (5.0 to 9.7) 122	7.2 (4.7 to 10.1) 97	7.4 (5.5 to 9.9) 93	7.6 (5.5 to 10.1) 100	0.1500
Monocytes, % Median (IQR) Missing data	0.62 (0.45 to 0.95) 398	0.63 (0.45 to 1.05) 118	0.56 (0.41 to 0.92) 96	0.64 (0.48 to 0.99) 88	0.61 (0.45 to 0.90) 96	0.1220
Basophils, % Median (IQR) Missing data	0.03 (0.01–0.07) 399	0.03 (0.02–0.08) 118	0.03 (0.01–0.07) 95	0.03 (0.01–0.07) 88	0.03 (0.01–0.06) 98	0.7491
Eosinophils, % Median (IQR) Missing data	0.08 (0.02–0.19) 401	0.10 (0.03–0.21) 119	0.08 80.03–0.20) 96	0.07 (0.02–0.17) 88	0.07 (0.02–0.16) 98	0.0237
Thromboctyes, G/L Median (IQR) Missing data	215 (183–257) 85	226 (188–272) 27	211 (179–250) 15	218 (185–253) 21	208 (176–255) 22	0.0026
Total cholesterol, mmol/L Median (IQR) Missing data	4.9 (4.1–5.7) 49	4.9 (4.1–5.8) 13	4.9 (4.0–5.8) 13	4.9 (4.1–5.7) 11	4.7 (4.0–5.6) 12	0.1602
HDL, mmol/L Median (IQR) Missing data	1.13 (0.69–1.60) 65	1.14 (0.93–1.40) 20	1.14 (0.96–1.40) 15	1.13 (0.94–1.40) 14	1.10 (0.93–1.33) 16	0.2852
LDL, mmol/L Median (IQR) Missing data	308 (2.34–3.84) 68	3.07 (2.32–3.94) 20	3.13 (2.40–3.82) 16	3.08 (2.37–3.81) 16	2.99 (2.26–3.81) 16	0.7418
Triglyceride Median (IQR) Missing data	1.02 (0.69–1.60) 58	1.08 (0.72–1.70) 17	1.05 (0.69–1.68) 15	1.00 (0.66–1.51) 13	0.99 (0.69–1.48) 13	0.2165
Creatinine, μmol/L Median (IQR) Missing data	76 (65–91) 0	75 (64–88) 0	76 (66–91) 0	76 (65–92) 0	77 (66–92) 0	0.3484
NT-ProBNP, ng/L Median (IQR) Missing data	414 (131–1436) 195	386 (118–1462) 60	355 (129–1196) 41	433 (132–1627) 42	544 (159–1633) 52	0.0390
eGFR, mL/min Median (IQR) Missing data	90.9 (73.5–108.8) 5	92.1 (75.8–110.9) 1	91.0 (73.5–107.2) 1	89.8 (73.5–108.3) 2	90.7 (70.5–108.6) 1	0.4791
Renal failure (eGFR <60), n (%)	214/1708	47/441	53/395	56/423	58/449	0.5828

* Kruskal-Wallis or Chi-2 tests (Fisher exact test for liver history). Abbreviations: Anti-apoA-I IgG—autoantibodies against apolipoprotein A-I; SBP—systolic blood pressure; DBP—diastolic blood pressure; IQR—interquartile range; HDL—high-density lipoprotein; LDL—low-density lipoprotein; NT-ProBNP—N-terminal pro-brain natriuretic peptide; eGFR—estimated glomerular filtration rate.

4. Discussion

The key finding of the present study is that anti-apoA-I IgG was predictive of incident MACE independently of the GRACE score 2.0, one of the strongest prognostic stratification tools currently validated in ACS. Furthermore, the prognostic accuracy of the GRACE score was slightly improved (+2%) by adding anti-apoA-I IgGs to it, according to C-statistics. Because of the important controversy regarding the reliability of C-statistics comparison and reclassification statistics (net reclassification index and integrated discrimination index) in embedded models due to the important false positive findings [34,35], we did not perform further statistical analyses to interpret this difference. Indeed, we feel that a 2% increase in C-statistics is unlikely to be clinically meaningful or to change current clinical practice in ACS. Nevertheless, these results strengthen the proof-of-principle that anti-apoA-I IgGs are predictive of a poor CV prognosis, as suggested earlier in the general population and in different high CV risk settings [11–17,20–22].

The second notable finding of this study is that the association between anti-apoA-I IgGs and MACE was not linear over the full range of anti-apoA-I IgG values. Indeed, the relationship was linear for the first 87% anti-apoA-I IgG values but started to decline for the remaining 23% highest anti-apoA-I IgG values. According to Cox regression models with splines confirming the significant nature of this non-linear relationship with MACE, we hypothesized that anti-apoA-I IgGs would mirror the U-shaped relationship between HDL-cholesterol and outcomes reported lately [27]. Nevertheless, there was no association between HDL-cholesterol and MACE or between HDL-cholesterol and anti-apoA-I IgGs in this study. Furthermore, due to the small number of events in this last 23% portion of anti-apoA-I IgG values, the confidence intervals of our regression model increased, preventing us from concluding on a possibly inverted U-shape relationship with MACE or the occurrence of a plateau at high anti-apoA-I IgG values. To the best of our knowledge, this is the first report of a non-linear association between anti-apoA-I IgG levels and MACE. Whether this non-linear association is specific for the ACS population is still elusive, but should be taken into account in future studies in order to avoid possible false negative findings related to the classic assumption of a proportional risk increase between anti-apoA-I IgG levels and hazards.

To further explore the possible associations between biological and clinical parameters with anti-apoA-I IgG levels in ACS, our linear regression analyses of anti-apoA-I IgG quartiles did not identify specific items markedly associated with anti-apoA-I IgG, and we did not reproduce the often reported association with previous MI history [11,20–22]. Interestingly, a trend was observed with the presence of valvular diseases, and an inverse relationship with the history of diabetes. This possible association with the presence of valvular diseases is in line with the concept supporting a causal role between humoral autoimmunity and different valvular diseases historically exemplified by rheumatic heart disease [6]. Nevertheless, as anti-apoA-I IgGs have never been shown to be associated with valvular diseases, these results open a new field of investigations, especially as we could not specify more in detail with which specific valvular disease these autoantibodies were associated. Furthermore, as anti-apoA-I IgGs were found to be higher in type 2 diabetic patients and associated with a CVD [17], the inverse association reported here between the prevalence of diabetes across anti-apoA-I IgG quartiles was unexpected and thus requires further confirmation. Of note, we did not retrieve any association between anti-apoA-I IgG and hs-CRP levels, despite previous reports, notably in the general population, where such association was retrieved [11]. Because anti-apoA-I IgG has been shown to promote various inflammatory cytokines production [10], including interleukin-6 regulating CRP liver production, this absence of association was unexpected. The reason for such difference is unknown and may be explained by differences related to the acute phase the samples of ACS patients were taken when compared to the stable situation of CoLaus individuals.

On the other hand, the inverse associations retrieved between these antibodies and lymphocyte and platelet counts are similar to what has been retrieved in HIV patients [36], and the possible causal nature of such associations is still under investigation. The inverse association between anti-apoA-I IgGs and eosinophil counts has not been reported. Whether patients with high anti-apoA-I IgG levels could be less prone to develop an allergic coronary syndrome (known as the Kounis syndrome) [37] should be further explored. The weak but significant positive anti-apoA-I IgG correlation with NT-proBNP, one of the strongest prognostic biomarkers of mitigated prognosis, may be related to the prognostic value previously ascribed to anti-apoA-I IgGs in different settings [11–17,20–22]. Because experimental evidence indicates that these autoantibodies could per se promote atherogenesis and atherothrombosis, as well as mitigate mouse survival by eliciting sterile inflammation through TLR2/4 and CD14 signaling [22–25], the present results reinforce the notion that anti-apoA-I IgGs could represent a future therapeutic target rather than a mere innocent bystander. Nevertheless, due to the non-linear relationship between these antibodies and MACE, knowing whether and how assessing anti-apoA-I IgG levels could affect current medical practice remains to be demonstrated.

The major limitation of this study relies on the fact that we did not assess other auto-antibodies of potential CV relevance, which is outside of the scope of the present work. The strength of the study relies on the fact that so far, it represents the largest and best characterized ACS cohort studying the prognostic value of anti-apoA-I IgGs in ACS.

5. Conclusions

In conclusion, anti-apoA-I IgGs are associated with incident MACE in ACS, independently of the GRACE 2.0 score, and with a nonlinear relationship to the hazards. Whether anti-apoA-I IgGs could provide incremental information over the existing risk stratification tool still needs to be determined.

Supplementary Materials: The following are available online at http://www.mdpi.com/2077-0383/8/7/1002/s1, Figure S1: Hazard ratio for the association between anti-apoA-I IgGs levels and specific endpoints of the composite MACE outcome assessed with Cox models accounting for the potential non linearity (with splines) of the relationship and adjusted for categories of the GRACE score, Table S1: Spearman Correlation coefficients between anti-apoA-I IgG and other continuous variables of interest

Author Contributions: N.V., S.P., R.K., and F.M. conceived and designed the study; methodology, N.V., S.P., C.C., R.K., and F.M.; software, C.C.; validation, N.V., S.P., C.C., B.G., J.V., L.R., D.C., S.C., D.N., N.R., S.W., S.H., W.Z., L.X., A.V.E., C.M., T.F.L., R.K., and F.M.; formal analysis, N.V., C.C., investigation, N.V., S.P., C.C., B.G., J.V., L.R., D.C., S.C., D.N., N.R., S.W., S.H., W.Z., L.X., A.V.E., C.M., T.F.L., R.K., and F.M.; resources, see funding; data curation, N.V., C.C.; writing—original draft preparation, N.V., S.P.; writing—review and editing, N.V., S.P., C.C., B.G., J.V., L.R., D.C., S.C., D.N., N.R., S.W., S.H., W.Z., L.X., A.V.E., C.M., T.F.L., R.K., and F.M. All authors have read and approved the final manuscript.

Funding: This work was supported by the Swiss National Science Foundation (SPUM 33CM30-124112 and 32473B_163271, 310030-163335 and 310030_1407736) and by the De Reuter Foundation (315112); the Leenaards Foundation (3698); the Swiss Heart Foundation; the Foundation Leducq and the Foundation for Cardiovascular Research—Zurich Heart House, Zurich. The SPUM consortium was also supported by Roche Diagnostics, Rotkreuz, Switzerland (providing the kits for high-sensitivity troponin T), Eli Lilly, Indianapolis (USA), AstraZeneca, Zug; Medtronic, Münchenbuchsee; Merck Sharpe and Dome (MSD), Lucerne; Sanofi-Aventis, Vernier; St. Jude Medical, Zurich (all Switzerland).

Conflicts of Interest: The authors declare no conflict of interest.

References

1. Hansson, G.K.; Hermansson, A. The immune system in atherosclerosis. *Nat. Immunol.* **2011**, *12*, 204–212. [CrossRef]
2. Ketelhuth, D.F.; Hansson, G.K. Modulation of Autoimmunity and Atherosclerosis—Common Targets and Promising Translational Approaches Against Disease. *Circ. J.* **2015**, *79*, 924–933. [CrossRef]
3. Ketelhuth, D.F.; Hansson, G.K. Adaptive Response of T and B Cells in Atherosclerosis. *Circ. Res.* **2016**, *118*, 668–678. [CrossRef]
4. Libby, P.; Tabas, I.; Fredman, G.; Fisher, E.A. Inflammation and its resolution as determinants of acute coronary syndromes. *Circ. Res.* **2014**, *114*, 1867–1879. [CrossRef]
5. Tsiantoulas, D.; Diehl, C.J.; Witztum, J.L.; Binder, C.J. B Cells and Humoral Immunity in Atherosclerosis. *Circ. Res.* **2014**, *114*, 1743–1756. [CrossRef]

6. Carapetis, J.R.; Beaton, A.; Cunningham, M.W.; Guilherme, L.; Karthikeyan, G.; Mayosi, B.M.; Sable, C.; Steer, A.; Wilson, N.; Wyber, R.; et al. Acute rheumatic fever and rheumatic heart disease. *Nat. Rev. Dis. Prim.* **2016**, *2*, 15084. [CrossRef]
7. Cunningham, M.W. Rheumatic Fever, Autoimmunity, and Molecular Mimicry: The Streptococcal Connection. *Int. Rev. Immunol.* **2014**, *33*, 314–329. [CrossRef]
8. Lazzerini, P.E.; Capecchi, P.L.; Laghi-Pasini, F.; Boutjdir, M. Autoimmune channelopathies as a novel mechanism in cardiac arrhythmias. *Nat. Rev. Cardiol.* **2017**, *14*, 521–535. [CrossRef]
9. Meier, L.A.; Binstadt, B.A. The Contribution of Autoantibodies to Inflammatory Cardiovascular Pathology. *Front. Immunol.* **2018**, *9*, 911. [CrossRef]
10. Satta, N.; Vuilleumier, N. Auto-Antibodies As Possible Markers and Mediators of Ischemic, Dilated, and Rhythmic Cardiopathies. *Curr. Drug Targets* **2015**, *16*, 342–360. [CrossRef]
11. Antiochos, P.; Marques-Vidal, P.; Virzi, J.; Pagano, S.; Satta, N.; Bastardot, F.; Hartley, O.; Montecucco, F.; Mach, F.; Waeber, G.; et al. Association between anti-apolipoprotein A-1 antibodies and cardiovascular disease in the general population. *Thromb. Haemost.* **2016**, *116*, 764–771.
12. Antiochos, P.; Marques-Vidal, P.; Virzi, J.; Pagano, S.; Satta, N.; Hartley, O.; Montecucco, F.; Mach, F.; Kutalik, Z.; Waeber, G.; et al. Impact of CD14 Polymorphisms on Anti-Apolipoprotein A-1 IgG-Related Coronary Artery Disease Prediction in the General PopulationHighlights. *Arter. Thromb. Vasc. Boil.* **2017**, *37*, 2342–2349. [CrossRef]
13. Antiochos, P.; Marques-Vidal, P.; Virzi, J.; Pagano, S.; Satta, N.; Hartley, O.; Montecucco, F.; Mach, F.; Kutalik, Z.; Waeber, G.; et al. Anti-Apolipoprotein A-1 IgG Predict All-Cause Mortality and Are Associated with Fc Receptor-Like 3 Polymorphisms. *Front. Immunol.* **2017**, *8*, 2234. [CrossRef]
14. Batuca, J.; Amaral, M.; Favas, C.; Justino, G.; Papoila, A.; Ames, P.; Alves, J. Antibodies against HDL Components in Ischaemic Stroke and Coronary Artery Disease. *Thromb. Haemost.* **2018**, *118*, 1088–1100. [CrossRef]
15. Batuca, J.R.; Amaral, M.C.; Favas, C.; Paula, F.S.; Ames, P.R.J.; Papoila, A.L.; Delgado Alves, J. Extended-Release Niacin Increases Anti-Apolipoprotein a-I Antibodies That Block the Antioxidant Effect of High-Density Lipoprotein-Cholesterol: The Explore Clinical Trial. *Br. J. Clin. Pharmacol.* **2017**, *83*, 1002–1010. [CrossRef]
16. Carbone, F.; Satta, N.; Montecucco, F.; Virzi, J.; Burger, F.; Roth, A.; Roversi, G.; Tamborino, C.; Casetta, I.; Seraceni, S.; et al. Anti-ApoA-1 IgG serum levels predict worse post-stroke outcomes. *Eur. J. Clin. Investig.* **2016**, *46*, 805–817. [CrossRef]
17. El-Lebedy, D.; Rasheed, E.; Kafoury, M.; Haleem, D.A.-E.; Awadallah, E.A.A.-S.; Ashmawy, I. Anti-apolipoprotein A-1 autoantibodies as risk biomarker for cardiovascular diseases in type 2 diabetes mellitus. *J. Diabetes its Complicat.* **2016**, *30*, 580–585. [CrossRef]
18. Keller, P.F.; Pagano, S.; Roux-Lombard, P.; Sigaud, P.; Rutschmann, O.T.; Mach, F.; Hochstrasser, D.; Vuilleumier, N. Autoantibodies against Apolipoprotein a-1 and Phosphorylcholine for Diagnosis of Non-St-Segment Elevation Myocardial Infarction. *J. Intern. Med.* **2012**, *271*, 451–462. [CrossRef]
19. Lagerstedt, J.O.; Dalla-Riva, J.; Marinkovic, G.; Del Giudice, R.; Engelbertsen, D.; Burlin, J.; Petrlova, J.; Lindahl, M.; Bernfur, K.; Melander, O.; et al. Anti-Apoa-I Igg Antibodies Are Not Associated with Carotid Artery Disease Progression and First-Time Cardiovascular Events in Middle-Aged Individuals. *J. Intern. Med.* **2019**, *285*, 49–58. [CrossRef]
20. Vuilleumier, N.; Bas, S.; Pagano, S.; Montecucco, F.; Guerne, P.-A.; Finckh, A.; Lovis, C.; Mach, F.; Hochstrasser, D.; Gabay, C.; et al. Anti-apolipoprotein A-1 IgG predicts major cardiovascular events in patients with rheumatoid arthritis. *Arthritis Rheum.* **2010**, *62*, 2640–2650. [CrossRef]
21. Vuilleumier, N.; Montecucco, F.; Spinella, G.; Pagano, S.; Bertolotto, M.B.; Pane, B.; Pende, A.; Galan, K.; Roux-Lombard, P.; Combescure, C.; et al. Serum levels of anti-apolipoprotein A-1 auto-antibodies and myeloperoxidase as predictors of major adverse cardiovascular events after carotid endarterectomy. *Thromb. Haemost.* **2013**, *109*, 706–715.
22. Vuilleumier, N.; Rossier, M.F.; Pagano, S.; Python, M.; Charbonney, E.; Nkoulou, R.; James, R.; Reber, G.; Mach, F.; Roux-Lombard, P. Anti-apolipoprotein A-1 IgG as an independent cardiovascular prognostic marker affecting basal heart rate in myocardial infarction. *Eur. Hear. J.* **2010**, *31*, 815–823. [CrossRef]

23. Montecucco, F.; Braunersreuther, V.; Burger, F.; Lenglet, S.; Pelli, G.; Carbone, F.; Fraga-Silva, R.; Stergiopulos, N.; Monaco, C.; Mueller, C.; et al. Anti-apoA-1 auto-antibodies increase mouse atherosclerotic plaque vulnerability, myocardial necrosis and mortality triggering TLR2 and TLR4. *Thromb. Haemost.* **2015**, *114*, 410–422.
24. Pagano, S.; Carbone, F.; Burger, F.; Roth, A.; Bertolotto, M.; Pane, B.; Spinella, G.; Palombo, D.; Pende, A.; Dallegri, F.; et al. Anti-apolipoprotein A-1 auto-antibodies as active modulators of atherothrombosis. *Thromb. Haemost.* **2016**, *116*, 554–564.
25. Pagano, S.; Satta, N.; Werling, D.; Offord, V.; De Moerloose, P.; Charbonney, E.; Hochstrasser, D.; Vuilleumier, N.; Roux-Lombard, P. Anti-apolipoprotein A-1 IgG in patients with myocardial infarction promotes inflammation through TLR2/CD14 complex. *J. Intern. Med.* **2012**, *272*, 344–357. [CrossRef]
26. Huang, W.; Fitzgerald, G.; Goldberg, R.J.; Gore, J.; McManus, R.H.; Awad, H.; Waring, M.E.; Allison, J.; Saczynski, J.S.; Kiefe, C.I.; et al. Performance of the GRACE Risk Score 2.0 Simplified Algorithm for Predicting 1-year Death Following Hospitalization for an Acute Coronary Syndrome in a Contemporary Multiracial Cohort. *Am. J. Cardiol.* **2016**, *118*, 1105–1110. [CrossRef]
27. Madsen, C.M.; Varbo, A.; Nordestgaard, B.G. Extreme high high-density lipoprotein cholesterol is paradoxically associated with high mortality in men and women: two prospective cohort studies. *Eur. Hear. J.* **2017**, *38*, 2478–2486. [CrossRef]
28. Klingenberg, R.; Aghlmandi, S.; Raber, L.; Gencer, B.; Nanchen, D.; Heg, D.; Carballo, S.; Rodondi, N.; Mach, F.; Windecker, S.; et al. Improved Risk Stratification of Patients with Acute Coronary Syndromes Using a Combination of Hstnt, Nt-Probnp and Hscrp with the Grace Score. *Eur. Heart J. Acute Cardiovasc. Care* **2018**, *7*, 129–138. [CrossRef]
29. Li, X.S.; Obeid, S.; Klingenberg, R.; Gencer, B.; Mach, F.; Räber, L.; Windecker, S.; Rodondi, N.; Nanchen, D.; Muller, O.; et al. Gut microbiota-dependent trimethylamine N-oxide in acute coronary syndromes: A prognostic marker for incident cardiovascular events beyond traditional risk factors. *Eur. Hear. J.* **2017**, *38*, 814–824. [CrossRef]
30. Gencer, B.; Montecucco, F.; Nanchen, D.; Carbone, F.; Klingenberg, R.; Vuilleumier, N.; Aghlmandi, S.; Heg, D.; Raber, L.; Auer, R.; et al. Prognostic Value of Pcsk9 Levels in Patients with Acute Coronary Syndromes. *Eur. Heart J.* **2016**, *37*, 546–553. [CrossRef]
31. Klingenberg, R.; Aghlmandi, S.; Liebetrau, C.; Räber, L.; Gencer, B.; Nanchen, D.; Carballo, D.; Akhmedov, A.; Montecucco, F.; Zoller, S.; et al. Cysteine-rich angiogenic inducer 61 (Cyr61): a novel soluble biomarker of acute myocardial injury improves risk stratification after acute coronary syndromes. *Eur. Hear. J.* **2017**, *38*, 3493–3502. [CrossRef] [PubMed]
32. Eilers, P.H.C.; Marx, B.D. Flexible smoothing with B -splines and penalties. *Stat. Sci.* **1996**, *11*, 89–121. [CrossRef]
33. Laaksonen, R.; Ekroos, K.; Sysi-Aho, M.; Hilvo, M.; Vihervaara, T.; Kauhanen, D.; Suoniemi, M.; Hurme, R.; März, W.; Scharnagl, H.; et al. Plasma ceramides predict cardiovascular death in patients with stable coronary artery disease and acute coronary syndromes beyond LDL-cholesterol. *Eur. Hear. J.* **2016**, *37*, 1967–1976. [CrossRef] [PubMed]
34. Demler, O.V.; Pencina, M.J.; D'Agostino, R.B. Misuse of DeLong test to compare AUCs for nested models. *Stat. Med.* **2012**, *31*, 2577–2587. [CrossRef] [PubMed]
35. Hilden, J.; Gerds, T.A. A Note on the Evaluation of Novel Biomarkers: Do Not Rely on Integrated Discrimination Improvement and Net Reclassification Index. *Stat. Med.* **2014**, *33*, 3405–3414. [CrossRef] [PubMed]
36. Satta, N.; Pagano, S.; Montecucco, F.; Gencer, B.; Mach, F.; Vuilleumier, N.; Aubert, V.; Barth, J.; Battegay, M.; Bernasconi, E.; et al. Anti-apolipoprotein A-1 autoantibodies are associated with immunodeficiency and systemic inflammation in HIV patients. *J. Infect.* **2018**, *76*, 186–195. [CrossRef] [PubMed]
37. Kounis, N.G. Coronary Hypersensitivity Disorder: The Kounis Syndrome. *Clin. Ther.* **2013**, *35*, 563–571. [CrossRef]

© 2019 by the authors. Licensee MDPI, Basel, Switzerland. This article is an open access article distributed under the terms and conditions of the Creative Commons Attribution (CC BY) license (http://creativecommons.org/licenses/by/4.0/).

Article

Relationship between HDL Cholesterol Efflux Capacity, Calcium Coronary Artery Content, and Antibodies against ApolipoproteinA-1 in Obese and Healthy Subjects

Nicolas Vuilleumier [1,2,†], Sabrina Pagano [1,2,*,†], Fabrizio Montecucco [3,4,5], Alessandra Quercioli [6], Thomas H. Schindler [7,8], François Mach [9], Eleonora Cipollari [10], Nicoletta Ronda [10] and Elda Favari [10]

1. Division of Laboratory Medicine, Diagnostic Department, Geneva University Hospitals, 1211 Geneva, Switzerland
2. Division of Laboratory Medicine, Department of Medical Specialties, Faculty of Medicine, University of Geneva, 1206 Geneva, Switzerland
3. First Clinic of Internal Medicine, Department of Internal Medicine, University of Genoa, 6 viale Benedetto XV, 16132 Genoa, Italy
4. Ospedale Policlinico San Martino, Genoa, 10 Largo Benzi, 16132 Genoa, Italy
5. Centre of Excellence for Biomedical Research (CEBR), University of Genoa, 9 viale Benedetto XV, 16132 Genoa, Italy
6. Division of Cardiology, "SS. Antonio e Biagio e Cesare Arrigo" Hospital, 6 via Venezia 16, 15121 Alessandria, Italy
7. Division of Nuclear Medicine—Cardiovascular Section, Department of Radiology and Radiological Science, School of Medicine, Johns Hopkins University, JHOC 3225, 601 N. Caroline Street, Baltimore, MD 21287, USA
8. Division of Cardiology, Department of Medicine, Johns Hopkins University, Baltimore, MD 21287, USA
9. Division of Cardiology, Cardiology Center, Geneva University Hospital, 1211 Geneva, Switzerland
10. Department of Food and Drug, University of Parma, Parco Area delle Scienze, 43124 Parma 27/A, Italy
* Correspondence: Sabrina.Pagano@hcuge.ch; Tel.: +41-2237-95321
† These authors contributed equally to this work.

Received: 25 June 2019; Accepted: 12 August 2019; Published: 15 August 2019

Abstract: Aims: To explore the associations between cholesterol efflux capacity (CEC), coronary artery calcium (CAC) score, Framingham risk score (FRS), and antibodies against apolipoproteinA-1 (anti-apoA-1 IgG) in healthy and obese subjects (OS). Methods and Results: ABCA1-, ABCG1-, passive diffusion (PD)-CEC and anti-apoA-1 IgG were measured in sera from 34 controls and 35 OS who underwent CAC score determination by chest computed tomography. Anti-apoA-1 IgG ability to modulate CEC and macrophage cholesterol content (MCC) was tested in vitro. Controls and OS displayed similar ABCG1-, ABCA1-, PD-CEC, CAC and FRS scores. Logistic regression analyses indicated that FRS was the only significant predictor of CAC lesion. Overall, anti-apoA-1 IgG were significantly correlated with ABCA1-CEC ($r = 0.48$, $p < 0.0001$), PD-CEC ($r = -0.33$, $p = 0.004$), and the CAC score ($r = 0.37$, $p = 0.03$). ABCA1-CEC was correlated with CAC score ($r = 0.47$, $p = 0.004$) and FRS ($r = 0.18$, $p = 0.29$), while PD-CEC was inversely associated with the same parameters (CAC: $r = -0.46$, $p = 0.006$; FRS: score $r = -0.40$, $p = 0.01$). None of these associations was replicated in healthy controls or after excluding anti-apoA-1 IgG seropositive subjects. In vitro, anti-apoA-1 IgG inhibited PD-CEC ($p < 0.0001$), increased ABCA1-CEC ($p < 0.0001$), and increased MCC ($p < 0.0001$). Conclusions: We report a paradoxical positive association between ABCA1-CEC and the CAC score, with the latter being inversely associated with PD in OS. Corroborating our clinical observations, anti-apoA-1 IgG enhanced ABCA1 while repressing PD-CEC, leading to MCC increase in vitro. These results indicate that anti-apoA-1 IgG have the potential to interfere with CEC and macrophage lipid metabolism, and may underpin paradoxical associations between ABCA1-CEC and cardiovascular risk.

Keywords: cholesterol efflux capacity; coronary artery calcium score; obesity; anti-apoA-1 IgG; autoantibodies

1. Introduction

Lately, impaired high-density lipoprotein (HDL) cholesterol efflux capacity (CEC) from macrophages, involving mainly the adenosine triphosphate (ATP)-binding cassette (ABC) transporter A1 (ABCA1) pathway, has gained considerable interest as a promising biomarker of atherosclerosis and cardiovascular (CV)-related risk, both in primary and secondary prevention settings [1–3], and served as proof of concept that HDL functional measures may provide improved CV risk stratification over standard lipid profile and other traditional CV risk factors [1–3]. Nevertheless, inverse and paradoxical positive associations between CEC and incident CV events have been reported in high-risk populations [4], and patients with metabolic syndrome have been shown to have an increased CEC, possibly due to increase pre-β-HDL levels [5,6]. To further fuel this controversy, no associations were retrieved between CEC and coronary artery calcium (CAC) score, either in the general population or in rheumatoid arthritis patients [3,7]. These contrasting observations relating CEC with CV risk and CAC score might be influenced by disease state or oxidative stress and immune-mediated inflammation.

Among factors influencing immune-mediated inflammation and related to lipid metabolism, we and others focused our investigations on antibodies against apolipoprotein A-1 (anti-apoA-1 IgG), the major protein fraction of HDL regulating ABCA1 activity, as a biomarker of poor general and CV outcomes, both in primary and secondary prevention settings [8–13]. In parallel, in vitro and animal studies have shown that anti-apoA-1 IgG could promote inflammation, atherogenesis, myocardial necrosis, and mice death through interaction with the Toll-like receptor (TLR) 2, 4 and the cluster of differentiation 14 (CD14) complex [14–18]. Accordingly, we demonstrated in obese but otherwise healthy subjects that anti-apoA-1 IgG were independent predictors of coronary artery lesion upon chest computed tomography [19]. Several groups have shown that these autoantibodies could also promote atherogenesis by affecting HDL anti-oxidant properties [20–24], and recently anti-apoA-1 IgG against the c-terminal part of the protein was found to be inversely associated with fibroblasts CEC [25].

In this pilot study, we explored whether specific pathways of HDL CEC (ABCG1-mediated, ABCA1-mediated, passive diffusion) could be differentially associated with the CAC score in healthy obese and non-obese participants, and whether anti-apoA-1 IgG are associated with a specific HDL CEC pathway modification on macrophages.

2. Experimental Section

2.1. Study Population and Design

The current investigation follows as sub-analysis of a previous published study [19], including 48 non-obese subjects (BMI < 30 kg/m^2) and 43 obese subjects (BMI ≥ 30 kg/m^2), without known traditional CV risk factors (such as arterial hypertension, smoking, and diabetes mellitus, anamnestic notion of variant angina, family history of premature coronary artery disease), clinically patent cardiovascular/systemic disease, or CV treatment (statins, any cardiac or vasoactive medication). From these 91 Subjects, 14 samples were missing from non-obese participants and 8 samples were missing from obese subjects, leaving 69 subjects available for HDL CEC determination. As reported earlier, "before inclusion in the cardiac perfusion assessment test, study participants underwent a complete visit, including a physical examination, electrocardiogram, blood pressure measurements, and blood puncture in a fasting state. Following inclusion, each study participant underwent multidetector computed tomography (MDCT) and CAC assessment. The study was approved by the University Hospitals of Geneva Institutional Review Board (protocol number 07-183), and each participant signed the approved informed consent form. This study has been conducted in compliance with the Declaration of Helsinki [19].

2.2. Study Endpoints

Two predetermined endpoints were considered for this pilot study. The primary endpoint was to test the ability of CEC to predict the presence of any coronary artery calcification on chest computed tomography (CT) scan, as described below. The secondary endpoint was to explore the possible association between anti-apoA-1 IgG and CEC. All CT scan data were assessed by two senior cardiologists blinded to the participants' biochemical data.

2.3. Assessment of Coronary Artery Calcification by Chest CTscan

In the first step of the perfusion assessment, a 64-slice multidetector computed tomography (MDCT) (64-sliceBiograph HiRez TruePoint Positron Emission Tomography/Computed Tomography (PET-CT) scanner, Siemens, Erlangen, Germany) was performed to determine the CAC score. The scanner was operated in the single slice mode with an image acquisition time of 100 ms and a Section thickness of 3 mm. Prospective electrocardiogram (ECG) triggering was done at 55% of the R-R interval. Contiguous slices to the apex of the heart were obtained. CAC was considered present if three or more contiguous pixels with a signal intensity of >130 Hounsfield Unit were identified. The size of the lesion was automatically calculated, and the CAC was scored using the Agatston algorithm. The CAC was computed across all lesions denoted within the left main, left anterior descending (LAD), left coronary circumflex (LCx), and right coronary artery (RCA), and the sum of all lesion scores yielded the total CAC score [26]. As recommended, we used the CAC scoring system in a binary fashion (CAC present or absent) [26,27]. Accordingly, any Agatston score above 0 was considered as a present CAC lesion [26,27].

2.4. Biochemical Analyses

A conventional lipid profile and high-sensitive C-Reactive Protein (hsCRP) profile were performed on routine autoanalyzers. Low-density lipoprotein (LDL) cholesterol levels were derived from the conventional Friedwald equation. Anti-apoA-1 IgG serum levels were measured as previously described [8,9,11–13]. Briefly, Maxi-Sorb plates (Nunc) were coated with purified, human-derived delipidated apoA-1 (20 µg/mL; 50 µL/well) for 1 h at 37 °C. After 3 washes with phosphate buffered saline (PBS)/2% bovine serum albumin (BSA; 100 µL/well), all wells were blocked for 1 h with 2% BSA at 37 °C. Samples were diluted 1 to 50 in PBS/2% BSA and incubated for 60 min. Additional patient samples at the same dilution were also added to an uncoated well to assess individual nonspecific binding. After 6 further washes, 50 µL/well of signal antibody (alkaline phosphatase-conjugated anti-human IgG; Sigma-Aldrich, Saint Louis, MO, USA) diluted 1:1000 in PBS/2% BSA solution were incubated for 1 h at 37 °C. After 6 more washes (150 µL/well) with PBS/2% BSA solution, the phosphatase substrate p-nitrophenyl phosphate disodium (50 µL/well; Sigma-Aldrich) dissolved in diethanolamine buffer (pH 9.8) was added. Each sample was tested in duplicate, and absorbance, determined as the optical density at 405 nm (OD405 nm), was measured after 20 min of incubation at 37 °C (FilterMax, Molecular Devices, San Jose, CA, USA). The corresponding nonspecific binding value was subtracted from the mean absorbance value for each sample. Anti-apoA-1 IgG seropositivity cutoff was predefined as previously validated and set at an optical density (OD) value of 0.6 and 37% of the positive control value, as described earlier [8,9,11–13]. OD values ranged from 0 to 1.3, and corresponding index values were between 0 and 84.9%.

Pre-β-HDL levels were measured using a quantitative enzyme-linked immunosorbent assay (ELISA) kit for pre-β-HDL detection in human plasma [28], following the manufacturer's instruction (Sekisui Diagnostic, Darmstadt, Germany). The coefficient of variation was 3.1% to 5.3% for individual analytical runs, and 4.9% to 9.1% between different analytical runs.

2.5. Serum HDL Cholesterol Efflux Capacity (CEC)

All serum samples were stored at −80 °C. The aliquots were slowly defrosted in ice. Four cholesterol efflux pathways were evaluated in cell cultures: total CEC (in J774 mouse macrophages treated with 0.3 mm cAMP for 18 h to up-regulate ABCA1 expression) [1]; passive diffusion (PD)-CEC (in J774 mouse macrophages, basal conditions) [29]; ABCA1-CEC (as a difference between total efflux and PD) [30]; ABCG1-CEC (in hABCG1-expressing Chinese hamster ovary cells, CHO-K1, as efflux difference between hABCG1-expressing and parent CHO-K1 control cells) [31]. In all determinations, cells were labeled with [1,2-^{3}H]-cholesterol in the presence of acyl CoA: cholesterol acyl tranferase (ACAT) inhibitor (2 µg/mL, Sandoz 58035) for 24 h. The efflux was promoted for 4 h (6 h for ABCG1-CEC) to 2% (v/v) serum samples. Serum HDL CEC was expressed as a percentage of the radioactivity released to the medium in 4 h (6 h for ABCG1-CEC) over the total radioactivity incorporated by cells. Control samples were run to confirm the responsiveness of cells. Background efflux, evaluated in the absence of acceptors, was subtracted from each samples value. Serum samples were determined in triplicate, while a standard pool of human serum from our laboratory (SN1) permitted correction for inter-assay variability and a second serum standard pool (SN2) was used to determine inter-assay variability [29].

2.6. Antibody Anti-apoA-1 Modulation of Cellular Cholesterol Efflux

The efflux process was evaluated in J774 cells as previously described. Following the ABCA1 upregulation with cyclic adenosine monophosphate (cAMP) we incubated cells, during the efflux time (4 h), with 25 µg/mL of anti-apoA-1 IgG. The efflux was promoted to either 10 µg/mL of apo-A-1 or 2% (v/v) of human serum standard [32].

2.7. Measurement of Intracellular Cholesterol Content

J774 cells were cultured in 10% fetal calf serum (FCS) in RPMI cell medium at 37 °C in 5% CO_2. To perform the experiments, cells were seeded in 24-well plates at a density of 2×10^5 cells/well for 24 h. Cells were exposed for 24 h to either 10% (v/v) whole serum and control IgG or anti-apoA-1 IgG 40 µg/mL. Cellular cholesterol content before and after serum exposure was measured as previously described [33]. Briefly, at the end of the experiment, cell monolayers were washed with Phosphat-Buffered Salin (PBS) and lysed in 0.5 mL of a 1% sodium cholate solution in water supplemented with 10 U/mL DNase. Cholesterol was than measured fluorimetrically using the Amplex Red Cholesterol Assay Kit (Molecular Probes, Eugene, OR, USA) as described by the manufacturer. The amount of cholesterol in each well was measured by comparison with a cholesterol standard curve included in each experiment. An aliquot of the cell lysates was also taken to measure cell protein by a modified Lowry method [34]. Cell cholesterol content after exposure of cells to serum and immunoglobulin G (IgG) was expressed as µg of cholesterol /mg protein.

2.8. Antibody Anti-apoA-1 Modulation of Membrane Free Cholesterol: Assay of Cholesterol Oxidase

Cholesterol oxidase treatment was essentially as previously described [35]. Briefly, cells were labelled with 3 µCi/mL [3H] cholesterol. Cholesterol oxidase (1 U/mL) was added, and cells were incubated for 4 h. Lipid was extracted with isopropanol, and radioactive cholesterol and cholestenone were separated using thin-layer chromatography and quantified.

2.9. Anti-apoA-1 IgG Modulation of Cellular Cholesterol Esterification: ACAT Activity

Cells were treated as previously described for the measurement of intracellular cholesterol content. Cholesterol esterification was evaluated as the incorporation of radioactivity into cellular cholesteryl esters after addition of [14C]-oleate-albumin complex [36]. At the end of incubation, cells were washed with PBS and lipids were extracted with hexane/isopropanol (3:2). The extracted lipids were separated by thin layer chromatography (TLC) (isoctane/diethyl ether/acetic acid, 75:25:2, v/v/v). Cholesterol radioactivity in the spots was determined by counting liquid scintillation.

2.10. Measurement of Free Cholesterol Content in Cell Supernatant

Free cholesterol in supernatant was measured using the fluorometric method Amplex® Red Cholesterol Assay Kit (Molecular Probe, Eugene, OR, USA) following the manufacturer's instruction. The experiments were conducted without an ACAT inhibitor.

2.11. Statistical Analyses

Analyses were performed using Statistica software (StatSoft, Tulsa, OK, USA). Normally, distributed clinical data are presented as mean ± standard deviation (SD), whereas data following a non-normal distribution are presented as median and interquartile range (IQR). Accordingly, means were compared using bilateral student T test, and medians were compared with the Mann-Whitney-test. Proportions were compared using two-tailed exact Fischer test. Correlations between variables were assessed using Spearman test. Due to the limited sample size, adjusted analyses were limited to the different forms of cholesterol efflux analyzed and the 10-year Framingham risk score (FRS) for coronary heart disease risk prediction [37] (allowing the aggregation of all traditional CV risk factors within a single continuous variable). Cholesterol efflux and intracellular cholesterol data were expressed as mean ± SD. Treatment groups were compared using an unpaired two tailed Student's t-test. Because of the explorative nature of this work with predefined endpoints, Bonferroni correction was not applied. Logistic regression analyses were used to assess associations between the presence of any CAC lesion and CEC, or anti-apoA-1 IgG as variables. In this model, the presence of any CAC lesion was set as the dependent variable. Associations are presented as odds ratios (OR) and corresponding 95% confidence intervals (95% CI). A p value below 0.05 was considered as statistically significant.

3. Results

3.1. Study Characteristics

Baseline demographic characteristics are presented in Table 1. As shown in this table, obese participants had lower levels of HDL, but higher triglycerides, hsCRP, and pre-βHDL levels when compared to non-obese subjects. No significant differences were found between control and obese subjects in terms of CEC mediated by ABCG1, ABCA1, or PD pathways and total (ABCA1 + PD) CEC, and no difference was observed between these two groups for the CAC score, the number of coronary artery lesions, or the presence of any lesion identified in chest CT scans. On the other hand, obese participants tended to have higher anti-apoA-1 IgG levels and a significantly higher anti-apoA-1 IgG positivity rate. Interestingly, anti-apoA-1 IgG positive participants ($n = 6$) were only observed in the obese subjects. No other significant association with any of the other parameters tested was retrieved (Table 1).

Table 1. Baseline demographic and biological characteristics.

	Overall (n = 69)	Obese (n = 35)	Non-Obese (n = 34)	* p Value
Age, mean (+/−SD)	44 (11.4)	44.6 (12.0)	43.5 (11.0)	0.61
Male Gender; n (%)	42 (60.8)	23 (65.7)	19 (55.8)	0.46
Weight in Kg, mean (+/−SD)	97.5 (26.37)	116 (19.3)	74.4 (12.5)	<0.001
Height in cm, mean (+/−SD)	74 (9.3)	173 (9.4)	172.2 (9.5)	0.56
BMI in kg/m², mean (+/−SD)	32.2 (8.8)	39.0 (7.1)	25.1 (3.0)	<0.001
Framingham risk score; median (IQR; range)	1 (0.9–4; 0.9–16)	1 (0.9–6; 0.9–16)	1 (0.9–3; 0.9–15)	0.13
FRS > 10%; n (%)	7 (10.1)	6 (17)	1 (2.9)	0.10
Chest CT:				
Total CAC score, mean (+/−SD)	4.9 (15.8)	4.65 (14.8)	5 (17)	0.75
Number of CAC lesions, mean (+/−SD)	0.28 (0.75)	0.28 (0.71)	0.26 (0.79)	0.74
Presence of any CAC lesion; n (%)	10 (14.4)	6 (17.1)	4 (11.7)	0.73

Table 1. Cont.

	Overall (n = 69)	Obese (n = 35)	Non-Obese (n = 34)	* p Value
Lipid profile:				
Total cholesterol in mg/dL, mean (+/−SD)	198.9 (36.4)	198.5 (41.4)	202.1 (31.5)	0.69
LDL cholesterol in mg/dL mean (+/−SD)	127.9 (32.2)	126.9 (36.9)	133.9 (26.7)	0.42
HDL cholesterol in mg/dL, mean (+/−SD)	47.1 (13.0)	42.1 (11.5)	50.4 (13.9)	0.04
Triglycerides in mg/dl, mean (+/−SD)	97.5 (73.2)	122.8 (70.8)	83.0 (67.0)	<0.001
Pre-β-HDL in µg/mL	42.6 (16.62)	49.2 (17.26)	35.6 (12.8)	<0.001
Cholesterol Efflux Capacity:				
ABCG1-mediated, mean (+/−SD)	4.04 (1.2)	4.04 (0.89)	3.85 (1.49)	0.96
ABCA1-mediated, mean (+/−SD)	3.98 (1.49)	4.18 (1.34)	3.67 (1.61)	0.18
Total, mean (+/−SD)	13.73 (1.55)	13.5 (1.57)	13.80 (1.57)	0.64
Passive diffusion, mean (+/−SD)	9.75 (2.05)	9.5 (1.9)	9.70 (2.16)	0.19
Anti-apoA-1 IgG OD; median (IQR; range)	0.31 (0.18–0.43; 0–1.3)	0.33 (0.2–0.48; 0.1–1.3)	0.26 (0.16–0.38; 0–0.56)	0.05
Anti-apoA-1 positivity, n (%)	6 (8.7)	6 (17.1)	0 (0)	0.02
hsCRP, median (IQR; range)	2.9 (0.9–6; 0.9–26.1)	5 (2.5–8.7; 0.9–26.1)	1 (0.9–3; 0.32–11)	<0.0001

Note: * p values were obtained by comparing obese versus non-obese subjects. For normally distributed parameters, p values were computed according to student t-test and for non-parametric parameters U-Mann Whitney test was used. Proportions were compared using two-tailed exact Fischer test. Abbreviations: SD = standard deviation; IQR = interquartile range; BMI = Body mass index; FRS = Framingham Risk Score; CT = Computed tomography; CAC = Coronary artery calcium; LDL = Low-density lipoprotein; HDL = High-density lipoprotein; ABCG1 = ATP-binding cassette subfamily G member 1; ABCA1 = ATP-binding cassette transporter 1; OD = Optical density; hsCRP = High-sensitivity C-reactive Protein.

3.2. Associations between CEC Pathways, CAC Score, Framingham Risk Score, Pre-Beta-HDL, and hsCRP Levels

ABCG1 CEC was not found to be associated with CAC score or FRS, nor with pre-β-HDL levels in any of the study groups (Table 2). ABCA1 CEC was found to be positively associated with CAC score and the number of CAC lesions in overall subjects ($r = 0.26$, $p = 0.02$), and the strength of this association seemed to increase in obese subjects ($r = 0.47$, $p = 0.004$), whereas it was lost in non-obese participants (Table 2). The only significant association retrieved between CEC and pre-β-HDL levels was between pre-β-HDL and PD-CEC ($r = 0.39$, $p = 0.02$), which was only observed in obese subjects. Furthermore, FRS correlated positively with ABCA1-CEC in overall and control subjects, but not in obese participants, whereas we observed an inverse correlation between PD-CEC, CAC score, and FRS in overall and obese participants, but not in non-obese subjects (Table 2). HsCRP levels did not correlate with any forms of CEC in any subgroups (Table 2), nor with the FRS ($r = 0.09$; $p = 0.43$), with CAC score ($r = -0.04$; $p = 0.75$), or with the number of CAC lesions ($r = -0.03$; $p = 0.77$). This absence of correlation between hsCRP, FRS, the CAC score, and the CAC lesion was unchanged in obese and non-obese individuals (data not shown). Because of the significant correlation shown between CAC score and anti-apoA-1 IgG [19], as well as between CAC score/lesions and ABCA1- and PD-CEC, we performed a sensitivity analysis after excluding anti-apoA-1 IgG positive patients ($n = 6$) in the obese subgroup in order to see if excluding high anti-apoA-1 IgG values could change the associations between different forms of CEC with FRS, number of CAC lesions, and CAC score, as well as with pre-β-HDL. Due to the non-significant aforementioned associations with hsCRP levels and any forms of CEC, hsCRP was not considered in our sensitivity analyses. As shown in Table 2, after excluding anti-apoA-1 IgG seropositive subjects, the correlations between CAC score/lesion and ABCA1-CEC and PD-CEC remained significant and of the same order of magnitude, whereas the significant association between FRS and ABCA1 was lost. The correlation between FRS and PD-CEC was unchanged. This does not point to anti-apoA-1 IgG as a key determinant of the associations retrieved between CAC score, ABCA1-CEC, and PD-CEC. Finally, if logistic regression analyses indicated that ABCA1-CEC, PD-CEC, and FRS were all significant predictors of CAC lesion in univariate analysis, the FRS was found to be the only independent predictor of CAC lesion in the overall cohort (Table 3). Given the limited study sample size, these analyses were not performed in the obese and non-obese subgroups separately.

Table 2. Correlations between CEC Pathways, CAC Score, Framingham Risk Score, Pre-Beta-HDL, and hsCRP Levels.

Correlations	Overall Subjects (n = 69)		Obese Subjects (n = 35)		Non-Obese Subjects (n = 34)	
	R Value (Spearman)	p Value	R Value (Spearman)	p Value	R Value (Spearman)	p Value
ABCG1-mediated CEC vs.:						
CAC score	0.02	0.84	−0.03	0.82	0.10	0.57
Nr of CAC lesions	0.02	0.86	−0.04	0.81	0.09	0.61
Framingham RS	−0.01	0.90	0.06	0.70	−0.09	0.59
Pre-β-HDL	−0.22	0.09	−0.19	0.27	−0.23	0.18
hsCRP	0.007	0.94	0.04	0.77	0.22	0.20
ABCA1-mediated CEC vs.:						
CAC score	0.26	0.02	0.47	0.004	0.05	0.76
Nr of CAC lesions	0.26	0.02	0.47	0.004	0.04	0.80
Framingham RS	0.30	0.01	0.18	0.29	0.40	0.01
Pre-β-HDL	−0.08	0.48	−0.15	0.38	−0.25	0.15
hsCRP	0.05	0.44	0.04	0.79	−0.12	0.47
Passive Diffusion vs.:						
CAC score	−0.30	0.01	−0.46	0.006	−0.10	0.53
Nr of CAC lesions	−0.29	0.01	−0.45	0.006	−0.09	0.60
Framingham RS	−0.33	0.006	−0.40	0.01	−0.22	0.21
Pre-β-HDL	0.15	0.20	0.39	0.02	0.06	0.72
hsCRP	−0.01	0.90	0.06	0.63	0.17	0.35

Correlations after excluding anti-apoA-1 IgG seropositive individuals in obese participants (n = 29)		
Correlations	R Value (Spearman)	p Value
ABCG1-mediated CEC vs.:		
CAC score	0.01	0.72
Nr of CAC lesions	0.01	0.97
Framingham RS	0.06	0.72
Pre-β HDL	−0.13	0.49
ABCA1-mediated CEC vs.:		
CAC score	0.39	0.03
Nr of CAC lesions	0.39	0.03
Framingham RS	0.24	0.20
Pre-β HDL	−0.14	0.48
Passive diffusion vs.:		
CAC score	−0.39	0.04
Nr of CAC lesions	−0.39	0.04
Framingham RS	−0.45	0.009
Pre-β-HDL	0.41	0.03

Table 3. Logistic regression analyses for CAC lesion prediction

Continuous Predictors	Univariate Analyses			Multivariate Analyses		
	Odds Ratio	95% CI	p	Odds Ratio	95% CI	p
ABCG1 CEC	1.10	0.64–1.91	0.72	1.02 *	0.51–2.05	0.94
ABCA1 CEC	1.17	1.17–3.13	0.009	1.38 **	0.67–2.86	0.38
Passive diffusion	0.56	0.32–0.97	0.03	0.78 ***	0.37–1.54	0.44
Framingham RS	1.31	1.13	0.004	1.27 ****	1.09–1.49	0.002

Note: * Adjusted for ABCA1 CEC, Passive diffusion and FRS; ** adjusted for ABCG1-CEC, passive diffusion, and FRS; *** adjusted for ABCG1-CEC and ABCA1-CEC, and FRS; **** adjusted for ABCA1, ABCG1, and passive diffusion.

3.3. Associations between Anti-apoA-1 IgGs, Specific CEC Pathways, Pre-β-HDL, and hsCRP Levels

Spearman correlations showed a positive association between ABCA1-CEC and anti-apoA-1 IgG ($r = 0.48$, $p < 0.001$) and a negative association between PD-CEC and these antibodies ($r = -0.33$; $p = 0.004$; Table 4) on our overall participants samples. The anti-apoA-1 IgG association with ABCA1-CEC remained significant for the obese and non-obese subgroups. Nevertheless, if the association with PD-CEC was unchanged in non-obese subjects, it was lost in obese participants (Table 4). No associations were retrieved between pre-β-HDL serum levels and anti-apoA-1 IgG, nor between these antibodies and hsCRP (data not shown).

Table 4. Spearman's correlations between anti-apoA-1 IgGs, Specific CEC Pathways, Pre-β-HDL, and hsCRP levels

	Overall Subjects ($n = 69$)		Obese Subjects ($n = 35$)		Non-Obese Subjects ($n = 34$)	
Correlations	R value	p value	R value	p value	R value	p value
Anti-apoA-1 IgG (OD) vs.:						
CAC score	0.21	0.09	0.37	0.03	0.01	0.91
Nr of CAC lesions	0.21	0.09	0.37	0.03	0.02	0.88
Framingham Risk Score	−0.03	0.81	−0.03	0.82	−0.11	0.51
ABCG1-mediated CEC	−0.16	0.17	−0.23	0.18	−0.10	0.54
ABCA1-mediated CEC	0.48	0.00002	0.49	0.002	0.44	0.007
Passive diffusion	−0.33	0.004	−0.29	0.09	−0.36	0.03
Pre-β-HDL	0.17	0.89	−0.04	0.81	−0.06	0.9

3.4. Anti-apoA-1 IgG-Mediated Modulation of Cellular Cholesterol Efflux

To further evaluate the possible impact of anti-apoA-1 IgG on CEC, we measured ABCA1-CEC and PD-CEC using standard serum or apoA-1 in J774 macrophages without cAMP stimulation in presence of anti-apoA-1 antibodies during the efflux time. As shown in Figure 1, cells incubated with the antibody showed an increased ABCA1-mediated efflux to the acceptors (4.28 ± 0.29 vs. 6.45 ± 0.45 $p < 0.0001$ to apoA1 as acceptor; 1.16 ± 0.15 vs. 1.81 ± 0.19 $p < 0.0001$ to human serum as acceptor). Conversely, the presence of the anti-apoA-1 IgG reduced the passive cellular efflux diffusion to both apoA-1 and human serum (0.77 ± 0.10 vs. 0.48 ± 0.11 $p < 0.0001$; 6.49 ± 0.05 vs. 4.67 ± 0.66 $p < 0.0001$).

3.5. Modulation of Anti-apoA-1 IgGs on Intracellular and Membrane Cholesterol Content

To understand the net effect of such dual modification of anti-apoA-1 IgGs on cellular cholesterol balance, we measured the impact of anti-apoA-1 IgGs on intracellular cholesterol content according to a previously validated method. As shown in Figure 2, anti-apoA-1 IgGs induced a significant increase in intracellular cholesterol content compared to control IgGs and baseline conditions.

To better investigate the possible anti-apoA-1 IgG-mediated passive diffusion inhibition, we look at the free cholesterol content in the membrane of J774 cells after four hour anti-apoA-1 IgG stimulation. As shown in Figure 3, anti-apoA-1 IgGs significantly reduced the free cholesterol membrane content when compared to the control IgGs or to the untreated condition. In the same experimental conditions, we observed a significant reduction of the free cholesterol content in the supernatant counterpart in the presence of anti-apoA-1 IgGs (Figure 4). Since ACAT is the key enzyme for the cholesterol esterification and its cellular redistribution, we measured the ACAT activity in the presence of anti-apoA-1 IgGs or control antibodies. As shown in Figure 5, anti-apoA-1 IgGs (and not the control IgGs) significantly increased ACAT activity.

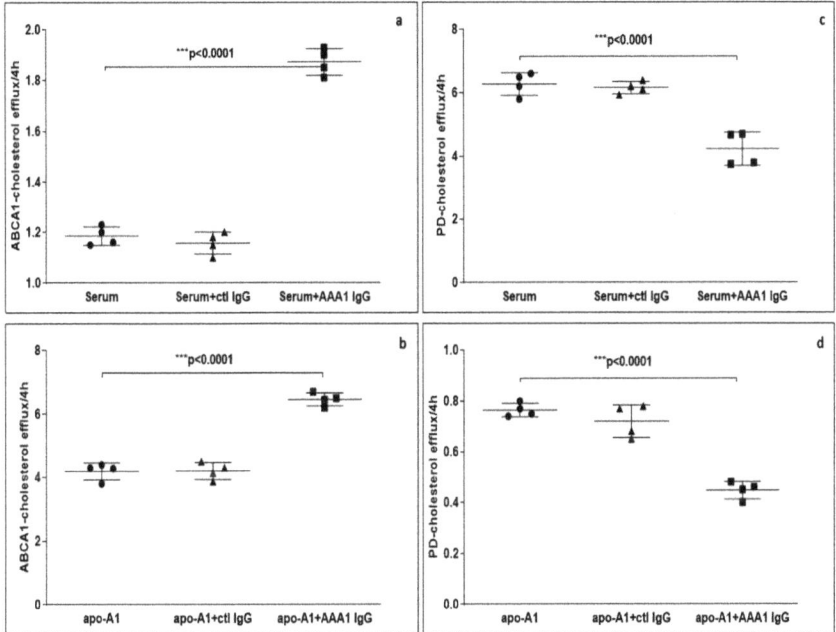

Figure 1. Cholesterol efflux from J774 to various acceptors in presence of anti-apoA-1 (AAA1) IgG. (**a,b**) The cells were activated with cAMP to measure the specific efflux from ABCA1 to serum or apoA-1 as acceptors after 4 h incubation. Data are expressed as the mean ± SD of the measurements done in triplicate and repeated four times ($n = 4$). Statistical differences were determined by one-way analysis of variance (one-way ANOVA), with $p \leq 0.05$ being considered significant. (**c,d**) The cells were not activated in order to quantify the passive cellular efflux diffusion to both apoA-1 or human serum. The percentage of cholesterol efflux is shown. Data are expressed as the mean ± SD of the measurements done in triplicate and repeated four times ($n = 4$). Statistical differences were determined by one-way analysis of variance (one-way ANOVA), with $p \leq 0.05$ being considered significant.

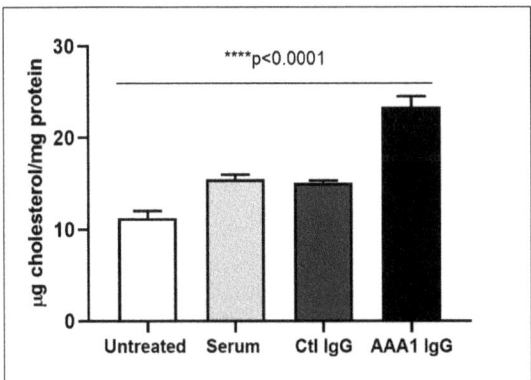

Figure 2. Anti-apoA-1 IgG increase the intracellular cholesterol content. Cell cholesterol content after exposure of cells to serum, anti-apoA-1 IgG, or control IgG, and was expressed as µg of cholesterol /mg total protein. Data are expressed as the mean ± SD of the measurements done in triplicate and repeated three times ($n = 3$). Statistical differences were determined by one-way analysis of variance (one-way ANOVA), with $p \leq 0.05$ being considered significant.

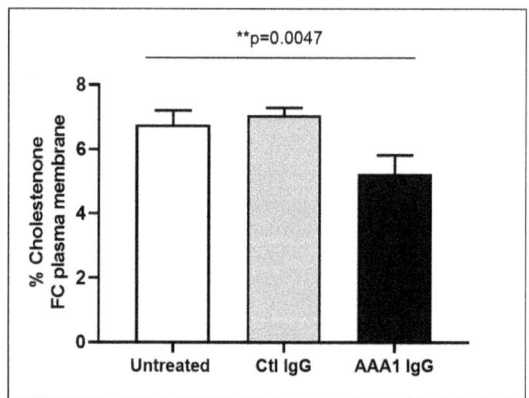

Figure 3. Free cholesterol in membrane. Monolayers were labeled with 3 µCi/mL [3H] cholesterol for 24 h in RPMI medium with 10% FCS, control IgGs or anti-apoA-1 IgGs 40 µg/mL. Cells were then washed and incubated with 1 U/mL cholesterol oxidase enzyme in Dulbecco's Phosphate Buffered Saline (DPBS) for 4 h at 37 °C. Data are expressed as the mean ± SD of the measurements done in triplicate and repeated three times ($n = 3$). Statistical differences were determined by one-way analysis of variance (one-way ANOVA), with $p \leq 0.05$ being considered significant.

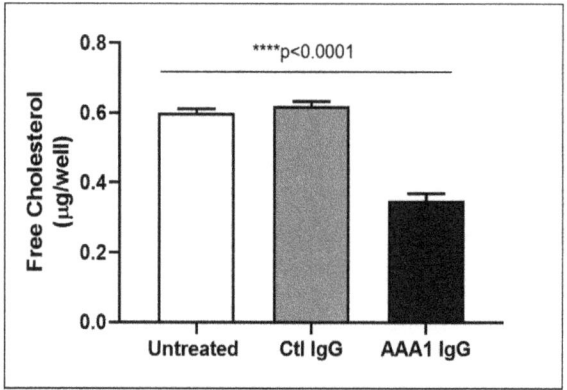

Figure 4. Free cholesterol in the cell supernatant. Free cholesterol content in J774 supernatant after exposure of cells to serum, anti-apoA-1 IgGs or control IgGs was expressed as µg of free cholesterol per well. Data are expressed as the mean ± SD of the measurements done in triplicate and repeated three times ($n = 3$). Statistical differences were determined by one-way analysis of variance (one-way ANOVA), with $p \leq 0.05$ being considered significant.

Taken together, these results indicate that the presence of anti-apoA-1 IgG on J774 macrophages enhances ACAT activity leading to increased intracellular esterified cholesterol pools, and decreases membrane free-cholesterol content. Such decrease in membrane free-cholesterol content reduce the cholesterol gradient, leading to a decrease in passive diffusion PD followed by the reduction of free-cholesterol content in cell supernatants. In this context, the increase in ABCA1-efflux in the presence of anti-apoA-1 IgG can be interpreted as the consequence of increased intracellular lipid pools.

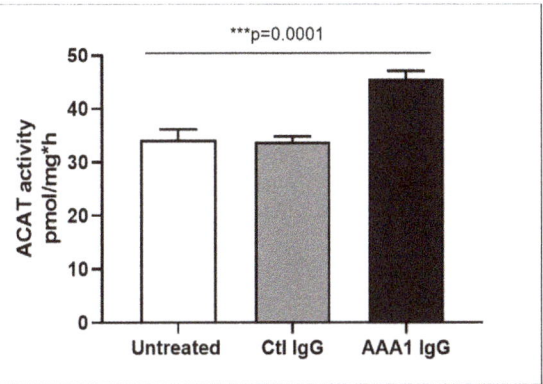

Figure 5. ACAT activity. Cells were incubated for 24 h in RPMI medium with 10% FCS, control IgG or anti-apoA-1 IgG 40 µg/mL. Monolayers underwent a second incubation (4 h) in the presence of [1-14C]-oleic acid albumin complex. Data are expressed as the mean ± SD of the measurements done in triplicate and repeated three times ($n = 3$). Statistical differences were determined by one-way analysis of variance (one-way ANOVA), with $p \leq 0.05$ being considered significant.

4. Discussion

In this study, we extend previous findings by demonstrating that on top of being significantly associated with higher CAC [19], anti-apoA-1 IgG levels are inversely associated with PD-CEC and positively with ABCA1-CEC in healthy obese subjects, despite the absence of ABCA1-CEC, ABCG1-CEC, and PD-CEC differences between obese and control participants. This apparently counterintuitive clinical observation derived from this small case-control study prompted us to evaluate the possible direct impact of anti-apoA-1 IgG on normal serum and apoA-1 CEC on macrophages, and on macrophage cholesterol content.

Corroborating our clinical observations, these in vitro investigations indicated that, per se, these autoantibodies could at the same time inhibit PD-CEC and potentiate ABCA1-CEC. PD efflux consists of the simple diffusion of free/unesterified cholesterol molecules across the membrane through the aqueous phase following their concentration gradient between the membrane and HDL/apoA-1, and accounts for the majority of total macrophage efflux (up to 80%) in normocholesterolemic conditions [38,39]. On the other hand, ABCA1-mediated efflux is an active pathway facilitating the transfer of unesterified cholesterol specifically to lipid-poor apoA-1, and represents up to 60% of total macrophage cholesterol efflux when activated by increased intracellular cholesterol load [38–40]. For this reason, determining the net effect of simultaneously blocking and activating these two important HDL CEC pathways at the cellular level was the key to providing a sound interpretation of our clinical and functional in vitro observations. As anti-apoA-1 IgG leads to increased intracellular cholesterol content in macrophages after 24 h of stimulation, we currently postulate that the anti-apoA-1 IgG overdriving mechanism consists of ACAT stimulation, leading to intracellular esterified cholesterol accumulation, decreasing the amount of membrane free-cholesterol available, and thus diminishing the cholesterol gradient requested for PD-CEC. In this context, the increase in ABCA1 efflux can be ascribed to a feedback loop aiming at preventing cellular cholesterol accumulation. Nevertheless, other possibilities cannot be formally excluded, and the exact mechanisms by which anti-apoA-1 IgG stimulates ACAT activity are still elusive. Several non-exclusive hypotheses are discussed below. Firstly, one can postulate an upstream mechanism, where anti-apoA-1 IgG could, per se, stimulate ACAT activity. With ACAT being an endoplasmic reticulum intracellular protein, an indirect effect may be evoked. Indeed, as ACAT deficiency has been shown to increase TLR4 expression and function in hepatocytes stellate cells [41], we cannot exclude the fact that anti-apoA-1 IgG-mediated TLR2/TLR4 stimulation may interfere with ACAT activity, although this hypothesis is currently devoid from of experimental evidence.

Alternatively, as ACAT activation is known to occur in response to cholesterol overload [42], any factor increasing either the intracellular cholesterol synthesis or uptake could be susceptible to increase ACAT activity. Therefore, knowing whether anti-apoA-1 IgG could affect HGMCoA reductase or LDL-receptor expression and function is currently under investigation.

Alternatively, we can consider the anti-apoA-1 IgG-mediated ACAT stimulation as a downstream consequence of mechanisms related to PD-CEC inhibition. Given the complexity of PD-CEC, where desorption of free cholesterol molecules from the cell plasma membrane into the surrounding aqueous phase is the rate-limiting step [38–40], there are several possibilities, including those in direct relation with apoA-1 [40]. Firstly, by stabilizing lecithin-cholesterol acyltransferase (LCAT) activity, apoA-1 maintains a favourable cholesterol gradient. One could hypothesize that anti-apoA-1 IgG may decrease LCAT stability, attenuating the cholesterol gradient and followed by PD-CEC attenuation. Secondly, as apoA-1 can also facilitate the maintenance of such gradient, either by acting as an acceptor for cholesterol molecules diffusing away from the cell surface, or by mediating the process of membrane microsolubilization, we can also envisage an antibody-mediated interference of one of these acceptor-related processes [38]. Indeed, as c-terminal (c-ter) part of apoA-1 is known to mediate membrane microsolubilization important for PD-CEC to occur [43], and because anti-apoA-1 IgG in humans is biased against the c-ter part of apoA-1 [44], we cannot rule-out a direct and antagonistic effect of anti-apoA-1 IgG on PD-CEC. Finally, as mentioned above, the increased ABCA1-CEC in the presence of anti-apoA-1 IgG is likely to represent a homeostatic feedback loop. Nevertheless, as ABCA1 activity is regulated by apoA-1's interaction with an epitope on the c-terminal part of apoA-1 [45], we cannot formally rule out a direct agonistic effect of these antibodies on ABCA1, or an indirect effect mediated through the antibody-induced macrophage interleukin-6 (IL-6) production known to increase ABCA1 efflux [46]. Despite these current mechanistic knowledge gaps requiring further detailed investigations, these results indicate that anti-apoA-1 IgG-induced ACAT stimulation followed by PD-CEC inhibition represent another mechanism by which these auto-antibodies can promote atherogenesis.

The second notable finding of the present study is the observation that in obese participants, macrophage ABCA1-CEC is positively associated with coronary atherosclerosis burden quantified by the CAC score, whereas no such association was retrieved in healthy controls. Although somehow contrasting with data published in primary and secondary prevention settings [1–3], these results are in line with previous observations in metabolic syndrome and rheumatoid arthritis [4–6], and lend weight to the notion that depending on disease-related systemic factors, the associations between macrophage ABCA1-CEC and coronary atherosclerosis burden are not necessarily straightforward. As metabolic syndrome patients share many pathophysiological similarities with obese patients, the existence of such a paradoxical association was suspected, but still remained to be demonstrated. Furthermore, providing PD-CEC results in parallel with ABCA1-CEC was found to be instrumental in facilitating the interpretation of these results. Indeed, the inverse association between CAC score and macrophage PD-CEC suggest that the lower the PD efflux, the higher the level of intracellular cholesterol, and thus atherosclerosis burden. In this context, the anti-apoA-1 IgG-dependent ABCA1-CEC increase, in fact, may not sufficiently compensate for the PD inhibition ascribed to anti-apoA-1 IgG. Furthermore, as our in vitro data demonstrated a substantial influence of anti-apoA-1 IgG on PD-CEC through ACAT stimulation, we performed a sensitivity analysis excluding anti-apoA-1 IgG seropositive subjects in the obese subgroup, which yielded similar results overall. Taken together, these results indicate that considering ABCA1-CEC without considering other concomitant functional effluxes may lead to erroneous conclusions. These results also suggest that the presence of anti-apoA-1 IgG may represent a new systemic factor susceptible to altering the usual association between ABCA1-CEC and CV risk.

This study has several limitations. The first one relates to the low number of subjects enrolled in this study, raising possible power issues when it comes to the plausibility of non-significant findings. Nevertheless, despite this power limitation, the significant findings derived from this cohort enabled us to generate hypotheses that were tested and validated in vitro, providing valuable insights to correctly interpret counterintuitive associations. A second limitation of this study is that we were not able to decipher the exact molecular mechanisms by which anti-apoA-1 IgG could stimulate ACAT activity and inhibit PD-CEC, and we did not explore their possible ability to modulate key cholesterol homeostasis regulating genes, which is currently under investigation (Pagano S. et al., under review). A further limitation is the fact that due to limited human sample availability, we could not evaluate the effects of human purified apoA-1 IgG in this study. Nevertheless, as we previously demonstrated that the commercial anti-human apoA-1 IgGs used in this study were promoting the same effects as human-purified IgG fraction containing high levels of these autoantibodies in vitro [13,18], and because these autoantibodies have been consistently used in different validated and published animal and in vitro studies [11–19], we expect similar impacts on in vitro lipid metabolism to occur by using human purified anti-apoA-1 IgG, even if not formally demonstrated. Finally, being focused on atherogenesis and foam cell formation, we did not evaluate whether the anti-apoA-1 IgG effects on lipid metabolism could be reproduced by other cell types, such as hepatocytes, and did not measure the scavenger B-I-related CEC.

5. Conclusions

In conclusions, the results of this study indicate that in healthy obese subjects there is a paradoxical positive association between CAC score and ABCA1-CEC, and a negative association between PD-CEC and CAC score. These associations could be partly explained by the propensity of anti-apoA-1 IgG to inhibit PD-mediated cholesterol efflux through ACAT stimulation and promote foam cell formation, notwithstanding an enhancing effect on ABCA1-mediated cholesterol efflux. Although preliminary, these results highlight the importance of simultaneously considering the different CEC pathways and using a translational approach for proper ABCA1-CEC interpretation.

Author Contributions: Conceptualization, N.V., S.P., and E.F.; methodology, S.P., E.C., and A.Q.; validation, F.M. and T.H.S.; formal analysis, N.V. and F.M.; investigation, T.H.S.; data curation, T.H.S. and A.Q.; writing—original draft preparation, N.V. and E.F.; writing—review and editing, N.V., S.P., E.F., N.R., and F.M.; supervision, F.M.; funding acquisition, N.V.

Funding: This work was supported by a grant from the Leenaards Foundation, grant number 3698 to N.V., by the Swiss National Science Foundation, grant number 310030-163335 to N.V., and by the De Reuter Foundation, grant number 315112 to N.V.

Acknowledgments: None.

Conflicts of Interest: The authors declare no conflict of interest.

References

1. Khera, A.V.; Cuchel, M.; de la Llera-Moya, M.; Rodrigues, A.; Burke, M.F.; Jafri, K.; French, B.C.; Phillips, J.A.; Mucksavage, M.L.; Wilensky, R.L.; et al. Cholesterol Efflux Capacity, High-Density Lipoprotein Function, and Atherosclerosis. *N. Engl. J. Med.* **2011**, *364*, 127–135. [CrossRef] [PubMed]
2. Khera, A.V.; Demler, O.V.; Adelman, S.J.; Collins, H.L.; Glynn, R.J.; Ridker, P.M.; Rader, D.J.; Mora, S. Cholesterol Efflux Capacity, High-Density Lipoprotein Particle Number, and Incident Cardiovascular Events: An Analysis from the Jupiter Trial (Justification for the Use of Statins in Prevention: An Intervention Trial Evaluating Rosuvastatin). *Circulation* **2017**, *135*, 2494–2504. [CrossRef] [PubMed]
3. Rohatgi, A.; Khera, A.; Berry, J.D.; Givens, E.G.; Ayers, C.R.; Wedin, K.E.; Neeland, I.J.; Yuhanna, I.S.; Rader, D.R.; de Lemos, J.A.; et al. Hdl Cholesterol Efflux Capacity and Incident Cardiovascular Events. *N. Engl. J. Med.* **2014**, *371*, 2383–2393. [CrossRef] [PubMed]
4. Li, X.M.; Tang, W.H.; Mosior, M.K.; Huang, Y.; Wu, Y.; Matter, W.; Gao, V.; Schmitt, D.; Didonato, J.A.; Fisher, E.A.; et al. Paradoxical Association of Enhanced Cholesterol Efflux with Increased Incident Cardiovascular Risks. *Arterioscler. Thromb. Vasc. Biol.* **2013**, *33*, 1696–1705. [CrossRef] [PubMed]
5. Lucero, D.; Sviridov, D.; Freeman, L.; Lopez, G.I.; Fassio, E.; Remaley, A.T.; Schreier, L. Increased Cholesterol Efflux Capacity in Metabolic Syndrome: Relation with Qualitative Alterations in Hdl and Lcat. *Atherosclerosis* **2015**, *242*, 236–242. [CrossRef] [PubMed]

6. Nestel, P.; Hoang, A.; Sviridov, D.; Straznicky, N. Cholesterol Efflux from Macrophages Is Influenced Differentially by Plasmas from Overweight Insulin-Sensitive and-Resistant Subjects. *Int. J. Obes.* **2012**, *36*, 407–413. [CrossRef] [PubMed]
7. Ormseth, M.J.; Yancey, P.G.; Yamamoto, S.; Oeser, A.M.; Gebretsadik, T.; Shintani, A.; Linton, M.F.; Fazio, S.; Davies, S.S.; Roberts, L.J., II; et al. Net Cholesterol Efflux Capacity of Hdl Enriched Serum and Coronary Atherosclerosis in Rheumatoid Arthritis. *IJC Metab. Endocr.* **2016**, *13*, 6–11. [CrossRef] [PubMed]
8. Antiochos, P.; Marques-Vidal, P.; Virzi, J.; Pagano, S.; Satta, N.; Bastardot, F.; Hartley, O.; Montecucco, F.; Mach, F.; Waeber, G.; et al. Association between Anti-Apolipoprotein a-1 Antibodies and Cardiovascular Disease in the General Population. Results from the Colaus Study. *Thromb. Haemost.* **2016**, *116*, 764–771. [CrossRef]
9. Antiochos, P.; Marques-Vidal, P.; Virzi, J.; Pagano, S.; Satta, N.; Hartley, O.; Montecucco, F.; Mach, F.; Kutalik, Z.; Waeber, G.; et al. Anti-Apolipoprotein a-1 Igg Predict All-Cause Mortality and Are Associated with Fc Receptor-Like 3 Polymorphisms. *Front. Immunol.* **2017**, *8*, 437. [CrossRef]
10. El-Lebedy, D.; Rasheed, E.; Kafoury, M.; Abd-El Haleem, D.; Awadallah, E.; Ashmawy, I. Anti-Apolipoprotein a-1 Autoantibodies as Risk Biomarker for Cardiovascular Diseases in Type 2 Diabetes Mellitus. *J. Diabetes Complic.* **2016**, *30*, 580–585. [CrossRef] [PubMed]
11. Vuilleumier, N.; Bas, S.; Pagano, S.; Montecucco, F.; Guerne, P.A.; Finckh, A.; Lovis, C.; Mach, F.; Hochstrasser, D.; Roux-Lombard, P.; et al. Anti-Apolipoprotein a-1 Igg Predicts Major Cardiovascular Events in Patients with Rheumatoid Arthritis. *Arthr. Rheum.* **2010**, *62*, 2640–2650. [CrossRef] [PubMed]
12. Vuilleumier, N.; Montecucco, F.; Spinella, G.; Pagano, S.; Bertolotto, M.; Pane, B.; Pende, A.; Galan, K.; Roux-Lombard, P.; Combescure, C.; et al. Serum Levels of Anti-Apolipoprotein a-1 Auto-Antibodies and Myeloperoxidase as Predictors of Major Adverse Cardiovascular Events after Carotid Endarterectomy. *Thromb. Haemost.* **2013**, *109*, 706–715. [CrossRef] [PubMed]
13. Vuilleumier, N.; Rossier, M.F.; Pagano, S.; Python, M.; Charbonney, E.; Nkoulou, R.; James, R.; Reber, G.; Mach, F.; Roux-Lombard, P. Anti-Apolipoprotein a-1 Igg as an Independent Cardiovascular Prognostic Marker Affecting Basal Heart Rate in Myocardial Infarction. *Eur. Heart J.* **2010**, *31*, 815–823. [CrossRef] [PubMed]
14. Mannic, T.; Satta, N.; Pagano, S.; Python, M.; Virzi, J.; Montecucco, F.; Frias, M.A.; James, R.W.; Maturana, A.D.; Rossier, M.F.; et al. Cd14 as a Mediator of the Mineralocorticoid Receptor-Dependent Anti-Apolipoprotein a-1 Igg Chronotropic Effect on Cardiomyocytes. *Endocrinology* **2015**, *156*, 4707–4719. [CrossRef] [PubMed]
15. Montecucco, F.; Braunersreuther, V.; Burger, F.; Lenglet, S.; Pelli, G.; Carbone, F.; Fraga-Silva, R.; Stergiopulos, N.; Monaco, C.; Mueller, C.; et al. Anti-Apoa-1 Auto-Antibodies Increase Mouse Atherosclerotic Plaque Vulnerability, Myocardial Necrosis and Mortality Triggering Tlr2 and Tlr4. *Thromb. Haemost.* **2015**, *114*, 410–422. [CrossRef]
16. Montecucco, F.; Vuilleumier, N.; Pagano, S.; Lenglet, S.; Bertolotto, M.; Braunersreuther, V.; Pelli, G.; Kovari, E.; Pane, B.; Spinella, G.; et al. Anti-Apolipoprotein a-1 Auto-Antibodies Are Active Mediators of Atherosclerotic Plaque Vulnerability. *Eur. Heart J.* **2011**, *32*, 412–421. [CrossRef] [PubMed]
17. Pagano, S.; Carbone, F.; Burger, F.; Roth, A.; Bertolotto, M.; Pane, B.; Spinella, G.; Palombo, D.; Pende, A.; Dallegri, F.; et al. Anti-Apolipoprotein a-1 Auto-Antibodies as Active Modulators of Atherothrombosis. *Thromb. Haemost.* **2016**, *116*, 554–564. [CrossRef]
18. Pagano, S.; Satta, N.; Werling, D.; Offord, V.; de Moerloose, P.; Charbonney, E.; Hochstrasser, D.; Roux-Lombard, P.; Vuilleumier, N. Anti-Apolipoprotein a-1 Igg in Patients with Myocardial Infarction Promotes Inflammation through Tlr2/Cd14 Complex. *J. Intern. Med.* **2012**, *272*, 344–357. [CrossRef]
19. Quercioli, A.; Montecucco, F.; Galan, K.; Ratib, O.; Roux-Lombard, P.; Pagano, S.; Mach, F.; Schindler, T.H.; Vuilleumier, N. Anti-Apolipoprotein a-1 Igg Levels Predict Coronary Artery Calcification in Obese but Otherwise Healthy Individuals. *Mediat. Inflamm.* **2012**, *2012*, 243158. [CrossRef]
20. Ahmed, M.M.; Elserougy, E.M.; Al, G., II; Fikry, I.M.; Habib, D.F.; Younes, K.M.; Salem, N.A. Anti-Apolipoprotein a-I Antibodies and Paraoxonase 1 Activity in Systemic Lupus Erythematosus. *EXCLI J.* **2013**, *12*, 719–732.
21. Ames, P.R.; Matsuura, E.; Batuca, J.R.; Ciampa, A.; Lopez, L.L.; Ferrara, F.; Iannaccone, L.; Alves, J.D. High-Density Lipoprotein Inversely Relates to Its Specific Autoantibody Favoring Oxidation in Thrombotic Primary Antiphospholipid Syndrome. *Lupus* **2010**, *19*, 711–716. [CrossRef] [PubMed]

22. Batuca, J.R.; Amaral, M.C.; Favas, C.; Paula, F.S.; Ames, P.R.J.; Papoila, A.L.; Delgado Alves, J. Extended-Release Niacin Increases Anti-Apolipoprotein a-I Antibodies That Block the Antioxidant Effect of High-Density Lipoprotein-Cholesterol: The Explore Clinical Trial. *Br. J. Clin. Pharmacol.* **2017**, *83*, 1002–1010. [CrossRef] [PubMed]
23. Batuca, J.R.; Ames, P.R.; Amaral, M.; Favas, C.; Isenberg, D.A.; Delgado Alves, J. Anti-Atherogenic and Anti-Inflammatory Properties of High-Density Lipoprotein Are Affected by Specific Antibodies in Systemic Lupus Erythematosus. *Rheumatology* **2009**, *48*, 26–31. [CrossRef] [PubMed]
24. Srivastava, R.; Yu, S.; Parks, B.W.; Black, L.L.; Kabarowski, J.H. Autoimmune-Mediated Reduction of High-Density Lipoprotein-Cholesterol and Paraoxonase 1 Activity in Systemic Lupus Erythematosus-Prone Gld Mice. *Arthritis Rheum.* **2011**, *63*, 201–211. [CrossRef] [PubMed]
25. Dullaart, R.P.F.; Pagano, S.; Perton, F.G.; Vuilleumier, N. Antibodies against the C-Terminus of Apoa-1 Are Inversely Associated with Cholesterol Efflux Capacity and Hdl Metabolism in Subjects with and without Type 2 Diabetes Mellitus. *Int. J. Mol. Sci.* **2019**, *20*. [CrossRef]
26. Villines, T.C.; Hulten, E.A.; Shaw, L.J.; Goyal, M.; Dunning, A.; Achenbach, S.; Al-Mallah, M.; Berman, D.S.; Budoff, M.J.; Cademartiri, F.; et al. Prevalence and Severity of Coronary Artery Disease and Adverse Events among Symptomatic Patients with Coronary Artery Calcification Scores of Zero Undergoing Coronary Computed Tomography Angiography: Results from the Confirm (Coronary Ct Angiography Evaluation for Clinical Outcomes: An International Multicenter) Registry. *J. Am. Coll. Cardiol.* **2011**, *58*, 2533–2540. [CrossRef] [PubMed]
27. Greenland, P.; Alpert, J.S.; Beller, G.A.; Benjamin, E.J.; Budoff, M.J.; Fayad, Z.A.; Foster, E.; Hlatky, M.A.; Hodgson, J.M.; Kushner, F.G.; et al. 2010 Accf/Aha Guideline for Assessment of Cardiovascular Risk in Asymptomatic Adults: A Report of the American College of Cardiology Foundation/American Heart Association Task Force on Practice Guidelines. *J. Am. Coll. Cardiol.* **2010**, *56*, e50–e103. [CrossRef]
28. Miyazaki, O.; Kobayashi, J.; Fukamachi, I.; Miida, T.; Bujo, H.; Saito, Y. A New Sandwich Enzyme Immunoassay for Measurement of Plasma Pre-Beta1-Hdl Levels. *J. Lipid Res.* **2000**, *41*, 2083–2088.
29. Vigna, G.B.; Satta, E.; Bernini, F.; Boarini, S.; Bosi, C.; Giusto, L.; Pinotti, E.; Tarugi, P.; Vanini, A.; Volpato, S.; et al. Flow-Mediated Dilation, Carotid Wall Thickness and Hdl Function in Subjects with Hyperalphalipoproteinemia. *Nutr. Metab. Cardiovasc. Dis. NMCD* **2014**, *24*, 777–783. [CrossRef]
30. Favari, E.; Ronda, N.; Adorni, M.P.; Zimetti, F.; Salvi, P.; Manfredini, M.; Bernini, F.; Borghi, C.; Cicero, A.F. Abca1-Dependent Serum Cholesterol Efflux Capacity Inversely Correlates with Pulse Wave Velocity in Healthy Subjects. *J. Lipid Res.* **2013**, *54*, 238–243. [CrossRef]
31. Ronda, N.; Favari, E.; Borghi, M.O.; Ingegnoli, F.; Gerosa, M.; Chighizola, C.; Zimetti, F.; Adorni, M.P.; Bernini, F.; Meroni, P.L. Impaired Serum Cholesterol Efflux Capacity in Rheumatoid Arthritis and Systemic Lupus Erythematosus. *Ann. Rheum. Dis.* **2014**, *73*, 609–615. [CrossRef] [PubMed]
32. Favari, E.; Calabresi, L.; Adorni, M.P.; Jessup, W.; Simonelli, S.; Franceschini, G.; Bernini, F. Small Discoidal Pre-Beta1 Hdl Particles Are Efficient Acceptors of Cell Cholesterol Via Abca1 and Abcg1. *Biochemistry* **2009**, *48*, 11067–11074. [CrossRef] [PubMed]
33. Zimetti, F.; Adorni, M.P.; Ronda, N.; Gatti, R.; Bernini, F.; Favari, E. The Natural Compound Berberine Positively Affects Macrophage Functions Involved in Atherogenesis. *Nutr. Metab. Cardiovasc. Dis. NMCD* **2015**, *25*, 195–201. [CrossRef] [PubMed]
34. Markwell, M.A.; Haas, S.M.; Bieber, L.L.; Tolbert, N.E. A Modification of the Lowry Procedure to Simplify Protein Determination in Membrane and Lipoprotein Samples. *Anal. Biochem.* **1978**, *87*, 206–210. [CrossRef]
35. Kellner-Weibel, G.; de La Llera-Moya, M.; Connelly, M.A.; Stoudt, G.; Christian, A.E.; Haynes, M.P.; Williams, D.L.; Rothblat, G.H. Expression of Scavenger Receptor Bi in Cos-7 Cells Alters Cholesterol Content and Distribution. *Biochemistry* **2000**, *39*, 221–229. [CrossRef] [PubMed]
36. Brown, M.S.; Ho, Y.K.; Goldstein, J.L. The Cholesteryl Ester Cycle in Macrophage Foam Cells. Continual Hydrolysis and Re-Esterification of Cytoplasmic Cholesteryl Esters. *J. Biol. Chem.* **1980**, *255*, 9344–9352. [PubMed]
37. Wilson, P.W.; D'Agostino, R.B.; Levy, D.; Belanger, A.M.; Silbershatz, H.; Kannel, W.B. Prediction of Coronary Heart Disease Using Risk Factor Categories. *Circulation* **1998**, *97*, 1837–1847. [CrossRef] [PubMed]
38. Adorni, M.P.; Zimetti, F.; Billheimer, J.T.; Wang, N.; Rader, D.J.; Phillips, M.C.; Rothblat, G.H. The Roles of Different Pathways in the Release of Cholesterol from Macrophages. *J. Lipid Res.* **2007**, *48*, 2453–2462. [CrossRef] [PubMed]

39. Phillips, M.C. Molecular Mechanisms of Cellular Cholesterol Efflux. *J. Biolog. Chem.* **2014**, *289*, 24020–24029. [CrossRef]
40. Phillips, M.C.; Gillotte, K.L.; Haynes, M.P.; Johnson, W.J.; Lund-Katz, S.; Rothblat, G.H. Mechanisms of High Density Lipoprotein-Mediated Efflux of Cholesterol from Cell Plasma Membranes. *Atherosclerosis* **1998**, *137*, S13–S17. [CrossRef]
41. Tomita, K.; Teratani, T.; Suzuki, T.; Shimizu, M.; Sato, H.; Narimatsu, K.; Usui, S.; Furuhashi, H.; Kimura, A.; Nishiyama, K.; et al. Acyl-Coa:Cholesterol Acyltransferase 1 Mediates Liver Fibrosis by Regulating Free Cholesterol Accumulation in Hepatic Stellate Cells. *J. Hepatol.* **2014**, *61*, 98–106. [CrossRef] [PubMed]
42. Chang, T.Y.; Chang, C.C.; Cheng, D. Acyl-Coenzyme A:Cholesterol Acyltransferase. *Annu. Rev. Biochem.* **1997**, *66*, 613–638. [CrossRef] [PubMed]
43. Gillotte, K.L.; Zaiou, M.; Lund-Katz, S.; Anantharamaiah, G.M.; Holvoet, P.; Dhoest, A.; Palgunachari, M.N.; Segrest, J.P.; Weisgraber, K.H.; Rothblat, G.H.; et al. Apolipoprotein-Mediated Plasma Membrane Microsolubilization. Role of Lipid Affinity and Membrane Penetration in the Efflux of Cellular Cholesterol and Phospholipid. *J. Biolog. Chem.* **1999**, *274*, 2021–2028. [CrossRef] [PubMed]
44. Pagano, S.; Gaertner, H.; Cerini, F.; Mannic, T.; Satta, N.; Teixeira, P.C.; Cutler, P.; Mach, F.; Vuilleumier, N.; Hartley, O. The Human Autoantibody Response to Apolipoprotein a-I Is Focused on the C-Terminal Helix: A New Rationale for Diagnosis and Treatment of Cardiovascular Disease? *PLoS ONE* **2015**, *10*, e0132780. [CrossRef] [PubMed]
45. Zannis, V.I.; Chroni, A.; Krieger, M. Role of Apoa-I, Abca1, Lcat, and Sr-Bi in the Biogenesis of Hdl. *J. Mol. Med.* **2006**, *84*, 276–294. [CrossRef] [PubMed]
46. Frisdal, E.; Lesnik, P.; Olivier, M.; Robillard, P.; Chapman, M.J.; Huby, T.; Guerin, M.; Le Goff, W. Interleukin-6 Protects Human Macrophages from Cellular Cholesterol Accumulation and Attenuates the Proinflammatory Response. *J. Biolog. Chem.* **2011**, *286*, 30926–30936. [CrossRef] [PubMed]

© 2019 by the authors. Licensee MDPI, Basel, Switzerland. This article is an open access article distributed under the terms and conditions of the Creative Commons Attribution (CC BY) license (http://creativecommons.org/licenses/by/4.0/).

Article

Anti-ApoA-1 IgGs in Familial Hypercholesterolemia Display Paradoxical Associations with Lipid Profile and Promote Foam Cell Formation

Sabrina Pagano [1,2,*], Alessandra Magenta [3], Marco D'Agostino [3], Francesco Martino [4], Francesco Barilà [5], Nathalie Satta [1,2], Miguel A. Frias [1,2], Annalisa Ronca [6], François Mach [7], Baris Gencer [7], Elda Favari [6] and Nicolas Vuilleumier [1,2]

1. Division of Laboratory Medicine, Department of Genetics and Laboratory Medicine, Geneva University Hospital, 4 rue Gabrielle-Perret-Gentil, 1205 Geneva, Switzerland; Nathalie.Satta@unige.ch (N.S.); Miguel.Frias@unige.ch (M.A.F.); nicolas.vuilleumier@hcuge.ch (N.V.)
2. Department of Internal Medicine Specialities, Medical Faculty, Geneva University, 1 rue Michel Servet, 1206 Geneva, Switzerland
3. Fondazione Luigi Maria Monti, Istituto Dermopatico dell' Immacolata-IRCCS, Experimental Immunology Laboratory, Via dei Monti di Creta 104, 00167 Rome, Italy; ale.magenta@gmail.com (A.M.); marcodagostino86@hotmail.it (M.D.)
4. Department of Pediatrics, Sapienza University of Rome, Viale Regina Elena 324, 00161 Rome, Italy; Francesco.Martino@uniroma1.it
5. Department of Cardiovascular, Respiratory, Nephrological, Anesthesiological and Geriatrical Sciences, Sapienza University of Rome, Viale del Policlinico 155, 00161 Rome, Italy; Francesco.Barilla@uniroma1.it
6. Department of Food and Drug, University of Parma, Parco Area delle Scienze 27/A, 43124 Parma, Italy; annalisa.ronca@studenti.unipr.it (A.R.); elda.favari@unipr.it (E.F.)
7. Division of Cardiology, Geneva University Hospital, 1205 Geneva, Switzerland; Francois.Mach@hcuge.ch (F.M.); Baris.Gencer@hcuge.ch (B.G.)
* Correspondence: Sabrina.Pagano@hcuge.ch

Received: 5 November 2019; Accepted: 18 November 2019; Published: 21 November 2019

Abstract: Aims: Anti-Apolipoprotein A-1 autoantibodies (anti-ApoA-1 IgG) promote atherogenesis via innate immune receptors, and may impair cellular cholesterol homeostasis (CH). We explored the presence of anti-ApoA-1 IgG in children (5–15 years old) with or without familial hypercholesterolemia (FH), analyzing their association with lipid profiles, and studied their in vitro effects on foam cell formation, gene regulation, and their functional impact on cholesterol passive diffusion (PD). Methods: Anti-ApoA-1 IgG and lipid profiles were measured on 29 FH and 25 healthy children. The impact of anti-ApoA-1 IgG on key CH regulators (SREBP2, HMGCR, LDL-R, ABCA1, and miR-33a) and foam cell formation detected by Oil Red O staining were assessed using human monocyte-derived macrophages. PD experiments were performed using a validated THP-1 macrophage model. Results: Prevalence of high anti-ApoA-1 IgG levels (seropositivity) was about 38% in both study groups. FH children seropositive for anti-ApoA-1 IgG had significant lower total cholesterol LDL and miR-33a levels than those who were seronegative. On macrophages, anti-ApoA-1 IgG induced foam cell formation in a toll-like receptor (TLR) 2/4-dependent manner, accompanied by NF-kB- and AP1-dependent increases of SREBP-2, LDL-R, and HMGCR. Despite increased ABCA1 and decreased mature miR-33a expression, the increased ACAT activity decreased membrane free cholesterol, functionally culminating to PD inhibition. Conclusions: Anti-ApoA-1 IgG seropositivity is frequent in children, unrelated to FH, and paradoxically associated with a favorable lipid profile. In vitro, anti-ApoA-1 IgG induced foam cell formation through a complex interplay between innate immune receptors and key cholesterol homeostasis regulators, functionally impairing the PD cholesterol efflux capacity of macrophages.

Keywords: anti-apolipoprotein A-1 IgG; familial hypercholesterolemia; cholesterol homeostasis; foam cells; miR-33a; TLR2/4; passive diffusion

1. Introduction

Humoral autoimmunity has recently been shown to represent a dual mediator of atherogenesis and cardiovascular diseases (CVD) by modulating three main pathways, including inflammation, coagulation, and foam cell formation [1,2]. Among autoantibodies of interest in CVD, the interest in antibodies against apolipoprotein A-1 (anti-ApoA-1 IgG) appears to be gaining momentum. Indeed, three recent studies derived from a large multicenter general population cohort demonstrated that anti-ApoA-1 IgGs were an independent cardiovascular (CV) risk factor predictive of poor prognosis [3–5] similarly to what had been reported previously and more recently in high CV risk populations [6–11]. In parallel, translational studies pointed to these autoantibodies as mediators of atherogenesis, promoting atherosclerosis, myocardial necrosis, and mice death through toll-like receptors (TLR) 2,4 and CD14 signaling [12–14]. Interestingly, inverse associations were regularly noted between anti-ApoA-1 IgG, total cholesterol, high-density lipoprotein (HDL), and low-density lipoprotein (LDL) levels [3,13,15,16], potentially suggesting that these antibodies could also interfere with cholesterol metabolism in addition to their established pro-inflammatory and pro-thrombotic properties [13,14,17]. Such a hypothesis has been further supported by two recent observations. The first showed that antibodies directed against the c-terminal part of ApoA-1 in type 2 diabetes were associated with decreased cholesterol efflux capacity (CEC) of fibroblasts [18]. The second reported that anti-ApoA-1 IgG levels were inversely associated with PD- and positively associated with ABCA1-CEC in healthy obese subjects, with the capacity, to enhance ABCA1-CEC in vitro while repressing PD-CEC, leading to foam cell formation [19], the hallmark of atherosclerosis [20]. Nevertheless, the exact molecular mechanisms underlying these drastic cellular phenotype modifications induced by anti-apoA-1 IgG are still elusive, and likely to imply major changes in the gene and protein expression levels of key regulators of cellular cholesterol homeostasis; whether such effect could be reproduced in human macrophages is currently unknown [20–23].

Among these, 3-hydroxy-3-methylglutaryl CoA reductase (HMGCR) and low-density lipoprotein receptor (LDL-R) are known as key cholesterol homeostasis regulatory elements leading to the developments of statins, the most efficient and commonly used pharmacological molecules available to prevent atherosclerosis-related complications [21]. HMGCR and LDLR are controlled by the sterol regulatory element binding protein 2 (SREBP2) encoded by *SREBF2* gene, which contains microRNAs 33a (miR-33a) in its intronic sequence and reduces cholesterol efflux via repression of the adenosine triphosphate (ATP)-binding cassette transporter A1 (ABCA1), impairing HDL biogenesis [22,23]. Upon *SREBF2* gene activation, both LDLR and HMGCR are upregulated by SREBP2, enhancing LDL uptake and increasing intracellular cholesterol synthesis. In response to *SREBF2* gene transcription, miR-33a expression will increase, inhibiting ABCA-1 cholesterol efflux [22,23]. Therefore, miR-33a and SREBP2 act synergistically to efficiently increase cellular cholesterol levels by inhibiting cellular cholesterol efflux and increasing lipid uptake and intracellular synthesis, respectively. While previous studies suggested that miR-33 may represent therapeutic target for the treatment of cardiovascular disease [24,25], a recent work showed that miR-33 deletion in mice results in dyslipidemia, obesity, and insulin resistance [26], assigning a role for miR-33 that is much more complex than what has been considered so far. Such complexity has also been illustrated in humans by the fact that miR-33 levels are surprisingly elevated in pediatric patients affected by familial hypercholesterolemia (FH) [27].

Since foam cells can be generated by the uncontrolled uptake of unmodified LDL via LDL-R [28,29], any factor influencing LDLR, HMGCR, or miR-33a expression will efficiently modulate lipid uptake, cellular cholesterol synthesis, and cholesterol efflux, and therefore affect atherogenesis.

Taken together, these observations suggest that anti-ApoA-1 IgG may reorient the lipids from the plasmatic compartment toward the intracellular lipid pools, potentially explaining the paradoxical associations frequently retrieved between anti-apoA-1 IgG and lipid profile despite an increased

CV risk. If so, such a hypothesis would imply modulation of key regulators of cellular cholesterol homeostasis and foam cell formation, including miR-33.

Therefore, in this translational study, we explored the associations between anti-ApoA-1 IgG, lipid profile, and miR-33a levels in FH children as an extreme and optimal human dyslipidemia phenotype to perform such explorations and compared whether such associations could be retrieved in age- and gender-matched controls. Fueled by these clinical observations, we then dissected the in vitro mechanisms by which anti-ApoA-1 IgG could lead to foam cell formation. Our results showed that associations between anti-ApoA-1 IgG, lipid profile, and miR-33 levels were only retrieved in FH children. Furthermore, our in vitro studies using human macrophages provided the first molecular insights on how these antibodies could act as novel endogenous cholesterol homeostasis disruptors.

2. Experimental Section

2.1. Patients

In this study, we considered 29 children (14 males and 15 females) aged 5–15 affected by FH and referred to the Center of Clinical Lipid Research, Department of Pediatrics, Sapienza University of Rome [27]. Children were classified as FH on the basis of the presence of a first-degree relative with hypercholesterolemia (total cholesterol (TC) >95th age and sex-specific percentile) according to the MEDPED criteria [30]. In the same center, a control group of 25 healthy children (11 males, 14 female) was recruited with a BMI (body mass index) appropriate for their age and gender, as matched for age with the FH group.

Exclusion criteria consisted of acute or chronic disease or infection (connective tissue disease, hypothyroidism, renal disease, malignancy, clinical evidence of CVD, diabetes mellitus, hypertension or metabolic syndrome), autoimmune disease, or the use of medication potentially affecting growth and development, associated with a history of alcohol consumption and smoking (when appropriate), immunosuppressive drugs, non-steroidal, anti-inflammatory, lipid-lowering drugs, and/or vitamin supplements.

At first visit, anthropometric data were measured (body weight and waist, hip, and arm circumferences). Weight was measured using an electronic scale (Soehnle, Murrhardt, Germany) and standing height was measured with the Harpenden Stadiometer (Holtain, Crymych, UK). Systolic and diastolic blood pressure was measured using a random zero sphygmomanometer (Hawksley & Sons Ltd., Lancing, UK); the mean of three measurements was used in the analysis. BMI was calculated as weight/height^2 (kg/m^2).

Informed written consent was obtained from all the participants. The study was in conformity with the ethical guidelines of the Declaration of Helsinki, and was reviewed and cleared by the Ethical Committee of Sapienza University of Rome.

2.2. Plasma Samples and Blood Analyses

Venous blood samples (10 mL) were collected in EDTA-containing tubes from 12 h fasted subjects. Blood was then centrifuged (1200g for 10 min at 4 °C). Supernatant was then collected and centrifuged (2000g for 10 min at 4 °C). Plasma samples were stored at −80 °C and were thawed on ice before use. Plasma concentrations of lipoprotein and apolipoproteins were determined as previously described [27]. LDL cholesterol was calculated using Friedewald's equation.

2.3. Assessment of Anti-ApoA-1 IgG Levels

Anti-ApoA-1 IgG were measured as previously described [7,9,13,14,17]. Briefly, MaxiSorp plates (Nunc^{TM}, city, Roskilde, Denmark) were coated with purified, human-derived delipidated apolipoprotein A-1 (20 µg/mL; 50 µL/well) for 1 h at 37 °C. After being washed, all wells were blocked for 1 h with 2% bovine serum albumin (BSA) in a phosphate-buffered solution (PBS) at 37 °C. FH samples were also added to a non-coated well in order to assess individual non-specific binding.

After six washing cycles, 50 µL/well of the alkaline phosphatase-conjugated anti-human IgG was added (Sigma-Aldrich, St Louis, MO, USA), it was diluted at 1:1000 in a PBS/BSA 2% solution, and this was added and incubated for 1 h at 37 °C. After washing six more times, phosphatase substrate p-nitrophanylphosphate disodium (Sigma-Aldrich) dissolved in a diethanolamine buffer (pH 9.8) was added and incubated for 30 min at 37 °C. Optical density (OD) was determined at 405 nm (Filtermax 3, Molecular Devices™, San Jose, CA, USA). and each sample was tested in duplicate. Corresponding non-specific binding was subtracted from mean OD for each sample. The specificity of detection was assessed using conventional saturation tests by Western blot analysis.

As previously described, elevated levels of anti-ApoA-1 IgG (seropositivity) were defined by an OD cutoff of OD > 0.64 corresponding to the 97.5th percentile of a reference population. In order to limit the impact of interassay variation, we further calculated an index consisting of the ratio between sample net absorbance and the positive control net absorbance × 100. The index value corresponding to the 97.5th percentile of the normal distribution was 37. Accordingly, to be considered as positive (presenting elevated anti-apoA-1 IgG levels), samples had to display both an absorbance value of OD > 0.64 and an index value ≥37%. [7,9,13,14,17].

2.4. Reagents

RPMI-1640 medium, fetal bovine serum (FBS), PBS free of Ca^{2+} and Mg^{2+}, L-glutamine, penicillin, and streptomycin were obtained from Gibco BRL-Life Technologies (Rockville, MD, USA). Interferon-gamma (IFN-γ) was from Roche (Mannheim, Germany).

Affinity purified goat polyclonal anti-human ApoA-1 IgG (ref. 11AG2) was obtained from Academy Bio-Medical Company (Houston, TX, USA) and goat control IgG were from Meridian Life Science (ref. A66200H) (Saco, ME, USA). Ultrapure lipopolysaccharide (LPS) from *Escherichia coli* was purchased from Alexis Enzo Life Sciences (Lausen, Switzerland). Blocking anti-human TLR4 (clone HTA 125) antibody, anti-human TLR2 antibody (clone TL2.1), and matched isotype control antibodies were from Biolegend (San Diego, CA, USA). Blocking anti-human TLR2 (clone TL2.5) and blocking anti-human CD14 antibodies were from InvivoGen (San Diego, CA, USA). SP600125 (c-Jun N-terminal kinase (JNK) inhibitor and BAY11-7082 (IkB-α inhibitor) were from InvivoGen. San Diego, CA, USA)

2.5. Human Monocyte-Derived Macrophage Preparation

Human monocytes were isolated from buffy coats obtained from healthy donors in the Geneva Hospital Blood Transfusion Center (Geneva, Switzerland) and differentiated into macrophages by 24 hours incubation with IFN-γ (500 U/mL) in a complete RPMI-1640 culture medium (10% heat-inactivated FCS, 50 µg/mL streptomycin, 50 U/mL penicillin, 2 mM L-glutamine) as previously described [14,17]. Macrophage preparation consisted of >90% CD68+ cells as assessed by flow cytometry.

When indicated, anti-TLR4, anti-TLR2 (TL2.5/2.1), and anti CD14 blocking antibodies, as well as isotype-matched control mAb (10 µg/mL) and specific pharmacological inhibitors, were added 30 min before stimulation with IgG [14,17].

2.6. Protein Purification and Western Blot Analysis

Macrophages (2×10^6) were lysed at 4 °C for 20 min with radioimmunoprecipitation assay (RIPA) lysis buffer with added protease inhibitors (Complete tablets, mini, Roche diagnostics, Mannheim, Germany) and phosphatase inhibitors (Halt Protease and Phosphatase Inhibitor Cocktail, Thermo Scientific, Waltham, MA, USA) and then samples were centrifuged at 14,000 rpm at 4 °C for 15 min. Protein concentration was determined by Bradford protein assay (Biorad, Hercules, CA, USA). To increase protein concentration, samples were loaded onto Amicon Ultra-0.5 centrifugal filter devices 3K (Merck Millipore, Darmstadt, Germany) following the manufacturer's instructions. Forty micrograms of total protein extract were resolved by 8% or 10% polyacrylamide gel electrophoresis under reducing conditions, and transferred to a polyvinylidene difluoride (PVDF) membrane

(Immobilon, Millipore IPVH 00010). Membranes were incubated with the following antibodies: anti-SREBP2 (rabbit polyclonal, ref. 10007663, Cayman chemical, Ann Arbor, MI, Stati Uniti), anti-LDL Receptor (rabbit monoclonal, ref. ab52818, Abcam, Cambridge, UK), anti-HMGCR (rabbit monoclonal, ref. ab174830, Abcam), anti-SCAP (Rabbit polyclonal, ref. ab125186 Abcam), anti-ABCA1 (rabbit polyclonal, ref. NB400-105, Novus Biologicals, Centennial, CO, USA), anti-β-actin (mouse monoclonal, ref. ab8226, Abcam). Horseradish peroxidase-conjugated antisera from Dako (Glostrup, Denmark), were used to reveal primary antibody binding, with detection by BM Chemiluminescence Blotting Substrate (POD) from Roche Diagnostics (Mannheim, Germany). Relative protein levels were measured using densitometric analysis with ImageJ software (1.48v, Java 1.5.0_20, 64-bit). Results are expressed in arbitrary units.

2.7. Lipid Uptake by Oil Red O Staining

Monocytes (3×10^5) were plated onto Lab-Tek chamber slide system (Nunc, Roskilde, Denmark) and IFN-γ (500 U/mL) was added to RPMI supplemented with 10% FCS for 24 h at 37 °C. After that time, cells were washed once with PBS and exposed, or not, for 24 h to 20 μg/mL LDL from Academy Bio-Medical Company (Houston, TX, USA). One hour before adding LDL, cells were incubated with 40 μg/mL of goat polyclonal anti-human apolipoprotein A-1 or control IgG, shown to be the optimal dose to elicit a pro-inflammatory response [7,9,13,14,17]. Cells were examined for lipid inclusion by Oil Red O staining. Briefly, cells were incubated with 10% formalin (Sigma-Aldrich, St Louis, MO, USA) for 30 min at room temperature, incubated with Oil Red O solution (Sigma- Aldrich, St Louis, MO, USA) for 15 min, and cells were then counterstained with hematoxylin (Sigma- Aldrich, St Louis, MO, USA). Aquatex (Merck Millipore, Darmstadt, Germany) was used as mounting media. Images were acquired with a microscope Zeiss Axioskop 2 plus. We used a 40× objective (Plan-Neofluar 40× /0.75 Ph2) for all images. The images were collected using the AxioVision 4.8.1.0 softwares (Zeiss, Oberkochen, Germania). Quantification of the lipid content (Oil Red O staining) per cell, identified by hematoxylin blue nuclear staining, was performed using Definiens Developer XD2 (Cambridge, MA, USA). Results are expressed as mean granule area in arbitrary units.

2.8. RNA Extraction

Macrophage cells (1×10^6) were subjected to RNA extraction using a total RNA purification RNeasy Micro kit (Qiagen, Hilden, Germany) according to the manufacturer's protocol. RNA from plasma (200 μL) was extracted using a Total RNA Purification Plus kit (Norgen Biotek, Thorold, ON, Canada) and RNA from macrophage supernatant (200 μL) was extracted using TRIzol LS (Ambion, Life-Technologies, Austin, TX, USA) as an internal control, with 10 fmol cel-miR-39a (Qiagen, Hilden, Germany) spiked into each plasma and supernatant sample after adding lysis buffer or TRIzol LS. We then followed the manufacturer's protocol for RNA extraction. RNA was quantified using NanoDrop 2000 software (NanoDrop Products, Thermo Fisher Scientific, Wilmington, DE, USA.).

2.9. miRNA Quantitative PCR

miRNA quantification was done using a TaqMan®MicroRNA Reverse Trancription Kit (Applied Biosystems) for reverse transcription reactions, according to the manufacturer's instructions. TaqMan Universal PCR master mix (Life Technologies, Carlsbad, CA, USA) was used to quantify miRNA levels in plasma, macrophage cells, and in the supernatant counterpart with a 7900HT SDS Fast System instrument (Applied Biosystems, Foster City, CA, USA). Primers for *miR-33a, cel-miR-39a* were obtained from Applied Biosystems (Foster City, CA, USA). *miR-33a* level in plasma and in cell supernatant were normalized to the spiked *cel-miR-39a* whereas *miR-33a* expression levels in each sample from macrophage cells were normalized to RNU6B, miR-16, Z30 controls (Applied Biosystems, Foster City, CA, USA). geNorm was used to determine the most stable genes from the set of control tested genes since geNorm calculates the gene expression stability measure (M value) for a control gene as the average pairwise variation for that gene with all other tested control genes [31]. Data were analyzed with the 7900HT SDS Software v2.3 (Applied Biosystems, Foster City, CA, USA). Relative expressions of *miR-33a* were calculated using the comparative threshold cycle values (C_T) method ($2^{-\Delta\Delta Ct}$) [32].

2.10. Quantitative PCR

For mRNA quantification, cDNA was synthesized using the High Capacity RNA to cDNA kit (Life Technologies, Carlsbad, CA, USA). Quantitative re al-time PCR was performed in duplicate using TaqMan Universal Master mix II, no UNG on the StepOne Plus Real-Time PCR System (Thermo Fisher Scientific, Waltham, MA, USA). The mRNA levels were normalized to Gapdh (hs99999905_m1) as a housekeeping gene. The following primers were used: pri-miR-33a (hs03293451), Abca1 (hs01059137_m1), and Srebf-2 (hs01081784_m1). All primers were obtained from (Applied Biosystems, Foster City, CA, USA). Data were analyzed with the 7900HT SDS Software v2.3 (Applied Biosystems, Foster City, CA, USA). Relative expression of mRNA was calculated using the comparative threshold cycles values (C_T) method ($2^{-\Delta\Delta Ct}$) [32].

2.11. Anti-ApoA-1 IgG Modulation of Membrane Free Cholesterol. Assay of Cholesterol Oxidase

Cholesterol oxidase treatment was essentially as previously described [33]. Briefly, cells were labeled with 3 µCi/mL [3 H] cholesterol. Cholesterol oxidase (1 U/mL) was added, and cells were incubated for 4 hours. Lipid was extracted with isopropanol, and radioactive cholesterol and cholestenone were separated using thin-layer chromatography and quantified.

2.12. ACAT Activity Assessment

Cholesterol esterification was evaluated as the incorporation of radioactivity into cellular cholesteryl esters after addition of [14C]-oleate-albumin complex [34]. At the end of incubation, cells were washed with PBS and lipids were extracted with hexane/isopropanol (3:2). The extracted lipids were separated by TLC (isooctane/diethyl ether/acetic acid, 75:25:2 (v/v/v)). Cholesterol radioactivity in the spots was determined by liquid scintillation counting.

2.13. Measurement of Free Cholesterol Content in Cell Supernatant

Free cholesterol in supernatant was measured using the fluorometric method using the Cholesterol Quantitation Assay Kit by Abcam (ab65359, Cambridge, UK) following the manufacturer's instruction. The experiments were conducted without ACAT inhibitors.

2.14. Passive Diffusion Analysis

We used a validated model of THP-1 monocytes cultured in RPMI 1640 medium supplemented with 10% FBS at 37 °C in 5% CO_2 [35]. To perform the experiments for passive diffusion, cells were seeded in 24-well plates at a density of 5×10^5 cells/well in the presence of 100 ng/mL PMA for 72 h to allow differentiation into macrophages. Cells were labeled with (1,2-3H)-cholesterol in the presence of an ACAT inhibitor (2 µg/mL, Sandoz 58035 by Sandoz, Holzkirchen, Germany) for 24 h. In order to measure passive diffusion, the key cholesterol efflux regulator in non-dyslipidemic macrophages [36], the experiment was performed in the absence of 10% whole serum, but cells were supplemented with 0.2% albumin to avoid any active ABCA1- or ABCG1-mediated cholesterol efflux. Cholesterol measurement was expressed as a percentage of the radioactivity released to the medium in 4 h over the total radioactivity incorporated by cells. Control samples were run to confirm the responsiveness of cells. Background efflux, evaluated in the absence of acceptors, was subtracted from each samples value.

2.15. Statistics

Continuous variables were expressed as median (interquartile range (IQR)) and comparisons between two groups were performed using a non-parametric Mann–Whitney U test, unless stated otherwise. Correlation analysis was carried out using a Spearman test. Analyses were performed with Statistica package (version13.5.0.17, TIBCO Software inc., Palo Alto, CA, USA). Statistical significance was defined at $p < 0.05$. For the in vitro results, statistical analysis was performed using the parametric unpaired Student's *t*-test using GraphPad Prism (version 7.0., GraphPad SoftwareSan Diego, CA, USA). Statistical significance was defined as $p < 0.05$.

3. Results

3.1. Anti-ApoA-1 IgG Associations with Lipid Profile in FH Children

Baseline, clinical, and biological characteristics of FH and healthy children are described in Table 1. As expected, FH children had higher levels of total cholesterol, LDL cholesterol, apoB particles, HDL cholesterol, and miR-33a than their age-matched controls. No other differences in baseline clinical characteristics were observed between these two groups. Median anti-ApoA-1 IgG were similar in these two groups and a similar prevalence of high anti-ApoA-1 IgG levels was retrieved in both groups (38.8% and 37.9%, respectively).

In FH children, there were no significant differences between anti-ApoA-1 IgG positive and anti-ApoA-1 IgG negative children in terms of age, gender, and overweight-associated features, with the possible exception of BMI and waist circumference, which tended to be higher in ApoA-1 IgG positive FH children than seronegative ones. Anti-ApoA-1 IgG positive subjects showed significantly lower median levels of LDL cholesterol, total cholesterol, and miR-33a, with a marginal trend for apoB particles when compared to anti-ApoA1 IgG negative subjects (Table 2). Significant correlations were only observed between anti-ApoA-1 IgG and miR-33a levels ($r = 0.42$, $p = 0.02$).

In age-matched healthy children, the aforementioned differences between anti-ApoA-1 IgG seropositive and seronegative individuals related to lipid profile and miR-33a levels were not observed (Table 3). In contrasting to the observation in FH children, we observed that BMI, waist, and arm circumference in healthy children were lower in subjects positive for anti-ApoA-1 IgG when compared to seronegative children.

Table 1. Baseline demographic and biological characteristics of subjects according to familial hypercholesterolemia (FH) status.

Characteristic	Healthy Children ($n = 25$)	FH Children ($n = 29$)	* p-Value
Age, year (IQR)	8.5 (7–11)	9 (6–11)	0.89
Sex (male %)	45	48.2	0.88
BMI (kg/m^2)	19 (16.6–20.2)	17.8 (15.6–20)	0.55
Systolic BP *, mmHg (IQR)	105 (95–110)	110 (100–120)	0.14
Diastolic BP, mmHg (IQR)	60 (60–70)	60 (60–70)	0.75
Waist circumference, cm (IQR)	61.5 (53–69)	60 (53–68)	0.94
Hip circumference, cm (IQR)	66 (61–76)	64 (57–73)	0.85
Arm circumference, cm (IQR)	20 (18–22)	20 (17–23)	0.83
Total cholesterol (mg/dL)	152.5 (138–167.5)	231 (213–280)	<0.0001
LDL cholesterol (mg/dL)	91.4 (79.5–103)	156 (137.6–211.6)	<0.0001
HDL cholesterol (mg/dL)	52 (43.5–58.5)	57 (51–66.5)	0.02
Triglycerides (mg/dL)	56.5 (45.5–63)	64 (47–75)	0.17
Apolipoprotein B (mg/dL)	65 (49–81.5)	93 (86–134)	<0.0001
Anti-ApoA-1 IgG (OD value)	0.53 (0.4–0.77)	0.55 (0.5–0.68)	1
Anti-ApoA-1 IgG positivity (%)	38.8	37.9	0.9
miR-33a ($2^{\wedge}$-ΔCt)	0.05 (0.03–0.2)	0.4 (0.19–1)	<0.0001

* p-values were obtained by comparing FH positive versus FH negative subjects, p-values were calculated using the Mann–Whitney U test. BMI, body mass index. BP, blood pressure. IQR: interquartile range.

Table 2. Baseline demographic and biological characteristics of FH children according to anti-ApoA-1 IgG status.

Characteristic	Healthy Children (n = 25)	Anti-ApoA-1 Negative (n = 15)	Anti-ApoA-1 Positive (n = 10)	* p-Value
Age, year (IQR)	8.5 (7–11)	10 (7–11)	7 (7–9)	0.08
Sex (male %)	45	42.8	50	0.28
BMI (kg/m^2)	19 (16.6–20.2)	19.2 (17–20.2)	17.2 (15.4–19.5)	0.02
Systolic BP *, mmHg (IQR)	105 (95–110)	100 (90–105)	110 (100–120)	0.7
Diastolic BP, mmHg (IQR)	60 (60–70)	60 (60–75)	62.5 (60–70)	0.53
Waist circumference, cm (IQR)	61.5 (53–69)	66 (51–69)	61 (54–64)	0.03
Hip circumference, cm (IQR)	66 (61–76)	70 (61–76)	66 (61–67)	0.09
Arm circumference, cm (IQR)	20 (18–22)	20 (17–22)	19.5 (18–21)	0.01
Total cholesterol (mg/dL)	151 (138.5–167.5)	147 (135–167)	163 (148–168)	0.27
LDL cholesterol (mg/dL)	91.4 (79.5–103)	89.4 (72–95)	101.8 (87–108)	0.11
HDL cholesterol (mg/dL)	52 (43.5–58.5)	50.5 (39–55)	56 (47–62)	0.12
Triglycerides (mg/dL)	56.5 (45.5–63)	58.5 (47–70)	55 (41–58)	0.97
Apolipoprotein B (mg/dL)	65 (49–81.5)	53.5 (40–66)	79.5 (69–85)	0.19
miR-33a (2^-ΔCt)	0.05 (0.03–0.25)	0.1 (0.03–0.26)	0.05 (0.04–0.1)	0.4

* p-values were obtained by comparing anti-ApoA-1 IgG positive versus anti-ApoA-1 IgG negative subjects, p-values were calculated using the Mann–Whitney U test. IQR: interquartile range.

Table 3. Baseline demographic and biological characteristics of healthy children according to anti-ApoA-1 IgG status.

Characteristic	FH Children (n = 29)	Anti-ApoA-1 Negative (n = 18)	Anti-ApoA-1 Positive (n = 11)	* p-Value
Age, year (IQR)	9 (6–11)	8 (7–10)	10 (6–13)	0.45
Sex (male %)	48	54.5	72.7	0.08
BMI (kg/m^2)	17.8 (15.6–20)	16.8 (15.3–18.2)	19.3 (17.4–21.9)	0.05
Systolic BP*, mmHg (IQR)	110 (100–120)	110 (100–115)	110 (100–125)	0.39
Diastolic BP, mmHg (IQR)	60 (60–70)	62.5 (60–70)	60 (60–70)	0.58
Waist circumference, cm (IQR)	60 (53–68)	56.5 (53–63)	66 (61–70)	0.05
Hip circumference, cm (IQR)	64 (57–73)	72 (64–80)	62.5 (57–71)	0.10
Arm circumference, cm (IQR)	20 (17–23)	22 (22–23)	19 (17–22)	0.14
Total cholesterol (mg/dL)	213 (213–280)	263 (219–305)	209 (188–240)	0.04
LDL cholesterol (mg/dL)	156.6 (137.6–211.6)	191.2 (144.6–228)	124.4 (113.8–168.8)	0.04
HDL cholesterol (mg/dL)	57 (51–66.5)	61 (55–67)	55 (50–58)	0.24
Triglycerides mg/dL)	64 (47–75)	60 (47–66)	66 (35–78)	0.35
Apolipoprotein B (mg/dL)	93 (86–134)	109 (91–143)	86 (76–103)	0.05
miR-33a (2^-Ct)	0.41 (0.19–1.03)	0.85 (0.36–1.2)	0.28 (0.14–0.5)	0.04

* p-values were obtained by comparing anti-ApoA-1 IgG positive versus anti-ApoA-1 IgG negative subjects, p-values were calculated using the Mann–Whitney U test. IQR: interquartile range.

3.2. Lower Level of miR-33a in Anti-ApoA-1 IgG-Treated Human Macrophages

To further explore a possible causal link between the presence of anti-ApoA-1 antibodies, lower miR-33a levels, and a more favorable lipid profile in FH children, we first tested the ability of polyclonal anti-ApoA-1 IgG or control IgG to modulate the in vitro miR-33a production in human monocyte-derived macrophages (HMDM).

After a time course of 4, 8, 16, and 24 h of antibody exposure, miR-33a levels were detected by quantitative PCR, as shown in Figure 1a. Anti-ApoA-1 IgG, and not control IgG, was found to decrease the miR-33a production in vitro on HMDM. As shown in Figure 1a, when compared to baseline conditions and control IgG, anti-ApoA-1 IgG (40 µg/mL) induced a slight but consistent decrease of cellular miR-33a levels during each time point, with the exception of 4 h stimulation, and was statistically significant at 24 hours ($p < 0.01$). As shown in Figure 1b, this anti-ApoA-1 IgG-induced miR-33a downregulation was associated with a decreased miR-33a level at 24 h in the cell supernatant. Up to now, most of the anti-ApoA-1 IgG pro-atherogenic effects have been shown to be mediated by TLR2/TLR4/CD14 complex engagement followed by nuclear factor (NF)-kB and activator protein (AP)-1 pathway activation [4,14,17], and we examined the role TLR2/4, CD14, NF-kB, and AP-1 in the anti-ApoA-1 IgG-mediated downregulation of miR-33a. Accordingly, we pretreated these cells with blocking antibodies against TLR2, TLR4, and the co-receptor CD14, and added chemical inhibitors BAY11-7082 (5 µM) or SP600125 (20 µM), specific to NF-kB and AP-1, respectively, 30 minutes before treatment with polyclonal anti-ApoA-1 IgG for 24 h. As shown in Figure 1c, none of the used inhibitors modulated the anti-ApoA-1 IgG effect on miR-33a, suggesting that the anti-ApoA-1 IgG-dependent miR-33a regulation was independent of TLR2/4/CD14 complex signaling.

We then looked at the primary miR-33a (pri-miR-33a) precursor transcript that, after complex processing, gives the mature miR-33a, as well as at the SREBP2 mRNA, since miR-33a is located in the intronic region of *SREBP2* gene [23,37] and we checked whether they would follow the same regulation as for miR-33a. Surprisingly, HMDM exposure to anti-ApoA-1 IgG for 24 hours significantly increased pri-miR-33a levels as well as SREBP2 mRNA (Figure 2a,b).

These results suggest that despite the fact that the levels of primary miR-33a transcript are increased by anti-ApoA-1 IgG, the mature intracellular miR-33a form is nevertheless decreased by anti-ApoA-1 IgG exposure. Current ongoing investigations to dissect the exact mechanisms underlying this discrepancy indicate that miR-33a stability issues rather than maturation or nuclear export defects may be involved, but no clear-cut mechanisms have been identified so far.

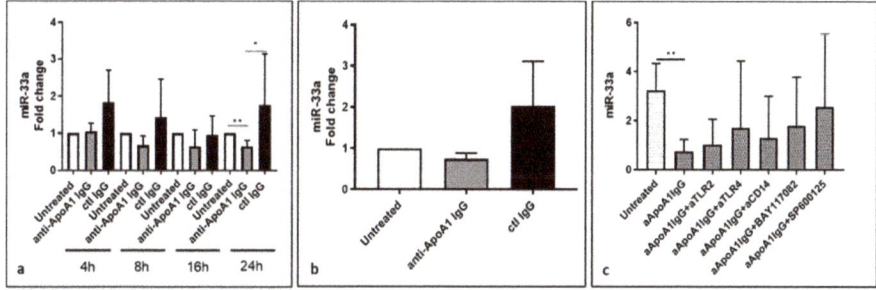

Figure 1. Lower levels of intracellular and extracellular miR-33a in anti-ApoA-1 IgG-treated human macrophage. Cells were treated with anti-ApoA-1 IgG or control antibodies for 4, 8, 16, and 24 h and RT-PCR was performed to determine (**a**) intracellular miR-33a levels as well as (**b**) miR-33a in the supernatant counterpart after 24 h anti-ApoA-1 IgG or control IgG treatment. (**c**) Blocking antibodies against TLR2/4 were used as well as inhibitors of NF-kB and AP-1 to investigate the anti-ApoA-1 IgG mediated miR-33a downregulation. In (**a**,**b**), data are expressed as fold change expression of the mean ± SD of miR-33a calculated by ΔΔCT method of five independent experiments ($n = 5$) and values were normalized to untreated condition, while in panel (**c**), data are expressed as miR-33a quantity ($2^{-\Delta\Delta Ct}$). *p*-values were calculated using Student's *t*-test. Panel a, * $p = 0.015$, ** $p = 0.0012$. Panel b, the difference is not statistically significant, anti-ApoA-1 IgG vs. untreated $p = 0.11$, anti-ApoA-1 IgG vs. ctl control IgG $p = 0.28$. (**c**), ** $p = 0.0017$.

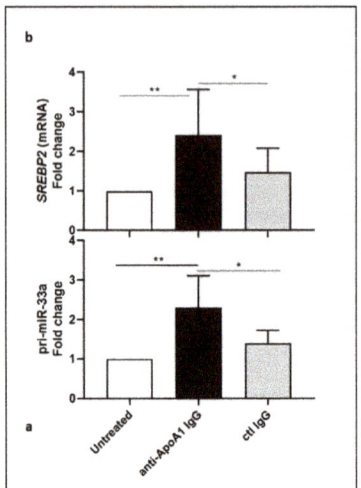

Figure 2. Primary miR-33a and SREBP2 mRNA are upregulated by anti-ApoA-1 IgG. Cells were treated with anti-ApoA-1 IgG or control antibodies for 24 h and RT-PCR was performed to determine pri-miR-33a and SREBP2 mRNA levels (**a**,**b**). Data are expressed as fold change expression of the mean ± SD of pri-miR-33a or SREBP2 mRNA calculated by ΔΔCT method of five independent experiments (*n* = 5) and values were normalized to untreated condition. *p*-values were calculated using the Student's *t*-test. (**a**), * *p* = 0.032, ** *p* = 0.0031; (**b**), * *p* = 0.043, ** *p* = 0.007.

3.3. Anti-ApoA-1 IgGs Modulate Cholesterol-Regulating Proteins in Macrophages

We further investigated the possible modulation, by anti-ApoA1 IgG, of SREBP2 transcription factor known to bind and activate *LDL-R* and *HMGCR* gene transcription [37–39] involved in cholesterol uptake and biosynthesis respectively. After 24 h of stimulation, anti-ApoA-1 IgG treatment increased the levels of SREBP2, HMGCR, and LDL-R protein when compared to the control IgG (Figure 3). As shown in Figure 3, only anti-ApoA-1 IgGs and not the control IgGs were found to increase the expression of the SCAP protein (SREBP cleavage-activating protein), responsible for cleaving and activating SREBP2 [40].

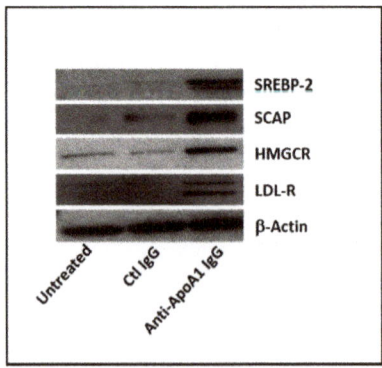

Figure 3. Anti-apoA-1 IgGs interfere with key regulators of cholesterol metabolism in macrophages. Anti-ApoA-1 IgGs induced the expression of the key regulators of the cholesterol pathway after 24 h as evidenced by Western blot. One of four representative Western blot is shown (*n* = 4).

3.4. The Impact of Anti-ApoA-1 IgG on Cholesterol-Regulating Proteins Is Mediated by TLR2 and TLR4

Furthermore, we examined the role of TLR members as well as CD14, NF-kB, and AP-1 in the anti-ApoA-1 IgG-mediated upregulation of SREBP-2, LDL-R, and HMGCR on HMDM. Blocking both TLR2 and TLR4, NF-kB, and AP-1, as previously described, induced a significant reduction in the SREBP-2, LDL-R, and HMGCR protein expression (Figure 4b–d), while only blocking AP-1 significantly decreased anti-ApoA-1 IgG-induced HMGCR protein expression (Figure 4d).

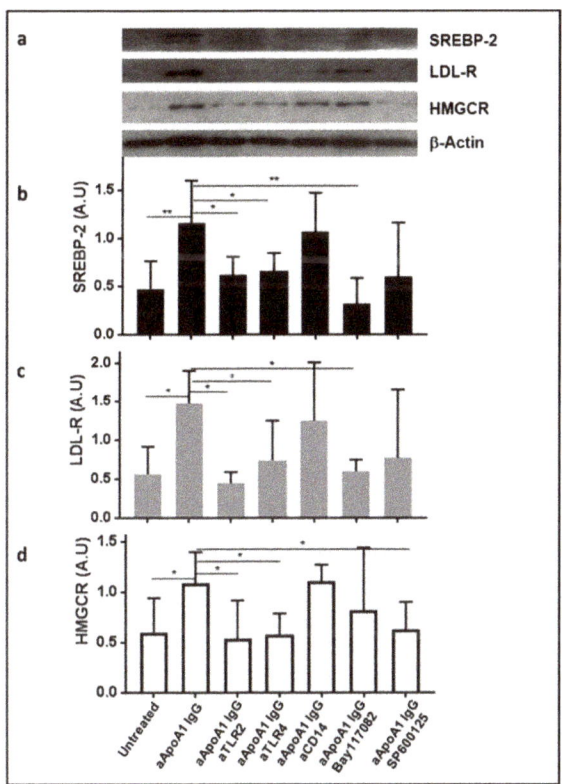

Figure 4. The impact of anti-ApoA-1 IgG on cholesterol metabolism is mediated by TLR2, TLR4, Nf-KB, and AP-1 transcription factors. Western blot-derived results showed that anti-ApoA-1 IgG-mediated upregulation of SREBP2, LDL-R, and HMGCR on human monocyte-derived macrophages (HMDM) is significantly reduced by blocking TLR2 and TLR4 with specific blocking antibodies. Blocking NF-kB and AP-1 using the specific inhibitors BAY117082 and SP600125 respectively provided similar results. (**a**) One of five representative Western blots is shown. (**b–d**) Data are expressed as mean ± SD of band intensity volume/actin intensity volume from five different experiments ($n = 5$). p-values were calculated using the Student's t-test: * $p < 0.05$, ** $p < 0.01$.

3.5. Anti-ApoA-1 IgGs Promote LDL Uptake and Foam Cell Formation

The complexity underlined by the previous observations indicating that i) anti-ApoA-1 IgG are associated with a more favorable lipid profile in FH children but lower miR-33a levels and ii) that these antibodies could increase both pro-atherogenic (HMGCR increase) and anti-atherogenic pathways (SREBP-2, LDLR) in vitro prompted us to evaluate the global result of exposing human macrophages to anti-ApoA-1 IgG in terms of foam cell formation. For this purpose, HMDM were treated with or without LDL (20 µg/mL) in the presence or absence of anti-ApoA-1 IgG (40 µg/mL) or control IgG (40 µg/mL) for 24 hours and followed by Oil Red O staining of HMDM. As shown in Figure 5, anti-ApoA-1 IgG exposure without LDL induced a modest but significant increase in cellular lipid content when compared to control IgG (Figure 5b). Furthermore, in the presence of LDL, the effect of anti-ApoA-1 IgG was enhanced (Figure 5a,b). In presence of LDL, the anti-ApoA-1 IgG effect on foam cell formation was abrogated by all the inhibitors used, with the exception of AP-1 inhibitor, which nevertheless provided a close to significant inhibition (Figure 5b). Taken together, these results indicate that anti-ApoA-1 IgG promotes foam cell formation through partly TLR2/4-dependent pathways. Of note, this anti-ApoA-1 IgG-induced foam cell formation was specifically enhanced by native LDL because in presence of oxidized LDL, anti-ApoA-1 IgG increases foam cell formation to the same extent as anti-ApoA-1 IgG alone (Supplemental Figure S1).

Importantly, given the fact that LDL per se substantially induced foam cell formation regardless of the concomitant presence of anti-ApoA-1 IgG, the rest of our experimental procedures were carried out without LDL in order to assess the effects specifically ascribed to anti-ApoA-1 IgGs.

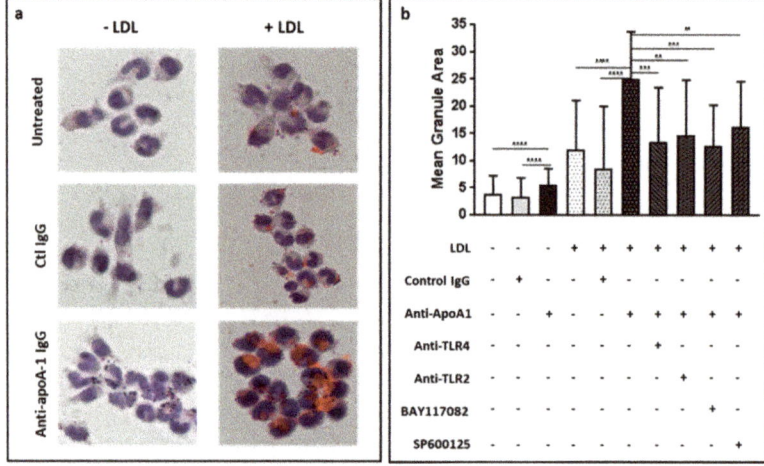

Figure 5. Anti-ApoA-1 IgG promote LDL uptake and foam cell formation mediated by TLR2/4, NF-kB, and AP1. (**a**) HMDM were treated for 24 h with anti-ApoA-1 IgG or ctl IgG in the presence or absence of native LDL (20 µg/mL). Cells were stained with Oil Red O to highlight the lipid uptake. (**b**) Blocking antibodies to TLR2/4 were used as well as inhibitors to NF-kB and AP-1 to try to inhibit the anti-ApoA-1 IgG mediated LDL uptake as evidenced by Oil Red O staining quantification as the mean granule area per cell. Oil Red O was quantified as indicated in the method section. Results are expressed in arbitrary units as mean ± SD of four independent experiments ($n = 4$), p-values were calculated using the Student's t-test: ** $p < 0.01$, *** $p < 0.001$, **** $p < 0.0001$.

3.6. Anti-ApoA-1 IgGs Upregulate ABCA1

Because ABCA1 is a known target of miR-33a [23], we expected that the anti-ApoA-1 IgG-induced downregulation of miR-33a would increase ABCA1 expression. As shown in Figure 6, anti-ApoA-1 IgG increased both the mRNA (Figure 6a) and protein levels of ABCA1 (Figure 6b) in HMDM.

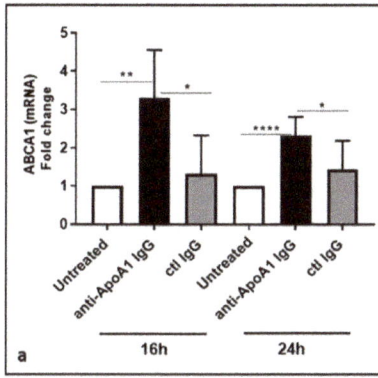 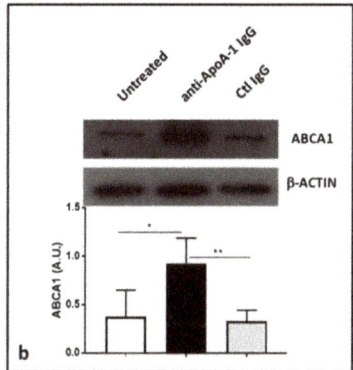

Figure 6. Anti-ApoA-1 IgGs upregulate ABCA1. ABCA1 was increased after 16 and 24 h of anti-ApoA-1 IgG exposure to HMDM at the mRNA level as revealed by RT-PCR (**a**) as well as at the protein level after 24 h anti-apoA-1 IgG stimulation, as revealed by Western blot analysis (**b**). In panel a, data are expressed as fold change expression of the mean ± SD of ABCA1 calculated by ΔΔCT method of six independent experiments ($n = 6$) and values were normalized to untreated condition. *p*-values were calculated using the Student's *t*-test: * $p < 0.05$, ** $p < 0.01$, **** $p < 0.0001$. (**b**) One of four representative Western blots is shown. Data are the mean ± SD of band intensity volume/actin intensity volume of four independent experiments ($n = 4$). *p*-values were calculated using the Student's *t*-test: * $p = 0.02$, ** $p = 0.005$.

3.7. ACAT Activity and Cellular Cholesterol Distribution in Macrophages Following Anti-ApoA1 IgG Treatment

Because ABCA1 levels are known to be upregulated as part of a homeostatic feedback loop in presence of intracellular lipid overload [36], the previous experiments did not allow us to conclude whether there was a direct or indirect effect of anti-ApoA-1 IgG on ABCA1 levels. Nevertheless, as the net effect of anti-ApoA-1 IgGs was an increase in foam cell formation (which is usually associated with lower ABCA1 levels), we primarily suspected that the ABCA1 increase in response to anti-ApoA-1 IgG was part of a negative feedback mechanism. In that case, the exposure of anti-ApoA-1 IgG on HMDM should primarily reorient the free cholesterol present in the plasmatic membrane toward esterified cholesterol intracellular pools, which is mediated by the acyl coenzyme A: cholesterol acyltransferase (ACAT) [41]. To validate this hypothesis, we assessed the impact of anti-ApoA-1 IgGs on the ACAT activity and protein levels on HMDM. As shown in Figure 7a, anti-ApoA-1 IgGs increased ACAT activity, while the protein levels remained unchanged (Supplemental Figure S2).

As shown in Figure 7b, this anti-ApoA-1 IgG- induced ACAT activation was accompanied by a significant reduction of free cholesterol content in membrane and a non-significant reduction of the free cholesterol content in the cell supernatant counterpart (Figure 7c).

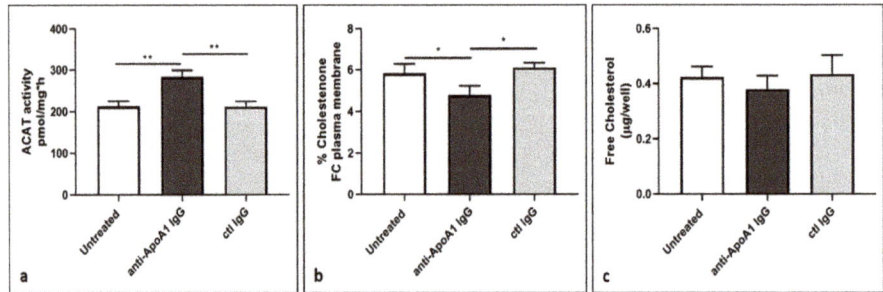

Figure 7. Effect of anti-ApoA-1 IgG on cholesterol distribution. (**a**) Cells were incubated for 24 hours in RPMI medium with 10% FCS, control IgGs, or anti-ApoA-1 IgG. Monolayers underwent a second incubation (4 h) in the presence of [1-14C]-oleic acid albumin complex. Data are expressed as the mean ± SD of the measurements done in triplicate and repeated on three HMDM donors ($n = 3$). Statistical differences were determined by Student's t-test: ** $p \leq 0.0036$. (**b**) HMDM were labelled with 3 μCi/mL [3H] cholesterol for 24 h in RPMI medium with 10% FCS, control IgGs or anti-ApoA-1 IgGs. Cells were then washed and incubated with 1 U/mL cholesterol oxidase enzyme in DPBS for 4 hours at 37 °C. Data are expressed as the mean ± SD of the measurements done in triplicate and repeated on three HMDM donors ($n = 3$), p-values were calculated using the Student's t-test: * $p < 0.05$. (**c**) Free cholesterol content in HMDM supernatant after exposure of cells to anti-ApoA-1 IgGs or control IgGs was expressed as μg of free cholesterol per well. Data are expressed as the mean ± SD of the measurements done on five HMDM donors ($n = 6$). The differences between groups were not statistically significant, anti-ApoA-1 IgG vs. untreated $p = 0.12$; anti-ApoA-1 IgG vs. ctl IgG $p = 0.14$.

3.8. Anti-ApoA-1 IgG Impact on Cholesterol Passive Diffusion

Cholesterol passive diffusion (also known as aqueous diffusion) represents the free cholesterol exchange occurring between cells membranes and plasma by a passive process driven by free cholesterol concentration gradients towards HDL/ApoA-1, independently of any efflux pumps [36] In normocholesterolemic conditions, passive diffusion has been shown to account for up to 80% total cholesterol macrophage efflux [42]. Given the decrease in membrane free cholesterol induced by anti-ApoA-1 IgG in response to the ACAT redistribution within intracellular pools, we expected these antibodies to decrease passive diffusion due to the decrease in the free cholesterol gradient. For this purpose, we used a validated model of human macrophage derived from THP-1 monocytic cell line [35] as our human primary macrophage model was not validated for such kind of experiments. Cells were treated for 24 h with anti-ApoA 1 IgGs or control IgGs in the absence of serum but in presence of 0.2% albumin to avoid any active ABCA1-, ABCG1- or SR-B1-mediated cholesterol efflux. As shown in Figure 8, anti ApoA-1 IgGs significantly decreased the passive diffusion.

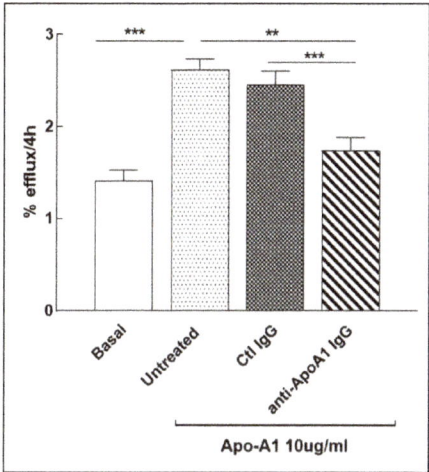

Figure 8. Anti-ApoA-1 IgGs inhibit passive diffusion. THP-1 cells were treated for 24 h with anti-ApoA-1 IgGs or ctl IgGs and passive diffusion was analyzed and expressed as percentage of the radioactivity released to the medium in 4 h over the total radioactivity incorporated by cells. p-values were calculated using the Student's t-test: ** $p = 0.010$, *** $p < 0.001$.

4. Discussion

In this study, we report, for the first time, that the prevalence of high anti-ApoA-1 IgG levels in children devoid of any concomitant autoimmune diseases is substantial, reaching 38%, and seemed unrelated to familial hypercholesterolemia. Such a prevalence is similar to what has been reported in adults in secondary prevention or equivalent CV risk-associated conditions [4,7,43], and close to double of what has been observed in a general adult population [3], where these antibodies were shown to be associated with a worse overall and CV prognosis [3,5].

The second notable finding of this study is that FH positive children with high anti-ApoA-1 IgG levels displayed a more favorable lipid profile consisting of lower total and LDL cholesterol levels, when compared to children who tested negative for these autoantibodies. These results complement previous observations [3,13,15,16] showing that high anti-ApoA-1 IgG levels were associated with lower total, LDL, and miR-33a levels, despite being associated with increased CVD risk, according to previous reports [3,13,15,16]. This clinical observation derived from a limited number of pediatric FH patients (and not retrieved on age and gender matched controls), prompted us to evaluate the ability of anti-ApoA-1 IgG to promote in vitro foam cell development on human primary macrophages, and to determine the molecular mechanisms involved, which led to the following new mechanistic findings.

If the role of the TLR2/4/CD14 complex has been well documented accounting for the cytokine-dependent pro-inflammatory, pro-thrombotic, and pro-arrythmogenic biological properties ascribed to anti-ApoA-1 IgG [9,14,17], this is the first demonstration that these antibodies promote foam cell formation on human macrophages by the same innate immune receptors. Our results demonstrate that anti-ApoA-1 IgGs upregulate the expression of key cholesterol homeostasis-related proteins, such as SREBP2, LDL-R, and HMGCR in a TLR2/4 complex-dependent manner, leading to both increased LDL uptake and intracellular cholesterol synthesis, independently of oxidized LDL uptake. We considered this cellular cholesterol overload as the key driver of ACAT stimulation [44] leading to the esterification of free cholesterol excess into intracellular cholesterol pools (lipid droplet accumulation) and decreased the amount of plasma membrane free cholesterol, diminishing the free cholesterol gradient requested for passive diffusion, a key macrophage cholesterol efflux regulator in normolipidemic conditions [36]. Our observations are in line with previous observations demonstrating that TLR4 and TLR2 agonists can upregulate SREBP2, HMGCR, and LDL-R expression and promote foam cell formation independently of oxidized-LDL uptake [29,45,46].

In this context, the anti-ApoA-1 IgG-induced increase of the anti-atherogenic ABCA1 protein expression, and decrease in pro-atherogenic miR-33a has to be understood as the reflection of the activation of a compensatory but insufficient homeostatic feedback loop to limit intracellular cholesterol overload for two main reasons. First of all, our results indicate that miR-33a expression is independent of TLR2/TLR4 signaling, whereas TLR2 and 4 ligands have been reported to downregulate miR-33a in the case of direct signaling [47]. Secondly, any increase in intracellular cholesterol accumulation is known to downregulate miR-33a expression which, in turn, will alleviate miR-33a-mediated ABCA1 inhibition, leading to an increase in ABCA1 protein expression and ABCA1 efflux independently of TLR stimulation [23,48]. Taken together, these results indicate that in our experimental context, changes in miR-33a and ABCA1 expression have to be considered as the effectors of a negative feedback loop aimed at limiting anti-ApoA-1 IgG-induced intracellular lipid accumulation without being able to fully prevent it, as a common feature of negative feedback loops [49]. Nevertheless, our results indicate that in presence of anti-ApoA-1 IgG, the usual negative feedback mechanism activated by high intracellular cholesterol levels to reduce LDLR seems to be blunted [49]. The exact mechanisms underlying this observation are currently unknown and warrant further investigations.

Despite these remaining questions, to the best of our knowledge, this is the first thorough experimental demonstration unravelling the detailed mechanisms by which anti-ApoA-1 IgGs could promote atherogenesis by acting as disruptors of macrophage cholesterol homeostasis by increasing both LDL uptake and intracellular cholesterol synthesis independently of oxidized LDL. As such, our results show that anti-ApoA-1 IgGs have at least two antagonist mechanisms of action to those of statins in the sense that the former act as HMGCR and ACAT activators, whereas statins act as HMGCR and ACAT inhibitors [50] Although, concordant with our clinical data, knowing whether these mechanisms could explain the clinical associations retrieved between anti-ApoA-1 IgG and lipid profile remains to be determined. Our current understanding of these possible pathophysiological mechanism is summarized in Supplemental Figure S3.

This study has several limitations. It is worth mentioning the very limited size of our FH pediatric sample used in this translational study, due to the low prevalence of FH [51] and the issues related to pediatric blood sampling. Despite this power limitation, the significant associations derived from this cohort were instrumental in providing guidance for the in vitro experiments which, in turn, provided results that could explain the reported clinical associations, albeit not providing a formal demonstration. Another limitation resides in the unresolved paradox raised by the fact that anti-ApoA-1 IgG can decrease mature miR-33a levels despite increasing levels of miR-33a precursor. The reason underlying such an observation is under active investigation, as we currently suspect a possible effect of anti-ApoA-1 IgG on miRNA decay that we did not investigate due to highly elusive mechanisms underlying miRNA stability [52,53], which could not be resolved within the scope of the present work. The third limitation is that we did not consider ABCA1 efflux in this work because all our experiments related to protein regulation were done in normocholesterolemic conditions without additional LDL, where passive diffusion is believed the main functional driver of the cholesterol efflux in macrophages [36]. Nevertheless, as ABCA1 levels were increased in response to anti-ApoA-1 IgG as part of the negative feedback loop (see above), we expect ABCA1 cholesterol efflux to be concomitantly increased, but not sufficiently to inhibit anti-ApoA-1 IgG-induced foam cell formation. Lastly, due to limited human sample availability, we could not evaluate the effect of human purified anti-ApoA-1 IgG in this study. Nevertheless, as we previously demonstrated that the commercial anti-human ApoA-1 IgG used in this study promoted the same effects as the human-purified IgG fraction containing high levels of these autoantibodies in vitro [9,17], and because these autoantibodies have been consistently used in different validated and published animal and in vitro studies [9,12–14,17], we expect a similar impact on in vitro lipid metabolism to occur by using human purified anti-ApoA-1 IgG, even if not formally demonstrated.

In conclusion, the results of this hypothesis-generating translational study show that the prevalence of ApoA-1 IgG seropositivity is frequent in children, unrelated to FH, and surprisingly associated with a favorable lipid profile in FH but not in controls. Furthermore, anti-ApoA-1 IgG were found to induce foam cell formation in vitro through a complex interplay between innate immune receptors and key cholesterol homeostasis regulators, functionally impairing the cholesterol efflux capacity of macrophages. The clinical implications of these findings are unclear, but may suggest that the presence of circulating anti-ApoA-1 IgG could not only interfere with some beneficial effects of lipid-lowering drugs in humans, but could also potentially decrease the contribution of the lipid profile to the individual's CV risk in current CV risk stratification tools. If so, these antibodies may be considered as an additional CV risk enhancer to the list of the factors already considered as such in the latest dyslipidemia management guidelines [54]. These hypotheses are currently devoid of any experimental evidence and await further investigations.

Supplementary Materials: The following are available online at http://www.mdpi.com/2077-0383/8/12/2035/s1, Figure S1: Anti-ApoA-1 IgGs don't promote oxLDL uptake, Figure S2: ACAT-1 expression is not modulated by anti-ApoA-1 IgG, Figure S3: Summary of the anti-ApoA-1 IgG-mediated foam cell formation.

Author Contributions: Conceptualization, S.P., N.V., A.M., E.F.; methodology, S.P., M.D., A.M., N.S., A.R.; software, S.P.; formal analysis, S.P., E.F., M.D., M.A.F.; investigation, F.B., F.M. (Francesco Martino); resources, F.B., F.M. (Francesco Martino); data curation, S.P., N.V.; writing—original draft preparation, S.P., N.V.; writing—review and editing S.P., N.V., A.M., M.A.F., F.M. (François Mach), B.G., E.F.; supervision, F.M. (François Mach); project administration, S.P., N.V.; funding acquisition, N.V.

Funding: This work was supported by a grant from the Leenaards Foundation (grant number 3698 to N.V.) by the Swiss National Science Foundation (grant number 310030-163335 to N.V.) and by the De Reuter Foundation (grant number 315112 to N.V.).

Conflicts of Interest: The authors declare no conflicts of interest.

References

1. Libby, P.; Tabas, I.; Fredman, G.; Fisher, E.A. Inflammation and Its Resolution as Determinants of Acute Coronary Syndromes. *Circ. Res.* **2014**, *114*, 1867–1879. [CrossRef]
2. Satta, N.; Vuilleumier, N. Auto-Antibodies as Possible Markers and Mediators of Ischemic, Dilated, and Rhythmic Cardiopathies. *Curr. Drug Targets* **2015**, *16*, 342–360. [CrossRef]
3. Antiochos, P.; Marques-Vidal, P.; Virzi, J.; Pagano, S.; Satta, N.; Bastardot, F.; Hartley, O.; Montecucco, F.; Mach, F.; Waeber, G.; et al. Association between Anti-Apolipoprotein a-1 Antibodies and Cardiovascular Disease in the General Population. Results from the Colaus Study. *Thromb. Haemost.* **2016**, *116*, 764–771. [CrossRef]
4. Antiochos, P.; Marques-Vidal, P.; Virzi, J.; Pagano, S.; Satta, N.; Hartley, O.; Montecucco, F.; Mach, F.; Kutalik, Z.; Waeber, G.; et al. Impact of Cd14 Polymorphisms on Anti-Apolipoprotein a-1 Igg-Related Coronary Artery Disease Prediction in the General Population. *Arterioscler. Thromb. Vasc. Biol.* **2017**, *37*, 2342–2349. [CrossRef]
5. Antiochos, P.; Marques-Vidal, P.; Virzi, J.; Pagano, S.; Satta, N.; Hartley, O.; Montecucco, F.; Mach, F.; Kutalik, Z.; Waeber, G.; et al. Anti-Apolipoprotein a-1 Igg Predict All-Cause Mortality and Are Associated with Fc Receptor-Like 3 Polymorphisms. *Front. Immunol.* **2017**, *8*, 437. [CrossRef]
6. Carbone, F.; Satta, N.; Montecucco, F.; Virzi, J.; Burger, F.; Roth, A.; Roversi, G.; Tamborino, C.; Casetta, I.; Seraceni, S.; et al. Anti-Apoa-1 Igg Serum Levels Predict Worse Poststroke Outcomes. *Eur. J. Clin. Investig.* **2016**, *46*, 805–817. [CrossRef]
7. Vuilleumier, N.; Bas, S.; Pagano, S.; Montecucco, F.; Guerne, P.A.; Finckh, A.; Lovis, C.; Mach, F.; Hochstrasser, D.; Roux-Lombard, P.; et al. Anti-Apolipoprotein a-1 Igg Predicts Major Cardiovascular Events in Patients with Rheumatoid Arthritis. *Arthritis Rheum.* **2010**, *62*, 2640–2650. [CrossRef] [PubMed]
8. Vuilleumier, N.; Montecucco, F.; Spinella, G.; Pagano, S.; Bertolotto, M.; Pane, B.; Pende, A.; Galan, K.; Roux-Lombard, P.; Combescure, C.; et al. Serum Levels of Anti-Apolipoprotein a-1 Auto-Antibodies and Myeloperoxidase as Predictors of Major Adverse Cardiovascular Events after Carotid Endarterectomy. *Thromb. Haemost.* **2013**, *109*, 706–715. [CrossRef] [PubMed]

9. Vuilleumier, N.; Rossier, M.F.; Pagano, S.; Python, M.; Charbonney, E.; Nkoulou, R.; James, R.; Reber, G.; Mach, F.; Roux-Lombard, P. Anti-Apolipoprotein a-1 Igg as an Independent Cardiovascular Prognostic Marker Affecting Basal Heart Rate in Myocardial Infarction. *Eur. Heart J.* **2010**, *31*, 815–823. [CrossRef] [PubMed]
10. Anderson, J.L.C.; Pagano, S.; Virzi, J.; Dullaart, R.P.F.; Annema, W.; Kuipers, F.; Bakker, S.J.L.; Vuilleumier, N.; Tietge, U.J.F. Autoantibodies to Apolipoprotein a-1 as Independent Predictors of Cardiovascular Mortality in Renal Transplant Recipients. *J. Clin. Med.* **2019**, *8*, 948. [CrossRef]
11. Vuilleumier, N.; Pagano, S.; Combescure, C.; Gencer, B.; Virzi, J.; Raber, L.; Carballo, D.; Carballo, S.; Nanchen, D.; Rodondi, N.; et al. Non-Linear Relationship between Anti-Apolipoprotein a-1 Iggs and Cardiovascular Outcomes in Patients with Acute Coronary Syndromes. *J. Clin. Med.* **2019**, *8*, 1002. [CrossRef] [PubMed]
12. Montecucco, F.; Braunersreuther, V.; Burger, F.; Lenglet, S.; Pelli, G.; Carbone, F.; Fraga-Silva, R.; Stergiopulos, N.; Monaco, C.; Mueller, C.; et al. Anti-Apoa-1 Auto-Antibodies Increase Mouse Atherosclerotic Plaque Vulnerability, Myocardial Necrosis and Mortality Triggering Tlr2 and Tlr4. *Thromb. Haemost.* **2015**, *114*, 410–422. [CrossRef] [PubMed]
13. Montecucco, F.; Vuilleumier, N.; Pagano, S.; Lenglet, S.; Bertolotto, M.; Braunersreuther, V.; Pelli, G.; Kovari, E.; Pane, B.; Spinella, G.; et al. Anti-Apolipoprotein a-1 Auto-Antibodies Are Active Mediators of Atherosclerotic Plaque Vulnerability. *Eur. Heart J.* **2011**, *32*, 412–421. [CrossRef] [PubMed]
14. Pagano, S.; Carbone, F.; Burger, F.; Roth, A.; Bertolotto, M.; Pane, B.; Spinella, G.; Palombo, D.; Pende, A.; Dallegri, F.; et al. Anti-Apolipoprotein a-1 Auto-Antibodies as Active Modulators of Atherothrombosis. *Thromb. Haemost.* **2016**, *116*, 554–564. [CrossRef]
15. Bridge, S.H.; Pagano, S.; Jones, M.; Foster, G.R.; Neely, D.; Vuilleumier, N.; Bassendine, M.F. Autoantibody to Apolipoprotein a-1 in Hepatitis C Virus Infection: A Role in Atherosclerosis? *Hepatol. Int.* **2018**, *12*, 17–25. [CrossRef]
16. Quercioli, A.; Montecucco, F.; Galan, K.; Ratib, O.; Roux-Lombard, P.; Pagano, S.; Mach, F.; Schindler, T.H.; Vuilleumier, N. Anti-Apolipoprotein a-1 Igg Levels Predict Coronary Artery Calcification in Obese but Otherwise Healthy Individuals. *Mediat. Inflamm.* **2012**, *2012*, 243158. [CrossRef]
17. Pagano, S.; Satta, N.; Werling, D.; Offord, V.; de Moerloose, P.; Charbonney, E.; Hochstrasser, D.; Roux-Lombard, P.; Vuilleumier, N. Anti-Apolipoprotein a-1 Igg in Patients with Myocardial Infarction Promotes Inflammation through Tlr2/Cd14 Complex. *J. Intern. Med.* **2012**, *272*, 344–357. [CrossRef]
18. Dullaart, R.P.F.; Pagano, S.; Perton, F.G.; Vuilleumier, N. Antibodies against the C-Terminus of Apoa-1 Are Inversely Associated with Cholesterol Efflux Capacity and Hdl Metabolism in Subjects with and without Type 2 Diabetes Mellitus. *Int. J. Mol. Sci.* **2019**, *20*, 732. [CrossRef]
19. Vuilleumier, N.; Pagano, S.; Montecucco, F.; Quercioli, A.; Schindler, T.H.; Mach, F.; Cipollari, E.; Ronda, N.; Favari, E. Relationship between Hdl Cholesterol Efflux Capacity, Calcium Coronary Artery Content, and Antibodies against Apolipoproteina-1 in Obese and Healthy Subjects. *J. Clin. Med.* **2019**, *8*, E1225. [CrossRef]
20. Hansson, G.K.; Hermansson, A. The Immune System in Atherosclerosis. *Nat. Immunol.* **2011**, *12*, 204–212. [CrossRef]
21. Goldstein, J.L.; Brown, M.S. A Century of Cholesterol and Coronaries: From Plaques to Genes to Statins. *Cell* **2015**, *161*, 161–172. [CrossRef] [PubMed]
22. Marquart, T.J.; Allen, R.M.; Ory, D.S.; Baldan, A. Mir-33 Links Srebp-2 Induction to Repression of Sterol Transporters. *Proc. Natl. Acad. Sci. USA* **2010**, *107*, 12228–12232. [CrossRef] [PubMed]
23. Najafi-Shoushtari, S.H.; Kristo, F.; Li, Y.; Shioda, T.; Cohen, D.E.; Gerszten, R.E.; Naar, A.M. Microrna-33 and the Srebp Host Genes Cooperate to Control Cholesterol Homeostasis. *Science* **2010**, *328*, 1566–1569. [CrossRef] [PubMed]
24. Rayner, K.J.; Sheedy, F.J.; Esau, C.C.; Hussain, F.N.; Temel, R.E.; Parathath, S.; van Gils, J.M.; Rayner, A.J.; Chang, A.N.; Suarez, Y.; et al. Antagonism of Mir-33 in Mice Promotes Reverse Cholesterol Transport and Regression of Atherosclerosis. *J. Clin. Investig.* **2011**, *121*, 2921–2931. [CrossRef] [PubMed]
25. Rotllan, N.; Ramirez, C.M.; Aryal, B.; Esau, C.C.; Fernandez-Hernando, C. Therapeutic Silencing of Microrna-33 Inhibits the Progression of Atherosclerosis in Ldlr$^{-/-}$Mice—Brief Report. *Arterioscler. Thromb. Vasc. Biol.* **2013**, *33*, 1973–1977. [CrossRef]

26. Price, N.L.; Singh, A.K.; Rotllan, N.; Goedeke, L.; Wing, A.; Canfran-Duque, A.; Diaz-Ruiz, A.; Araldi, E.; Baldan, A.; Camporez, J.P.; et al. Genetic Ablation of Mir-33 Increases Food Intake, Enhances Adipose Tissue Expansion, and Promotes Obesity and Insulin Resistance. *Cell Rep.* **2018**, *22*, 2133–2145. [CrossRef]
27. Campagna, F.; Martino, F.; Bifolco, M.; Montali, A.; Martino, E.; Morrone, F.; Antonini, R.; Cantafora, A.; Verna, R.; Arca, M. Detection of Familial Hypercholesterolemia in a Cohort of Children with Hypercholesterolemia: Results of a Family and DNA-Based Screening. *Atherosclerosis* **2008**, *196*, 356–364. [CrossRef]
28. Moore, K.J.; Kunjathoor, V.V.; Koehn, S.L.; Manning, J.J.; Tseng, A.A.; Silver, J.M.; McKee, M.; Freeman, M.W. Loss of Receptor-Mediated Lipid Uptake Via Scavenger Receptor a or Cd36 Pathways Does Not Ameliorate Atherosclerosis in Hyperlipidemic Mice. *J. Clin. Investig.* **2005**, *115*, 2192–2201. [CrossRef]
29. Ye, Q.; Chen, Y.; Lei, H.; Liu, Q.; Moorhead, J.F.; Varghese, Z.; Ruan, X.Z. Inflammatory Stress Increases Unmodified Ldl Uptake Via Ldl Receptor: An Alternative Pathway for Macrophage Foam-Cell Formation. *Inflamm. Res. Off. J. Eur. Histamine Res. Soc. [et al.]* **2009**, *58*, 809–818. [CrossRef]
30. Haase, A.; Goldberg, A.C. Identification of People with Heterozygous Familial Hypercholesterolemia. *Curr. Opin. Lipidol.* **2012**, *23*, 282–289. [CrossRef]
31. Vandesompele, J.; De Preter, K.; Pattyn, F.; Poppe, B.; Van Roy, N.; De Paepe, A.; Speleman, F. Accurate Normalization of Real-Time Quantitative Rt-Pcr Data by Geometric Averaging of Multiple Internal Control Genes. *Genome Biol.* **2002**, *3*, RESEARCH0034. [CrossRef] [PubMed]
32. Livak, K.J.; Schmittgen, T.D. Analysis of Relative Gene Expression Data Using Real-Time Quantitative Pcr and the 2(-Delta Delta C(T)) Method. *Methods* **2001**, *25*, 402–408. [CrossRef] [PubMed]
33. Kellner-Weibel, G.; de La Llera-Moya, M.; Connelly, M.A.; Stoudt, G.; Christian, A.E.; Haynes, M.P.; Williams, D.L.; Rothblat, G.H. Expression of Scavenger Receptor Bi in Cos-7 Cells Alters Cholesterol Content and Distribution. *Biochemistry* **2000**, *39*, 221–229. [CrossRef] [PubMed]
34. Brown, M.S.; Ho, Y.K.; Goldstein, J.L. The Cholesteryl Ester Cycle in Macrophage Foam Cells. Continual Hydrolysis and Re-Esterification of Cytoplasmic Cholesteryl Esters. *J. Biol. Chem.* **1980**, *255*, 9344–9352. [PubMed]
35. Barberio, M.D.; Kasselman, L.J.; Playford, M.P.; Epstein, S.B.; Renna, H.A.; Goldberg, M.; DeLeon, J.; Voloshyna, I.; Barlev, A.; Salama, M.; et al. Cholesterol Efflux Alterations in Adolescent Obesity: Role of Adipose-Derived Extracellular Vesical Micrornas. *J. Transl. Med.* **2019**, *17*, 232. [CrossRef] [PubMed]
36. Phillips, M.C. Molecular Mechanisms of Cellular Cholesterol Efflux. *J. Biol. Chem.* **2014**, *289*, 24020–24029. [CrossRef] [PubMed]
37. Rayner, K.J.; Suarez, Y.; Davalos, A.; Parathath, S.; Fitzgerald, M.L.; Tamehiro, N.; Fisher, E.A.; Moore, K.J.; Fernandez-Hernando, C. Mir-33 Contributes to the Regulation of Cholesterol Homeostasis. *Science* **2010**, *328*, 1570–1573. [CrossRef]
38. Karunakaran, D.; Thrush, A.B.; Nguyen, M.A.; Richards, L.; Geoffrion, M.; Singaravelu, R.; Ramphos, E.; Shangari, P.; Ouimet, M.; Pezacki, J.P.; et al. Macrophage Mitochondrial Energy Status Regulates Cholesterol Efflux and Is Enhanced by Anti-Mir33 in Atherosclerosis. *Circ. Res.* **2015**, *117*, 266–278. [CrossRef]
39. Niesor, E.J.; Schwartz, G.G.; Perez, A.; Stauffer, A.; Durrwell, A.; Bucklar-Suchankova, G.; Benghozi, R.; Abt, M.; Kallend, D. Statin-Induced Decrease in Atp-Binding Cassette Transporter A1 Expression Via Microrna33 Induction May Counteract Cholesterol Efflux to High-Density Lipoprotein. *Cardiovasc. Drugs Ther.* **2015**, *29*, 7–14. [CrossRef]
40. Brown, M.S.; Goldstein, J.L. A Proteolytic Pathway That Controls the Cholesterol Content of Membranes, Cells, and Blood. *Proc. Natl. Acad. Sci. USA* **1999**, *96*, 11041–11048. [CrossRef]
41. Buhman, K.F.; Accad, M.; Farese, R.V. Mammalian Acyl-Coa:Cholesterol Acyltransferases. *Biochim. Biophys. Acta* **2000**, *1529*, 142–154. [CrossRef]
42. Adorni, M.P.; Zimetti, F.; Billheimer, J.T.; Wang, N.; Rader, D.J.; Phillips, M.C.; Rothblat, G.H. The Roles of Different Pathways in the Release of Cholesterol from Macrophages. *J. Lipid. Res.* **2007**, *48*, 2453–2462. [CrossRef] [PubMed]
43. Pruijm, M.; Schmidtko, J.; Aho, A.; Pagano, S.; Roux-Lombard, P.; Teta, D.; Burnier, M.; Vuilleumier, N. High Prevalence of Anti-Apolipoprotein/a-1 Autoantibodies in Maintenance Hemodialysis and Association with Dialysis Vintage. *Ther. Apher. Dial. Off. Peer-Rev. J. Int. Soc. Apher. Jpn. Soc. Apher. Jpn. Soc. Dial. Ther.* **2012**, *16*, 588–594. [CrossRef] [PubMed]

44. Chang, T.Y.; Chang, C.C.; Cheng, D. Acyl-Coenzyme A:Cholesterol Acyltransferase. *Annu. Rev. Biochem.* **1997**, *66*, 613–638. [CrossRef] [PubMed]
45. Li, L.C.; Varghese, Z.; Moorhead, J.F.; Lee, C.T.; Chen, J.B.; Ruan, X.Z. Cross-Talk between Tlr4-Myd88-Nf-Kappab and Scap-Srebp2 Pathways Mediates Macrophage Foam Cell Formation. *Am. J. Physiol. Heart Circ. Physiol.* **2013**, *304*, H874–H884. [CrossRef]
46. Li, Y.J.; Zhu, P.; Liang, Y.; Yin, W.G.; Xiao, J.H. Hepatitis B Virus Induces Expression of Cholesterol Metabolism-Related Genes Via Tlr2 in Hepg2 Cells. *World J. Gastroenterol.* **2013**, *19*, 2262–2269. [CrossRef]
47. Lai, L.; Azzam, K.M.; Lin, W.C.; Rai, P.; Lowe, J.M.; Gabor, K.A.; Madenspacher, J.H.; Aloor, J.J.; Parks, J.S.; Naar, A.M.; et al. Microrna-33 Regulates the Innate Immune Response Via Atp Binding Cassette Transporter-Mediated Remodeling of Membrane Microdomains. *J. Biol. Chem.* **2016**, *291*, 19651–19660. [CrossRef]
48. Horie, T.; Ono, K.; Horiguchi, M.; Nishi, H.; Nakamura, T.; Nagao, K.; Kinoshita, M.; Kuwabara, Y.; Marusawa, H.; Iwanaga, Y.; et al. Microrna-33 Encoded by an Intron of Sterol Regulatory Element-Binding Protein 2 (Srebp2) Regulates Hdl in Vivo. *Proc. Natl. Acad. Sci. USA* **2010**, *107*, 17321–17326. [CrossRef]
49. Afonso, M.S.; Machado, R.M.; Lavrador, M.S.; Quintao, E.C.R.; Moore, K.J.; Lottenberg, A.M. Molecular Pathways Underlying Cholesterol Homeostasis. *Nutrients* **2018**, *10*, 760. [CrossRef]
50. Kam, N.T.; Albright, E.; Mathur, S.; Field, F.J. Effect of Lovastatin on Acyl-Coa: Cholesterol O-Acyltransferase (Acat) Activity and the Basolateral-Membrane Secretion of Newly Synthesized Lipids by Caco-2 Cells. *Biochem. J.* **1990**, *272*, 427–433. [CrossRef]
51. Akioyamen, L.E.; Genest, J.; Shan, S.D.; Reel, R.L.; Albaum, J.M.; Chu, A.; Tu, J.V. Estimating the Prevalence of Heterozygous Familial Hypercholesterolaemia: A Systematic Review and Meta-Analysis. *BMJ Open* **2017**, *7*, e016461. [CrossRef] [PubMed]
52. Ameres, S.L.; Horwich, M.D.; Hung, J.H.; Xu, J.; Ghildiyal, M.; Weng, Z.; Zamore, P.D. Target Rna-Directed Trimming and Tailing of Small Silencing Rnas. *Science* **2010**, *328*, 1534–1539. [CrossRef] [PubMed]
53. Ruegger, S.; Grosshans, H. Microrna Turnover: When, How, and Why. *Trends Biochem. Sci.* **2012**, *37*, 436–446. [CrossRef] [PubMed]
54. Mach, F.; Baigent, C.; Catapano, A.L.; Koskinas, K.C.; Casula, M.; Badimon, L.; Chapman, M.J.; De Backer, G.G.; Delgado, V.; Ference, B.A.; et al. Scientific Document. 2019 Esc/Eas Guidelines for the Management of Dyslipidaemias: Lipid Modification to Reduce Cardiovascular Risk. *Eur. Heart J.* **2019**. [CrossRef] [PubMed]

© 2019 by the authors. Licensee MDPI, Basel, Switzerland. This article is an open access article distributed under the terms and conditions of the Creative Commons Attribution (CC BY) license (http://creativecommons.org/licenses/by/4.0/).

Article

Autoantibodies to Apolipoprotein A-1 as Independent Predictors of Cardiovascular Mortality in Renal Transplant Recipients

Josephine L.C. Anderson [1], Sabrina Pagano [2,3], Julien Virzi [2,3], Robin P.F. Dullaart [4], Wijtske Annema [1,5], Folkert Kuipers [1,6], Stephan J.L. Bakker [7], Nicolas Vuilleumier [2,†] and Uwe J.F. Tietge [1,8,9,*,†]

1. Department of Pediatrics, University Medical Center Groningen, University of Groningen, 1205 Groningen, The Netherlands
2. Division of Laboratory Medicine, Department of Genetics and Laboratory Medicine, Geneva University Hospital, 1205 Geneva, Switzerland
3. Department of Medical Specialties, Faculty of Medicine, Geneva University, 1205 Geneva, Switzerland
4. Department of Endocrinology, University Medical Center Groningen, University of Groningen, 9713 GZ Groningen, The Netherlands
5. Institute of Clinical Chemistry, University Hospital of Zurich and University of Zurich, 8006 Zurich, Switzerland
6. Laboratory Medicine, University Medical Center Groningen, University of Groningen, 9713 GZ Groningen, The Netherlands
7. Department of Nephrology, University Medical Center Groningen, University of Groningen, 9713 GZ Groningen, The Netherlands
8. Division of Clinical Chemistry, Department of Laboratory Medicine H5, Karolinska Institutet, 14183 Stockholm, Sweden
9. Clinical Chemistry, Karolinska University Laboratory, Karolinska University Hospital, SE-141 86 Stockholm, Sweden
* Correspondence: uwe.tietge@ki.se; Tel.: +46-852-481-361
† These authors contributed equally to this work.

Received: 16 May 2019; Accepted: 27 June 2019; Published: 29 June 2019

Abstract: Renal transplant recipients (RTRs) are known to have a high cardio-vascular disease (CVD) burden only partly explained by traditional CVD risk factors. The aim of this paper was therefore to determine: i) the prognostic value of autoantibodies against apoA-1 (anti-apoA-1 IgG) for incidence of CVD mortality, all-cause mortality and graft failure in RTR. Four hundred and sixty two (462) prospectively included RTRs were followed for 7.0 years. Baseline anti-apoA-1 IgG were determined and associations with incidence of CVD mortality (n = 48), all-cause mortality (n = 92) and graft failure (n = 39) were tested. Kaplan–Meier analyses demonstrated significant associations between tertiles of anti-apoA-1 IgG and CVD mortality (log rank test: p = 0.048). Adjusted Cox regression analysis showed a 54% increase in risk for CVD mortality for each anti-apoA-1 IgG levels standard deviation increase (hazard ratio [HR]: 1.54, 95% Confidence Interval [95%CI]: 1.14–2.05, p = 0.005), and a 33% increase for all-cause mortality (HR: 1.33; 95%CI: 1.06–1.67, p = 0.01), independent of CVD risk factors, renal function and HDL function. The association with all-cause mortality disappeared after excluding cases of CVD specific mortality. The sensitivity, specificity, positive predictive value, and negative predictive value of anti-apoA-1 positivity for CVD mortality were 18.0%, 89.3%, 17.0%, and 90.0%, respectively. HDL functionality was not associated with anti-apoA-1 IgG levels. This prospective study demonstrates that in RTR, anti-apoA-1 IgG are independent predictors of CVD mortality and are not associated with HDL functionality.

Keywords: biomarker; anti-apolipoprotein A-1 antibodies; renal transplant recipient; HDL function; prognosis; cholesterol

1. Introduction

Impaired kidney function is a major risk factor for cardiovascular diseases (CVD) through all stages of renal dysfunction, amounting to a substantial 40-fold increased risk of CVD mortality in end-stage renal disease (ESRD) patients [1,2]. Even after renal transplantation the CVD risk remains four to six times higher in age-adjusted analyses, with half of all deaths of renal transplant recipients (RTRs) being attributable to a CVD origin [3]. Traditional CVD risk factors, such as the ones aggregated in the Framingham risk score (FRS), are known to be of little use in CVD risk prediction in RTRs [4,5]. Thus, accurate CVD risk stratification in RTRs represents an unmet clinical need in this constantly increasing patient population.

Autoantibodies against apoA-1 (anti-apoA-1 IgG) represent a recently identified biomarker with a high potential to predict increased CVD risk. Increased levels of these antibodies are associated with a pro-atherogenic lipid profile, a systemic pro-inflammatory state [6–8], as well as high-density lipoprotein (HDL) dysfunction [9,10], and were shown to be associated with increased CVD risk and poorer prognosis in high-risk patients, as well as in the general population [7,11–14]. When administered to mouse models of atherosclerosis, anti-apoA-1 IgG enhanced atherogenesis, myocardial necrosis and premature death indicating that anti-apoA-1 IgG have the potential to serve as a causative biomarker for CVD [15–17]. A previous study showed that ESRD patients have a high prevalence of elevated anti-apoA-1 IgG levels, which was associated with dialysis vintage, and was a major determinant of cardiovascular outcomes in these patients [18]. The reasons underpinning such association in ESRD are still elusive, but may suggest that a prolonged exposure to the uremic milieu, characterized by increased oxidative stress and inflammation, could increase apoA-1 immunogenicity leading to an anti-apoA-1 IgG response. Accordingly, as RTR are exposed to a uremic milieu prior to transplantation, they could also constitute a particularly risk-prone group to such a humoral autoimmunity phenomenon, despite receiving immunosuppressive treatment.

The aim of our present study was to determine: i) the prognostic value of anti-apoA-1 IgG for incidence of CVD specific mortality, overall mortality and graft failure in RTR and ii) to delineate the relationship of anti-apoA-1 IgG with apoA-1 levels and HDL functionality.

2. Experimental Section

2.1. Study Design and Study Population

This study included all adult RTR who visited the University Medical Centre Groningen (UMCG) outpatient clinic between August 2001 and July 2003 with a functioning renal graft for at least 1 year. Of 847 eligible patients, 606 consented to participate in the overall study. Exclusion criteria consisted of congestive heart failure, malignant disease other than cured skin cancer, as well as endocrine abnormalities other than diabetes, or suspected acute infection upon inclusion, indicated by a CRP value above 15 mg/L. This way 477 patients were initially included in the present study; serum was available from 462 RTRs, in which subsequently anti-apoA-1 IgG levels were measured. All relevant patient characteristics were obtained from the "Groningen Renal Transplant Database". Patients were followed for a period of 7 years, and no patients were lost during follow-up. More detailed definitions of the characteristics of the database patients' baseline characteristics, as well as the routine laboratory methods used have been previously described [19,20]. The study was approved by the local Medical Ethics Committee (METc2001/039), and is in accordance with the Declaration of Helsinki and Principles of the Declaration of Istanbul as outlined in the 'Declaration of Istanbul on Organ Trafficking and Transplant Tourism'.

2.2. Outcome Measures

The main outcome measure in this study is the level of anti-apoA-1 IgG. The primary endpoints consisted in CVD mortality, all-cause mortality and graft failure. As previously reported [21,22]. graft failure was defined as return to dialysis or re-transplantation. Cause of death was obtained by linking the number on the death certificate to the primary cause of death, as coded by a physician from the Central Bureau of Statistics. CVD mortality was defined as deaths in which the principal cause of death was cardiovascular in nature, using ICD-9 codes 410–447. The secondary endpoint was a possible association between anti-apoA-1 IgG levels and apoA-1 levels, as well as a key HDL functionality, namely macrophage cholesterol efflux capacity (CEC).

2.3. Sensitivity Analyses

Sensitivity analyses were performed, in which the association of anti-apoA-1 IgG with non CVD-mortality was assessed, in order to assess the specificity of anti-apoA-1 with CVD.

2.4. Determination of Anti-apoA-1 IgG

Anti-apoA-1 IgG were measured using RTR serum aliquots stored at −80 °C, as previously described [6–8,11–13]. The experiments demonstrating the specificity of our assay against the native and unmodified form of apoA-1 are available in the Table 1.

Table 1. Baseline characteristics according to gender stratified tertiles of anti-apoA-1 IgG.

Characteristic	Gender Stratified Tertiles of Anti-apoA-1 Levels			p-Value
	First (n = 154)	Second (n = 154)	Third (n = 154)	
Recipient demographics				
Anti-apoA-1 IgG, AU (OD$_{405}$ nm)	0.15 (0.11–0.19)	0.31 (0.26–0.36)	0.64 (0.52–0.82)	<0.001
Recipient demographics				
Age, years	50.5 (41.6–59.4)	53.4 (44.7–61.1)	52.1 (44.0–60.8)	0.12
Male gender, n (%)	84 (55)	84 (55)	84 (55)	1.00
Current smoking, n (%)	28 (18)	33 (21)	22 (14)	0.26
Previous smoking, n (%)	70 (46)	67 (44)	72 (47)	0.85
Metabolic syndrome, n (%)	83 (57)	94 (65)	84 (60)	0.34
Body composition				
BMI, kg/m^2	26.1 ± 4.3	26.1 ± 4.2	25.9 ± 4.2	0.86
Lipid Profile				
Total cholesterol, mmol/L	5.6 ± 1.0	5.7 ± 0.9	5.7 ± 1.3	0.49
LDL cholesterol, mmol/L	3.5 ± 1.0	3.6 ± 0.8	3.6 ± 1.2	0.82
HDL cholesterol, mmol/L	1.1 ± 0.3	1.3 ± 0.3	1.1 ± 0.3	0.55
Apolipoprotein A-I, g/L	1.6 ± 0.3	1.6 ± 0.3	1.6 ± 0.3	0.75
Triglycerides, mmol/L	2.1 (1.3–2.7)	2.1 (1.4–2.5)	2.2 (1.4–2.7)	0.22
Cholesterol efflux percentage	7.3 (6.2–8.4)	7.5 (6.3–8.3)	7.6 (6.5–8.9)	0.11
Use of statins, n (%)	79 (51)	88 (57)	74 (48)	0.27
Cardiovascular disease history				
History of MI, n (%)	12 (7)	8 (5)	20 (13)	0.047
TIA/CVA, n (%)	5 (3)	8 (5)	9 (6)	0.54
Blood pressure				
Systolic blood pressure, mmHg	152.2 ± 23.9	151.0 ± 21.4	154.1 ± 22.0	0.47
Diastolic blood pressure, mmHg	89.8 (± 9.8)	89.0 (± 9.5)	90.1 (± 10.0)	0.59
Use of ACE inhibitors, n (%)	55 (36)	49 (32)	58 (38)	0.55
Use of β-blockers, n (%)	90 (58)	93 (60)	95 (61)	0.84
Use of diuretics, n (%)	59 (38)	75 (49)	63 (41)	0.16
Number of antihypertensive drugs, n (%)	2 (1–3)	2 (1–3)	2 (1–3)	0.16

Table 1. Cont.

Characteristic	Gender Stratified Tertiles of Anti-apoA-1 Levels			p-Value
	First (n = 154)	Second (n = 154)	Third (n = 154)	
Glucose homeostasis				
Glucose, mmol/L	4.9 (4.1–5.0)	4.8 (4.1–5.0)	4.8 (4.1–5.1)	0.69
Insulin, µmol/L	11.3 (8.7–16.5)	10.6 [7.8–14.8)	11.4 (7.6–15.4)	0.16
HbA1c, %	6.3 (5.8–6.9)	6.3 (5.8–7.0)	6.4 (5.7–7.1)	0.47
HOMA-IR	3.1 (1.7–3.6)	2.7 (1.5–3.4)	2.8 (1.5–3.4)	0.21
Post-Tx diabetes mellitus, n (%)	24 (15)	29 (19)	29 (19)	0.69
Use of anti-diabetic drugs, n (%)	17 (11)	25 (16)	21 (14)	0.41
Use of insulin, n (%)	7 (5)	9 (6)	13 (8)	0.36
Inflammation				
hsCRP, mg/L	3.4 (0.7–4.4)	3.3 (0.9–4.1)	3.3 (1.0–4.2)	0.43
Framingham risk score	17.2 (7.6–27.3)	20.8 (9.6–32.9)	20.7 (8.6–31.3)	0.28
Donor demographics				
Age, years	37.6 (23.0–50.0)	37.2 (23.8–50.0)	37.1 (23.0–51.3)	0.76
Male gender, n (%)	76 (49)	90 (59)	85 (56)	0.24
Living kidney donor, n (%)	28 (18)	17 (11)	15 (10)	0.06
(Pre)transplant history				
Dialysis time, months	34.8 (12.0–48.3)	37.0 (14.8–51.0)	33.6 (12.8–45.0)	0.44
Primary renal disease				
Primary glomerular disease, n (%)	35 (23)	40 (26)	52 (34)	0.08
Glomerulonephritis, n (%)	11 (7)	6 (4)	12 (8)	0.32
Tubulo-interstitial disease, n (%)	33 (21)	24 (16)	17 (11)	0.05
Polycystic renal disease, n (%)	26 (17)	31 (20)	24 (16)	0.56
Dysplasia and hypoplasia, n (%)	7 (5)	8 (5)	2 (1)	0.15
Renovascular disease, n (%)	11 (7)	12 (8)	6 (4)	0.32
Diabetic nephropathy, n (%)	3 (2)	2 (1)	9 (6)	0.04
Other or unknown cause, n (%)	28 (18)	31 (20)	32 (21)	0.84
Immunosuppressive medication				
Daily prednisolone dose, mg	9.2 (7.5–10.0)	9.1 (7.5–10.0)	9.1 (7.5–10.0)	0.33
Calcineurin inhibitors, n (%)	120 (78)	126 (82)	124 (81)	0.68
Proliferation inhibitors, n (%)	124 (81)	109 (71)	108 (70)	0.07
Renal allograft function				
Creatinine clearance, mL/min	47.3 ± 14.6	48.2 ± 15.9	46.5 ± 16.1	0.62
Urinary protein excretion, g/24 h	0.3 (0.0–0.3)	0.2 (0.1–0.2)	0.4 (0.1–0.4)	0.07

Normally distributed continuous variables are presented as mean ± SD. Continuous variables with a skewed distribution are presented as median [25th to 75th percentile]. Categorical data are summarized by n (%), and differences were tested by chi-squared test. MI, myocardial infarction; TIA, transient ischemic attack; CVA, cerebrovascular accident; ACE, angiotensin-converting enzyme; Tx, transplantation.

2.5. Determination of HDL Function

To determine HDL-mediated CEC, a previously published method was used [21,22]. For further details see the Table 1.

2.6. Statistical Analysis

In order to eliminate bias due to gender specific differences in levels of anti-apoA-1 IgG renal transplant recipients were divided into gender-stratified tertiles based on levels of anti-apoA-1 IgG. This was done by first dividing the group into males and females, then computing tertiles based on the levels of anti-apoA-1 IgG, and subsequently merging the groups back together. Differences between baseline characteristics were tested. For continuous variables with a skewed distribution differences were tested by Kruskal–Wallis test. Differences for normally distributed continuous variables were tested by one-way analysis of variance followed by Bonferroni post-hoc test. Differences in categorical data were tested by chi-squared test.

Thereafter, multivariable linear regression analysis was performed to evaluate which variables predict levels of anti-apoA-1. Baseline characteristics with a *p*-value of ≤0.2 between tertiles of anti-apoA-1 IgG were first fitted into a univariate linear regression. Variables that had a significant association with anti-apoA-1 IgG in a univariate analysis were then entered into a multivariate linear regression.

The association of anti-apoA-1 IgG levels with the primary endpoints was assessed by the log-rank test and by Cox proportional hazards regression. Kaplan-Meier curve analyses were performed across anti-apoA-1 IgG tertiles and according to anti-apoA-1 IgG seropositivity, based upon a predefined and validated anti-apoA-1 IgG cut-off value (an OD value >0.64 and a percentage of the positive control above 37%) [7,8,11–14,18]. Differences were assessed using a log-rank test. Cox regression analyses were used to calculate hazard ratios (HR) and reported with their 95% confidence intervals (95%CI). Univariate and multivariate Cox regression analyses were performed per standard deviation (SD, 0.316) increase of anti-apoA-1 IgG levels, and according to anti-apoA-1 IgG seropositivity. Multivariate analyses were performed using different models, taking into account traditional CV risk factors, renal function, HDL functionality, and all the variables that had significant association with anti-apoA-1 levels in linear regression. Schoenfeld residuals test was used to test the proportional hazard assumption for the outcomes of CVD mortality, all-cause mortality and graft failure for analysis per standard deviation increase ($p = 0.18$, $p = 0.20$ and $p = 0.69$ respectively) and for analysis with seropositivity ($p = 0.38$, $p = 0.09$ and $p = 0.77$ respectively), and was found not to be violated. Sensitivity (SN), specificity (SP), positive predictive and negative predictive values (PPV and NPV, respectively) for anti-apoA-1 IgG seropositivity were computed. *p*-values <0.05 were considered statistically significant. All statistical analyses were performed using the Statistical Package for the Social Sciences version 24 (IBM SPSS, IBM Corporation, Armonk, NY, USA) and GraphPad Prism version 6.0 (GraphPad Software, San Diego, CA, USA).

3. Results

3.1. Baseline Demographic Characteristics

In this longitudinal follow-up study the levels of anti-apoA-1 IgG were measured in 462 RTR. Of these patients, 92 (20%) died within the follow-up of 7 years, 48 of these from a confirmed CVD cause, as determined by ICD-9 codes 410–447 (10% of included patients, 52% of all recorded deaths). A total of 39 (8%) patients experienced graft failure. Overall, the prevalence of high levels of anti-apoA-1 IgG (anti-apoA-1 IgG seropositivity according to a previously defined cut-off value based on data from the general population) was 11.5 % (53/462). [7,11–13,18] In order to better explore the architecture of anti-apoA-1 IgG in RTR, patients were divided into gender-stratified tertiles of anti-apoA-1 IgG, with median values of 0.15 (range 0–0.24), 0.31 (range 0.24–0.49), and 0.64 (range 0.50–2.09) for the first, second and third tertile, respectively (Table 1). Analyses between tertiles showed a significant difference for the history of myocardial infarction (MI), which was most common in patients with the highest levels of anti-apoA-1 IgG ($p = 0.047$), as well as for diabetic nephropathy ($p = 0.04$) as the primary renal disease. A trend was also observed for tubulo-interstitial disease ($p = 0.05$), which is characterised by acute or chronic inflammation of the renal tubules and interstitium, and primary glomerular disease ($p = 0.08$), which covers a group of conditions in which there is primary injury in the glomerulus [23].

When participating RTR were stratified according to anti-apoA-1 IgG seropositivity, there was again a significant association with a higher prevalence of previous MI ($p = 0.02$), glomerular disease ($p = 0.03$) and tubulo-interstitial disease ($p = 0.03$) as the primary renal disease (Table 1). Furthermore, anti-apoA-1 IgG seropositive patients tended to have received grafts from older donors ($p = 0.05$) and showed a higher urinary protein excretion ($p = 0.04$, Supplementary Table S1). Importantly, cholesterol efflux capacity as central HDL function metric did not differ between patients seropositive for anti-apoA-1 IgG and those seronegative for these antibodies (Supplementary Table S1).

Subsequently, univariate and thereafter multivariate linear regression was performed to deduce which variables are independently associated with levels of anti-apoA-1 IgG (Table 2). A positive, independent association was seen between anti-apoA-1 IgG and a history of myocardial infarction ($\beta = 0.103$, $p = 0.026$) and primary glomerular disease ($\beta = 0.116$, $p = 0.016$). A negative association was seen with tubule-interstitial disease ($\beta = -0.106$, $p = 0.028$). No significant relationship between anti-apoA-1 IgG and the metric of HDL function, CEC, was discernible, nor with concentrations of HDL-C or with apoA-1. There was also no association with immunosuppressive drugs, either individually or combined.

Table 2. Multivariate linear regression for baseline characteristics that are significantly associated with anti-apoA-1 IgG in a univariate linear regression.

Characteristics	Unstandardized Coefficient	95% CI	Standardized Coefficient	p-Value
Primary glomerular disease	0.086	0.016–0.156	0.116	0.016
History of MI	0.121	0.015–0.227	0.103	0.026
Tubulo-interstitial disease	−0.0.96	−0.182–−0.010	−0.106	0.028

Variables are listed in decreasing order of strength of association. $R^2 = 0.043$ (Cox & Snell). Model x^2: 49.8; $p < 0.001$.

3.2. Association with Incidence of CVD Mortality, All-Cause Mortality, and Graft Failure

As shown in Figure 1, Kaplan Meier curves showed a significant association of tertiles of anti-apoA-1 IgG with CVD mortality ($p = 0.048$), but not with all-cause mortality ($p = 0.22$) or graft failure ($p = 0.13$).

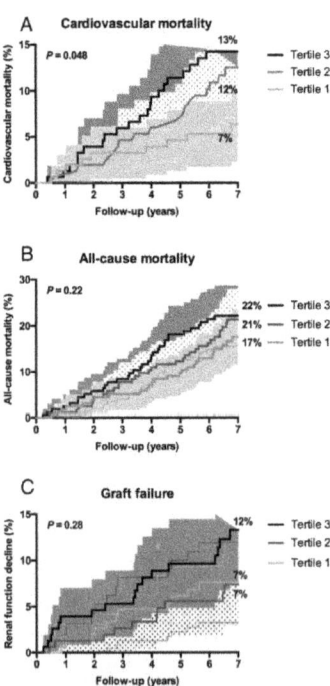

Figure 1. Higher levels of anti-apoA-1 IgG are associated with an increased incidence of cardiovascular mortality and all-cause mortality in renal transplant recipients. Kaplan-Meier curves depicting (**A**) cardiovascular mortality, (**B**) all-cause mortality, and (**C**) graft failure according to tertiles of anti-apoA-1 IgG. The corresponding P value was obtained from log-rank tests.

When Kaplan Meier curves were generated comparing anti-apoA-1 IgG seropositive versus anti-apoA-1 IgG seronegative RTR (Supplementary Figure S1), the same associations were retrieved, namely significance for CVD mortality ($p = 0.035$), but not for all-cause mortality or for incident graft failure.

At the pre-specified cut-off for anti-apoA-1 IgG positivity, sensitivity was 18.0% (95% CI: 9–32), specificity 89.3% (95% CI: 86–92), positive predictive value 17.0% (95% CI: 9–30) and negative predictive value 90.0% (95% CI: 87–92) for CVD-related deaths.

Finally, as shown in Table 3 Cox regression analyses showed that anti-apoA-1 IgG levels were significantly associated with CVD mortality in a model adjusted for age and gender (model 1, HR: 1.56, $p = 0.002$). This association remained significant, independent of adjustment either for the Framingham risk score (FRS, model 2, HR: 1.56, $p = 0.002$), eGFR (model 3, HR: 1.54, $p = 0.004$) or both parameters combined (model 4, HR: 1.54, $p = 0.004$), HDL CEC (model 5, HR: 1.54, $p = 0.003$), history of MI (model 6, HR: 1.45, $p = 0.0013$), primary renal disease (model 7, HR: 1.53, $p = 0.005$) and time between transplantation and baseline (model 8, HR: 1.56, $p = 0.002$). Importantly, FRS itself was not associated with CVD mortality in our RTR cohort (unadjusted HR: 1.00 [0.99–1.02], $p = 0.51$; age and gender adjusted HR: 1.00 [0.98–1.03], $p = 0.95$). For all-cause mortality, the same associations were retrieved (models 3 and 4; Table 3). On the other hand, no significant associations were detected between anti-apoA-1 IgG levels and incident graft failure (Table 3). In our sensitivity analyses there was no association between anti-apoA-1 IgG levels and non-CVD mortality, further supporting the specificity of the relationship between anti-apoA-1 IgG and CVD mortality (Table 3).

Table 3. Hazard ratios for cardiovascular mortality, all-cause mortality, and graft failure per one standard deviation increase of anti-apoA-1 IgG.

	CVD Mortality		All-Cause Mortality		Graft Failure		Non-CVD Mortality Sensitivity Analysis	
	HR [95%CI] per 1–SD Increase	p	HR [95%CI] per 1–SD Increase	p	HR [95%CI] per 1–SD Increase	p	HR [95%CI] per 1–SD Increase	p
Model 1	1.56 [1.17–2.07]	0.002	1.36 [1.09–1.70]	0.007	1.17 [0.93–1.48]	0.18	1.41 [0.95–2.09]	0.09
Model 2	1.56 [1.17–2.08]	0.002	1.36 [1.09–1.70]	0.007	1.18 [0.94–1.49]	0.16	1.41 [0.94–2.11]	0.09
Model 3	1.54 [1.15–2.06]	0.004	1.32 [1.05–1.67]	0.017	1.14 [0.91–1.42]	0.26	1.39 [0.92–2.08]	0.11
Model 4	1.54 [1.15–2.07]	0.004	1.32 [1.04–1.66]	0.020	1.15 [0.92–1.44]	0.22	1.39 [0.92–2.10]	0.12
Model 5	1.54 [1.16–2.05]	0.003	1.36 [1.09–1.71]	0.007	1.19 [0.94–1–50]	0.15	1.44 [0.98–2.11]	0.06
Model 6	1.45 [1.08–1.94]	0.013	1.32 [1.05–1.66]	0.016	1.14 [0.90–1.45]	0.27	1.39 [0.90–2.14]	0.14
Model 7	1.53 [1.14–2.05]	0.005	1.33 [1.06–1.67]	0.016	1.17 [0.93–1.48]	0.18	1.45 [0.96–2.20]	0.08
Model 8	1.56 [1.18–2.09]	0.002	1.37 [1.09–1.17]	0.07	1.17 [0.93–1.48]	0.18	1.47 [0.98–2.23]	0.07

Model 1: adjustment for recipient age and gender; model 2: model 1 + adjustment for FRS; model 3: model 1 + adjustment for eGFR; model 4: model 1 + adjustment for FRS and eGFR; model 5: model 1 + adjustment for cholesterol efflux capacity; model 6: model 1 + adjustment for history of MI; model 7: model 1 + adjustment for primary renal disease; model 8: model 1 + adjustment for time between transplantation and baseline. In sensitivity analysis non-CVD deaths were used as endpoint. One standard deviation is equivalent to 0.316. HR: Hazard ratios; FRS: Framingham risk score.

When Cox regression analyses were performed according to anti-apoA-1 IgG seropositivity, the aforementioned associations remained unchanged, at the exception of all-cause mortality which remained significant after adjusting for renal function, and for which the association became close to significance after adjusting for previous MI on top of age and gender (Supplementary Table S2). Again, no association between anti-apoA-1 IgG levels and graft failure could be observed (Supplementary Table S2).

4. Discussion

The novel finding of this prospective study is that anti-apoA-1 IgGs are an independent predictor of CVD mortality in a RTR cohort with a follow-up of 7 years. Our observations indicate that traditional CVD risk factors were presently not associated with CVD mortality. Furthermore, no association was observed between anti-apoA-1 IgG and non-CVD mortality in the preformed sensitivity analysis. This reinforces both the possible clinical relevance and the CVD specificity of the present findings. Indeed, to the best of our knowledge and at the exception of renal function markers [3–5,21,22], no specific biomarkers of CVD outcome independent of renal function have been identified so far in RTR. Considering the absence of currently validated tools for CVD risk prediction in RTR, a rule-out test with a 90% NPV could conceivably be of clinical interest as a first step in the field of CV risk stratification in these patients. In this context, we hypothesize that a simple standard follow-up could be particularly well adapted to RTR patients with low anti-apoA-1 IgG values. Further validation studies are now required to challenge this hypothesis before any clinical recommendations can be made.

The second notable finding of this study is that anti-apoA-1 IgG were not associated with graft failure, nor with HDL CEC. Although further reinforcing the specificity between anti-apoA-1 IgG and CV outcomes, these results were somehow unexpected, as anti-apoA-1 IgG have been previously shown to be associated with impaired HDL CEC [9,10], lately reported as being an independent predictor of incident graft failure [21]. The reasons for such differences are still elusive, most likely numerous, and possibly related to pathophysiological differences between CVD and graft atherosclerosis. Indeed, rupture of vulnerable atherosclerotic plaques is known to underlie most cases of acute CVD events, while this is not thought to play an important role in chronic transplant vasculopathy-induced graft failure, where progressive arteriolar luminal narrowing due to the intimal accumulation of degenerating smooth muscle-like cells and adventitial fibrosis represent the major pathogenic processes [23]. Furthermore, another explanation could lie in the fact that RTR represent a unique patient population in terms of oxidative stress exposure and persistent loss of HDL function, when compared to e.g. systemic lupus erythematosus patients [10] or dyslipidaemic subjects with preserved renal function [9]. Finally due to the heterogeneity of methodological protocols underlying the numerous unstandardized HDL functional assays, we cannot exclude that an analytical difference between our HDL assay and those from other groups could undermine the present observation [24–26]. Therefore, further studies are warranted to determine if this absence of correlation between anti-apoA-1 IgG and HDL functionality in RTR is intrinsically disease-specific.

Thirdly, this study strengthens previous observations and provides the first insights of the anti-apoA-1 IG architecture in RTR. Indeed, the association between these antibodies with previous MI has been consistently reported across different populations with preserved renal function [7,8,11,13]. Reproducing this association reinforces the notion that a previous acute coronary event is an important acquired factor, that could, together with niacin therapy [9] and genetic determinants [27], contribute to better understand the reasons underlying the existence of anti-apoA-1 IgG in individuals without overt signs of clinical autoimmunity. In this context, we report for the first time specific associations with primary glomerular disease and tubulo-interstitial disease as primary renal diseases possibly associated to the existence of anti-apoA-1 IgG.

Lastly, the somewhat lower than expected prevalence of anti-apoA-1 IgG seropositivity retrieved presently (11.5%) when compared to maintenance hemodialysis patients and the general population (20%) [8,18], is worth a comment, as we would have expected an increased prevalence as previously reported in all other clinical situations with a high CV risk [7,8,10,12–14,18]. A conceivable explanation for this observation might be that RTRs are under chronic immunosuppressive medication, known to improve features of autoimmunity and thus decrease autoantibody levels. The trend toward a decrease in the prevalence of proliferation inhibitors along the increasing anti-apoA-1 IgG tertiles may lend weight to this hypothesis and warrants further investigations.

Although the results of the present study lend further weight to the growing body of evidence indicating that humoral autoimmunity contributes to CVD, we could not explore the mechanisms by which anti-apoA-1 IgG levels may associate with CVD in RTR. So far, previous animal and in-vitro studies showed that anti-apoA-1 IgG could be active mediators of atherogenesis, inducing myocardial necrosis and death in mice through toll-like receptors (TLR) [2,4] and CD14 heterodimer signaling [6,7,11,15–17,28]. Since these deleterious effects could potentially be amended by immunomodulation therapies, either using a specific apoA-1 mimetic peptide or intravenous immunoglobulins, anti-apoA-1 IgGs have been proposed as emergent therapeutic targets [11,29]. In accordance with these in vitro and animal experiments, a functional CD14 polymorphism was recently shown to be a strong modulator of anti-apoA-1 IgG-related CVD risk prediction in the general population [12]. As CD14 expressing monocytes display higher TLR2 and 4 expression in RTR [30], knowing whether CD14 and/or TLR2/4 polymorphisms together with the presence of anti-apoA-1 IgG could further improve prognosis assessment in RTR remains to be investigated. Given the important pathophysiological differences between atherogenesis and transplant vasculopathy [23], knowing whether the aforementioned molecular mechanisms could also explain the increased CV risk ascribed to these antibodies in RTR remains unknown and constitutes an important limitation of the present study. Further, despite the relatively large number of included RTR in this adequately powered study, the number of events was still somewhat low, leading to restricted possibilities with regards to statistical analysis. Since this investigation was also carried out in a single center, further validation of our findings in a larger multicentre cohort appears desirable. In addition, it would be interesting to analyse whether RTR with high anti-apoA-1 IgG titers show a differential response to an intervention with cardiovascular treatment strategies. A further limitation resides in the fact that we did not measure other autoantibodies of possible CV relevance, such as auto-antibodies to β2 glycoprotein I domain I and IV, cardiolipin, heat-shock protein 60, and to phosphorylcholine. Because anti-apoA-1 IgG were shown to display the strongest and independent prognostic accuracy for major adverse cardiovascular events in non-autoimmune settings when compared to the aforementioned auto-antibodies [31], we focused our work specifically on this class of antibodies. Knowing whether the present association could be reproduced with other auto-antibodies remains to be shown. Also, before utilizing anti-apoA-1 IgGs as clinical biomarker, it would be interesting to screen kidney graft donors to learn, whether intra-individual variability in titers has a potential impact on outcomes after transplantation.

In conclusion, we report anti-apoA-1 IgG as a novel prognostic biomarker for CVD mortality in RTR, independent of traditional CVD risk factors and HDL functionality. These data indicate that anti-apoA-1 IgG holds potential as a clinical biomarker for CVD risk stratification in RTR patients, a high CVD risk population with altered functionality of the immune system. Further investigations to define the potential usefulness of anti-apoA-1 IgG assessments in clinical decision making are required, as well as studies to delineate the intrinsic pathophysiological pathways that these antibodies activate to subsequently result in an increased CVD risk in RTR.

Supplementary Materials: Supplementary materials can be found at http://www.mdpi.com/2077-0383/8/7/948/s1. Table S1: Baseline characteristics according to seropositivity of anti-apoA-1 IgG, Table S2: Hazard ratios for cardiovascular disease mortality, all-cause mortality, and graft failure by seropositivity of anti-apoA1 IgG, Figure S1: Seropositivity of anti-apoA-1 IgG is associated with increased cardiovascular mortality in renal transplant recipients.

Author Contributions: Conceptualization, N.V. and U.J.F.T.; Data curation, J.L.C.A., S.P., J.V. and W.A.; Formal analysis, J.L.C.A. and W.A.; Methodology, S.J.L.B., N.V. and U.J.F.T.; Supervision, R.P.F.D., F.K., S.J.L.B. and U.J.F.T.; Writing—original draft, J.L.C.A. and U.J.F.T.; Writing—review & editing, S.P., J.V., R.P.F.D., W.A., F.K., S.J.L.B. and N.V.

Funding: The TxL-IRI Biobank and Cohort Study was financially supported by the Dutch Kidney Foundation (grant C00.1877). This work was also supported by the Leenaards Foundation, and the Swiss National Science Foundation (grant number 310030-163335).

Acknowledgments: For this study, we made use of samples and data of the TransplantLines Insulin Resistance and Inflammation (TxL-IRI) Biobank and Cohort Study.

Conflicts of Interest: The authors declare no conflict of interest.

References

1. Chen, J.; Budoff, M.J.; Reilly, M.P.; Yang, W.; Rosas, S.E.; Rahman, M.; Zhang, X.; Roy, J.A.; Lustigova, E.; Nessel, L.; et al. Coronary artery calcification and risk of cardiovascular disease and death among patients with chronic kidney disease. *JAMA Cardiol.* **2017**, *2*, 635–643. [CrossRef] [PubMed]
2. Go, A.S.; Chertow, G.M.; Fan, D.; McCulloch, C.E.; Hsu, C. Chronic kidney disease and the risks of death, cardiovascular events, and hospitalization. *N. Engl. J. Med.* **2004**, *351*, 1296–1305. [CrossRef] [PubMed]
3. Oterdoom, L.H.; de Vries, A.P.J.; van Ree, R.M.; Gansevoort, R.T.; van Son, W.J.; van der Heide, J.J.H.; Navis, G.; de Jong, P.E.; Gans, R.O.B.; Bakker, S.J.L. N-terminal pro-B-type natriuretic peptide and mortality in renal transplant recipients versus the general population. *Transplantation* **2009**, *87*, 1562–1570. [CrossRef] [PubMed]
4. Israni, A.K.; Snyder, J.J.; Skeans, M.A.; Peng, Y.; Maclean, J.R.; Weinhandl, E.D.; Kasiske, B.L. PORT investigators predicting coronary heart disease after kidney transplantation: Patient outcomes in renal transplantation (PORT) Study. *Am. J. Transplant.* **2010**, *10*, 338–353. [CrossRef]
5. Foster, M.C.; Weiner, D.E.; Bostom, A.G.; Carpenter, M.A.; Inker, L.A.; Jarolim, P.; Joseph, A.A.; Kusek, J.W.; Pesavento, T.; Pfeffer, M.A.; et al. Filtration markers, cardiovascular disease, mortality, and kidney outcomes in stable kidney transplant recipients: The FAVORIT trial. *Am. J. Transplant.* **2017**, *17*, 2390–2399. [CrossRef] [PubMed]
6. Pagano, S.; Satta, N.; Werling, D.; Offord, V.; de Moerloose, P.; Charbonney, E.; Hochstrasser, D.; Roux-Lombard, P.; Vuilleumier, N. Anti-apolipoprotein A-1 IgG in patients with myocardial infarction promotes inflammation through TLR2/CD14 complex. *J. Intern. Med.* **2012**, *272*, 344–357. [CrossRef] [PubMed]
7. Vuilleumier, N.; Bas, S.; Pagano, S.; Montecucco, F.; Guerne, P.-A.; Finckh, A.; Lovis, C.; Mach, F.; Hochstrasser, D.; Roux-Lombard, P.; et al. Anti-apolipoprotein A-1 IgG predicts major cardiovascular events in patients with rheumatoid arthritis. *Arthritis Rheum.* **2010**, *62*, 2640–2650. [CrossRef] [PubMed]
8. Antiochos, P.; Marques-Vidal, P.; Virzi, J.; Pagano, S.; Satta, N.; Bastardot, F.; Hartley, O.; Montecucco, F.; Mach, F.; Waeber, G.; et al. Association between anti-apolipoprotein A-1 antibodies and cardiovascular disease in the general population. Results from the CoLaus study. *Thromb. Haemost.* **2016**, *116*, 764–771.
9. Batuca, J.R.; Amaral, M.C.; Favas, C.; Paula, F.S.; Ames, P.R.J.; Papoila, A.L.; Delgado Alves, J. Extended-release niacin increases anti-apolipoprotein A-I antibodies that block the antioxidant effect of high-density lipoprotein-cholesterol: The EXPLORE clinical trial. *Br. J. Clin. Pharmacol.* **2017**, *83*, 1002–1010. [CrossRef]
10. Batuca, J.R.; Ames, P.R.J.; Amaral, M.; Favas, C.; Isenberg, D.A.; Delgado Alves, J. Anti-atherogenic and anti-inflammatory properties of high-density lipoprotein are affected by specific antibodies in systemic lupus erythematosus. *Rheumatology* **2009**, *48*, 26–31. [CrossRef]
11. Vuilleumier, N.; Rossier, M.F.; Pagano, S.; Python, M.; Charbonney, E.; Nkoulou, R.; James, R.; Reber, G.; Mach, F.; Roux-Lombard, P. Anti-apolipoprotein A-1 IgG as an independent cardiovascular prognostic marker affecting basal heart rate in myocardial infarction. *Eur. Heart J.* **2010**, *31*, 815–823. [CrossRef] [PubMed]
12. Antiochos, P.; Marques-Vidal, P.; Virzi, J.; Pagano, S.; Satta, N.; Hartley, O.; Montecucco, F.; Mach, F.; Kutalik, Z.; Waeber, G.; et al. Impact of CD14 polymorphisms on anti-apolipoprotein A-1 IGG-related coronary heart disease prediction in the general population. *Atherosclerosis* **2017**, *263*, e45. [CrossRef]
13. Vuilleumier, N.; Montecucco, F.; Spinella, G.; Pagano, S.; Bertolotto, M.; Pane, B.; Pende, A.; Galan, K.; Roux-Lombard, P.; Combescure, C.; et al. Serum levels of anti-apolipoprotein A-1 auto-antibodies and myeloperoxidase as predictors of major adverse cardiovascular events after carotid endarterectomy. *Thromb. Haemost.* **2013**, *109*, 706–715. [PubMed]
14. El-Lebedy, D.; Rasheed, E.; Kafoury, M.; Abd-El Haleem, D.; Awadallah, E.; Ashmawy, I. Anti-apolipoprotein A-1 autoantibodies as risk biomarker for cardiovascular diseases in type 2 diabetes mellitus. *J. Diabetes Complications* **2016**, *30*, 580–585. [CrossRef] [PubMed]
15. Montecucco, F.; Vuilleumier, N.; Pagano, S.; Lenglet, S.; Bertolotto, M.; Braunersreuther, V.; Pelli, G.; Kovari, E.; Pane, B.; Spinella, G.; et al. Anti-Apolipoprotein A-1 auto-antibodies are active mediators of atherosclerotic plaque vulnerability. *Eur. Heart J.* **2011**, *32*, 412–421. [CrossRef] [PubMed]
16. Montecucco, F.; Braunersreuther, V.; Burger, F.; Lenglet, S.; Pelli, G.; Carbone, F.; Fraga-Silva, R.; Stergiopulos, N.; Monaco, C.; Mueller, C.; et al. Anti-apoA-1 auto-antibodies increase mouse atherosclerotic plaque vulnerability, myocardial necrosis and mortality triggering TLR2 and TLR4. *Thromb. Haemost.* **2015**, *114*, 410–422. [PubMed]

17. Pagano, S.; Carbone, F.; Burger, F.; Roth, A.; Bertolotto, M.; Pane, B.; Spinella, G.; Palombo, D.; Pende, A.; Dallegri, F.; et al. Anti-apolipoprotein A-1 auto-antibodies as active modulators of atherothrombosis. *Thromb. Haemost.* **2016**, *116*, 554–564.
18. Pruijm, M.; Schmidtko, J.; Aho, A.; Pagano, S.; Roux-Lombard, P.; Teta, D.; Burnier, M.; Vuilleumier, N. High prevalence of anti-apolipoprotein/A-1 autoantibodies in maintenance hemodialysis and association with dialysis vintage. *Ther. Apher. Dial.* **2012**, *16*, 588–594. [CrossRef]
19. van Ree, R.M.; de Vries, A.P.J.; Oterdoom, L.H.; The, T.H.; Gansevoort, R.T.; Homan van der Heide, J.J.; van Son, W.J.; Ploeg, R.J.; de Jong, P.E.; Gans, R.O.B.; et al. Abdominal obesity and smoking are important determinants of C-reactive protein in renal transplant recipients. *Nephrol. Dial. Transplant.* **2005**, *20*, 2524–2531. [CrossRef]
20. de Vries, A.P.J.; Bakker, S.J.L.; van Son, W.J.; van der Heide, J.J.H.; Ploeg, R.J.; The, H.T.; de Jong, P.E.; Gans, R.O.B. Metabolic syndrome is associated with impaired long-term renal allograft function; not all component criteria contribute equally. *Am. J. Transplant.* **2004**, *4*, 1675–1683. [CrossRef]
21. Annema, W.; Dikkers, A.; Freark de Boer, J.; Dullaart, R.P.F.; Sanders, J.-S.F.; Bakker, S.J.L.; Tietge, U.J.F. HDL Cholesterol efflux predicts graft failure in renal transplant recipients. *J. Am. Soc. Nephrol.* **2016**, *27*, 595–603. [CrossRef] [PubMed]
22. Leberkühne, L.J.; Ebtehaj, S.; Dimova, L.G.; Dikkers, A.; Dullaart, R.P.F.; Bakker, S.J.L.; Tietge, U.J.F. The predictive value of the antioxidative function of HDL for cardiovascular disease and graft failure in renal transplant recipients. *Atherosclerosis* **2016**, *249*, 181–185. [CrossRef] [PubMed]
23. Mitchell, R.N.; Libby, P. Vascular remodeling in transplant vasculopathy. *Circ. Res.* **2007**, *100*, 967–978. [CrossRef] [PubMed]
24. Campise, M.; Bamonti, F.; Novembrino, C.; Ippolito, S.; Tarantino, A.; Cornelli, U.; Lonati, S.; Cesana, B.M.; Ponticelli, C. Oxidative stress in kidney transplant patients. *Transplantation* **2003**, *76*, 1474–1478. [CrossRef] [PubMed]
25. Kopecky, C.; Haidinger, M.; Birner-Grünberger, R.; Darnhofer, B.; Kaltenecker, C.C.; Marsche, G.; Holzer, M.; Weichhart, T.; Antlanger, M.; Kovarik, J.J.; et al. Restoration of renal function does not correct impairment of uremic HDL properties. *J. Am. Soc. Nephrol.* **2015**, *26*, 565–575. [CrossRef] [PubMed]
26. Triolo, M.; Annema, W.; Dullaart, R.P.F.; Tietge, U.J.F. Assessing the functional properties of high-density lipoproteins: an emerging concept in cardiovascular research. *Biomark. Med.* **2013**, *7*, 457–472. [CrossRef] [PubMed]
27. Antiochos, P.; Marques-Vidal, P.; Virzi, J.; Pagano, S.; Satta, N.; Hartley, O.; Montecucco, F.; Mach, F.; Kutalik, Z.; Waeber, G.; et al. Anti-apolipoprotein A-1 IgG predict all-cause mortality and are associated with FC receptor-like 3 polymorphisms. *Front. Immunol.* **2017**, *8*, 437. [CrossRef] [PubMed]
28. Rossier, M.F.; Pagano, S.; Python, M.; Maturana, A.D.; James, R.W.; Mach, F.; Roux-Lombard, P.; Vuilleumier, N. Antiapolipoprotein A-1 IgG chronotropic effects require nongenomic action of aldosterone on L-type calcium channels. *Endocrinology* **2012**, *153*, 1269–1278. [CrossRef]
29. Pagano, S.; Gaertner, H.; Cerini, F.; Mannic, T.; Satta, N.; Teixeira, P.C.; Cutler, P.; Mach, F.; Vuilleumier, N.; Hartley, O. The Human autoantibody response to apolipoprotein A-I is focused on the C-terminal helix: A new rationale for diagnosis and treatment of cardiovascular disease? *PLoS ONE* **2015**, *10*, e0132780. [CrossRef]
30. Hosseinzadeh, M.; Nafar, M.; Ahmadpoor, P.; Noorbakhsh, F.; Yekaninejad, M.S.; Niknam, M.H.; Amirzargar, A. Increased expression of toll-like receptors 2 and 4 in renal transplant recipients that develop allograft dysfunction: A cohort study. *Iran J. Immunol.* **2017**, *14*, 24–34.
31. Vuilleumier, N. Head-to-head comparison of auto-antibodies for cardiovascular outcome prediction after myocardial infarction: A prospective study. *J. Clinic. Experiment. Cardiol.* **2011**, *2*, 169. [CrossRef]

© 2019 by the authors. Licensee MDPI, Basel, Switzerland. This article is an open access article distributed under the terms and conditions of the Creative Commons Attribution (CC BY) license (http://creativecommons.org/licenses/by/4.0/).

Article

Estimating the Level of Carbamoylated Plasma Non-High-Density Lipoproteins Using Infrared Spectroscopy

Sigurd E. Delanghe [1], Sander De Bruyne [2], Linde De Baene [2], Wim Van Biesen [1], Marijn M. Speeckaert [1,3] and Joris R. Delanghe [2,*]

1. Department of Nephrology, Ghent University Hospital, 9000 Ghent, Belgium; Sigurd.Delanghe@ugent.be (S.E.D.); Wim.Vanbiesen@ugent.be (W.V.B.); Marijn.Speeckaert@ugent.be (M.M.S.)
2. Department of Clinical Chemistry, Ghent University Hospital, 9000 Ghent, Belgium; Sander.Debruyne@uzgent.be (S.D.B.); Linde.Debaene@ugent.be (L.D.B.)
3. Research Foundation-Flanders (FWO), 1000 Brussels, Belgium
* Correspondence: Joris.Delanghe@ugent.be; Tel.: +32-9332-2956; Fax: +32-9332-3659

Received: 4 May 2019; Accepted: 29 May 2019; Published: 31 May 2019

Abstract: Background: The increased cardiovascular morbidity and mortality observed in chronic kidney disease (CKD) patients can be partly explained by the presence of carbamoylated lipoproteins. Lipid profiles can be determined with infrared spectroscopy. In this paper, the effects of carbamoylation on spectral changes of non-high-density lipoproteins (non-HDL) were studied. Methods: In the present study, fasting serum samples were obtained from 84 CKD patients (CKD stage 3–5: $n = 37$ and CKD stage 5d (hemodialysis): $n = 47$) and from 45 healthy subjects. In vitro carbamoylation of serum lipoproteins from healthy subjects was performed using increasing concentrations of potassium cyanate. Lipoprotein-containing pellets were isolated by precipitation of non-HDL. The amount of carbamoylated serum non-HDL was estimated using attenuated total reflectance-Fourier transform infrared (ATR-FTIR) spectroscopy, followed by soft independent modelling by class analogy analysis. Results: Carbamoylation resulted in a small increase of the amide I band (1714–1589 cm^{-1}) of the infrared spectroscopy (IR) spectrum. A significant difference in the amide II/amide I area under the curves (AUC) ratio was observed between healthy subjects and CKD patients, as well as between the two CKD groups (non-dialysis versus hemodialysis patients). Conclusions: ATR-FTIR spectroscopy can be considered as a novel method to detect non-HDL carbamoylation.

Keywords: carbamoylation; chronic kidney disease; lipoproteins; infrared spectroscopy

1. Introduction

Carbamoylation is a post-translational modification, playing a role in chronic kidney disease (CKD), comparable to the role of glycation in diabetes mellitus [1,2]. This non-enzymatic reaction is characterized by a covalent binding of isocyanic acid to the amino group (either the epsilon amino group of lysine residus or the N-terminal amino group) of amino acids, polypetides, and (lipo)proteins. This post-translational molecular modification contributes to the molecular ageing of proteins. Isocyanic acid is formed continuously in an equilibrium reaction with urea or by myeloperoxidase (MPO). MPO is an enzyme (present in e.g., neutrophils, monocytes, and some tissue macrophages) that catalyzes the oxidation of thiocyanate in the presence of hydrogen peroxide, producing isocyanate at inflammation sites (e.g., atherosclerotic plaques) [3]. Plasma isocyanic acid concentrations increase with a declining kidney function [4].

An increased cardiovascular morbidity and mortality (10–30 times higher than in the general population) has been reported in the end-stage renal disease (ESRD) population [5], which can be partly attributed to the influence of carbamoylated lipoproteins [6]. In renal failure, dyslipidemia contributes to a worsening kidney function. Proteinuria is accompanied by a marked elevation of low-density lipoproteins (LDL) [7]. Due to its carbamoylation, LDL can exert its prothrombotic [8] and atherosclerosis-prone effects by stimulating an increased adhesion of monocytes to endothelial cells [9], by inducing endothelial dysfunction [6] and endothelial mitotic cell death [10], and by promoting smooth muscle proliferation [11]. Carbamoylated LDL (cLDL) has been identified as the most abundant modified LDL isoform in human blood, which is also present in healthy individuals [12]. It is generated by carbamoylation of apolipoprotein B, the protein component of the LDL particle [13]. CKD is also associated with decreased serum high-density lipoprotein (HDL) concentrations. Carbamoylation of HDL leads to a loss of the atheroprotective function of HDL, illustrated by an impaired ability to promote cholesterol efflux from macrophages [14].

As carbamoylation is of major clinical importance, practical biomarkers for assessing carbamoylation and lipoprotein carbamoylation, in particular, are needed. In the present study, we explored the possibilities of infrared (IR) spectroscopy to assess non-HDL carbamoylation. The advantages of IR spectroscopy to determine lipid profiles were already applied in the past [15]. In this paper, the effects of in vitro carbamoylation on spectral changes of non-HDL were studied. Furthermore, non-HDL carbamoylation was investigated in healthy subjects, in non-dialysis (CKD stage 3–5) as well as in hemodialysis patients (CKD stage 5d).

2. Materials and Methods

2.1. Study Participants

The control group consisted of 45 healthy subjects (median age: 28 years, interquartile range (IQR): 24–33 years), whereas the patient group consisted of 84 CKD patients (CKD stage 3–5: $n = 37$, median age: 70 years, IQR: 56–75 years, and CKD stage 5d (hemodialysis): $n = 47$, median age: 67 years, IQR: 56–75 years) of the Department of Nephrology, Ghent University Hospital. The approval of this study was granted by the Ethical committee of the Ghent University Hospital (EC/2015/0932).

2.2. In Vitro Carbamoylation of Lipids

In vitro carbamoylation of lipids was achieved by adding increasing volumes of 0.5 mol/L potassium cyanate (KOCN) solution (Sigma–Aldrich, St. Louis, MO, USA) in a phosphate buffered salt (PBS) solution (0.1 mol/L, pH 8.0) (Sigma–Aldrich, MO, USA) to 1000 µL serum of healthy subjects. Serum samples were carbamoylated using increasing concentrations of KOCN: 0 mmol/L, 20 mmol/L, 50 mmol/L, 80 mmol/L and 100 mmol/L. In vitro carbamoylation was carried out for 48 h at 37 °C (these reaction conditions warrant a completeness of the reaction). Proof of carbamoylation was obtained by verifying the electrophoretic mobility of lipoprotein fractions on a lipoprotein agarose electrophosis (Hydragel 7, Sebia, Lisses, France) using a semi-automated HYDRASYS instrument (Sebia, Lisses, France). The separated lipoproteins were stained with a lipid-specific Sudan black stain. The excess of stain was removed with an alcoholic solution. The resulting electropherogram was evaluated visually.

2.3. In Vitro Oxidation

Oxidative stress is involved in the exacerbation of disease burden in CKD patients. In vitro oxidation of serum was performed to reveal potential influences on the infrared spectrum. Ten samples from the serum pool at the laboratory of clinical biology of the University Hospital in Ghent were randomly selected. One milliliter of each sample was pooled. Before the oxidation process, non-HDL were precipitated 5 times. According to a modified version of the method used by Coffey et al. [16], oxidized non-HDLs were prepared by dialyzing 4 mL of the serum pool against 400 mL isotonic saline (0.15 mol/L NaCl (VWR International, Haasrode, Belgium) dissolved in distilled water) containing 60 µmol/L CuSO4 (copper(II)sulphate pentahydrate, Merck Eurolab, Leuven, Belgium) during one hour. The dialysate was then changed to isotonic saline containing 0.5 mmol/L EDTA (BDH Chemicals,

Poole, England) and dialysis continued during one hour with changes of dialysate every 15 min. After the oxidation process, non-HDL were precipitated 5 times from the oxidized serum pool.

2.4. Precipitation Procedure

All serum samples were centrifugated during 10 min at 3000× g. A precipitation reaction was performed, in which non-HDL fats very-low-density lipoprotein (VLDL), intermediate-density lipoprotein (IDL), lipoprotein(a), LDL and chylomicrons) were precipitated. 20 µL of a 13 mmol/L sodium phosphotungstate hydrate solution (Sigma Aldrich St Louis, MO, USA) and 5 µL of 2 mol/L $MgCl_2$ (E. Merck KG, Darmstadt, Germany) were added to 200 µL serum. After vortexing, the samples were centrifuged (10 min, 6000× g) (centrifuge 54515 D, Eppendorf, Hamburg, Germany) [17]. The precipitate was subsequently dried for 48 h in an incubator. Completeness of the precipitation reaction was assessed by lipid electrophoresis of the serum pre- and post-precipitation (5 samples were precipitated in triplicate). The formed pellet was ground prior to analysis with attenuated total reflectance-Fourier transform infrared (ATR-FTIR) spectroscopy.

2.5. ATR-FTIR Analysis

All spectra were obtained by a Perkin Elmer Two ATR-FTIR spectrometer with the ATR accesory and spectrum 10 software (Perkin Elmer, Waltham, MA, USA). Before and after each analysis, the 50 mm ZnSE crystal was thoroughly cleaned with an alcoholic solution (Dax alcoliquid, Dialex biomedica, Sweden). A background scan was taken after the complete evaporation of the alcoholic solution. The lipoprotein powder was placed in contact with the surface of the crystal until complete covering. The pressure applied to the sample was standardised at 100 gauche to obtain a good contact between the sample and the crystal.

Three peaks were investigated: the carbonyl peak, the peak of the amide I band and the peak of the amide II band. Within-run coefficients of variation (CV) and between-run CV were calculated. The area under the curves (AUC) of the amide I and amide II bands were obtained by auto-labeling of the peaks in the Perkin Elmer 10 software.

Spectra were analyzed using the software program SIMCA version 14.1 (Umetrics, Sartorius Stedim Biotech, Umeå, Sweden). SIMCA (soft independent modeling of class analogy) was used to identify the spectral changes due to carbamoylation of lipoproteins. By applying various spectral filters, the noise was eliminated and the region of interest was selected. Data were normalized using the standard normal variate (SNV) method and were converted to their second derivative. The Savitsky-Golay algorithm allowed smoothing of the spectrum.

2.6. Routine Laboratory Measurements

After overnight fasting, blood samples were collected and centrifuged (10 min, 3000× g). Urea, creatinine, albumin, triglycerides, total and HDL-cholesterol concentrations were assayed using commercial reagents on a Cobas 8000 analyzer (Roche, Mannheim, Germany) [18]. The serum concentration of apolipoprotein B was determined by immunonephelometry on a Behring BN II nephelometer (Siemens, Marburg, Germany) [19]. The LDL-cholesterol concentration was estimated using the Friedewald-formula [20]. The estimated glomerular filtration rate (eGFR) was calculated with the Chronic Kidney Disease Epidemiology Collaboration (CKD-EPI) formula [21].

2.7. Statistics

Statistical analyses were performed using MedCalc (MedCalc, Mariakerke, Belgium). Normality of distributions was tested using the D'Agostino Pearson test. Data are expressed as median ± IQR or mean ± standard deviation (SD). Differences between patient groups were assessed using the Student's t-test and the Kruskall–Wallis test. The effect of the biological parameters on the spectrum was evaluated using a multiple linear regression model. A p-value < 0.05 was considered a priori to be statistically significant.

3. Results

3.1. In Vitro Carbamoylation

In vitro carbamoylation of lipoproteins in serum of healthy subjects was demonstrated by agarose gel electrophoresis, which showed a progressive increase in electrophoretic mobility of lipoproteins with increasing KOCN concentrations. The effectiveness of the precipitation reaction was illustrated by the disappearance of the LDL and VLDL fraction on agarose gel electrophoresis. Serum samples with the highest concentrations of KOCN showed a less efficient precipitation, probably due to an altered protein structure, which could interfere with the precipitation process.

Carbamoylation was further investigated with ATR-FTIR spectroscopy. Figure 1 presents the different IR spectra of the pellet of precipitated lipoproteins, urea, and KOCN. The visual spectrum of urea and KOCN did not interfere with the IR spectra of the precipitated lipoproteins. The between-run and within-run CVs for the detection of the carbonyl band (4.5% and 4.9%), the amide I band (4.9% and 8.4%) and the amide II band (5.8% and 8.1%) were low.

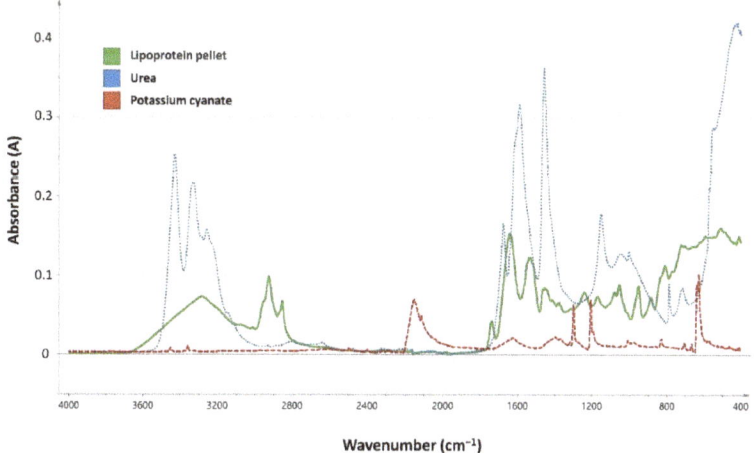

Figure 1. Infrared spectra of precipitated lipoproteins (green), urea (blue) and potassium cyanate (red).

Using the software package SIMCA 14.1, it was possible to differentiate non-carbamoylated from in vitro carbamoylated non-HDL. The data set was centered, normalized and fitted, and a loading line was formed from the cleaned data. We focused on the fingerprint region (1500–600 cm^{-1}) and the amide I and amide II region (1700–1500 cm^{-1}). Carbamoylation resulted in a small increase in the amide I band (1714–1589 cm^{-1}) of the spectrum (Figure 2). The data set was reduced to the amide I band, normalized and the second derivative was taken before fitting. Moreover, the in vitro experiments showed a diminishing amide II band/amide I area under the curve (AUC) ratio with increasing KOCN concentrations.

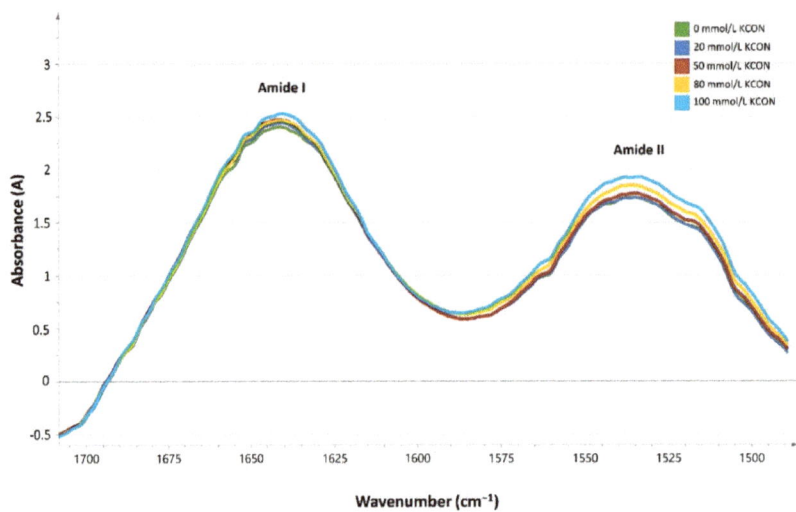

Figure 2. Absorbance spectra of in vitro carbamoylated lipids, adding increasing concentrations of potassium cyanate (0 mmol/L, 20 mmol/L, 50 mmol/L and 100 mmol/L) to serum of healthy subjects.

In vitro oxidation revealed an increased absorption in the amide I and amide II band ($p < 0.01$). However, the amide II/amide I ratio remained the same before and after in vitro oxidation (ratio 0.55).

3.2. In Vivo Samples

Table 1 describes the general characteristics of the healthy subjects and the CKD patients. In the in vivo part of the study, the findings of the in vitro study were compared with the CKD patients' samples. The same spectral filters were applied and the amide I band was selected. There was a clear distinction between the various groups (healthy subjects, patients with CKD stage 3–5 and CKD 5d patients).

Table 1. General characteristics of the healthy subjects and the chronic kidney disease patients.

	Healthy Subjects	CKD Stage 3, 4 or 5	CKD Stage 5d	p
N	45	37	47	
Median age (years)	28 (24–33)	70 (56–75)	67 (56–75)	<0.0001
% diabetes mellitus	0	30	32	
Urea (mmol/L)	8.9 (7.8–9.9)	25.0 (18.7–37.5)	32.8 (28.5–40.1)	<0.0001
Creatinine (µmol/L)	72.5 (62.8–80.0)	160.9 (138.6–243.8)	627.6 (474.7–774.4)	<0.0001
eGFR (mL/min/1.73 m^2)	>90	31.0 ± 13.6	<15	<0.0001
Albumin (g/L)	47.3 (45.0–50.0)	42.2 (40.3–44.5)	40.0 (36.2–42.9)	<0.0001
Total cholesterol (mmol/L)	4.8 (4.3–5.5)	4.7 (3.7–5.5)	4.3 (3.7–5.8)	NS
HDL cholesterol (mmol/L)	1.6 (1.3–2.0)	1.3 (1.0–1.6)	1.1 (0.9–1.5)	<0.0001
LDL cholesterol (mmol/L)	2.7 (2.3–3.1)	2.3 (1.8–3.1)	2.4 (1.7–3.2)	NS
Triglycerides (mmol/L)	1.0 (0.8–1.3)	1.7 (1.1–2.3)	1.4 (1.1–2.4)	=0.0002
Apolipoprotein B (g/L)	0.8 (0.7–1.0)	0.8 (0.7–1.0)	0.8 (0.7–0.9)	NS

NS = not significant.

In addition, a significant difference in the amide II/amide I AUC ratio was observed between the healthy subjects and the CKD groups ($p < 0.0001$) (Figure 3), as well as between the two CKD groups (non-dialysis versus hemodialysis patients). A negative correlation was observed between the amide II/amide I AUC ratio and the serum urea concentration ($r = -0.63$, $p < 0.0001$).

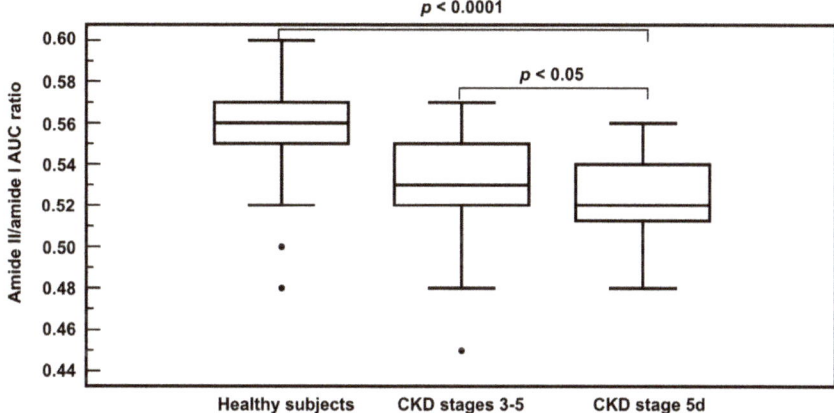

Figure 3. Amide II/amide I AUC ratio among the different study groups.

Multiple regression analysis with the amide II/amide I AUC ratio as a dependent variable revealed that the serum urea concentration and the serum apolipoprotein B concentration were the main predictors (Table 2).

Table 2. Multiple regression model with the amide II/amide I area under the curves (AUC) ratio as dependent variable.

	Variable	β (Standard Error)	p
Amide II/amide I AUC ratio, $r^2 = 0.54$, $p < 0.001$	Triglycerides (mmol/L)	−0.001819 (0.001367)	0.1858
	Apolipoprotein B (g/L)	−0.0306 (0.006277)	<0.0001
	Creatinine (μmol/L)	−0.00001067 (0.000006989)	0.1293
	Urea (mmol/L)	−0.0006622 (0.0001436)	<0.0001
	Age (years)	−0.0002128 (0.00008219)	0.0108

4. Discussion

In the present study, we have demonstrated for the first time the detection of carbamoylated non-HDL using ATR-FTIR spectroscopy. More specifically, in vitro carbamoylation of non-HDL induced structural changes, which were clearly visible in the mid-IR spectrum of the lipid pellets. The amide I band, containing mainly C=O stretching vibrations of protein peptide bonds was identified as the relevant region. In the clinical part of this study, significant differences at the amide I band were observed between healthy subjects, patients with CKD stage 3–5 and hemodialysis patients (CKD stage 5d). The amide I band depends on the secondary structure of the backbone and is, therefore, the amide vibration, which is most commonly used for secondary structure analysis [22]. The amide II mode is the out-of-phase combination of the N−H in plane bend and the C≡N stretching vibration with smaller contributions from the C=O in plane bend and the C≡C and N≡C stretching vibrations. Although the protein secondary structure and frequency correlate less straightforward than for the amide I vibration, the amide II band provides valuable structural information [23]. Previous studies have attributed the amide I band to apolipoproteins [24,25]. The change in the amide I band can be expected as the carbamoylation process alters the protein component of LDL, namely apolipoprotein B [13].

Building on the in vitro model of carbamoylation, we showed that increasing serum KOCN concentrations resulted in a reduced amide II/amide I AUC ratio. These amide bands in the IR spectrum take part in the adding of the carbamoyl group on the amino group of the epsilon-amino group of lysine and the terminal amino groups. The amide II/amide I AUC ratio reflects the observed spectral changes due to carbamoylation. Significant differences of the amide II/amide I AUC ratio were observed between healthy subjects, patients with advanced stages of CKD and hemodialysis patients. As expected, this ratio showed a negative correlation with the serum urea concentration. This result is supported by earlier findings, showing a similar regression coefficient between % carbamoylated albumin and blood urea concentrations in ESRD subjects [26].

As demonstrated in the multiple regression model, age was identified as a minor predictor of the amide II/amide I AUC ratio in comparison with apolipoprotein B and urea. Tissue accumulation of carbamoylated proteins may be considered as a general hallmark of ageing, linking cumulative metabolic alterations and age-related complications. In addition to the association with carbamoylation, many other nonenzymatic posttranslational modifications occur during the biological life of proteins, leading to protein molecular ageing [27].

A limitation of the present study is the fact that we did not perform liquid chromatography tandem-mass spectrometry (LC-MS/MS) to objectify the amount of carbamoylation. This technique has already been used for the detection of carbamoylated albumin [26], but not for carbamoylated non-HDL. However, the relationship between the amide II/amide I AUC ratio and the KOCN concentration, as well as its relationship with the serum urea concentration are very suggestive for the carbamoylation process of non-HDL. A potential confounder could be the effect of diabetes mellitus, as the lysine residues are susceptible to both carbamoylation and glycation by glucose [28]. However, previous work of our research group showed no significant changes in the amide I and amide II band after in vitro glycation of keratins in nail powder. After incubation of nail powders with respectively 1 mL of 0.9% sodium chloride solution, 5% glucose solution and 10% glucose solution, a clear difference in the area under the infrared peak was observed at wavenumber 1047 cm^{-2}, a region characterized by a characteristic carbohydrate absorption. This band was used for measuring the degree of keratin glycation. No influence of glycation products was observed on the amide I and amide II bands [29]. In the CKD stage 3-5 group and in the CKD 5d group, there were respectively 29.7% and 31.3% diabetics as compared to 0% in the group of healthy subjects. In addition, also oxidation may modify lipoproteins in a similar way as it has been demonstrated by amino acid analysis that modification of lysine residues occurs during LDL oxidation [30]. However as demonstrated in our in vitro experiments, oxidation had no important influence on the reported results.

5. Conclusions

Carbamoylation is involved in the pathogenesis of various diseases (atherosclerosis, kidney diseases, autoimmune diseases, infections, and thrombus formation) and has been identified as an important risk factor for mortality in dialysis patients or in those with accelerated atherogenesis. The development of novel tools to determine the degree of post-translational modification-derived products is a demanding task. At this moment, the majority of potential carbamoylation biomarkers can only be assessed by rather complex analytical methods, hampering their use in clinical practice [2].

In the present study, we have demonstrated that ATR-FTIR spectroscopy is an easy-to-use, reagent-free, and cost-effective method. It is a non-destructive technique, consuming only a small amount of sample [31]. Spectral changes of non-HDL were observed depending on a declining kidney function. So, ATR-FTIR can be regarded as a new method for identification of carbamoylated non-HDL in CKD patients.

Author Contributions: Conceptualization, J.R.D.; analysis, S.E.D., S.D.B. and L.D.B.; writing, review and editing, S.E.D., W.V.B. and M.M.S.

Funding: This research was funded by an Assistant Academic Personnel Grant from the Ghent University (S.E.D.) and by a Senior Clinical Researcher Grant from the Research Foundation Flanders (FWO) (M.M.S.). The APC was funded by the Ghent University Hospital.

Conflicts of Interest: The authors declare no conflict of interest.

References

1. Jaisson, S.; Pietrement, C.; Gillery, P. Carbamylation-derived products: Bioactive compounds and potential biomarkers in chronic renal failure and atherosclerosis. *Clin. Chem.* **2011**, *57*, 1499–1505. [CrossRef]
2. Delanghe, S.; Delanghe, J.R.; Speeckaert, R.; Van Biesen, W.; Speeckaert, M.M. Mechanisms and consequences of carbamoylation. *Nat. Rev. Nephrol.* **2017**, *13*, 580–593. [CrossRef]
3. Wang, Z.; Nicholls, S.J.; Rodriguez, E.R.; Kummu, O.; Hörkkö, S.; Barnard, J.; Reynolds, W.F.; Topol, E.J.; DiDonato, J.A.; Hazen, S.L. Protein carbamylation links inflammation, smoking, uremia and atherogenesis. *Nat. Med.* **2007**, *13*, 1176–1184. [CrossRef] [PubMed]
4. Nilsson, L.; Lundquist, P.; Kagedal, B.; Larsson, R. Plasma cyanate concentrations in chronic renal failure. *Clin. Chem.* **1996**, *42*, 482–483.
5. Foley, R.N.; Parfrey, P.S.; Sarnak, M.J. Clinical epidemiology of cardiovascular disease in chronic renal disease. *Am. J. Kidney Dis.* **1998**, *32*, S112–S119. [CrossRef] [PubMed]
6. Speer, T.; Owala, F.O.; Holy, E.W.; Zewinger, S.; Frenzel, F.L.; Stähli, B.E.; Razavi, M.; Triem, S.; Cvija, H.; Rohrer, L.; et al. Carbamylated low-density lipoprotein induces endothelial dysfunction. *Eur. Heart J.* **2014**, *35*, 3021–3032. [CrossRef]
7. Vaziri, N.D. Causes of dysregulation of lipid metabolism in chronic renal failure. *Semin. Dial.* **2009**, *22*, 644–651. [CrossRef]
8. Holy, E.W.; Akhmedov, A.; Speer, T.; Camici, G.G.; Zewinger, S.; Bonetti, N.; Beer, J.H.; Lüscher, T.F.; Tanner, F.C. Carbamylated low-density lipoproteins induce a prothrombotic state via LOX-1: Impact on arterial thrombus formation in vivo. *J. Am. Coll. Cardiol.* **2016**, *68*, 1664–1676. [CrossRef]
9. Apostolov, E.O.; Shah, S.V.; Ok, E.; Basnakian, A.G. Carbamylated low-density lipoprotein induces monocyte adhesion to endothelial cells through intercellular adhesion molecule-1 and vascular cell adhesion molecule-1. *Arterioscler. Thromb. Vasc. Biol.* **2007**, *27*, 826–832. [CrossRef]
10. Apostolov, E.O.; Ray, D.; Alobuia, W.M.; Mikhailova, M.V.; Wang, X.; Basnakian, A.G.; Shah, S.V. Endonuclease G mediates endothelial cell death induced by carbamylated LDL. *Am. J. Physiol. Heart Circ. Physiol.* **2011**, *300*, H1997–H2004. [CrossRef]
11. Asci, G.; Basci, A.; Shah, S.V.; Basnakian, A.; Toz, H.; Ozkahya, M.; Duman, S.; Ok, E. Carbamylated low-density lipoprotein induces proliferation and increases adhesion molecule expression of human coronary artery smooth muscle cells. *Nephrology (Carlton)* **2008**, *13*, 480–486. [CrossRef]
12. Apostolov, E.O.; Shah, S.V.; Ok, E.; Basnakian, A.G. Quantification of carbamylated LDL in human sera by a new sandwich ELISA. *Clin. Chem.* **2005**, *51*, 719–728. [CrossRef]
13. Apostolov, E.O.; Ray, D.; Savenka, A.V.; Shah, S.V.; Basnakian, A.G. Chronic uremia stimulates LDL carbamylation and atherosclerosis. *J. Am. Soc. Nephrol.* **2010**, *21*, 1852–1857. [CrossRef]
14. Holzer, M.; Gauster, M.; Pfeifer, T.; Wadsack, C.; Fauler, G.; Stiegler, P.; Koefeler, H.; Beubler, E.; Schuligoi, R.; Heinemann, A.; et al. Protein carbamylation renders high-density lipoprotein dysfunctional. *Antioxid. Redox Signal.* **2011**, *14*, 2337–2346. [CrossRef]
15. Krilov, D.; Balarin, M.; Kosovic, M.; Gamulin, O.; Brnjas-Kraljevic, J. FT-IR spectroscopy of lipoproteins—A comparative study, Spectrochim. *Acta A Mol. Biomol. Spectrosc.* **2009**, *74*, 701–706. [CrossRef]
16. Coffey, M.D.; Coe, R.A.; Colles, S.M.; Chisolm, G.M. In vitro cell injury by oxidized low density lipoprotein involves lipid hydroperoxide-induced formation of alkoxyl, lipid, and peroxyl radicals. *J. Clin. Investig.* **1995**, *96*, 1866–1873. [CrossRef] [PubMed]
17. Burstein, M.; Scholnick, H.R.; Morfin, R. Rapid method for the isolation of lipoproteins from human serum by precipitation with polyanions. *J. Lipid Res.* **1970**, *11*, 583–595.
18. Roche Diagnostics, Cobas 8000 Modular Series: Intelligent Lab Power (2013). Available online: http://www.cobas.be (accessed on 30 May 2019).

19. Fink, P.; Roemer, M.; Haeckel, R.; Fateh-Moghadam, A.; Delanghe, J.; Gressner, A.M.; Dubs, R.W. Measurement of proteins with the Behring nephelometer. A multicenter evaluation. *J. Clin. Chem. Clin. Biochem.* **1989**, *27*, 261–276.
20. Friedewald, W.T.; Levy, R.I.; Fredrickson, D.S. Estimation of the concentration of low-density lipoprotein cholesterol in plasma, without use of the preparative ultracentrifuge. *Clin. Chem.* **1972**, *18*, 499–502. [PubMed]
21. Levey, A.S.; Stevens, L.A.; Schmid, C.H.; Zhang, Y.L.; Castro, A.F., 3rd; Feldman, H.I.; Kusek, J.W.; Eggers, P.; Van Lente, F.; Greene, T.; et al. A new equation to estimate glomerular filtration rate. *Ann. Intern. Med.* **2009**, *150*, 604–612. [CrossRef]
22. Barth, A. Infrared spectroscopy of proteins. *Biochim. Biophys. Acta* **2007**, *1767*, 1073–1101. [CrossRef]
23. Oberg, K.A.; Ruysschaert, J.M.; Goormaghtigh, E. The optimization of protein secondary structure determination with infrared and circular dichroism spectra. *Eur. J. Biochem.* **2004**, *271*, 2937–2948. [CrossRef]
24. Fernandez-Higuero, J.A.; Salvador, A.M.; Martin, C.; Milicua, J.C.; Arrondo, J.L. Human LDL structural diversity studied by IR spectroscopy. *PLoS ONE* **2014**, *9*, e92426. [CrossRef]
25. Liu, K.Z.; Man, A.; Dembinski, T.C.; Shaw, R.A. Quantification of serum apolipoprotein B by infrared spectroscopy. *Anal. Bioanal. Chem.* **2007**, *387*, 1809–1814. [CrossRef]
26. Berg, A.H.; Drechsler, C.; Wenger, J.; Buccafusca, R.; Hod, T.; Kalim, S.; Ramma, W.; Parikh, S.M.; Steen, H.; Friedman, D.J.; et al. Carbamylation of serum albumin as a risk factor for mortality in patients with kidney failure. *Sci. Transl. Med.* **2013**, *5*, 175ra29. [CrossRef]
27. Gorisse, L.; Pietrement, C.; Vuiblet, V.; Schmelzer, C.E.; Köhler, M.; Duca, L.; Debelle, L.; Fornès, P.; Jaisson, S.; Gillery, P. Protein carbamylation is a hallmark of aging. *Proc. Natl. Acad. Sci. USA* **2016**, *113*, 1191–1196. [CrossRef]
28. Nicolas, C.; Jaisson, S.; Gorisse, L.; Tessier, F.J.; Niquet-Leridon, C.; Jacolot, P.; Pietrement, C.; Gillery, P. Carbamylation is a competitor of glycation for protein modification in vivo. *Diabetes Metab.* **2018**, *44*, 160–167. [CrossRef]
29. Coopman, R.; Van de Vyver, T.; Kishabongo, A.S.; Katchunga, P.; Van Aken, E.H.; Cikomola, J.; Monteyne, T.; Speeckaert, M.M.; Delanghe, J.R. Glycation in human fingernail clippings using ATR-FTIR spectrometry, a new marker for the diagnosis and monitoring of diabetes mellitus. *Clin. Biochem.* **2017**, *50*, 62–67. [CrossRef]
30. Steinbrecher, U.P. Oxidation of human low density lipoprotein results in derivatization of lysine residues of apolipoprotein B by lipid peroxide decomposition products. *J. Biol. Chem.* **1987**, *262*, 3603–3608.
31. De Bruyne, S.; Speeckaert, M.M.; Delanghe, J.R. Applications of mid-infrared spectroscopy in the clinical laboratory setting. *Crit. Rev. Clin. Lab. Sci.* **2018**, *55*, 1–20. [CrossRef]

© 2019 by the authors. Licensee MDPI, Basel, Switzerland. This article is an open access article distributed under the terms and conditions of the Creative Commons Attribution (CC BY) license (http://creativecommons.org/licenses/by/4.0/).

Review

HDL and LDL: Potential New Players in Breast Cancer Development

Lídia Cedó [1,2], Srinivasa T. Reddy [3], Eugènia Mato [1,5], Francisco Blanco-Vaca [1,2,4,*] and Joan Carles Escolà-Gil [1,2,4,*]

1. Institut d'Investigacions Biomèdiques (IIB) Sant Pau, Sant Quintí 77, 08041 Barcelona, Spain; lcedo@santpau.cat (L.C.); emato@santpau.cat (E.M.)
2. CIBER de Diabetes y Enfermedades Metabólicas Asociadas (CIBERDEM), Monforte de Lemos 3-5, 28029 Madrid, Spain
3. Department of Molecular and Medical Pharmacology, David Geffen School of Medicine, University of California, Los Angeles, CA 90095-1736, USA; sreddy@mednet.ucla.edu
4. Departament de Bioquímica i Biologia Molecular, Universitat Autònoma de Barcelona, Av. de Can Domènech 737, 08193 Cerdanyola del Vallès, Spain
5. CIBER de Bioingeniería, Biomateriales y Nanomedicina (CIBER-BBN), Monforte de Lemos 3-5, 28029 Madrid, Spain
* Correspondence: fblancova@santpau.cat (F.B.-V.); jescola@santpau.cat (J.C.E.-G.); Tel.: +34-935537588 (F.B.-V. & J.C.E.-G.); Fax: +34-935537589 (F.B.-V. & J.C.E.-G.)

Received: 29 May 2019; Accepted: 12 June 2019; Published: 14 June 2019

Abstract: Breast cancer is the most prevalent cancer and primary cause of cancer-related mortality in women. The identification of risk factors can improve prevention of cancer, and obesity and hypercholesterolemia represent potentially modifiable breast cancer risk factors. In the present work, we review the progress to date in research on the potential role of the main cholesterol transporters, low-density and high-density lipoproteins (LDL and HDL), on breast cancer development. Although some studies have failed to find associations between lipoproteins and breast cancer, some large clinical studies have demonstrated a direct association between LDL cholesterol levels and breast cancer risk and an inverse association between HDL cholesterol and breast cancer risk. Research in breast cancer cells and experimental mouse models of breast cancer have demonstrated an important role for cholesterol and its transporters in breast cancer development. Instead of cholesterol, the cholesterol metabolite 27-hydroxycholesterol induces the proliferation of estrogen receptor-positive breast cancer cells and facilitates metastasis. Oxidative modification of the lipoproteins and HDL glycation activate different inflammation-related pathways, thereby enhancing cell proliferation and migration and inhibiting apoptosis. Cholesterol-lowering drugs and apolipoprotein A-I mimetics have emerged as potential therapeutic agents to prevent the deleterious effects of high cholesterol in breast cancer.

Keywords: Breast cancer; cholesterol; 27-hydroxycholesterol; HDL; LDL; cholesterol-lowering therapies

---

1. Introduction

Breast cancer is the third most common cancer overall, with an estimated incidence of 1.7 million cases in 2016 and a 29% increase in incident cases between 2006 and 2016. Moreover, breast cancer was the fifth leading cause of cancer deaths for both sexes in 2016 and the primary cause of death for women [1]. A substantial proportion of the worldwide burden of cancer could be prevented; however, improved primary prevention of cancer requires identification of risk markers [2]. Reproductive, hormonal factors, and unhealthy lifestyles that trigger obesity are considered significant risk factors for breast cancer [3]. Obesity represents a potentially modifiable risk factor that could increase the risk of breast cancer in women [4,5]. The biological association between obesity and disease risk, at least in part, may be related to circulating lipid levels and tissue lipid metabolism [6].

Cancer cells show specific alterations in different aspects of lipid metabolism, which can affect the availability of structural lipids for the synthesis of membranes, contribution of lipids to energy homeostasis, and lipid signaling functions, including the activation of inflammation-related pathways. All these changes are related to important cellular processes, including cell growth, proliferation, differentiation, and motility [7]. The interplay among cholesterol, lipoproteins, proinflammatory signaling pathways, and tumor development has mainly been studied in breast cancer cells and experimental models in vivo. Furthermore, in humans, both benign and malignant proliferation of breast tissue were associated with changes in plasma lipid and lipoprotein levels [8], despite that epidemiological data on the association between lipoproteins and breast cancer showed inconclusive results [9–11]. This article reviews the progress to date in research on the role of cholesterol and its main lipoprotein transporters, the low-density and high-density lipoproteins (LDL and HDL), on breast cancer development, mainly focusing on recent findings in human trials and those obtained in experimental models of breast cancer. PubMed was searched comprehensively with combinations of the keyword Breast Cancer and the rest of keywords related with cholesterol and lipoproteins.

2. Association of Cholesterol in Breast Cancer Risk: Clinical and Epidemiological Studies

Study of the relationship between serum cholesterol levels and risk of cancer is of special interest and has sparked debate, especially with the expansion of lipid-modifying therapies and more aggressive cholesterol goals to reduce the risk of cardiovascular events [12]. However, different studies have produced divergent results. Indeed, one study found that total cholesterol was associated with the risk of breast cancer [13], but others failed in finding such an association [14–18], or they even found that total cholesterol was inversely associated with the risk of breast cancer [19].

Since cholesterol is mainly transported by LDL and HDL, several clinical trials have associated them with breast cancer. A clinical study in which the lipid profile was assessed in women with breast cancer showed that LDL cholesterol (LDL-C) levels at diagnosis was a prognostic factor of breast tumor progression. A systemic LDL-C level above 117 mg dL^{-1} was found to be a predictive factor of tumor stage, and it was positively associated with worse prognosis because of a higher histological grade, higher proliferative rate, and more advanced clinical stage [20] (Table 1). Moreover, patients with LDL-C above 144 mg dL^{-1} were also prone to have lymph node metastasis [20]. More importantly, a Mendelian randomization study found that genetically raised LDL-C was associated with a higher risk of breast cancer [11]. However, other meta-analyses and prospective studies found no association between LDL-C and breast cancer risk [9,10,16,21]; some trials even found that LDL-C or non-HDL were inversely associated with the risk of breast cancer [14,22] (Table 1).

Concerning HDL-C, discordant results were also found. One prospective study with a follow-up time of 11.5 years found an inverse association between HDL-C and breast cancer risk [19], and retrospectively collected clinical data showed that decreased HDL-C levels had a significant association with worse overall survival in breast cancer patients [23] (Table 1). In contrast, a Mendelian randomization study showed that raised HDL-C increased the risk of estrogen receptor (ER)-positive breast cancer [11] (Table 1). It should also be noted that other studies failed to find any association between HDL-C and breast cancer risk [10,21,24] or survival [24]. Moreover, controversy also exists when considering the menopausal status of patients (Table 1). Some studies have found that low HDL-C among premenopausal women increased breast cancer risk [9,25,26], while others found that low HDL-C was associated with an increased postmenopausal risk of breast cancer [16,27].

Table 1. Clinical and epidemiological studies linking low-density lipoprotein cholesterol (LDL-C) and high-density lipoprotein cholesterol (HDL-C) levels to breast cancer risk.

Reference	Year	Study Design	Participants	Main Findings
Nowak et al. [11]	2018	Mendelian randomization	>400,000	Raised LDL-C increased the risk of breast cancer (OR = 1.09 (1.02–1.18)) and ER-positive breast cancer (OR = 1.14 (1.05–1.24)). Raised HDL-C increased the risk of ER-positive breast cancer (OR = 1.13 (1.01–1.26)).
Ni et al. [16]	2015	Meta-analysis	1,189,635	Inverse association between HDL-C and breast cancer risk among postmenopausal women (RR = 0.45 (0.64–0.93)). No association in premenopausal women. No association between LDL-C and breast cancer risk.
Touvier et al. [9]	2015	Meta-analysis	1,489,484	Inverse association between HDL-C and breast cancer risk among premenopausal women (HR = 0.77 (0.31–0.67)). No association in postmenopausal women. No association between LDL-C and breast cancer risk.
Borgquist et al. [21]	2016	Prospective	5281	No evident associations between LDL-C or HDL-C and breast cancer incidence.
Chandler et al. [10]	2016	Prospective	15,602	No association between LDL-C or HDL-C and breast cancer risk.
His et al. [19]	2014	Prospective	7557	HDL-C was inversely associated with breast cancer risk (HR 1 mmol L^{-1} increment = 0.48 (0.28–0.83)).
Rodrigues dos Santos et al. [20]	2014	Prospective	244	Systemic levels of LDL-C correlated positively with tumor size (Spearman's r = 0.199, p = 0.002).
Kucharska-Newton et al. [25]	2008	Prospective	7575	Modest association of low HDL-C (<50 mg dL^{-1}) with breast cancer among premenopausal women (HR = 1.67 (1.06–2.63)). No association in postmenopausal women.
Furberg et al. [27]	2004	Prospective	30,546	The risk of postmenopausal breast cancer was reduced in women in the highest quartile of HDL-C (>1.64 mmol L^{-1}) compared with women in the lowest quartile (<1.20 mmol L^{-1}; RR = 0.73 (0.55–0.95)). No association was found in premenopausal women.
Li et al. [23]	2017	Retrospective	1044	Decreased HDL-C levels showed significant association with worse overall survival (HR = 0.528 (0.302–0.923)).
Li et al. [28]	2018	Case–control	Total: 3537 Cases: 1054 Controls: 2483	The levels of LDL-C and HDL-C were lower in breast cancer patients than controls (p < 0.001).
His et al. [24]	2017	Case–control	Total: 1626 Cases: 583 Controls: 1043	No association between LDL-C or HDL-C and breast cancer risk or survival.
Martin et al. [14]	2015	Case–control	Total: 837 Cases: 279 Controls: 558	HDL-C was positively associated (75th vs. 25th percentile: 23% higher, p = 0.05) and non-HDL-C was negatively associated (75th vs. 25th percentile: 19% lower, p = 0.03) with breast cancer risk.
Llanos et al. [22]	2012	Case–control	Total: 199 Cases: 97 Controls: 102	Increasing levels of LDL-C were inversely associated with breast cancer risk (OR = 0.41 (0.21–0.81)). Lower levels of HDL-C were associated with a significant increase in breast cancer risk (OR = 1.99 (1.06–3.74)).
Yadav et al. [29]	2012	Case–control	Total: 139 Cases: 69 Controls: 70	Postmenopausal breast cancer patients had higher LDL-C levels (p < 0.001) and lower HDL-C levels (p = 0.025) than controls. No significant changes in premenopausal women.
Kim et al. [26]	2009	Case–control	Total: 2070 Cases: 690 Controls: 1380	Protective effect of HDL-C on breast cancer was only observed among premenopausal women (OR = 0.49 (0.33–0.72) for HDL-C ≥ 60 vs. <50 mg dL^{-1} (p < 0.01)).

Table 1. *Cont.*

Reference	Year	Study Design	Participants	Main Findings
Owiredu et al. [30]	2009	Case–control	Total: 200 Cases: 100 Controls: 100	Increased LDL-C levels in postmenopausal breast cancer patients vs. controls ($p < 0.05$). No significant changes in premenopausal women. No changes in HDL-C levels between cases and controls.
Michalaki et al. [31]	2005	Case–control	Total: 100 Cases: 56 Controls: 44	A decrease in HDL-cholesterol was observed in patients with breast cancer vs. controls ($p < 0.05$).

LDL-C = low-density lipoprotein cholesterol, HDL-C = high-density lipoprotein cholesterol, ER = estrogen receptor, OR = odds ratio, RR = risk ratio, and HR = hazard ratio. Between brackets, 95% confidence interval.

In summary, although some studies failed to find associations between lipoproteins and breast cancer, the results of some large clinical trials seem to point to a direct association between LDL-C and breast cancer risk as well as an inverse association between HDL-C and breast cancer risk. It is important to note that clinical or methodological differences in the design of the studies, including variation in geographic regions, menopausal status, number of cases, or follow-up length, could explain the discrepancies found in these studies (summarized in Table 1). For this reason, basic scientific research can contribute to determining potential underlying mechanisms that may explain these associations [12].

3. Hypercholesterolemia and Breast Cancer

Diet and obesity are important risk factors for breast cancer development [5,32]. High cholesterol intake was found to be positively associated with the risk of breast cancer, mainly among postmenopausal women [33,34]. To address interactions between body weight and dietary fat intake on subsequent mammary tumor development, a study was performed in which female murine mammary tumor virus (MMTV)-transforming growth factor α (TGFα) mice consumed a moderately high-fat diet [35]. The MMTV promoter specifically directs expression to the mammary epithelium [36], obtaining a model that recapitulates human breast cancer progression from early hyperplasia to malignant breast carcinoma [37]. These mice exhibited mammary tumor latency inversely related to their body fat, suggesting that body fat may be the mediating factor of the effect of a high-fat diet on mammary tumor development [35]. Moreover, the expression of a number of proteins associated with leptin and apoptosis signaling pathways were also affected by diet in the mammary tumors of these animals [38].

Some studies have specifically addressed the role of dietary cholesterol in the regulation of tumor progression in different experimental mouse models of breast cancer. Llaverias et al. studied the role of a high-fat/high-cholesterol (HFHC) diet administration in MMTV polyoma middle T (PyMT) oncogene transgenic mice and found that the HFHC diet accelerated and enhanced tumor progression in these mice [39]. Plasma cholesterol levels were reduced during tumor development but not prior to its initiation, providing new evidence for an increased utilization of cholesterol by tumors and for its role in tumor formation [39]. Another group administered an HFHC diet to female immunodeficient mice implanted orthotopically with MDA-MB-231 cells and found that diet induced angiogenesis and accelerated breast tumor growth in this model of breast cancer [40].

The role of dyslipidemia in breast cancer growth and metastasis was also explored in hypercholesterolemic apolipoprotein E knockout mice (apoE$^{-/-}$) fed an HFHC diet and injected with non-metastatic Met-1 and metastatic Mvt-1 mammary cancer cells derived from PyMT mice and c-Myc/vegf tumor explants, respectively [41]. The apoE glycoprotein is a structural component of all lipoprotein particles other than LDL, and it acts as a ligand of lipoprotein receptors and participates in the uptake of lipids into cells. The absence of apoE leads to the accumulation of cholesterol and triglycerides in plasma [42]. ApoE$^{-/-}$ mice exhibited increased tumor growth and displayed a greater number of spontaneous metastases to the lungs. The results in tumor growth were only observed when an HFHC diet was administered to the mice, not when they were fed a standard chow diet [41]. Therefore, although the uptake of cholesterol via apoE was blocked, other adipocyte apoE-independent receptors, such as LDL receptor (LDLR) [43], may be involved in the cholesterol uptake by cancer cells. Moreover, the phosphoinositide 3-kinase (PI3K)/Akt pathway, involved in proinflammatory and cell proliferation signals, was found to be one mediator of the tumor-promoting activity of hypercholesterolemia [41].

Whereas the selection of HFHC diets for these studies reflects current dietary trends, this approach has not allowed an evaluation of the specific effect of cholesterol on tumor biology [44]. To directly address this question, PyMT mice were administered a high-cholesterol diet from weaning and developed palpable tumors earlier than mice on a control chow diet were [45]. High-cholesterol diet administration to mice injected with different breast cancer cell lines (human breast cancer HTB20 and MDA-MB-231, and the mouse breast cancer cell line 4 T1) also promoted breast tumor growth. Tumors of animals in the high-cholesterol diet group showed a higher proliferative ratio than those from chow-fed mice, and lung metastasis was increased [46].

4. 27-Hydroxycholesterol and Breast Cancer

Estrogen receptor α-induced signal transduction controls the growth of most breast cancers [47]. 27-hydroxycholesterol (27-HC), one of the most prevalent oxysterols, was identified as an endogenous selective ER modulator (SERM) and liver X receptor (LXR) agonist [48]. This oxysterol is generated enzymatically from cholesterol by the P450 enzyme sterol 27-hydroxylase CYP27A1. CYP27A1 is abundant in the liver, but it is also expressed in the intestine, vasculature, brain, and macrophages. 27-HC is mainly transported in association with HDL and LDL, primarily in the esterified form [49]. Regarding its catabolism, 27-HC is hydroxylated by oxysterol 7α-hydroxylase CYP7B1, which is also abundant in the liver [50].

The first evidence for 27-HC's role in breast cancer began with studies that found that it stimulated the growth of ER-positive MCF-7 cells but not that of ER-negative MCF-10 cells. The effect of a concentration of 1–2 µM of 27-HC was similar to that of 1–2 nM of 17β-estradiol [51]. The proliferative role of 27-HC in vitro on MCF-7 cells was also confirmed by others, who also reported that 27-HC increased tumor growth in vivo in PyMT mice and in murine or human cancer cell xenografts [45,52]. 27-HC was also found to hasten metastasis to the lungs, an effect that implicated LXR activation [45]. 27-HC also hastened myeloid immune cell functions, as it was found that this oxysterol increased the number of polymorphonuclear neutrophils and γδ T cells as well as decreased cytotoxic CD8$^+$ T cells within tumors and metastatic lesions [53].

In human breast cancer tissue, 27-HC concentration was found to increase because of decreased catabolism, since *CYP7B1* gene expression was downregulated, whereas *CYP27A1* remained unchanged. Moreover, increased *CYP7B1* mRNA was correlated with better survival [52]. Consistently, Nelson et al. found increased CYP27A1 protein expression in higher grade tumors [45]. Nevertheless, the first prospective epidemiological study on prediagnosis of circulating 27-HC and breast cancer risk showed an inverse association between blood 27-HC and breast cancer risk among postmenopausal women. The authors hypothesized that 27-HC-associated inhibition of estradiol–ER binding outweighed 27-HC's agonistic effect in human breast cancer [54].

Unlike humans, mice do not normally become severely hypercholesterolemic when fed an HFHC diet [44]. To circumvent this limitation, breast cancer cells were implanted in mice in which the mouse *Apoe* gene was replaced with the human *APOE3* allele, which codes for the most frequent human isoform. The animals on an HFHC diet exhibited both increased cholesterol and 27-HC in plasma as well as promotion of larger tumors, effects that were partially reversed by treatment with the CYP27A1 inhibitor GW273297X [45].

Several studies investigated the potential mechanisms involved in 27-HC-induced breast cancer development. First, 27-HC inhibited p53 protein and activity in MCF-7 cells via ER. The oxysterol increased p53 regulator mouse double minute 2 (MDM2) levels and enhanced interaction between p53 and MDM2, suggesting that 27-HC proliferation depended on MDM2-mediated p53 degradation. Interestingly, estradiol, the main physiological endogenous ligand for ER, which had similar effects to 27-HC on cell proliferation, had no effect on p53 activity; this demonstrates that 27-HC may contribute to ER-positive breast cancer progression via different mechanisms compared with known estrogens [55]. Another study found that 27-HC increased Myc protein stability (a critical oncogene that can promote proliferation, migration, and invasion of cancer cells) by reducing its dephosphorylation and ubiquitination for proteasomal degradation [56]. Signal transducer and activator of transcription (STAT)-3 is an important transcription factor that can target c-Myc, vascular endothelial growth factor (VEGF), cyclin D1, matrix metalloproteinase (MMP) 2, and MMP9 to promote the development of cancer involving tumor proliferation, invasion, metastasis, and angiogenesis [57]. 27-hydroxycholesterol induced activation of STAT-3, which promoted the angiogenesis of breast cancer cells via proinflammatory-related reactive oxygen species (ROS)/STAT-3/VEGF signaling [58]. Moreover, it induced the epithelial–mesenchymal transition (EMT) [59], a mechanism that promotes migration and invasion, via STAT-3/MMP9 and STAT-3/EMT [60], in both ER-positive and ER-negative breast cancer cells. Furthermore, 27-HC causes greater macrophage infiltration and exacerbation of inflammation in the setting of hypercholesterolemia [61], thereby providing a link between inflammation and cancer development. Collectively, mechanisms involved in 27-HC-promoted progression of breast cancer are complex. Therefore, seeking effective measures to prevent 27-HC-caused pathogenicity is difficult, and further studies should be carried out with an emphasis on deeply investigating the potential mechanisms involved in 27-HC breast cancer promotion [58].

The discovery of 27-HC as an endogenous ER ligand that promotes ER-positive breast tumor growth could help explain why some breast cancer patients are resistant to aromatase inhibitors [62]. In this way, 27-HC may act as an alternate estrogenic ligand in a low-estrogen environment [63]. Assessments of 27-HC or their metabolic enzymes' abundance in tumors could aid in personalizing hormone-based therapy [64].

5. Low-Density Lipoprotein and Breast Cancer

Proliferating cancer cells have an increased cholesterol need. Increased LDLR expression was demonstrated in breast cancer tissue to increase the uptake of LDL-C from the bloodstream [65]. In vitro, LDLR gene and protein expression was found increased in ER-negative MDA-MB-231 cells in contrast to ER-positive MCF-7 cells [66,67]. Accordingly, LDL-C mainly promoted proliferation [68–70] and migration [46,71] in ER-negative cells, but this was not evident in ER-positive cell lines. This difference between the two cell types corresponded to a greater ability of ER-negative cells to take up, store, and utilize exogenous cholesterol because of the increased activity of acyl-CoA:cholesterol acyltransferase 1 (ACAT1) [68]. The Women's Intervention Nutrition Study (WINS) found that a low-fat diet mainly extended relapse-free survival in women with ER-negative breast cancer [72]. At least in part, that ER-negative breast cancer cells differentially uptake and store cholesterol may explain the differential effect of a low-fat diet on human breast cancer recurrence [68]. Another study found that LDL-C also induced proliferation in ER-positive BT-474 breast cancer cells [46]. This discrepancy could be because BT-474 cells usually express the Her2 (ErbB2) receptor [73]; furthermore, high plasma LDL-C levels were found to be associated with Her2-positive breast cells [20]. It is noteworthy that the Her2-positive and triple-negative subtypes are the most aggressive breast cancers [74].

Beyond in vitro studies, tumors from breast cancer cells with high LDLR expression (murine MCNeuA (Her2-positive) and human MDA-MB-231 (triple-negative), respectively) have been incrementally grown in immunocompetent (LDLR$^{-/-}$ and apoE$^{-/-}$) and immunodeficient (Rag1$^{-/-}$/LDLR$^{-/-}$ and Rag1$^{-/-}$/apoE$^{-/-}$) mouse models of hyperlipidemia with increasing serum LDL concentrations. Importantly, silencing LDLR in the tumor cells reduced tumor growth [67].

Finally, in human samples, LDLR and ACAT1 were also found to be increased in Her2-positive and triple-negative tumors compared with luminal A tumors. Her2-positive and triple-negative tumors were more cholesteryl ester-rich and had higher histological grades, Ki-67 expression, and tumor necrosis. Therefore, cholesteryl ester accumulation due to increased LDL-C internalization and esterification was associated with breast cancer proliferation [75]. In line with these findings, higher LDLR expression was found to be associated with a worse prognosis in patients who underwent systemic therapy [67]. Overall, elevated circulating LDL and breast cancer expression of LDLR have roles, at least in Her2-positive and triple-negative breast cancers, in disease progression and disease-free survival.

Oxidized Low-Density Lipoprotein and Breast Cancer

Lipid peroxidation is associated with carcinogenesis [76]. Lipid peroxidation metabolites cause structural alterations in DNA and decrease DNA repair capacity through their direct interaction with repair enzymes [77]. The oxidation of LDL affects both protein and lipid contents, resulting in the formation of peroxidation metabolites. Patients with breast cancer exhibited elevated serum levels of oxidized LDL (oxLDL) [78]. Moreover, serum oxLDL levels were associated with increased breast cancer risk [78]. Oxidized LDL was also reported to trigger pro-oncogenic signaling in MCF10A cells; concretely, cells treated with oxLDL showed a dose-dependent stimulation of proliferation mediated by stimulation of the microRNA miR-21, which, in turn, activated the related proinflammatory PI3K/Akt signaling pathways [79].

OxLDL lecithin-like receptor 1 (OLR1) is the main receptor for internalization of oxLDL. It is overexpressed in human breast cancer and positively correlates to tumor stage and grade [80]. A microarray analysis of hearts of *Olr1* KO mice compared with wild-type mice showed a reduction in the expression of nuclear factor κB (NF-κB) target genes involved in cellular transformation (regulation of apoptosis, proliferation, wound healing, defense response, immune response, and cell migration) as well as an inhibition of key enzymes involved in lipogenesis. The human breast cancer cell line HCC1143 showed increased *OLR1* expression compared with the normal mammary epithelial cell line MCF10A [81]. Forced overexpression of *OLR1* in both cell lines resulted in upregulation of NF-κB and its target pro-oncogenes involved in the inhibition of apoptosis (*BCL2*, *BCL2A1*, and *TNFAIP3*) and regulation of the cell cycle (*CCND2*) in HCC1143 cells. Moreover, upregulation of *OLR1* in breast cancer cell lines enhanced cell migration [81,82]. In line with these findings, *OLR1* depletion by siRNAs, or ORL1 inhibition by antibodies or a recombinant OLR1 protein, significantly suppressed the invasion and migration of breast cancer cells [81–83]. TBC1D3 is a hominoid-specific oncogene that also regulates migration of human breast cancer cells. TBC1D3 was found to stimulate the expression of *OLR1*, and this *TBC1D3*-induced *OLR1* expression was regulated by tumor necrosis factor α (TNFα)/NF-κB signaling [84]. Therefore, *OLR1* may function in special situations, such as obesity and chronic inflammation, to increase breast cancer susceptibility.

6. High-Density Lipoprotein and Breast Cancer

Controversy exists about the association between HDL-C levels and breast cancer risk, as detailed in Section 2. In the present section, experimental data evaluating the role of HDL in breast cancer development are reviewed. In vitro analyses have shown that HDL stimulated proliferation in both ER-positive [69,85] and ER-negative breast cancer cell lines [69] in a dose-dependent manner, but ER-negative cells showed a higher response [69]. Human HDL3 also induced migration and activated Akt and extracellular signal-regulated kinases (ERK)1/2 signal transduction pathways in both MCF7 and MDA-MB-231 cells [86].

The scavenger receptor class B type I (SR-BI) acts as an HDL receptor and mediates its cholesterol uptake in breast cancer cells [87]. The receptor SR-BI is abundantly expressed in human breast cancer tissue compared with adjacent normal tissue [88]. Moreover, high SR-BI expression was found related to tumor aggressiveness and poor prognosis in breast cancer [75,89,90], whereas knockdown of SR-BI in vitro attenuated Akt activation and inhibited breast cancer cell proliferation and migration [86]. Moreover, HDL-induced proliferation was blocked in transfected MCF-7 cells with a mutant, nonfunctional SR-BI [88]. Beyond in vitro studies, mice injected with SR-BI-knockdown breast cancer cells showed a decreased tumor burden, accompanied with reduced Akt and ERK1/2 activation, reduced angiogenesis, and increased apoptosis [86]. Therefore, cholesteryl ester entry via HDL-SR-BI and Akt signaling seems to play a critical role in the regulation of cellular proliferation and migration and tumor growth. SR-BI was also found to increase concomitantly with an increased number and size of tumors in PyMT mice fed an HFHC diet compared with those fed a chow diet. However, cholesterol was not found accumulated in the mammary tumors, suggesting that even if tumor cholesterol uptake was increased, cholesterol was probably metabolized to sustain a high level of proliferation [39].

Serum HDL particles contain either a single copy or multiple copies of apolipoprotein A-I (apoA-I), the most abundant HDL apolipoprotein [91]. Apolipoprotein A-I plays a role in promoting cholesterol release from cells; possesses anti-inflammatory, antioxidant, and antiapoptotic properties; and influences innate immunity [92]. The levels of apoA-I have normally been found to be inversely associated with breast cancer risk [19,93,94], although one study found that apoA-I was positively associated with breast cancer [21]. Our group showed that human apoA-I-containing HDL could not hinder breast tumor development in PyMT mice. While overexpression of human apoA-I reduced the levels of oxLDL, 27-HC levels were increased, which could promote tumor growth [95]. Concerning apoA-II, the second major protein constituent of HDL [96], our research group showed that human apoA-II-containing HDL increased the breast tumor burden in PyMT mice (Figure 1A) (unpublished results). These results may be related with the apoA-II-mediated alteration in HDL remodeling, decreased capacity to protect against LDL oxidative modification and its proinflammatory actions, and postprandial hyperlipidemia (Figure 1B) [97,98].

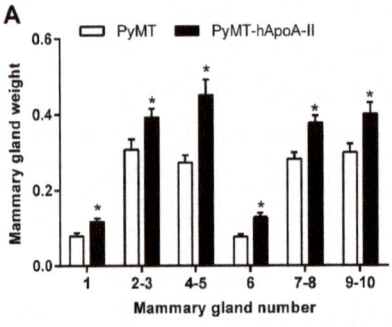 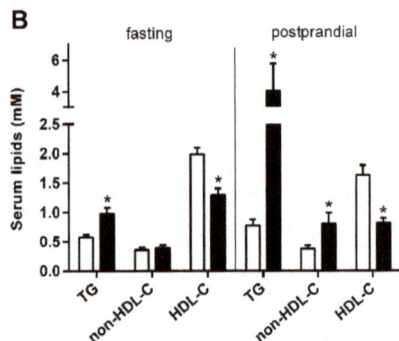

Figure 1. Effects of human apolipoprotein A-II (hApoA-II) overexpression on tumor development in polyoma middle T (PyMT) mice. PyMT mice were backcrossed with hApoA-II transgenic (TG) mice on a C57BL/6 background. The mice were maintained on a regular chow diet until 19 weeks of age, when they were euthanized, and the mammary glands were excised and weighed. Serum lipids were determined after an overnight fasting period and 3 h after a 0.15 mL dose of olive oil by oral gavage. A) Mammary gland weight. B) Serum lipid levels in fasting and postprandial conditions (TG = triglycerides, and HDL-C = high-density lipoprotein cholesterol). Values shown represent the mean ± SEM. A t-test was performed to determine the statistical significance between groups. * $p < 0.05$ vs. PyMT mice.

Dysfunctional High-Density Lipoprotein and Breast Cancer

Under conditions of oxidative stress, HDL can be oxidatively modified, and these modifications may have an effect on HDL function. Hypochlorite-oxidized HDL was found to stimulate cell proliferation, migration, invasion, and adhesion in vitro, involving the protein kinase C (PKC) pathway, which regulates numerous cellular responses including cell proliferation and the inflammatory response.

This modified HDL promoted breast cancer cell pulmonary and hepatic metastasis compared with normal HDL in vivo. Interestingly, in this study, normal HDL reduced the metastasis of MCF7 cells in the liver compared with control animals in which HDL was not injected [99].

In patients with type 2 diabetes mellitus (T2DM), HDL can be modified into dysfunctional glycated HDL and oxidized HDL [100]. Indeed, T2DM patients have a 20% increased risk of breast cancer compared with nondiabetic subjects [101]. In this context, diabetic HDL was found to have a stronger capability to promote cell proliferation, migration, and invasion of breast cancer cells through the Akt, ERK, and p38 mitogen-activated protein kinase (MAPK) pathways. These observations were also found in glycated and oxidized HDL produced in vitro, compared with normal HDL [102]. Pretreatment with diabetic, glycated, and oxidized HDL also promoted the metastasis capacity of breast cancer cells in vivo, and it increased their capacity of adhesion to human umbilical vein endothelial cells (HUVECs) and attachment to the extracellular matrix in vitro, compared with normal HDL. These effects mainly were due to elevated PCK activity, which, in turn, could stimulate secretion of integrins, which are important in promoting breast cancer metastasis [103]. Similarly, HDL isolated from patients with breast cancer complicated with T2DM promoted an increase in breast cancer cell adhesion to HUVECs and stimulated higher intercellular adhesion molecule 1 (ICAM-1) and vascular cell adhesion molecule 1 (VCAM-1) expression on the cell surfaces of breast cancer cells and HUVECs, along with the activation of PKC, compared with HDL isolated from breast cancer patients. However, in breast cancer patients complicated with T2DM, a lower expression of ICAM-I and VCAM-I was found in their tumor tissue, which may contribute to the metastasis of tumor cells [104]. Collectively, associations between T2DM and breast cancer could be attributed, in part, to alterations in HDL structure and composition and their proinflammatory actions.

7. Effects of Cholesterol-Lowering Therapies on Breast Cancer

The studies reviewed indicate that cholesterol and its main metabolite, 27-HC, may increase breast cancer development and metastasis. To address this, cholesterol-lowering drugs have emerged as potential therapies to reverse the deleterious effects of impaired cholesterol metabolism in breast cancer.

7.1. Statins

Statins are inhibitors of the enzyme hydroxy-methyl-glutaryl-coenzyme A reductase (HMGCR), which catalyzes the conversion of HMG-CoA to mevalonate, the rate-limiting step of cholesterol synthesis [105]. In humans, the effect of statins in cancer prevention and treatment remains controversial (Table 2). The use of lipid-lowering drugs, and more concretely, statins, was found to be associated with a reduced risk of breast cancer in older women [106]. Specifically, the use of lipophilic statins but not hydrophilic statins were found to significantly reduce the risk of breast cancer in Thai women [107]. Conversely, other studies, including a large Mendelian randomization study, failed to find a protective effect of statins against breast cancer risk [108–113], or they even found a positive association between long-term use of statins and increased risk of breast cancer [114]. In contrast, treatment with statins seems to have a more important effect in protecting against breast cancer recurrence and death [115–122]. Considering the type of statin, lipophilic statins were mainly found to be associated with a reduced risk of breast cancer recurrence or mortality [123–125], although hydrophilic statin use was also found to be associated with improved progression-free survival compared with no statin use in inflammatory breast cancer patients [126]. Taken together, HMGCR inhibitors do not seem to protect against breast cancer development, but statins, and more concretely, lipophilic statins, could be a good strategy for protecting against breast cancer recurrence and death.

Statins also exert antiproliferative and cytotoxic effects on breast cancer cells in vitro by increasing apoptosis, autophagy, and cell cycle arrest [127,128]. However, only lipophilic statins show anticancer activity [129], and the ER-negative phenotype seems to be more sensitive than those that overexpress ER [129,130]. ER-positive cell resistance to statin treatment is associated with high expression of cholesterol biosynthesis genes [131].

In vivo studies have also reported controversial results. Atorvastatin was able to reduce the level of circulating cholesterol, and it attenuated enhanced tumor growth and lung metastasis associated with an HFHC diet in a transgenic model in which the murine *Apoe* gene was replaced with the human *APOE3* allele and injected with ER-positive E0771 murine mammary cancer cells [45,53]. Moreover, simvastatin and fluvastatin treatments were found to inhibit tumor growth in mice inoculated with breast cancer cells [129,132], and fluvastatin was also found to reduce the metastatic burden in a murine breast cancer metastasis model [133]. The mechanisms of action of simvastatin included the inhibition of NF-κB transcription factor, which attenuated expression of antiapoptotic Bcl_{XL} and derepressed expression of the antiproliferative/proapoptotic tumor suppressor PTEN, which reduced the phosphorylation of Akt, resulting in decreased cancer cell proliferation and survival [132]. In contrast, statin treatment failed to reduce plasma cholesterol levels or tumor growth in mice injected with breast cancer cells on an HFHC diet [46] or other models of breast cancer in mice and rats [134]. An explanation for these negative results could be that mice are generally unresponsive to statins [135].

Finally, an interesting study investigated the biological effect of short-term lipophilic fluvastatin exposure on in situ and invasive breast cancer through paired tissue, blood, and imaging-based biomarkers in women with a diagnosis of ductal carcinoma in situ or stage 1 breast cancer. Fluvastatin exposure showed reduced tumor proliferation and increased apoptotic activity in high-grade breast cancer, concomitant with a reduction of cholesterol levels [136]. An upregulation of HMGCR was observed in breast cancer patients after two weeks of atorvastatin treatment, which was interpreted as activation of the negative feedback loop controlling cholesterol synthesis. Moreover, in tumors expressing HMGCR before treatment with atorvastatin, the proliferation marker Ki67 was found to be downregulated. In summary, these results suggested that HMGCR was targeted by statins in breast cancer cells in vivo, and that statins could have antiproliferative effects, mostly in HMGCR-positive breast cancers [137]. Importantly, atorvastatin was also found to decrease serum 27-HC and CYP27A1 expression in tumors of breast cancer patients [138].

Table 2. Clinical and epidemiological studies linking statin treatment to breast cancer risk.

Reference	Year	Study Design	Participants	Main Findings
Ference et al. [113]	2019	Mendelian randomization	654,783	Genetic inhibition of *HMGCR* did not affect breast cancer risk.
Islam et al. [109]	2017	Meta-analysis	121,399	There was no association between statin use and breast cancer risk.
Liu et al. [123]	2017	Meta-analysis	197,048	Significant protective effects of lipophilic statin use, but not hydrophilic statins, against cancer-specific mortality (HR = 0.57 (0.46–0.70)).
Mansourian et al. [116]	2016	Meta-analysis	124,669	Significant reduction in breast cancer recurrence (OR = 0.792 (0.735–0.853)) and death (OR = 0.849 (0.827–0.870)) among statin users.
Manthravadi et al. [124]	2016	Meta-analysis	75,684	Lipophilic statin use was associated with improved recurrence-free survival (HR = 0.72 (0.59–0.89)).
Wu et al. [119]	2015	Meta-analysis	144,830	There was a significantly negative association between prediagnosis statin use and breast cancer mortality (for overall survival: HR = 0.68 (0.54–0.84), and for disease-specific survival (HR = 0.72 (0.53–0.99)). There was also a significant inverse association between postdiagnosis statin use and breast cancer disease-specific survival (HR = 0.65 (0.43–0.98)). No significant association was detected between statin use and breast cancer risk.
Undela et al. [111]	2012	Meta-analysis	>2.4 million	Statin use and long-term statin use did not significantly affect breast cancer risk.
Bonovas et al. [108]	2005	Meta-analysis	327,238	Statin use did not significantly affect breast cancer risk.
Dale et al. [112]	2005	Meta-analysis	86,936	Statins did not reduce the incidence of breast cancer.

Table 2. Cont.

Reference	Year	Study Design	Participants	Main Findings
Borgquist et al. [115]	2017	Prospective	8010	Initiation of cholesterol-lowering medication in postmenopausal women with early stage, hormone receptor-positive invasive breast cancer during endocrine therapy was related to improved disease-free survival (HR = 0.79 (0.66–0.95)), breast cancer-free interval (HR = 0.76 (0.60–0.97)), and distant recurrence-free interval (HR = 0.74 (0.56–0.97)).
Murtola et al. [122]	2014	Prospective	31,236	Both postdiagnostic and prediagnostic statin uses were associated with a lowered risk of breast cancer death (HR = 0.46 (0.38–0.55) and HR = 0.54 (0.44–0.67), respectively).
Brewer et al. [126]	2013	Prospective	723	Hydrophilic statins were associated with significantly improved progression-free survival compared with no statin (HR = 0.49 (0.28–0.84)) in inflammatory breast cancer patients.
Ahern et al. [125]	2011	Prospective	18,769	Significant reduction in breast cancer recurrence among patients using simvastatin after 10 y of follow up (adjusted HR = 0.70 (0.57–0.86)).
Cauley et al. [106]	2003	Prospective	7528	Older women who used statins had a reduced risk of breast cancer (RR = 0.28 (0.09–0.86), adjusted for age and body weight) compared with nonusers.
Shaitelman et al. [120]	2017	Retrospective	869	Statin use was significantly associated with overall survival (HR = 0.10 (0.01–0.76)) in triple-negative breast cancer.
Smith et al. [121]	2017	Retrospective	6314	Prediagnostic statin use was associated with breast cancer-specific mortality (HR = 0.81 (0.68–0.96)). This reduction was greatest in statin users with ER-positive tumors (HR = 0.69 (0.55–0.85)).
Anothaisintawee et al. [107]	2016	Retrospective	15,718	Using lipophilic statins, but not hydrophilic statins, could significantly reduce the risk of breast cancer (risk difference = −0.0034 (−0.006,−0.001) lipophilic statin users vs. nonusers).
Mc Menamin et al. [139]	2016	Retrospective	15,140	There was no evidence of an association between statin use and breast cancer-specific death.
Sakellaki et al. [117]	2016	Retrospective	610	Statins may be linked to a favorable outcome in early breast cancer patients, especially in younger age groups (HR = 0.58 (0.36–0.94)).
Chae et al. [118]	2011	Retrospective	703	Significant reduction in breast cancer recurrence among patients who used statins (HR = 0.43 (0.26–0.70)). No association was found regarding overall survival.
Schairer et al. [110]	2018	Case–control	Total: 228,973 Cases: 30,004 Controls: 198,969	Statin use did not significantly affect breast cancer risk.
McDougall et al. [114]	2013	Case–control	Total: 2886 Cases: 916 IDC + 1068 ILC Controls: 902	Current users of statins for ≥10 y had increased risk of IDC (OR = 1.83 (1.14–2.93)) and ILC (OR = 1.97 (1.25–3.12)) compared with never users of statins.

OR = odds ratio, RR = risk ratio, HR = hazard ratio, y = years, IDC = invasive ductal carcinoma; and ILC = invasive lobular carcinoma. Between brackets, 95% confidence interval.

7.2. Ezetimibe

Ezetimibe is a drug that specifically targets intestinal Niemann-Pick C1-Like 1 (NPC1L1) and mediates the inhibition of intestinal sterol absorption [140]. Few studies have explored the effects of ezetimibe on breast cancer. However, considering that statins may have little effect on plasma cholesterol in mice [135], ezetimibe's action on breast cancer development is of interest. A study by Pelton et al. investigated the effects of ezetimibe administered in an HFHC diet on breast cancer development in an orthotopic breast tumor model, in which mice were implanted with MDA-MB-231 cells. Ezetimibe was able to reduce tumor volume, proliferation, and angiogenesis and increase apoptosis compared with the HFHC-fed mice, achieving similar results to those in mice fed a low-fat/low cholesterol (LFLC) diet. These results were accompanied with a reduction in circulating cholesterol levels, but intratumoral cholesterol levels remained unchanged [40].

To our knowledge, the effects of ezetimibe treatment on breast cancer risk or mortality have not been studied. Only Kobberø Lauridsen et al. explored the effects of genetic variants of *NPC1L1* (−133A>G and V1296V T>C), mimicking treatment with ezetimibe, on breast cancer risk. These researchers found that *NPC1L1* variants were not associated with the risk of breast cancer [141].

7.3. Phytosterols

Plant sterols, or phytosterols, lower serum LDL-C levels by reducing intestinal cholesterol absorption [142]. Several in vivo studies have tested the efficacy of dietary phytosterol in breast cancer development. Female severe combined immunodeficiency (SCID) mice supplemented with 2% phytosterols and injected with MDA-MB-231 cells exhibited a reduction in serum cholesterol, accompanied with a reduction in tumor size and metastasis to lymph nodes and lungs [143]. In ovariectomized athymic mice injected with MCF-7 cells, supplementation with β-sitosterol, the most common phytosterol, was also able to reduce tumor size [144]. Furthermore, phytosterol supplementation could decrease both the development of mammary hyperplastic lesions and tumor burden in PyMT mice fed an HFHC diet. This protective effect was not observed in mice fed an LFLC diet. A potential mechanism of action of phytosterol was the prevention of lipoprotein oxidation [145].

Numerous experimental in vitro studies showed that phytosterols functioned as anticancer compounds acting on host systems to affect tumor surveillance or on tumors to affect tumor cell biology. Mechanisms affecting the tumors include slowing of cell cycle progression, induction of apoptosis, inhibition of tumor metastasis, altered signal transduction, and activation of angiogenesis. Host influences comprise enhancing immune recognition of cancer, influencing hormonal-dependent growth of endocrine tumors, and altering cholesterol metabolism (reviewed in [146,147]).

7.4. Other Therapies

Fibrates are agonists of the peroxisome proliferator-activated receptor α (PPARα), which stimulate the expression of genes involved in fatty acid and lipoprotein metabolism, resulting in a shift from hepatic fat synthesis to fat oxidation. Fibrates are used as therapeutic agents for treating dyslipidemia [148]. A meta-analysis of 17 long-term, randomized, placebo-controlled trials found that fibrates had a neutral effect on breast cancer and other cancer outcomes [149].

The levels of HDL-C and apoA-I are inversely related to cardiovascular risk [150]. The beneficial effects of HDL have largely been attributed to apoA-I, and researchers have sought apoA-I mimetic peptides as therapeutic agents based on physical–chemical and biological properties [151]. To our knowledge, a study from our group was the only one to analyze the effects of apoA-I mimetics on breast cancer. In that study, the apoA-I mimetic peptide D-4F was administered to PyMT female mice, and the treatment significantly increased tumor latency and inhibited the development of tumors. D-4F was unable to reduce the levels of 27-HC in the tumors, but it decreased the plasma levels of oxLDL and prevented the oxLDL-mediated proliferative response in MCF-7 cells, suggesting that D-4F inhibited breast cancer by protecting against LDL oxidative modifications [95].

8. Concluding Remarks

Results of some large clinical studies indicate a direct association for LDL-C and an inverse association for HDL-C and breast cancer risk; however, these findings have not been reproduced in all epidemiological studies and are still debated. Basic research studies have determined the important role of cholesterol, especially the 27-HC metabolite, and its transporters in breast cancer development. Both LDL and HDL, and their modified forms (oxLDL and oxidized and glycated HDL), may promote breast cancer via several mechanisms. Investigations in breast cancer cells and experimental models in vivo have demonstrated an interplay among modified lipoproteins, proinflammatory signaling pathways, and breast cancer tumorigenic processes (summarized in Figure 2). Cholesterol can be esterified or metabolized to 27-HC, which has been hypothesized to be responsible for stimulating the proliferation of ER-positive breast cancer cells rather than cholesterol (Figure 2). Oxidized LDL as well as oxidized and glycated HDL induce different OLR1 and SR-BI downstream inflammation-related pathways, thereby inhibiting apoptosis and enhancing cell proliferation and migration. Therefore, considering the important role of cholesterol in breast cancer development, cholesterol-lowering drugs and apoA-I mimetics, which possess antioxidant and anti-inflammatory properties, could emerge as potential therapies for preventing the deleterious effects of high cholesterol in breast cancer. Lipophilic statins seem a good strategy for protecting against breast cancer recurrence and death. However, more studies in humans are necessary to evaluate the role of other therapies, such as ezetimibe, phytosterols or fibrates, on breast cancer risk and prognosis.

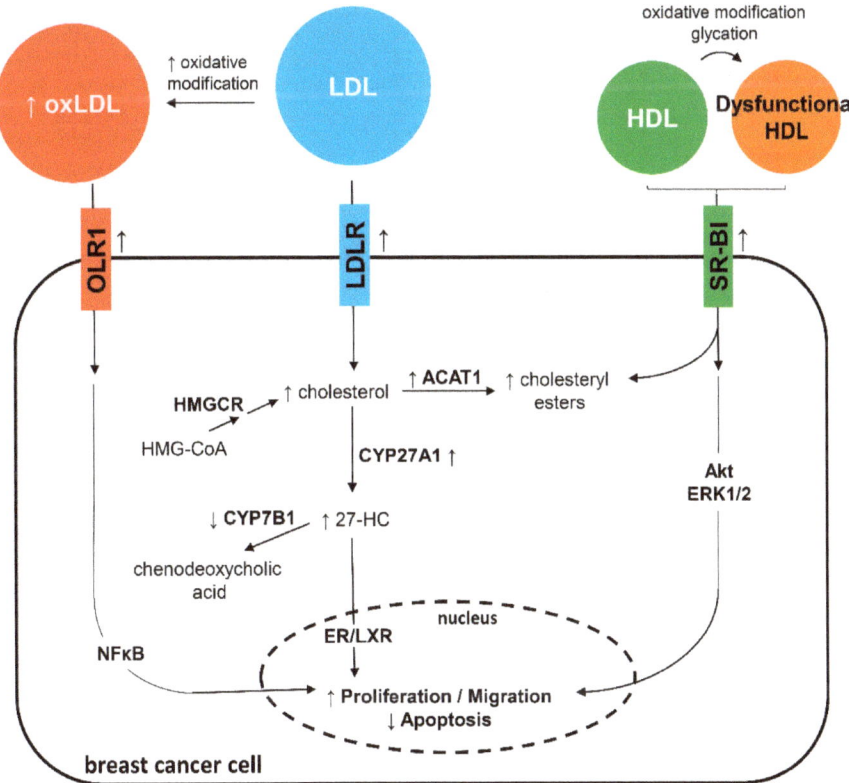

Figure 2. Mechanisms by which low-density lipoprotein (LDL), high-density lipoprotein (HDL), and their modified forms induce proliferation and migration and reduce apoptosis in breast cancer cells. OLR1 = OxLDL lecithin-like receptor 1, LDLR = LDL receptor, SR-BI = scavenger receptor class B type I, HMGCR = hydroxy-methyl-glutaryl-coenzyme A reductase, ACAT1 = acyl-CoA:cholesterol acyltransferase 1, 27-HC = 27-hydroxycholesterol, ERK1/2 = extracellular signal-regulated kinases $\frac{1}{2}$, NFκB = nuclear factor κB, and ER/LXR = estrogen receptor/liver X receptor.

Author Contributions: L.C., F.B.-V., and J.C.E.-G. wrote the manuscript. L.C. and J.C.E.-G. designed the figures and tables. E.M. and S.T.R. conducted a critical review of the manuscript and contributed to its final version.

Funding: This work was partly funded by the Instituto de Salud Carlos III and FEDER "Una manera de hacer Europa", including grants FIS 18/00164 (to F.B.-V.), FIS 16/00139 (to J.C.E-G.), and grant 12/C/2015 from La Fundació la Marató TV3 (to F.B-V.). CIBERDEM and CIBEROBN are Instituto de Salud Carlos III projects.

Conflicts of Interest: The authors declare no conflict of interest.

References

1. Global Burden of Disease Cancer Collaboration; Fitzmaurice, C.; Akinyemiju, T.F.; Al Lami, F.H.; Alam, T.; Alizadeh-Navaei, R.; Allen, C.; Alsharif, U.; Alvis-Guzman, N.; Amini, E.; et al. Global, Regional, and National Cancer Incidence, Mortality, Years of Life Lost, Years Lived With Disability, and Disability-Adjusted Life-years for 29 Cancer Groups, 1990 to 2016: A Systematic Analysis for the Global Burden of Disease Study. *JAMA Oncol.* **2018**, *4*, 1553–1568. [PubMed]
2. Jemal, A.; Bray, F.; Center, M.M.; Ferlay, J.; Ward, E.; Forman, D. Global cancer statistics. *CA Cancer J. Clin.* **2011**, *61*, 69–90. [CrossRef] [PubMed]
3. Torre, L.A.; Bray, F.; Siegel, R.L.; Ferlay, J.; Lortet-Tieulent, J.; Jemal, A. Global cancer statistics, 2012. *CA Cancer J. Clin.* **2015**, *65*, 87–108. [CrossRef] [PubMed]
4. Grundy, S.M. Metabolic complications of obesity. *Endocrine* **2000**, *13*, 155–165. [CrossRef]
5. Yung, R.L.; Ligibel, J.A. Obesity and breast cancer: Risk, outcomes, and future considerations. *Clin. Adv. Hematol. Oncol.* **2016**, *14*, 790–797. [PubMed]
6. Park, J.; Morley, T.S.; Kim, M.; Clegg, D.J.; Scherer, P.E. Obesity and cancer—mechanisms underlying tumour progression and recurrence. *Nat. Rev. Endocrinol.* **2014**, *10*, 455–465. [CrossRef] [PubMed]
7. Santos, C.R.; Schulze, A. Lipid metabolism in cancer. *FEBS J.* **2012**, *279*, 2610–2623. [CrossRef] [PubMed]
8. Lane, D.M.; Boatman, K.K.; McConathy, W.J. Serum lipids and apolipoproteins in women with breast masses. *Breast Cancer Res. Treat.* **1995**, *34*, 161–169. [CrossRef]
9. Touvier, M.; Fassier, P.; His, M.; Norat, T.; Chan, D.S.M.; Blacher, J.; Hercberg, S.; Galan, P.; Druesne-Pecollo, N.; Latino-Martel, P. Cholesterol and breast cancer risk: A systematic review and meta-analysis of prospective studies. *Br. J. Nutr.* **2015**, *114*, 347–357. [CrossRef]
10. Chandler, P.D.; Song, Y.; Lin, J.; Zhang, S.; Sesso, H.D.; Mora, S.; Giovannucci, E.L.; Rexrode, K.E.; Moorthy, M.V.; Li, C.; et al. Lipid biomarkers and long-term risk of cancer in the Women's Health Study. *Am. J. Clin. Nutr.* **2016**, *103*, 1397–1407. [CrossRef]
11. Nowak, C.; Ärnlöv, J. A Mendelian randomization study of the effects of blood lipids on breast cancer risk. *Nat. Commun.* **2018**, *9*, 3957. [CrossRef]
12. Jafri, H.; Alsheikh-Ali, A.A.; Karas, R.H. Baseline and on-treatment high-density lipoprotein cholesterol and the risk of cancer in randomized controlled trials of lipid-altering therapy. *J. Am. Coll. Cardiol.* **2010**, *55*, 2846–2854. [CrossRef] [PubMed]
13. Kitahara, C.M.; Berrington de González, A.; Freedman, N.D.; Huxley, R.; Mok, Y.; Jee, S.H.; Samet, J.M. Total cholesterol and cancer risk in a large prospective study in Korea. *J. Clin. Oncol. Off. J. Am. Soc. Clin. Oncol.* **2011**, *29*, 1592–1598. [CrossRef] [PubMed]
14. Martin, L.J.; Melnichouk, O.; Huszti, E.; Connelly, P.W.; Greenberg, C.V.; Minkin, S.; Boyd, N.F. Serum Lipids, Lipoproteins, and Risk of Breast Cancer: A Nested Case-Control Study Using Multiple Time Points. *J. Natl. Cancer Inst.* **2015**, *107*, djv032. [CrossRef] [PubMed]
15. Ha, M.; Sung, J.; Song, Y.-M. Serum total cholesterol and the risk of breast cancer in postmenopausal Korean women. *Cancer Causes Control* **2009**, *20*, 1055–1060. [CrossRef] [PubMed]
16. Ni, H.; Liu, H.; Gao, R. Serum Lipids and Breast Cancer Risk: A Meta-Analysis of Prospective Cohort Studies. *PLoS ONE* **2015**, *10*, e0142669. [CrossRef] [PubMed]
17. Bosco, J.L.F.; Palmer, J.R.; Boggs, D.A.; Hatch, E.E.; Rosenberg, L. Cardiometabolic factors and breast cancer risk in U.S. black women. *Breast Cancer Res. Treat.* **2012**, *134*, 1247–1256. [CrossRef] [PubMed]
18. Eliassen, A.H.; Colditz, G.A.; Rosner, B.; Willett, W.C.; Hankinson, S.E. Serum lipids, lipid-lowering drugs, and the risk of breast cancer. *Arch. Intern. Med.* **2005**, *165*, 2264–2271. [CrossRef] [PubMed]

19. His, M.; Zelek, L.; Deschasaux, M.; Pouchieu, C.; Kesse-Guyot, E.; Hercberg, S.; Galan, P.; Latino-Martel, P.; Blacher, J.; Touvier, M. Prospective associations between serum biomarkers of lipid metabolism and overall, breast and prostate cancer risk. *Eur. J. Epidemiol.* **2014**, *29*, 119–132. [CrossRef] [PubMed]
20. Rodrigues Dos Santos, C.; Fonseca, I.; Dias, S.; Mendes de Almeida, J.C. Plasma level of LDL-cholesterol at diagnosis is a predictor factor of breast tumor progression. *BMC Cancer* **2014**, *14*, 132. [CrossRef] [PubMed]
21. Borgquist, S.; Butt, T.; Almgren, P.; Shiffman, D.; Stocks, T.; Orho-Melander, M.; Manjer, J.; Melander, O. Apolipoproteins, lipids and risk of cancer. *Int. J. Cancer J. Int. Cancer* **2016**, *138*, 2648–2656. [CrossRef] [PubMed]
22. Llanos, A.A.; Makambi, K.H.; Tucker, C.A.; Wallington, S.F.; Shields, P.G.; Adams-Campbell, L.L. Cholesterol, lipoproteins, and breast cancer risk in African American women. *Ethn. Dis.* **2012**, *22*, 281–287. [PubMed]
23. Li, X.; Tang, H.; Wang, J.; Xie, X.; Liu, P.; Kong, Y.; Ye, F.; Shuang, Z.; Xie, Z.; Xie, X. The effect of preoperative serum triglycerides and high-density lipoprotein-cholesterol levels on the prognosis of breast cancer. *Breast Edinb. Scotl.* **2017**, *32*, 1–6. [CrossRef] [PubMed]
24. His, M.; Dartois, L.; Fagherazzi, G.; Boutten, A.; Dupré, T.; Mesrine, S.; Boutron-Ruault, M.-C.; Clavel-Chapelon, F.; Dossus, L. Associations between serum lipids and breast cancer incidence and survival in the E3N prospective cohort study. *Cancer Causes Control CCC* **2017**, *28*, 77–88. [CrossRef] [PubMed]
25. Kucharska-Newton, A.M.; Rosamond, W.D.; Mink, P.J.; Alberg, A.J.; Shahar, E.; Folsom, A.R. HDL-cholesterol and incidence of breast cancer in the ARIC cohort study. *Ann. Epidemiol.* **2008**, *18*, 671–677. [CrossRef] [PubMed]
26. Kim, Y.; Park, S.K.; Han, W.; Kim, D.-H.; Hong, Y.-C.; Ha, E.H.; Ahn, S.-H.; Noh, D.-Y.; Kang, D.; Yoo, K.-Y. Serum high-density lipoprotein cholesterol and breast cancer risk by menopausal status, body mass index, and hormonal receptor in Korea. *Cancer Epidemiol. Biomark. Prev. Publ. Am. Assoc. Cancer Res. Cosponsored Am. Soc. Prev. Oncol.* **2009**, *18*, 508–515. [CrossRef] [PubMed]
27. Furberg, A.-S.; Veierød, M.B.; Wilsgaard, T.; Bernstein, L.; Thune, I. Serum high-density lipoprotein cholesterol, metabolic profile, and breast cancer risk. *J. Natl. Cancer Inst.* **2004**, *96*, 1152–1160. [CrossRef] [PubMed]
28. Li, X.; Liu, Z.-L.; Wu, Y.-T.; Wu, H.; Dai, W.; Arshad, B.; Xu, Z.; Li, H.; Wu, K.-N.; Kong, L.-Q. Status of lipid and lipoprotein in female breast cancer patients at initial diagnosis and during chemotherapy. *Lipids Health Dis.* **2018**, *17*, 91. [CrossRef] [PubMed]
29. Yadav, N.K.; Poudel, B.; Thanpari, C.; Chandra Koner, B. Assessment of biochemical profiles in premenopausal and postmenopausal women with breast cancer. *Asian Pac. J. Cancer Prev. APJCP* **2012**, *13*, 3385–3388. [CrossRef]
30. Owiredu, W.K.B.A.; Donkor, S.; Addai, B.W.; Amidu, N. Serum lipid profile of breast cancer patients. *Pak. J. Biol. Sci.* **2009**, *12*, 332–338. [CrossRef]
31. Michalaki, V.; Koutroulis, G.; Syrigos, K.; Piperi, C.; Kalofoutis, A. Evaluation of serum lipids and high-density lipoprotein subfractions (HDL2, HDL3) in postmenopausal patients with breast cancer. *Mol. Cell. Biochem.* **2005**, *268*, 19–24. [CrossRef] [PubMed]
32. Kotepui, M. Diet and risk of breast cancer. *Contemp. Oncol. Pozn. Pol.* **2016**, *20*, 13–19. [CrossRef] [PubMed]
33. Hu, J.; La Vecchia, C.; de Groh, M.; Negri, E.; Morrison, H.; Mery, L.; Canadian Cancer Registries Epidemiology Research Group. Dietary cholesterol intake and cancer. *Ann. Oncol. Off. J. Eur. Soc. Med. Oncol.* **2012**, *23*, 491–500. [CrossRef] [PubMed]
34. Li, C.; Yang, L.; Zhang, D.; Jiang, W. Systematic review and meta-analysis suggest that dietary cholesterol intake increases risk of breast cancer. *Nutr. Res.* **2016**, *36*, 627–635. [CrossRef] [PubMed]
35. Cleary, M.P.; Grande, J.P.; Maihle, N.J. Effect of high fat diet on body weight and mammary tumor latency in MMTV-TGF-alpha mice. *Int. J. Obes. Relat. Metab. Disord. J. Int. Assoc. Study Obes.* **2004**, *28*, 956–962. [CrossRef] [PubMed]
36. Guy, C.T.; Cardiff, R.D.; Muller, W.J. Induction of mammary tumors by expression of polyomavirus middle T oncogene: A transgenic mouse model for metastatic disease. *Mol. Cell. Biol.* **1992**, *12*, 954–961. [CrossRef]
37. Lin, E.Y.; Jones, J.G.; Li, P.; Zhu, L.; Whitney, K.D.; Muller, W.J.; Pollard, J.W. Progression to malignancy in the polyoma middle T oncoprotein mouse breast cancer model provides a reliable model for human diseases. *Am. J. Pathol.* **2003**, *163*, 2113–2126. [CrossRef]
38. Dogan, S.; Hu, X.; Zhang, Y.; Maihle, N.J.; Grande, J.P.; Cleary, M.P. Effects of high-fat diet and/or body weight on mammary tumor leptin and apoptosis signaling pathways in MMTV-TGF-α mice. *Breast Cancer Res.* **2007**, *9*, R91. [CrossRef]

39. Llaverias, G.; Danilo, C.; Mercier, I.; Daumer, K.; Capozza, F.; Williams, T.M.; Sotgia, F.; Lisanti, M.P.; Frank, P.G. Role of cholesterol in the development and progression of breast cancer. *Am. J. Pathol.* **2011**, *178*, 402–412. [CrossRef]
40. Pelton, K.; Coticchia, C.M.; Curatolo, A.S.; Schaffner, C.P.; Zurakowski, D.; Solomon, K.R.; Moses, M.A. Hypercholesterolemia induces angiogenesis and accelerates growth of breast tumors in vivo. *Am. J. Pathol.* **2014**, *184*, 2099–2110. [CrossRef]
41. Alikhani, N.; Ferguson, R.D.; Novosyadlyy, R.; Gallagher, E.J.; Scheinman, E.J.; Yakar, S.; LeRoith, D. Mammary tumor growth and pulmonary metastasis are enhanced in a hyperlipidemic mouse model. *Oncogene* **2013**, *32*, 961–967. [CrossRef] [PubMed]
42. Zhang, S.H.; Reddick, R.L.; Piedrahita, J.A.; Maeda, N. Spontaneous hypercholesterolemia and arterial lesions in mice lacking apolipoprotein E. *Science* **1992**, *258*, 468–471. [CrossRef] [PubMed]
43. Constantinou, C.; Mpatsoulis, D.; Natsos, A.; Petropoulou, P.-I.; Zvintzou, E.; Traish, A.M.; Voshol, P.J.; Karagiannides, I.; Kypreos, K.E. The low density lipoprotein receptor modulates the effects of hypogonadism on diet-induced obesity and related metabolic perturbations. *J. Lipid Res.* **2014**, *55*, 1434–1447. [CrossRef] [PubMed]
44. McDonnell, D.P.; Park, S.; Goulet, M.T.; Jasper, J.; Wardell, S.E.; Chang, C.-Y.; Norris, J.D.; Guyton, J.R.; Nelson, E.R. Obesity, cholesterol metabolism, and breast cancer pathogenesis. *Cancer Res.* **2014**, *74*, 4976–4982. [CrossRef] [PubMed]
45. Nelson, E.R.; Wardell, S.E.; Jasper, J.S.; Park, S.; Suchindran, S.; Howe, M.K.; Carver, N.J.; Pillai, R.V.; Sullivan, P.M.; Sondhi, V.; et al. 27-Hydroxycholesterol links hypercholesterolemia and breast cancer pathophysiology. *Science* **2013**, *342*, 1094–1098. [CrossRef] [PubMed]
46. Dos Santos, C.R.; Domingues, G.; Matias, I.; Matos, J.; Fonseca, I.; de Almeida, J.M.; Dias, S. LDL-cholesterol signaling induces breast cancer proliferation and invasion. *Lipids Health Dis.* **2014**, *13*, 16. [CrossRef]
47. Jensen, E.V.; Jordan, V.C. The estrogen receptor: A model for molecular medicine. *Clin. Cancer Res. Off. J. Am. Assoc. Cancer Res.* **2003**, *9*, 1980–1989.
48. Umetani, M.; Shaul, P.W. 27-Hydroxycholesterol: The first identified endogenous SERM. *Trends Endocrinol. Metab.* **2011**, *22*, 130–135. [CrossRef]
49. Burkard, I.; von Eckardstein, A.; Waeber, G.; Vollenweider, P.; Rentsch, K.M. Lipoprotein distribution and biological variation of 24S- and 27-hydroxycholesterol in healthy volunteers. *Atherosclerosis* **2007**, *194*, 71–78. [CrossRef]
50. Russell, D.W. Oxysterol biosynthetic enzymes. *Biochim. Biophys. Acta BBA Mol. Cell Biol. Lipids* **2000**, *1529*, 126–135. [CrossRef]
51. Cruz, P.; Torres, C.; Ramírez, M.E.; Epuñán, M.J.; Valladares, L.E.; Sierralta, W.D. Proliferation of human mammary cancer cells exposed to 27-hydroxycholesterol. *Exp. Ther. Med.* **2010**, *1*, 531–536. [CrossRef] [PubMed]
52. Wu, Q.; Ishikawa, T.; Sirianni, R.; Tang, H.; McDonald, J.G.; Yuhanna, I.S.; Thompson, B.; Girard, L.; Mineo, C.; Brekken, R.A.; et al. 27-Hydroxycholesterol promotes cell-autonomous, ER-positive breast cancer growth. *Cell Rep.* **2013**, *5*, 637–645. [CrossRef] [PubMed]
53. Baek, A.E.; Yu, Y.-R.A.; He, S.; Wardell, S.E.; Chang, C.-Y.; Kwon, S.; Pillai, R.V.; McDowell, H.B.; Thompson, J.W.; Dubois, L.G.; et al. The cholesterol metabolite 27 hydroxycholesterol facilitates breast cancer metastasis through its actions on immune cells. *Nat. Commun.* **2017**, *8*, 864. [CrossRef] [PubMed]
54. Lu, D.-L.; Le Cornet, C.; Sookthai, D.; Johnson, T.S.; Kaaks, R.; Fortner, R.T. Circulating 27-Hydroxycholesterol and Breast Cancer Risk: Results From the EPIC-Heidelberg Cohort. *J. Natl. Cancer Inst.* **2019**, *111*, 365–371. [CrossRef] [PubMed]
55. Raza, S.; Ohm, J.E.; Dhasarathy, A.; Schommer, J.; Roche, C.; Hammer, K.D.P.; Ghribi, O. The cholesterol metabolite 27-hydroxycholesterol regulates p53 activity and increases cell proliferation via MDM2 in breast cancer cells. *Mol. Cell. Biochem.* **2015**, *410*, 187–195. [CrossRef] [PubMed]
56. Ma, L.-M.; Liang, Z.-R.; Zhou, K.-R.; Zhou, H.; Qu, L.-H. 27-Hydroxycholesterol increases Myc protein stability via suppressing PP2A, SCP1 and FBW7 transcription in MCF-7 breast cancer cells. *Biochem. Biophys. Res. Commun.* **2016**, *480*, 328–333. [CrossRef]
57. Hu, B.; Zhang, K.; Li, S.; Li, H.; Yan, Z.; Huang, L.; Wu, J.; Han, X.; Jiang, W.; Mulatibieke, T.; et al. HIC1 attenuates invasion and metastasis by inhibiting the IL-6/STAT3 signalling pathway in human pancreatic cancer. *Cancer Lett.* **2016**, *376*, 387–398. [CrossRef]

58. Zhu, D.; Shen, Z.; Liu, J.; Chen, J.; Liu, Y.; Hu, C.; Li, Z.; Li, Y. The ROS-mediated activation of STAT-3/VEGF signaling is involved in the 27-hydroxycholesterol-induced angiogenesis in human breast cancer cells. *Toxicol. Lett.* **2016**, *264*, 79–86. [CrossRef]
59. Torres, C.G.; Ramírez, M.E.; Cruz, P.; Epuñan, M.J.; Valladares, L.E.; Sierralta, W.D. 27-hydroxycholesterol induces the transition of MCF7 cells into a mesenchymal phenotype. *Oncol. Rep.* **2011**, *26*, 389–397.
60. Shen, Z.; Zhu, D.; Liu, J.; Chen, J.; Liu, Y.; Hu, C.; Li, Z.; Li, Y. 27-Hydroxycholesterol induces invasion and migration of breast cancer cells by increasing MMP9 and generating EMT through activation of STAT-3. *Environ. Toxicol. Pharmacol.* **2017**, *51*, 1–8. [CrossRef]
61. Umetani, M.; Ghosh, P.; Ishikawa, T.; Umetani, J.; Ahmed, M.; Mineo, C.; Shaul, P.W. The cholesterol metabolite 27-hydroxycholesterol promotes atherosclerosis via proinflammatory processes mediated by estrogen receptor alpha. *Cell Metab.* **2014**, *20*, 172–182. [CrossRef] [PubMed]
62. Kaiser, J. Cholesterol forges link between obesity and breast cancer. *Science* **2013**, *342*, 1028. [CrossRef] [PubMed]
63. DuSell, C.D.; Umetani, M.; Shaul, P.W.; Mangelsdorf, D.J.; McDonnell, D.P. 27-hydroxycholesterol is an endogenous selective estrogen receptor modulator. *Mol. Endocrinol.* **2008**, *22*, 65–77. [CrossRef] [PubMed]
64. Lee, W.-R.; Ishikawa, T.; Umetani, M. The interaction between metabolism, cancer and cardiovascular disease, connected by 27-hydroxycholesterol. *Clin. Lipidol.* **2014**, *9*, 617–624. [CrossRef] [PubMed]
65. Pires, L.A.; Hegg, R.; Freitas, F.R.; Tavares, E.R.; Almeida, C.P.; Baracat, E.C.; Maranhão, R.C. Effect of neoadjuvant chemotherapy on low-density lipoprotein (LDL) receptor and LDL receptor-related protein 1 (LRP-1) receptor in locally advanced breast cancer. *Braz. J. Med. Biol. Res. Rev. Bras. Pesqui. Medicas E Biol.* **2012**, *45*, 557–564. [CrossRef] [PubMed]
66. Stranzl, A.; Schmidt, H.; Winkler, R.; Kostner, G.M. Low-density lipoprotein receptor mRNA in human breast cancer cells: Influence by PKC modulators. *Breast Cancer Res. Treat.* **1997**, *42*, 195–205. [CrossRef] [PubMed]
67. Gallagher, E.J.; Zelenko, Z.; Neel, B.A.; Antoniou, I.M.; Rajan, L.; Kase, N.; LeRoith, D. Elevated tumor LDLR expression accelerates LDL cholesterol-mediated breast cancer growth in mouse models of hyperlipidemia. *Oncogene* **2017**, *36*, 6462–6471. [CrossRef]
68. Antalis, C.J.; Arnold, T.; Rasool, T.; Lee, B.; Buhman, K.K.; Siddiqui, R.A. High ACAT1 expression in estrogen receptor negative basal-like breast cancer cells is associated with LDL-induced proliferation. *Breast Cancer Res. Treat.* **2010**, *122*, 661–670. [CrossRef]
69. Rotheneder, M.; Kostner, G.M. Effects of low- and high-density lipoproteins on the proliferation of human breast cancer cells in vitro: Differences between hormone-dependent and hormone-independent cell lines. *Int. J. Cancer* **1989**, *43*, 875–879. [CrossRef]
70. Lu, C.-W.; Lo, Y.-H.; Chen, C.-H.; Lin, C.-Y.; Tsai, C.-H.; Chen, P.-J.; Yang, Y.-F.; Wang, C.-H.; Tan, C.-H.; Hou, M.-F.; et al. VLDL and LDL, but not HDL, promote breast cancer cell proliferation, metastasis and angiogenesis. *Cancer Lett.* **2017**, *388*, 130–138. [CrossRef]
71. Antalis, C.J.; Uchida, A.; Buhman, K.K.; Siddiqui, R.A. Migration of MDA-MB-231 breast cancer cells depends on the availability of exogenous lipids and cholesterol esterification. *Clin. Exp. Metastasis* **2011**, *28*, 733–741. [CrossRef] [PubMed]
72. Blackburn, G.L.; Wang, K.A. Dietary fat reduction and breast cancer outcome: Results from the Women's Intervention Nutrition Study (WINS). *Am. J. Clin. Nutr.* **2007**, *86*, s878–s881. [CrossRef] [PubMed]
73. Neve, R.M.; Chin, K.; Fridlyand, J.; Yeh, J.; Baehner, F.L.; Fevr, T.; Clark, L.; Bayani, N.; Coppe, J.-P.; Tong, F.; et al. A collection of breast cancer cell lines for the study of functionally distinct cancer subtypes. *Cancer Cell* **2006**, *10*, 515–527. [CrossRef] [PubMed]
74. Cornejo, K.M.; Kandil, D.; Khan, A.; Cosar, E.F. Theranostic and molecular classification of breast cancer. *Arch. Pathol. Lab. Med.* **2014**, *138*, 44–56. [CrossRef] [PubMed]
75. De Gonzalo-Calvo, D.; López-Vilaró, L.; Nasarre, L.; Perez-Olabarria, M.; Vázquez, T.; Escuin, D.; Badimon, L.; Barnadas, A.; Lerma, E.; Llorente-Cortés, V. Intratumor cholesteryl ester accumulation is associated with human breast cancer proliferation and aggressive potential: A molecular and clinicopathological study. *BMC Cancer* **2015**, *15*, 460. [CrossRef]
76. Sánchez-Pérez, Y.; Carrasco-Legleu, C.; García-Cuellar, C.; Pérez-Carreón, J.; Hernández-García, S.; Salcido-Neyoy, M.; Alemán-Lazarini, L.; Villa-Treviño, S. Oxidative stress in carcinogenesis. Correlation between lipid peroxidation and induction of preneoplastic lesions in rat hepatocarcinogenesis. *Cancer Lett.* **2005**, *217*, 25–32. [CrossRef]

77. Wiseman, H.; Halliwell, B. Damage to DNA by reactive oxygen and nitrogen species: Role in inflammatory disease and progression to cancer. *Biochem. J.* **1996**, *313 Pt 1*, 17–29. [CrossRef]
78. Delimaris, I.; Faviou, E.; Antonakos, G.; Stathopoulou, E.; Zachari, A.; Dionyssiou-Asteriou, A. Oxidized LDL, serum oxidizability and serum lipid levels in patients with breast or ovarian cancer. *Clin. Biochem.* **2007**, *40*, 1129–1134. [CrossRef]
79. Khaidakov, M.; Mehta, J.L. Oxidized LDL triggers pro-oncogenic signaling in human breast mammary epithelial cells partly via stimulation of MiR-21. *PLoS ONE* **2012**, *7*, e46973. [CrossRef]
80. Pucci, S.; Polidoro, C.; Greggi, C.; Amati, F.; Morini, E.; Murdocca, M.; Biancolella, M.; Orlandi, A.; Sangiuolo, F.; Novelli, G. Pro-oncogenic action of LOX-1 and its splice variant LOX-1Δ4 in breast cancer phenotypes. *Cell Death Dis.* **2019**, *10*, 53. [CrossRef]
81. Khaidakov, M.; Mitra, S.; Kang, B.-Y.; Wang, X.; Kadlubar, S.; Novelli, G.; Raj, V.; Winters, M.; Carter, W.C.; Mehta, J.L. Oxidized LDL receptor 1 (OLR1) as a possible link between obesity, dyslipidemia and cancer. *PLoS ONE* **2011**, *6*, e20277. [CrossRef] [PubMed]
82. Liang, M.; Zhang, P.; Fu, J. Up-regulation of LOX-1 expression by TNF-α promotes trans-endothelial migration of MDA-MB-231 breast cancer cells. *Cancer Lett.* **2007**, *258*, 31–37. [CrossRef] [PubMed]
83. Hirsch, H.A.; Iliopoulos, D.; Joshi, A.; Zhang, Y.; Jaeger, S.A.; Bulyk, M.; Tsichlis, P.N.; Shirley Liu, X.; Struhl, K. A transcriptional signature and common gene networks link cancer with lipid metabolism and diverse human diseases. *Cancer Cell* **2010**, *17*, 348–361. [CrossRef] [PubMed]
84. Wang, B.; Zhao, H.; Zhao, L.; Zhang, Y.; Wan, Q.; Shen, Y.; Bu, X.; Wan, M.; Shen, C. Up-regulation of OLR1 expression by TBC1D3 through activation of TNFα/NF-κB pathway promotes the migration of human breast cancer cells. *Cancer Lett.* **2017**, *408*, 60–70. [CrossRef] [PubMed]
85. Gospodarowicz, D.; Lui, G.M.; Gonzalez, R. High-density lipoproteins and the proliferation of human tumor cells maintained on extracellular matrix-coated dishes and exposed to defined medium. *Cancer Res.* **1982**, *42*, 3704–3713. [CrossRef]
86. Danilo, C.; Gutierrez-Pajares, J.L.; Mainieri, M.A.; Mercier, I.; Lisanti, M.P.; Frank, P.G. Scavenger receptor class B type I regulates cellular cholesterol metabolism and cell signaling associated with breast cancer development. *Breast Cancer Res.* **2013**, *15*, R87. [CrossRef] [PubMed]
87. Pussinen, P.J.; Karten, B.; Wintersperger, A.; Reicher, H.; McLean, M.; Malle, E.; Sattler, W. The human breast carcinoma cell line HBL-100 acquires exogenous cholesterol from high-density lipoprotein via CLA-1 (CD-36 and LIMPII analogous 1)-mediated selective cholesteryl ester uptake. *Biochem. J.* **2000**, *349*, 559–566. [CrossRef] [PubMed]
88. Cao, W.M.; Murao, K.; Imachi, H.; Yu, X.; Abe, H.; Yamauchi, A.; Niimi, M.; Miyauchi, A.; Wong, N.C.W.; Ishida, T. A mutant high-density lipoprotein receptor inhibits proliferation of human breast cancer cells. *Cancer Res.* **2004**, *64*, 1515–1521. [CrossRef]
89. Yuan, B.; Wu, C.; Wang, X.; Wang, D.; Liu, H.; Guo, L.; Li, X.-A.; Han, J.; Feng, H. High scavenger receptor class B type I expression is related to tumor aggressiveness and poor prognosis in breast cancer. *Tumor Biol.* **2016**, *37*, 3581–3588. [CrossRef]
90. Li, J.; Wang, J.; Li, M.; Yin, L.; Li, X.-A.; Zhang, T.-G. Up-regulated expression of scavenger receptor class B type 1 (SR-B1) is associated with malignant behaviors and poor prognosis of breast cancer. *Pathol. Res. Pract.* **2016**, *212*, 555–559. [CrossRef]
91. Lee-Rueckert, M.; Escola-Gil, J.C.; Kovanen, P.T. HDL functionality in reverse cholesterol transport—Challenges in translating data emerging from mouse models to human disease. *Biochim. Biophys. Acta Mol. Cell Biol. Lipids* **2016**, *1861*, 566–583. [CrossRef] [PubMed]
92. Mineo, C.; Shaul, P.W. Novel Biological Functions of High-Density Lipoprotein Cholesterol. *Circ. Res.* **2012**, *111*, 1079–1090. [CrossRef] [PubMed]
93. Huang, H.-L.; Stasyk, T.; Morandell, S.; Dieplinger, H.; Falkensammer, G.; Griesmacher, A.; Mogg, M.; Schreiber, M.; Feuerstein, I.; Huck, C.W.; et al. Biomarker discovery in breast cancer serum using 2-D differential gel electrophoresis/MALDI-TOF/TOF and data validation by routine clinical assays. *Electrophoresis* **2006**, *27*, 1641–1650. [CrossRef] [PubMed]
94. Chang, S.-J.; Hou, M.-F.; Tsai, S.-M.; Wu, S.-H.; Hou, L.A.; Ma, H.; Shann, T.-Y.; Wu, S.-H.; Tsai, L.-Y. The association between lipid profiles and breast cancer among Taiwanese women. *Clin. Chem. Lab. Med.* **2007**, *45*, 1219–1223. [CrossRef] [PubMed]

95. Cedó, L.; García-León, A.; Baila-Rueda, L.; Santos, D.; Grijalva, V.; Martínez-Cignoni, M.R.; Carbó, J.M.; Metso, J.; López-Vilaró, L.; Zorzano, A.; et al. ApoA-I mimetic administration, but not increased apoA-I-containing HDL, inhibits tumour growth in a mouse model of inherited breast cancer. *Sci. Rep.* **2016**, *6*, 36387. [CrossRef] [PubMed]
96. Blanco-Vaca, F.; Escolà-Gil, J.C.; Martín-Campos, J.M.; Julve, J. Role of apoA-II in lipid metabolism and atherosclerosis: Advances in the study of an enigmatic protein. *J. Lipid Res.* **2001**, *42*, 1727–1739. [PubMed]
97. Julve, J.; Escolà-Gil, J.C.; Rotllan, N.; Fiévet, C.; Vallez, E.; de la Torre, C.; Ribas, V.; Sloan, J.H.; Blanco-Vaca, F. Human apolipoprotein A-II determines plasma triglycerides by regulating lipoprotein lipase activity and high-density lipoprotein proteome. *Arterioscler. Thromb. Vasc. Biol.* **2010**, *30*, 232–238. [CrossRef]
98. Ribas, V.; Sánchez-Quesada, J.L.; Antón, R.; Camacho, M.; Julve, J.; Escolà-Gil, J.C.; Vila, L.; Ordóñez-Llanos, J.; Blanco-Vaca, F. Human apolipoprotein A-II enrichment displaces paraoxonase from HDL and impairs its antioxidant properties: A new mechanism linking HDL protein composition and antiatherogenic potential. *Circ. Res.* **2004**, *95*, 789–797. [CrossRef]
99. Pan, B.; Ren, H.; Lv, X.; Zhao, Y.; Yu, B.; He, Y.; Ma, Y.; Niu, C.; Kong, J.; Yu, F.; et al. Hypochlorite-induced oxidative stress elevates the capability of HDL in promoting breast cancer metastasis. *J. Transl. Med.* **2012**, *10*, 65. [CrossRef]
100. Kontush, A.; Chapman, M.J. Why is HDL functionally deficient in type 2 diabetes? *Curr. Diab. Rep.* **2008**, *8*, 51–59. [CrossRef]
101. Larsson, S.C.; Mantzoros, C.S.; Wolk, A. Diabetes mellitus and risk of breast cancer: A meta-analysis. *Int. J. Cancer* **2007**, *121*, 856–862. [CrossRef] [PubMed]
102. Pan, B.; Ren, H.; Ma, Y.; Liu, D.; Yu, B.; Ji, L.; Pan, L.; Li, J.; Yang, L.; Lv, X.; et al. High-density lipoprotein of patients with type 2 diabetes mellitus elevates the capability of promoting migration and invasion of breast cancer cells. *Int. J. Cancer* **2012**, *131*, 70–82. [CrossRef] [PubMed]
103. Pan, B.; Ren, H.; He, Y.; Lv, X.; Ma, Y.; Li, J.; Huang, L.; Yu, B.; Kong, J.; Niu, C.; et al. HDL of patients with type 2 diabetes mellitus elevates the capability of promoting breast cancer metastasis. *Clin. Cancer Res. Off. J. Am. Assoc. Cancer Res.* **2012**, *18*, 1246–1256. [CrossRef] [PubMed]
104. Huang, X.; He, D.; Ming, J.; He, Y.; Zhou, C.; Ren, H.; He, X.; Wang, C.; Jin, J.; Ji, L.; et al. High-density lipoprotein of patients with breast cancer complicated with type 2 diabetes mellitus promotes cancer cells adhesion to vascular endothelium via ICAM-1 and VCAM-1 upregulation. *Breast Cancer Res. Treat.* **2016**, *155*, 441–455. [CrossRef] [PubMed]
105. Goldstein, J.L.; Brown, M.S. Regulation of the mevalonate pathway. *Nature* **1990**, *343*, 425–430. [CrossRef] [PubMed]
106. Cauley, J.A.; Zmuda, J.M.; Lui, L.-Y.; Hillier, T.A.; Ness, R.B.; Stone, K.L.; Cummings, S.R.; Bauer, D.C. Lipid-lowering drug use and breast cancer in older women: A prospective study. *J. Womens Health* **2003**, *12*, 749–756. [CrossRef] [PubMed]
107. Anothaisintawee, T.; Udomsubpayakul, U.; McEvoy, M.; Lerdsitthichai, P.; Attia, J.; Thakkinstian, A. Effect of Lipophilic and Hydrophilic Statins on Breast Cancer Risk in Thai Women: A Cross-sectional Study. *J. Cancer* **2016**, *7*, 1163–1168. [CrossRef]
108. Bonovas, S.; Filioussi, K.; Tsavaris, N.; Sitaras, N.M. Use of statins and breast cancer: A meta-analysis of seven randomized clinical trials and nine observational studies. *J. Clin. Oncol. Off. J. Am. Soc. Clin. Oncol.* **2005**, *23*, 8606–8612. [CrossRef] [PubMed]
109. Islam, M.M.; Yang, H.-C.; Nguyen, P.-A.; Poly, T.N.; Huang, C.-W.; Kekade, S.; Khalfan, A.M.; Debnath, T.; Li, Y.-C.J.; Abdul, S.S. Exploring association between statin use and breast cancer risk: An updated meta-analysis. *Arch. Gynecol. Obstet.* **2017**, *296*, 1043–1053. [CrossRef]
110. Schairer, C.; Freedman, D.M.; Gadalla, S.M.; Pfeiffer, R.M. Lipid-lowering drugs, dyslipidemia, and breast cancer risk in a Medicare population. *Breast Cancer Res. Treat.* **2018**, *169*, 607–614. [CrossRef] [PubMed]
111. Undela, K.; Srikanth, V.; Bansal, D. Statin use and risk of breast cancer: A meta-analysis of observational studies. *Breast Cancer Res. Treat.* **2012**, *135*, 261–269. [CrossRef] [PubMed]
112. Dale, K.M.; Coleman, C.I.; Henyan, N.N.; Kluger, J.; White, C.M. Statins and cancer risk: A meta-analysis. *JAMA* **2006**, *295*, 74–80. [CrossRef] [PubMed]
113. Ference, B.A.; Ray, K.K.; Catapano, A.L.; Ference, T.B.; Burgess, S.; Neff, D.R.; Oliver-Williams, C.; Wood, A.M.; Butterworth, A.S.; Di Angelantonio, E.; et al. Mendelian Randomization Study of ACLY and Cardiovascular Disease. *N. Engl. J. Med.* **2019**, *380*, 1033–1042. [CrossRef]

114. McDougall, J.A.; Malone, K.E.; Daling, J.R.; Cushing-Haugen, K.L.; Porter, P.L.; Li, C.I. Long-Term Statin Use and Risk of Ductal and Lobular Breast Cancer among Women 55 to 74 Years of Age. *Cancer Epidemiol. Biomark. Prev.* **2013**, *22*, 1529–1537. [CrossRef] [PubMed]
115. Borgquist, S.; Giobbie-Hurder, A.; Ahern, T.P.; Garber, J.E.; Colleoni, M.; Láng, I.; Debled, M.; Ejlertsen, B.; von Moos, R.; Smith, I.; et al. Cholesterol, Cholesterol-Lowering Medication Use, and Breast Cancer Outcome in the BIG 1-98 Study. *J. Clin. Oncol. Off. J. Am. Soc. Clin. Oncol.* **2017**, *35*, 1179–1188. [CrossRef] [PubMed]
116. Mansourian, M.; Haghjooy-Javanmard, S.; Eshraghi, A.; Vaseghi, G.; Hayatshahi, A.; Thomas, J. Statins Use and Risk of Breast Cancer Recurrence and Death: A Systematic Review and Meta-Analysis of Observational Studies. *J. Pharm. Pharm. Sci. Publ. Can. Soc. Pharm. Sci. Soc. Can. Sci. Pharm.* **2016**, *19*, 72–81. [CrossRef] [PubMed]
117. Sakellakis, M.; Akinosoglou, K.; Kostaki, A.; Spyropoulou, D.; Koutras, A. Statins and risk of breast cancer recurrence. *Breast Cancer Dove Med. Press* **2016**, *8*, 199–205. [PubMed]
118. Chae, Y.K.; Valsecchi, M.E.; Kim, J.; Bianchi, A.L.; Khemasuwan, D.; Desai, A.; Tester, W. Reduced risk of breast cancer recurrence in patients using ACE inhibitors, ARBs, and/or statins. *Cancer Investig.* **2011**, *29*, 585–593. [CrossRef] [PubMed]
119. Wu, Q.-J.; Tu, C.; Li, Y.-Y.; Zhu, J.; Qian, K.-Q.; Li, W.-J.; Wu, L. Statin use and breast cancer survival and risk: A systematic review and meta-analysis. *Oncotarget* **2015**, *6*, 42988–43004. [CrossRef] [PubMed]
120. Shaitelman, S.F.; Stauder, M.C.; Allen, P.; Reddy, S.; Lakoski, S.; Atkinson, B.; Reddy, J.; Amaya, D.; Guerra, W.; Ueno, N.; et al. Impact of Statin Use on Outcomes in Triple Negative Breast Cancer. *J. Cancer* **2017**, *8*, 2026–2032. [CrossRef]
121. Smith, A.; Murphy, L.; Zgaga, L.; Barron, T.I.; Bennett, K. Pre-diagnostic statin use, lymph node status and mortality in women with stages I-III breast cancer. *Br. J. Cancer* **2017**, *117*, 588–596. [CrossRef] [PubMed]
122. Murtola, T.J.; Visvanathan, K.; Artama, M.; Vainio, H.; Pukkala, E. Statin use and breast cancer survival: A nationwide cohort study from Finland. *PLoS ONE* **2014**, *9*, e110231. [CrossRef] [PubMed]
123. Liu, B.; Yi, Z.; Guan, X.; Zeng, Y.-X.; Ma, F. The relationship between statins and breast cancer prognosis varies by statin type and exposure time: A meta-analysis. *Breast Cancer Res. Treat.* **2017**, *164*, 1–11. [CrossRef] [PubMed]
124. Manthravadi, S.; Shrestha, A.; Madhusudhana, S. Impact of statin use on cancer recurrence and mortality in breast cancer: A systematic review and meta-analysis: Breast cancer: A systematic review and meta-analysis. *Int. J. Cancer* **2016**, *139*, 1281–1288. [CrossRef] [PubMed]
125. Ahern, T.P.; Pedersen, L.; Tarp, M.; Cronin-Fenton, D.P.; Garne, J.P.; Silliman, R.A.; Sørensen, H.T.; Lash, T.L. Statin prescriptions and breast cancer recurrence risk: A Danish nationwide prospective cohort study. *J. Natl. Cancer Inst.* **2011**, *103*, 1461–1468. [CrossRef] [PubMed]
126. Brewer, T.M.; Masuda, H.; Liu, D.D.; Shen, Y.; Liu, P.; Iwamoto, T.; Kai, K.; Barnett, C.M.; Woodward, W.A.; Reuben, J.M.; et al. Statin use in primary inflammatory breast cancer: A cohort study. *Br. J. Cancer* **2013**, *109*, 318–324. [CrossRef] [PubMed]
127. Afzali, M.; Vatankhah, M.; Ostad, S.N. Investigation of simvastatin-induced apoptosis and cell cycle arrest in cancer stem cells of MCF-7. *J. Cancer Res. Ther.* **2016**, *12*, 725–730.
128. Alarcon Martinez, T.; Zeybek, N.D.; Müftüoğlu, S. Evaluation of the Cytotoxic and Autophagic Effects of Atorvastatin on MCF-7 Breast Cancer Cells. *Balk. Med. J.* **2018**, *35*, 256–262. [CrossRef] [PubMed]
129. Campbell, M.J.; Esserman, L.J.; Zhou, Y.; Shoemaker, M.; Lobo, M.; Borman, E.; Baehner, F.; Kumar, A.S.; Adduci, K.; Marx, C.; et al. Breast cancer growth prevention by statins. *Cancer Res.* **2006**, *66*, 8707–8714. [CrossRef]
130. Göbel, A.; Breining, D.; Rauner, M.; Hofbauer, L.C.; Rachner, T.D. Induction of 3-hydroxy-3-methylglutaryl-CoA reductase mediates statin resistance in breast cancer cells. *Cell Death Dis.* **2019**, *10*, 91. [CrossRef]
131. Kimbung, S.; Lettiero, B.; Feldt, M.; Bosch, A.; Borgquist, S. High expression of cholesterol biosynthesis genes is associated with resistance to statin treatment and inferior survival in breast cancer. *Oncotarget* **2016**, *7*, 59640–59651. [CrossRef] [PubMed]
132. Ghosh-Choudhury, N.; Mandal, C.C.; Ghosh-Choudhury, N.; Ghosh Choudhury, G. Simvastatin induces derepression of PTEN expression via NFkappaB to inhibit breast cancer cell growth. *Cell. Signal.* **2010**, *22*, 749–758. [CrossRef] [PubMed]

133. Vintonenko, N.; Jais, J.-P.; Kassis, N.; Abdelkarim, M.; Perret, G.-Y.; Lecouvey, M.; Crepin, M.; Di Benedetto, M. Transcriptome analysis and in vivo activity of fluvastatin versus zoledronic acid in a murine breast cancer metastasis model. *Mol. Pharmacol.* **2012**, *82*, 521–528. [CrossRef] [PubMed]
134. Lubet, R.A.; Boring, D.; Steele, V.E.; Ruppert, J.M.; Juliana, M.M.; Grubbs, C.J. Lack of efficacy of the statins atorvastatin and lovastatin in rodent mammary carcinogenesis. *Cancer Prev. Res. Phila.* **2009**, *2*, 161–167. [CrossRef] [PubMed]
135. Krause, B.R.; Princen, H.M. Lack of predictability of classical animal models for hypolipidemic activity: A good time for mice? *Atherosclerosis* **1998**, *140*, 15–24. [CrossRef]
136. Garwood, E.R.; Kumar, A.S.; Baehner, F.L.; Moore, D.H.; Au, A.; Hylton, N.; Flowers, C.I.; Garber, J.; Lesnikoski, B.-A.; Hwang, E.S.; et al. Fluvastatin reduces proliferation and increases apoptosis in women with high grade breast cancer. *Breast Cancer Res. Treat.* **2010**, *119*, 137–144. [CrossRef] [PubMed]
137. Bjarnadottir, O.; Romero, Q.; Bendahl, P.-O.; Jirström, K.; Rydén, L.; Loman, N.; Uhlén, M.; Johannesson, H.; Rose, C.; Grabau, D.; et al. Targeting HMG-CoA reductase with statins in a window-of-opportunity breast cancer trial. *Breast Cancer Res. Treat.* **2013**, *138*, 499–508. [CrossRef] [PubMed]
138. Kimbung, S.; Chang, C.-Y.; Bendahl, P.-O.; Dubois, L.; Thompson, J.W.; McDonnell, D.P.; Borgquist, S. Impact of 27-hydroxylase (CYP27A1) and 27-hydroxycholesterol in breast cancer. *Endocr. Relat. Cancer* **2017**, *24*, 339–349. [CrossRef]
139. Mc Menamin, Ú.C.; Murray, L.J.; Hughes, C.M.; Cardwell, C.R. Statin use and breast cancer survival: A nationwide cohort study in Scotland. *BMC Cancer* **2016**, *16*, 600. [CrossRef]
140. Cedó, L.; Blanco-Vaca, F.; Escolà-Gil, J.C. Antiatherogenic potential of ezetimibe in sitosterolemia: Beyond plant sterols lowering. *Atherosclerosis* **2017**, *260*, 94–96. [CrossRef]
141. Kobberø Lauridsen, B.; Stender, S.; Frikke-Schmidt, R.; Nordestgaard, B.G.; Tybjærg-Hansen, A. Using genetics to explore whether the cholesterol-lowering drug ezetimibe may cause an increased risk of cancer. *Int. J. Epidemiol.* **2017**, *46*, 1777–1785. [CrossRef] [PubMed]
142. Miettinen, T.A.; Puska, P.; Gylling, H.; Vanhanen, H.; Vartiainen, E. Reduction of serum cholesterol with sitostanol-ester margarine in a mildly hypercholesterolemic population. *N. Engl. J. Med.* **1995**, *333*, 1308–1312. [CrossRef] [PubMed]
143. Awad, A.B.; Downie, A.; Fink, C.S.; Kim, U. Dietary phytosterol inhibits the growth and metastasis of MDA-MB-231 human breast cancer cells grown in SCID mice. *Anticancer Res.* **2000**, *20*, 821–824. [PubMed]
144. Ju, Y.H.; Clausen, L.M.; Allred, K.F.; Almada, A.L.; Helferich, W.G. beta-Sitosterol, beta-Sitosterol Glucoside, and a Mixture of beta-Sitosterol and beta-Sitosterol Glucoside Modulate the Growth of Estrogen-Responsive Breast Cancer Cells In Vitro and in Ovariectomized Athymic Mice. *J. Nutr.* **2004**, *134*, 1145–1151. [CrossRef] [PubMed]
145. Llaverias, G.; Escolà-Gil, J.C.; Lerma, E.; Julve, J.; Pons, C.; Cabré, A.; Cofán, M.; Ros, E.; Sánchez-Quesada, J.L.; Blanco-Vaca, F. Phytosterols inhibit the tumor growth and lipoprotein oxidizability induced by a high-fat diet in mice with inherited breast cancer. *J. Nutr. Biochem.* **2013**, *24*, 39–48. [CrossRef] [PubMed]
146. Bradford, P.G.; Awad, A.B. Phytosterols as anticancer compounds. *Mol. Nutr. Food Res.* **2007**, *51*, 161–170. [CrossRef] [PubMed]
147. Blanco-Vaca, F.; Cedo, L.; Julve, J. Phytosterols in cancer: From molecular mechanisms to preventive and therapeutic potentials. *Curr. Med. Chem.* **2018**. [CrossRef] [PubMed]
148. Després, J.-P.; Lemieux, I.; Robins, S.J. Role of fibric acid derivatives in the management of risk factors for coronary heart disease. *Drugs* **2004**, *64*, 2177–2198. [CrossRef] [PubMed]
149. Bonovas, S.; Nikolopoulos, G.K.; Bagos, P.G. Use of Fibrates and Cancer Risk: A Systematic Review and Meta-Analysis of 17 Long-Term Randomized Placebo-Controlled Trials. *PLoS ONE* **2012**, *7*, e45259. [CrossRef]
150. Kwiterovich, P.O. The antiatherogenic role of high-density lipoprotein cholesterol. *Am. J. Cardiol.* **1998**, *82*, 13Q–21Q. [CrossRef]
151. Navab, M.; Anantharamaiah, G.M.; Reddy, S.T.; Hama, S.; Hough, G.; Grijalva, V.R.; Yu, N.; Ansell, B.J.; Datta, G.; Garber, D.W.; et al. Apolipoprotein A-I mimetic peptides. *Arterioscler. Thromb. Vasc. Biol.* **2005**, *25*, 1325–1331. [CrossRef] [PubMed]

© 2019 by the authors. Licensee MDPI, Basel, Switzerland. This article is an open access article distributed under the terms and conditions of the Creative Commons Attribution (CC BY) license (http://creativecommons.org/licenses/by/4.0/).

Article

Short-Term Cooling Increases Plasma ANGPTL3 and ANGPTL8 in Young Healthy Lean Men but Not in Middle-Aged Men with Overweight and Prediabetes

Laura G.M. Janssen [1,2], Matti Jauhiainen [3], Vesa M. Olkkonen [3], P.A. Nidhina Haridas [3], Kimberly J. Nahon [1,2], Patrick C.N. Rensen [1,2] and Mariëtte R. Boon [1,2,*]

1. Department of Medicine, Division of Endocrinology, Leiden University Medical Center, P.O. Box 9600, 2300 RC Leiden, The Netherlands
2. Einthoven Laboratory for Experimental Vascular Medicine, Leiden University Medical Center, P.O. Box 9600, 2300 RC Leiden, The Netherlands
3. Minerva Foundation Institute for Medical Research, Biomedicum 2U, 00290 Helsinki, Finland
* Correspondence: m.r.boon@lumc.nl; Tel.: +31-71-52-63078

Received: 3 July 2019; Accepted: 12 August 2019; Published: 14 August 2019

Abstract: Angiopoietin-like proteins (ANGPTLs) regulate triglyceride (TG)-rich lipoprotein distribution via inhibiting TG hydrolysis by lipoprotein lipase in metabolic tissues. Brown adipose tissue combusts TG-derived fatty acids to enhance thermogenesis during cold exposure. It has been shown that cold exposure regulates ANGPTL4, but its effects on ANGPTL3 and ANGPTL8 in humans have not been elucidated. We therefore investigated the effect of short-term cooling on plasma ANGPTL3 and ANGPTL8, besides ANGPTL4. Twenty-four young, healthy, lean men and 20 middle-aged men with overweight and prediabetes were subjected to 2 h of mild cooling just above their individual shivering threshold. Before and after short-term cooling, plasma ANGPTL3, ANGPTL4, and ANGPTL8 were determined by ELISA. In young, healthy, lean men, short-term cooling increased plasma ANGPTL3 (+16%, $p < 0.05$), ANGPTL4 (+15%, $p < 0.05$), and ANGPTL8 levels (+28%, $p < 0.001$). In middle-aged men with overweight and prediabetes, short-term cooling only significantly increased plasma ANGPTL4 levels (+15%, $p < 0.05$), but not ANGPTL3 (230 ± 9 vs. 251 ± 13 ng/mL, $p = 0.051$) or ANGPTL8 (2.2 ± 0.5 vs. 2.3 ± 0.5 µg/mL, $p = 0.46$). We show that short-term cooling increases plasma ANGPTL4 levels in men, regardless of age and metabolic status, but only overtly increases ANGPTL3 and ANGPTL8 levels in young, healthy, lean men.

Keywords: adipose tissue; ANGPTL3; ANGPTL4; ANGPTL8; lipid metabolism

---

1. Introduction

Increased plasma triacylglycerol (TG) levels are an independent risk factor for cardiovascular disease [1]. TG is either derived from dietary lipids or synthesized by the liver and white adipose tissue (WAT) from glucose and carried in the circulation within TG-rich lipoproteins (TRLs). TRLs can be hydrolysed by lipoprotein lipase (LPL) on endothelial cells to provide underlying oxidative tissues with fatty acids (FA) as fuel during increased energy demands (fasting or exercise) or to increase lipid storage in WAT during nutrient excess [2].

As energy needs can rapidly change, the LPL-mediated clearance of TRL-derived TG is under strict regulation of several factors including lipoprotein-associated apolipoproteins (e.g., APOC2 and APOC3) and angiopoietin-like proteins (ANGPTLs) [3]. ANGPTLs consist of a family of multifunctional glycoproteins, of which ANGPTL3, ANGPTL4, and ANGPTL8 inhibit LPL activity and work in concert to regulate lipoprotein metabolism [4]. Loss-of-function mutations in either one of these proteins are associated with a favourable lipid profile including lower TG levels in humans [5–8]. Moreover, deficiency for either one of these proteins in mouse models results in lower plasma TG levels, whereas overexpression leads to hypertriglyceridemia [9–13].

A recently identified novel player in TG metabolism is brown adipose tissue (BAT). Mouse studies have shown that activating BAT reduces plasma TG [14], mainly via LPL-mediated processing of TRLs [15] and alleviates dyslipidemia and atherosclerosis [16]. Therefore, BAT activation is currently widely investigated as potential treatment strategy aiming to improve cardiometabolic diseases in humans [17]. The main function of BAT is to generate heat by combustion of intracellular lipids to maintain core body temperature, and the LPL-dependent FA influx into activated brown adipocytes is required to replenish these intracellular lipid stores. The most potent physiological stimulus of BAT activation is cold exposure, which results in sympathetic activation of β-adrenergic receptors on brown adipocytes [18].

Cold exposure was previously shown to affect ANGPTL4 expression in adipose tissue in mice and to increase circulating ANGPTL4 levels in men [19,20]. The effects of cold exposure on ANGPTL3 and ANGPTL8 in humans have not been established as yet. We, therefore, investigated the effect of short-term cooling on plasma ANGPTL3 and ANGPTL8, in comparison with ANGPTL4, in relation to changes in lipid metabolism in young, healthy, lean men as well as middle-aged men with overweight and prediabetes.

2. Experimental Section

2.1. Study Design and Participants

In this study, blood samples of two clinical trials were used; one consisting of a cohort of young, healthy, lean men [21] (Dutch Trial Register 2473) and one consisting of a cohort of middle-aged men with overweight and prediabetes [22] (Clinicaltrials.gov NCT02291458). Both studies were approved by the Medical Ethical Committee of the LUMC and conducted in accordance with the principles of the revised Declaration of Helsinki (2013) and the Medical Research Involving Human Subjects Act (WMO). All subjects signed written informed consent prior to participation.

The study setup of the young, healthy, lean cohort was described in detail elsewhere [21]. In short, the study investigated the effect of cold exposure on brown adipose tissue and energy metabolism in 24 healthy, lean (BMI < 25 kg/m^2) men, aged 18–28 years, of white Caucasian ($n = 12$) and South Asian ($n = 12$) descent. Subjects were included between March 2013 and June 2013. After an overnight fast, body fat percentage was measured by dual-energy X-ray absorptiometry (iDXA, GE Healthcare, UK). A blood sample was collected at thermoneutrality and at the end of an individualized water-cooling protocol, and here analyzed for plasma ANGPTL3, ANGPTL4, ANGPTL8, and serum lipids. The individualized cooling protocol consisted of approximately 2 h mild cooling between 2 water-perfused blankets; 1 above and 1 beneath the study subject (Blanketrol III, Cincinnati Sub-Zero Products, Cincinnati, OH, USA). The individualized cooling protocol started at 32 °C and the water temperature was gradually decreased until shivering occurred, after which the temperature was increased by 3–4 °C to stop shivering. Hereafter, an [^{18}F]FDG-PET/CT scan was performed to quantify BAT volume and glucose uptake.

An elaborate description of the study setup of the middle-aged overweight prediabetic cohort can be found elsewhere [22]. Briefly, 20 middle-aged (40–55 years) white Caucasian ($n = 10$) and South Asian ($n = 10$) men with overweight or obesity (BMI 25–35 kg/m^2) and prediabetes were included in a randomized double-blind cross-over study, evaluating the effect of L-arginine on brown adipose tissue and energy metabolism. Subjects were included between October 2014 and June 2015. Prediabetes was defined according to ADA criteria as having either fasting plasma glucose levels between 5.6–6.9 mmol/L or plasma glucose levels 2 h after an oral glucose tolerance test between 7.8–11.1 mmol/L [23]. Two South Asian subjects used simvastatin 40 mg once daily. While subjects received either L-arginine or placebo for 6 weeks, in the current study only data after ingestion of the placebo were used. Placebo tablets consisted of a mixture of pregelatinized maize starch, microcrystalline cellulose, and magnesium stearate. Placebo supplements were divided over 3 gifts: after breakfast, lunch, and dinner. On the following day after placebo intake, after 4 h of fasting in the morning, a blood sample was collected under thermoneutral conditions. Hereafter, an individualized mild cooling protocol lasting approximately 2 h was initiated, after which another blood sample was collected. The individualized cooling protocol consisted of gradually lowering the water temperature until just above the shivering point of the subjects, here being wrapped in a water-perfused suit (ThermaWrap Universal 3166, MTRE Advanced Technologies, Yavne, Israel). In the current study, we assessed these blood samples for serum lipids and plasma ANGPTL3, ANGPTL4, and ANGPTL8 as well as plasma glucose and insulin. The following day after the cooling experiment, body fat percentage was determined with DXA (Discovery A, Hologic, Bedford, MA, USA).

In both the young, healthy, lean cohort and middle-aged overweight prediabetic cohort, subjects were instructed not to exercise more than 3 times per week and to refrain from exercise prior to the experimental day. In both study cohorts, subjects were also instructed to not change their dietary habits and to consume a standardized evening meal prior to the experimental day.

2.2. Serum and Plasma Analyses

Plasma samples of both the young, healthy, and lean and middle-aged overweight prediabetic cohorts were analysed for ANGPTL3 [24] and ANGPTL8 [25] by in-house developed ELISAs. For the ANGPTL8 ELISA, in brief, antibodies against multiple synthesized ANGPTL8 peptides were chosen from different parts of the ANGPTL8 protein molecule. Rabbit R355 antibodies against the ANGPTL8 amino acid region 54–68 were combined with the horseradish peroxidase-labelled capture antibody against the ANGPTL8 peptide region 182–196 (rabbit R360) [25]. Plasma samples were also analysed for ANGPTL4 with a commercial ELISA assay (R&D Systems, Minneapolis, MN, USA). The intra- and inter-assay coefficients of variation for ANGPTL3 and ANGPTL4 were <15% [24]. Precision or intra- and inter-assay CVs for ANGPTL8 were approximately 10%. Serum TG and free fatty acid (FFA) concentrations were determined with enzymatic kits in both the young, healthy, lean cohort (Roche Diagnostics, Woerden, the Netherlands and Wako Chemicals, Neuss, Germany, respectively) and the middle-aged overweight prediabetic cohort (ABX Pentra 400 autoanalyzer, HORIBA Medical, Montpellier, France). In the middle-aged overweight prediabetic cohort, plasma glucose was also measured with an enzymatic kit (ABX Pentra 400 autoanalyzer, HORIBA Medical, Montpellier, France) and plasma insulin via a commercially available radioimmunoassay kit (Human Insulin-specific Radioimmunoassay, Millipore Corporation, Burlington, MA, USA). Four young, healthy, lean subjects and 1 middle-aged subject with overweight and prediabetes were excluded from this study due to absent plasma samples.

2.3. Statistical Analysis

Statistical analyses were performed with SPSS Statistics version 25 for Windows (IBM, Armonk, NY, USA). Baseline characteristics were compared between cohorts and ethnicities with a two-tailed unpaired student's t-test, and with a two-way mixed-effect ANOVA in case temperature was an additional factor. A linear mixed-model analysis was performed with cohort, ethnicity, and temperature modelled as fixed effects, to investigate the effect of cold exposure on ANGPTLs both within and between cohorts and ethnicities. Temperature was additionally modelled as a random effect with intercepts and an unstructured covariance type. Correlations between changes in plasma ANGPTLs and serum lipids were performed using linear regression analysis and were assessed for interaction of ethnicity. Data are presented as mean ± SEM, unless stated otherwise. A p-value < 0.05 was considered statistically significant. No correction for multiple testing was applied.

3. Results

3.1. Clinical Characteristics

Clinical characteristics of the healthy lean cohort as well as effects of short-term cooling on serum lipids and other metabolic parameters have been described elsewhere [21]. In brief, subjects were 24 ± 1 years of age, had a BMI of 21.9 ± 0.4 kg/m^2, and a body fat percentage of 21.4 ± 1.2% (Table 1). Short-term cooling increased both serum TG (+0.22 ± 0.06 mmol/L, $p < 0.01$) and FFA levels (+0.19 ± 0.05 mmol/L, $p < 0.001$; Table 1). Similar observations were made when taking into account South Asian and white Caucasian ethnicities separately (Table S1).

Clinical characteristics for the middle-aged men with overweight and prediabetes have also been described in detail elsewhere [22]. Compared with the young, healthy, lean men, middle-aged overweight prediabetic subjects were older (47 ± 2 vs. 24 ± 1 years, $p < 0.001$), had a higher BMI (30.6 ± 0.8 vs. 21.9 ± 0.4 kg/m^2, $p < 0.001$) and a higher body fat percentage (30.9 ± 0.9 vs. 21.4 ± 1.2%, $p < 0.001$) (Table 1). In addition, compared with the young, healthy, lean men, FFA levels were lower (0.54 ± 0.04 vs. 0.84 ± 0.08 mmol/L, $p < 0.01$) and TG levels higher (1.56 ± 0.14 vs. 0.87 ± 0.10 mmol/L, $p < 0.001$) in middle-aged overweight prediabetic subjects (Table 1). Short-term cooling increased serum TG (+0.18 ± 0.04 mmol/L, $p < 0.001$) but not FFA levels (+0.06 ± 0.04 mmol/L, $p = 0.11$) in middle-aged men with overweight and prediabetes. Similar observations were made when taking into account South Asian and white Caucasian ethnicities separately (Table S1).

Table 1. Clinical characteristics. Data are mean ± SEM. *** $p < 0.001$, ** $p < 0.01$ middle-aged overweight prediabetic men vs. young, healthy, lean men. BMI = body mass index, FFA = free fatty acids, TG = triglycerides. Four healthy, young, lean subjects and one middle-aged overweight subject were excluded from the original cohorts due to absent plasma samples.

Clinical Characteristics	Young Healthy Lean Men (n = 20)	Middle-Aged Overweight Prediabetic Men (n = 19)
Age (years)	24 ± 1	47 ± 2 ***
Height (m)	1.79 ± 0.02	1.78 ± 0.01
Weight (kg)	70.6 ± 2.1	96.9 ± 2.9 ***
BMI (kg/m^2)	21.9 ± 0.4	30.6 ± 0.8 ***
Body fat percentage	21.4 ± 1.2	30.9 ± 0.9 ***
Thermoneutral TG (mmol/L)	0.87 ± 0.10	1.56 ± 0.14 ***
Cold-induced change TG (mmol/L)	+0.22 ± 0.06	+0.18 ± 0.04
Thermoneutral FFA (mmol/L)	0.84 ± 0.08	0.54 ± 0.04 **
Cold-induced change FFA (mmol/L)	+0.19 ± 0.05	+0.06 ± 0.04 $p = 0.053$

3.2. Short-Term Cooling Increases Plasma ANGPTL3 and ANGPTL8 in Young, Healthy, Lean Men but Not in Middle-Aged Men with Overweight and Prediabetes

In young, healthy, lean men, short-term cooling increased ANGPTL3 (142 ± 9 vs. 164 ± 11 ng/mL, +16%, $p < 0.05$; Figure 1A), ANGPTL4 (165 ± 19 vs. 190 ± 19 ng/mL, +15%, $p < 0.05$; Figure 1B), and ANGPTL8 levels (1.8 ± 0.3 vs. 2.3 ± 0.3 µg/mL, +28%, $p < 0.001$; Figure 1C).

In middle-aged men with overweight and prediabetes, short-term cooling also increased ANGPTL4 levels (193 ± 27 vs. 234 ± 31 ng/mL, +15%, $p < 0.001$; Figure 1E). In contrary to the young, healthy, lean men, short-term cooling did not overtly increase ANGPTL3 (230 ± 9 vs. 251 ± 13 ng/mL, $p = 0.051$; Figure 1D) or ANGPTL8 levels (2.2 ± 0.5 vs. 2.3 ± 0.5 µg/mL, $p = 0.46$; Figure 1F). The effect of cold on plasma ANGPTL8 levels was significantly different between the young, healthy, lean men and the middle-aged men with overweight and prediabetes (+28% vs. +3%, $p < 0.01$). Similar trends were observed for cold-induced ANGPTL4 and ANGPTL8 levels in both cohorts when taking into account South Asian and white Caucasian men separately. However, in case of ANGPTL3, plasma levels increased in white Caucasian but not South Asian middle-aged men with overweight and prediabetes (Figure S1).

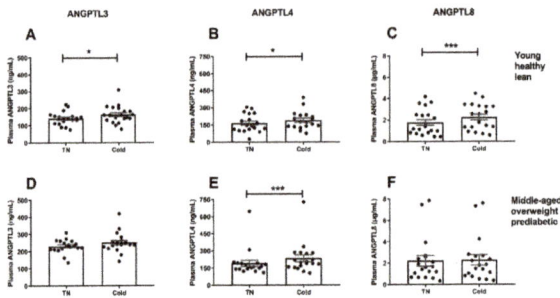

Figure 1. Effect of cold exposure on plasma ANGPTL3, ANGPTL4, and ANGPTL8 levels in young, healthy, lean men (**A–C**) and middle-aged men with overweight and prediabetes (**D–F**). Data are mean ± SEM. *** $p < 0.001$, * $p < 0.05$ cold vs. thermoneutrality (TN).

3.3. The Change in Plasma ANGPTL4 Negatively Correlates with the Change in Triglycerides after Short-Term Cooling in Young, Healthy, Lean Men

To further investigate whether the change in ANGPTLs during short-term cooling was related to the increase in serum lipids, we performed correlation analyses between the changes in the ANGPTLs and TG and FFA levels in both study cohorts.

Data of both ethnicities were pooled, as ethnic origin did not show interaction with any of the correlation analyses. We did not observe a correlation between the cold-induced response in either one of the ANGPTLs and FFA levels in the young, healthy, lean men or middle-aged men with overweight and prediabetes. In the young, healthy, lean men, the cold-induced response in ANGPTL4 negatively correlated with the cold-induced response in TG ($R^2 = 0.39$, $p < 0.01$; Figure 2B), whereas no correlations were observed for ANGPTL3 (Figure 2A) or ANGPTL8 (Figure 2C) levels. In addition, no correlations between the cold-induced response in ANGPTL3, ANGPTL4, and ANGPTL8 and TG levels were observed in the middle-aged men with overweight and prediabetes (Figure 2D–F). Of note, body fat percentage did not correlate with cold-induced changes in levels of either of the ANGPTLs (Figure S2). Moreover, body fat percentage did not affect any of the correlation analyses between cold-induced changes in ANGPTLs and TG or FFA levels.

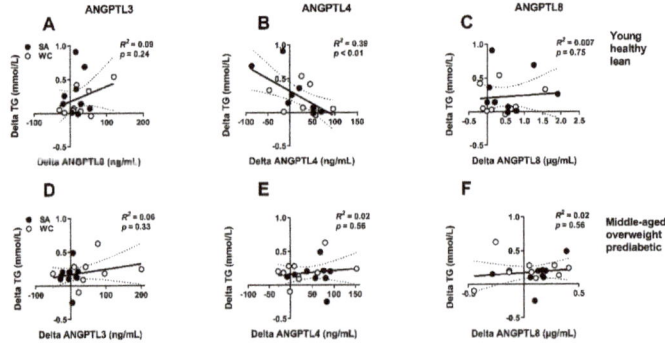

Figure 2. Correlation between cold-induced changes in serum triglyceride (TG) and plasma ANGPTL3, ANGPTL4, and ANGPTL8 levels in young, healthy, lean men (**A–C**) and middle-aged men with overweight and prediabetes (**D–F**). Dotted lines represent 95% confidence interval. TG = triglycerides. Black circles are South Asians (SA), white circles are white Caucasians (WC).

3.4. Changes in ANGPTLs are not Overtly Correlated to [^{18}F]FDG Uptake by BAT or Plasma Glucose or Insulin Levels after Short-Term Cooling

Short-term cooling not only increases TRL-derived FA uptake, but also glucose uptake to stimulate thermogenesis by BAT, the latter likely to enhance de novo lipogenesis [26]. ANGPTLs have a well-established role in TRL-derived FA uptake by metabolic tissues, but their interplay with glucose metabolism is more controversial [27]. We therefore evaluated cold-induced changes in levels of ANGPTLs in relation to [^{18}F]FDG uptake by BAT on PET/CT scan in both young, healthy, lean men and middle-aged men with overweight and prediabetes. We additionally assessed cold-induced changes in ANGPTL levels in relation to delta glucose, delta insulin levels, as well as HOMA-IR under insulin resistant conditions (i.e., in the cohort of middle-aged men with overweight and prediabetes).

Data of both ethnicities were pooled, as ethnic origin did not show interaction with any of the correlation analyses aside from ANGPTL8. We did not observe correlations between the cold-induced response in either one of the ANGPTLs and BAT volume, SUVmean, or BAT metabolic activity (BAT volume multiplied by SUVmean) in the young, healthy, lean men (Figure S3). In the middle-aged men with overweight and prediabetes, we only observed a negative correlation between the cold-induced change in ANGPTL4 levels and BAT volume, but no correlations between other ANGPTLs and BAT parameters (Figure S4). In addition, we did not observe correlations between the cold-induced response in either one of the ANGPTLs and delta glucose, delta insulin levels, or HOMA-IR in the middle-aged men with overweight and prediabetes (Figure S5).

4. Discussion

ANGPTLs are inhibitors of LPL activity and function in modulating TRLs that traffic between tissues depending on specific situational energy demands. During cold exposure, LPL-mediated hydrolysis of TRL-TG by BAT is enhanced to meet the increased FA demand to facilitate thermogenesis [15]. Here, we confirmed that short-term cooling increases plasma ANGPTL4 levels in both young, healthy, lean men and middle-aged men with overweight and prediabetes. In addition, we now show that cooling increases plasma ANGPTL3 and ANGPTL8 levels, but only in young, healthy, lean men. We propose that the elevated circulating ANGPTL3 and ANGPTL8 levels during short-term cooling represent a compensatory response aimed at preventing an ANGPTL4-promoted excessive lipid accumulation in oxidative tissues in young, healthy, lean men.

First, we show that short-term cooling increased plasma ANGPTL4 levels in both young, healthy, lean men and middle-aged men with overweight and prediabetes. We previously obtained serum ANGPTL4 levels in this same young, healthy, lean cohort, and these findings are in line albeit measured with a different ELISA [20]. Plasma ANGPTL4 levels also increased after 48 h of continuous mild cold exposure (16 °C) in young obese males [19]. In addition to cold exposure, both fasting and exercise increased circulating ANGPTL4 levels in humans [28–30]. Regulation of *Angptl4* expression is tissue-specific, as fasting upregulated *Angptl4* expression and impaired LPL activity in WAT in mice, thereby facilitating the uptake of TG-derived FA by tissues with an increased energy demand [28,31]. In line with this, in mice, cold exposure upregulated *Angptl4* expression and limited TRL-derived FA uptake in WAT, whereas *Angptl4* expression was downregulated and, as a consequence, TG-derived FA uptake was increased in BAT [19,31]. We previously hypothesized that during short-term cooling in humans, FAs derived from intracellular lipolysis bind to peroxisome proliferator-activated receptor-γ to stimulate ANGPTL4 expression in WAT, thereby increasing circulating ANGPTL4 levels [32]. This may subsequently limit TG-derived FA uptake by WAT and redirect TRLs towards active BAT for hydrolysis of TG [20]. In the current study, we also observed a negative correlation between the cold-induced changes in ANGPTL4 levels and TG levels in young, healthy, lean men. According to our hypothesis, a higher increase in plasma ANGPTL4 levels during short-term cooling possibly indicates more shuttling of TRLs away from WAT towards active BAT, which might be accompanied by enhanced TRL-TG hydrolysis by BAT that subsequently results in a less pronounced cold-induced increase in serum TG.

In addition to ANGPTL4, we now show that short-term cooling increased plasma ANGPTL3 levels in young, healthy, lean men. ANGPTL3 in humans is nearly exclusively expressed in the liver and inhibits LPL activity in metabolic tissues in an endocrine fashion [33,34]. In contrast to ANGPTL4, ANGPTL3 inhibits LPL activity and TG-derived FA uptake by oxidative tissues after (re)feeding, thereby promoting lipid storage in WAT. This was demonstrated by a study showing that *Angptl3*$^{-/-}$ mice are unable to suppress LPL activity specifically in oxidative tissues in a fed state, thereby increasing very-low density-lipoprotein (VLDL)-TG-derived FA uptake by oxidative tissues (skeletal muscle, heart, and BAT) and reducing VLDL-TG-derived FA uptake by WAT. As a consequence, plasma TG levels were markedly lower in *Angptl3*$^{-/-}$ mice compared with wild-type mice [35]. In addition to TG levels, plasma FFA and glycerol levels were lower in *Angptl3*$^{-/-}$ mice, likely due to impaired inhibition of lipolysis [36]. Lowering circulating TG levels via ANGPTL3 inactivation by antisense oligonucleotides or monoclonal antibodies is a promising treatment strategy to target dyslipidemia and cardiovascular disease, as this reduced atherosclerosis progression in mice and significantly improved lipid profile in subjects with dyslipidemia in phase I trials [7,37]. However, the effect of pharmacologically targeting ANGPTL3 on risk factors for cardiovascular disease in specific metabolically challenged subjects remains to be elucidated in future studies.

We also show that short-term cooling increases plasma ANGPTL8 levels in young, healthy, lean men. This is in line with mouse studies showing that cold exposure enhances expression of *Angptl8* in liver, BAT, and WAT, although circulating ANGPTL8 levels were not reported in these studies [31,38]. The expression of ANGPTL8 is enriched in liver and present to a lesser extent in WAT and BAT and is highly upregulated in both tissues after (re)feeding [12,39]. Similar to ANGPTL3, ANGPTL8 likely inhibits LPL activity and TG-derived FA uptake by oxidative tissues after feeding to promote lipid storage in WAT. This was evident from a study showing that *Angptl8*$^{-/-}$ mice have increased LPL activity only in oxidative tissues (heart and skeletal muscle) and not in WAT upon (re)feeding [11]. Moreover, Wang et al. [13] showed that *Angptl8*$^{-/-}$ mice have impaired uptake of VLDL-TG-derived FA by WAT in a fed state.

ANGPTL3 and ANGPTL8 share sequence homology, form a protein–protein complex, and need each other's presence to sufficiently regulate circulating lipid levels [12]. Zhang et al. [40] proposed the idea of an ANGPTL3-4-8 axis that ensures adequate distribution of TRL-TG to energy-demanding tissues in different nutritional states. In this model, during fasting, ANGPTL4 negatively regulates LPL activity in WAT to redirect TRL-TG towards other energy requiring tissues for hydrolysis, whereas upon feeding, ANGPTL3 and ANGPTL8 negatively regulate LPL activity in oxidative tissues to make TG-derived FA available for storage by WAT. In this context, we hypothesize that the increase in plasma ANGPTL4 during short-term cooling indicates inhibition of LPL activity in WAT to shuttle TRLs towards active BAT for uptake of TG-derived FA, whereas ANGPTL3 and ANGPTL8 redirect TRLs towards WAT to prevent the accumulation of excess lipids in active BAT in young, healthy, lean men.

It is tempting to speculate about the underlying mechanisms that increase plasma ANGPTL3 and ANGPTL8 during short-term cooling in young, healthy, lean men. Both ANGPTL3 and ANGPTL8 are regulated by the liver X receptor (LXR) [41–43]. Oxysterols are metabolites of cholesterol and are endogenous ligands of the LXR. As mouse studies have shown that cold exposure rapidly generates TRL-derived cholesterol-enriched remnants that are cleared by the liver [16], this might provide a source of cholesterol that enhances oxysterol formation and thereby stimulates LXR-induced expression of ANGPTL3 and ANGPTL8. Possibly, reduced formation of oxysterols as a consequence of decreased hepatic clearance of cholesterol-enriched remnants under insulin resistant conditions is involved in the absent response of ANGPTL3 and ANGPTL8 during short-term cooling in our cohort of middle-aged men with overweight and prediabetes [44]. Besides activation of the LXR, ANGPTL8 expression is upregulated in both hepatocytes and adipocytes by insulin [34,39,45]. As insulin release is stimulated during cold exposure [46], we speculate that insulin might contribute to the increased plasma ANGPTL8 levels during short-term cooling. Similar to other metabolic tissues, BAT becomes less sensitive to insulin during (pre)diabetes [47]. Therefore, we speculate that the absent cold-induced changes in

ANGPTL3 and ANGPTL8 in middle-aged men with overweight and prediabetes might reflect an attempt of the body to overcome impaired glucose uptake by insulin-resistant BAT. It should be noted that we did not observe correlations between cold-induced changes in ANGPTL3 or ANGPTL8 and changes in glucose or insulin levels to support these hypotheses. However, circulating levels of insulin may not reflect local signaling function in metabolic tissues, including its effects on ANGPTL8 secretion. We also did not observe overt correlations between cold-induced changes in ANGPTLs and [^{18}F]FDG uptake by BAT measured with PET/CT scan. However, this only reflects the uptake of glucose by BAT. Taking into account the LPL-inhibitory function of ANGPTLs, it would be highly interesting to specifically investigate cold-induced changes in ANGPTLs in relation to the uptake of (TRL-derived) FA by BAT in future studies. Of note, we observed an increase in FFA levels upon short-term cooling in the young, healthy, lean cohort but not in the middle-aged overweight prediabetic cohort. Interestingly, an increased fat mass is associated with an impaired FFA release from subcutaneous WAT [48]. This likely reflects impaired lipolysis, which may be mediated in part via catecholamine-resistance during obesity (reviewed in [49]). We therefore propose that reduced activity of the sympathetic nervous system in our cohort of overweight and obese men contributed to their unchanged FFA levels upon short-term cooling.

A strong aspect of our study is that we evaluated the effects of short-term cooling on ANGPTLs in both a cohort of healthy and metabolically challenged men. However, this study also had its limitations. For example, its cross-sectional design, as it would be interesting to investigate whether differences in cold-induced ANGPTL levels arise during metabolic changes within individuals over time. In addition, the study designs of both cohorts, although comparable to a certain extent, are not identical with respect to timing and fasting duration (which may explain lower baseline FFA levels in the middle-aged vs. young cohort). As the applied cooling protocols are also slightly different, we cannot exclude that variability in temperatures using individualized cooling protocols affected ANGPTL and lipid levels. Additionally, it is likely that a fasting state, maintained during the short-term cooling, partly contributed to an increase in circulating lipids and ANGPTL4, whereas this might have abolished the increase in ANGPTL3 and ANGPTL8. Importantly, we cannot exclude possible confounding in our analyses by a difference in age between both study cohorts, nor an effect of the placebo in the middle-aged overweight prediabetic cohort. Lastly, as the sample size of both studies is small, larger studies are warranted to confirm these observations.

5. Conclusions

In conclusion, we show that short-term cooling not only increases plasma ANGPTL4, but also plasma ANGPTL3 and ANGPTL8 levels in young, healthy, lean men. We propose that these ANGPTLs act in concert to facilitate TG partitioning between tissues in response to cold. While ANGPTL4 likely functions to shuttle TRLs away from WAT towards active thermogenic tissues during cold exposure, we suggest that ANGPTL3 and ANGPTL8 redirect TRLs away from thermogenic tissues to prevent excessive lipid accumulation. Whether these increases in circulating ANGPTLs reflect their ability to locally inhibit LPL-mediated TG hydrolysis and subsequent TG-derived FA uptake by metabolic tissues, remains to be elucidated.

Supplementary Materials: The following are available online at http://www.mdpi.com/2077-0383/8/8/1214/s1, Figure S1: Effect of cold exposure on plasma ANGPTL3, ANGPTL4 and ANGPTL8 levels in young healthy lean South Asian and white Caucasian men, and in South Asian and white Caucasian middle-aged men with overweight and prediabetes; Figure S2: Correlation between body fat percentage and cold-induced changes in plasma ANGPTL3, ANGPTL4 and ANGPTL8 levels in the young healthy lean and middle-aged overweight prediabetic cohorts; Figure S3: Correlation between cold-induced changes in plasma ANGPTL3, ANGPTL4 and ANGPTL8 levels and BAT volume, SUVmean and metabolic activity, i.e., volume*SUVmean, in young healthy lean men; Figure S4: Correlation between cold-induced changes in plasma ANGPTL3, ANGPTL4 and ANGPTL8 levels and BAT volume, SUVmean and metabolic activity, i.e., volume*SUVmean, in middle-aged men with overweight and prediabetes; Figure S5: Correlation between cold-induced changes in plasma ANGPTL3, ANGPTL4 and ANGPTL8 levels and cold-induced changes in plasma glucose and insulin levels and HOMA1-IR in middle-aged men with overweight and prediabetes; Table S1: Clinical characteristics per ethnicity.

Author Contributions: Conceptualization, M.R.B. and P.C.N.R.; methodology, L.G.M.J., M.J., V.M.O., P.A.N.H., K.J.N., and M.R.B.; validation, M.J., V.M.O., P.A.N.H., K.J.N., and M.R.B.; formal analysis, L.G.M.J. and M.R.B.; investigation, M.J., V.M.O., P.A.N.H., K.J.N., and M.R.B.; resources, M.J., V.M.O., P.A.N.H., and M.R.B.; data curation, L.G.M.J., K.J.N., and M.R.B.; writing—original draft preparation, L.G.M.J.; writing—review and editing, M.J., V.M.O., P.A.N.H., K.J.N., M.R.B., and P.C.N.R.; visualization, L.G.M.J., M.J., and M.R.B.; supervision, P.C.N.R.; project administration, L.G.M.J. and M.R.B.; funding acquisition, M.J., V.M.O., P.A.N.H., M.R.B., and P.C.N.R.

Funding: This research was funded by the Finnish Foundation for Cardiovascular Research, the Jane and Aatos Erkko Foundation, the Paavo Nurmi Foundation, the Novo-Nordisk Foundation, the Dutch Heart Foundation (2009T038), the Netherlands Organisation for Scientific Research (825.13.021) and the Dutch Diabetes Research Foundation (2015.81.1808).

Acknowledgments: Jari Metso, is acknowledged for expert technical assistance.

Conflicts of Interest: The authors declare no conflict of interest.

References

1. Nordestgaard, B.G.; Varbo, A. Triglycerides and cardiovascular disease. *Lancet* **2014**, *384*, 626–635. [CrossRef]
2. Kersten, S. Physiological regulation of lipoprotein lipase. *Biochim. Biophys. Acta* **2014**, *1841*, 919–933. [CrossRef] [PubMed]
3. Voshol, P.J.; Rensen, P.C.; van Dijk, K.W.; Romijn, J.A.; Havekes, L.M. Effect of plasma triglyceride metabolism on lipid storage in adipose tissue: Studies using genetically engineered mouse models. *Biochim. Biophys. Acta* **2009**, *1791*, 479–485. [CrossRef] [PubMed]
4. Dijk, W.; Kersten, S. Regulation of lipid metabolism by angiopoietin-like proteins. *Curr. Opin. Lipidol.* **2016**, *27*, 249–256. [CrossRef] [PubMed]
5. Peloso, G.M.; Auer, P.L.; Bis, J.C.; Voorman, A.; Morrison, A.C.; Stitziel, N.O.; Brody, J.A.; Khetarpal, S.A.; Crosby, J.R.; Fornage, M.; et al. Association of low-frequency and rare coding-sequence variants with blood lipids and coronary heart disease in 56,000 whites and blacks. *Am. J. Hum. Genet.* **2014**, *94*, 223–232. [CrossRef] [PubMed]
6. Romeo, S.; Pennacchio, L.A.; Fu, Y.; Boerwinkle, E.; Tybjaerg-Hansen, A.; Hobbs, H.H.; Cohen, J.C. Population-based resequencing of ANGPTL4 uncovers variations that reduce triglycerides and increase HDL. *Nat. Genet.* **2007**, *39*, 513–516. [CrossRef] [PubMed]
7. Dewey, F.E.; Gusarova, V.; Dunbar, R.L.; O'Dushlaine, C.; Schurmann, C.; Gottesman, O.; McCarthy, S.; Van Hout, C.V.; Bruse, S.; Dansky, H.M.; et al. Genetic and Pharmacologic Inactivation of ANGPTL3 and Cardiovascular Disease. *N. Engl. J. Med.* **2017**, *377*, 211–221. [CrossRef] [PubMed]
8. Dewey, F.E.; Gusarova, V.; O'Dushlaine, C.; Gottesman, O.; Trejos, J.; Hunt, C.; Van Hout, C.V.; Habegger, L.; Buckler, D.; Lai, K.M.; et al. Inactivating Variants in ANGPTL4 and Risk of Coronary Artery Disease. *N. Engl. J. Med.* **2016**, *374*, 1123–1133. [CrossRef] [PubMed]
9. Koishi, R.; Ando, Y.; Ono, M.; Shimamura, M.; Yasumo, H.; Fujiwara, T.; Horikoshi, H.; Furukawa, H. Angptl3 regulates lipid metabolism in mice. *Nat. Genet.* **2002**, *30*, 151–157. [CrossRef] [PubMed]
10. Koster, A.; Chao, Y.B.; Mosior, M.; Ford, A.; Gonzalez-DeWhitt, P.A.; Hale, J.E.; Li, D.; Qiu, Y.; Fraser, C.C.; Yang, D.D.; et al. Transgenic angiopoietin-like (angptl) 4 overexpression and targeted disruption of angptl4 and angptl3: Regulation of triglyceride metabolism. *Endocrinology* **2005**, *146*, 4943–4950. [CrossRef]
11. Fu, Z.; Abou-Samra, A.B.; Zhang, R. A lipasin/Angptl8 monoclonal antibody lowers mouse serum triglycerides involving increased postprandial activity of the cardiac lipoprotein lipase. *Sci. Rep.* **2015**, *5*, 18502. [CrossRef] [PubMed]
12. Quagliarini, F.; Wang, Y.; Kozlitina, J.; Grishin, N.V.; Hyde, R.; Boerwinkle, E.; Valenzuela, D.M.; Murphy, A.J.; Cohen, J.C.; Hobbs, H.H. Atypical angiopoietin-like protein that regulates ANGPTL3. *Proc. Natl. Acad. Sci. USA* **2012**, *109*, 19751–19756. [CrossRef] [PubMed]
13. Wang, Y.; Quagliarini, F.; Gusarova, V.; Gromada, J.; Valenzuela, D.M.; Cohen, J.C.; Hobbs, H.H. Mice lacking ANGPTL8 (Betatrophin) manifest disrupted triglyceride metabolism without impaired glucose homeostasis. *Proc. Natl. Acad. Sci. USA* **2013**, *110*, 16109–16114. [CrossRef] [PubMed]
14. Bartelt, A.; Bruns, O.T.; Reimer, R.; Hohenberg, H.; Ittrich, H.; Peldschus, K.; Kaul, M.G.; Tromsdorf, U.I.; Weller, H.; Waurisch, C.; et al. Brown adipose tissue activity controls triglyceride clearance. *Nat. Med.* **2011**, *17*, 200–205. [CrossRef] [PubMed]

15. Khedoe, P.P.; Hoeke, G.; Kooijman, S.; Dijk, W.; Buijs, J.T.; Kersten, S.; Havekes, L.M.; Hiemstra, P.S.; Berbee, J.F.; Boon, M.R.; et al. Brown adipose tissue takes up plasma triglycerides mostly after lipolysis. *J. Lipid Res.* **2015**, *56*, 51–59. [CrossRef] [PubMed]
16. Berbee, J.F.; Boon, M.R.; Khedoe, P.P.; Bartelt, A.; Schlein, C.; Worthmann, A.; Kooijman, S.; Hoeke, G.; Mol, I.M.; John, C.; et al. Brown fat activation reduces hypercholesterolaemia and protects from atherosclerosis development. *Nat. Commun.* **2015**, *6*, 6356. [CrossRef] [PubMed]
17. Ruiz, J.R.; Martinez-Tellez, B.; Sanchez-Delgado, G.; Osuna-Prieto, F.J.; Rensen, P.C.N.; Boon, M.R. Role of Human Brown Fat in Obesity, Metabolism and Cardiovascular Disease: Strategies to Turn Up the Heat. *Prog. Cardiovasc. Dis.* **2018**, *61*, 232–245. [CrossRef]
18. Hoeke, G.; Kooijman, S.; Boon, M.R.; Rensen, P.C.; Berbee, J.F. Role of Brown Fat in Lipoprotein Metabolism and Atherosclerosis. *Circ. Res.* **2016**, *118*, 173–182. [CrossRef]
19. Dijk, W.; Heine, M.; Vergnes, L.; Boon, M.R.; Schaart, G.; Hesselink, M.K.; Reue, K.; van Marken Lichtenbelt, W.D.; Olivecrona, G.; Rensen, P.C.; et al. ANGPTL4 mediates shuttling of lipid fuel to brown adipose tissue during sustained cold exposure. *Elife* **2015**, *4*, 1303. [CrossRef]
20. Nahon, K.J.; Hoeke, G.; Bakker, L.E.H.; Jazet, I.M.; Berbee, J.F.P.; Kersten, S.; Rensen, P.C.N.; Boon, M.R. Short-term cooling increases serum angiopoietin-like 4 levels in healthy lean men. *J. Clin. Lipidol.* **2018**, *12*, 56–61. [CrossRef]
21. Bakker, L.E.; Boon, M.R.; van der Linden, R.A.; Arias-Bouda, L.P.; van Klinken, J.B.; Smit, F.; Verberne, H.J.; Jukema, J.W.; Tamsma, J.T.; Havekes, L.M.; et al. Brown adipose tissue volume in healthy lean south Asian adults compared with white Caucasians: A prospective, case-controlled observational study. *Lancet Diabetes Endocrinol.* **2014**, *2*, 210–217. [CrossRef]
22. Boon, M.R.; Hanssen, M.J.W.; Brans, B.; Hulsman, C.J.M.; Hoeks, J.; Nahon, K.J.; Bakker, C.; van Klinken, J.B.; Havekes, B.; Schaart, G.; et al. Effect of L-arginine on energy metabolism, skeletal muscle and brown adipose tissue in South Asian and Europid prediabetic men: A randomised double-blinded crossover study. *Diabetologia* **2018**. [CrossRef]
23. American Diabetes, A. Standards of medical care in diabetes—2014. *Diabetes Care* **2014**, *37* (Suppl. 1), S14–S80. [CrossRef]
24. Robciuc, M.R.; Tahvanainen, E.; Jauhiainen, M.; Ehnholm, C. Quantitation of serum angiopoietin-like proteins 3 and 4 in a Finnish population sample. *J. Lipid Res.* **2010**, *51*, 824–831. [CrossRef]
25. Tikka, A.; Metso, J.; Jauhiainen, M. ANGPTL3 serum concentration and rare genetic variants in Finnish population. *Scand. J. Clin. Lab. Investig.* **2017**, *77*, 601–609. [CrossRef]
26. Labbe, S.M.; Caron, A.; Bakan, I.; Laplante, M.; Carpentier, A.C.; Lecomte, R.; Richard, D. In vivo measurement of energy substrate contribution to cold-induced brown adipose tissue thermogenesis. *FASEB J.* **2015**, *29*, 2046–2058. [CrossRef]
27. Davies, B.S.J. Can targeting ANGPTL proteins improve glucose tolerance? *Diabetologia* **2018**, *61*, 1277–1281. [CrossRef]
28. Kersten, S.; Mandard, S.; Tan, N.S.; Escher, P.; Metzger, D.; Chambon, P.; Gonzalez, F.J.; Desvergne, B.; Wahli, W. Characterization of the fasting-induced adipose factor FIAF, a novel peroxisome proliferator-activated receptor target gene. *J. Biol. Chem.* **2000**, *275*, 28488–28493. [CrossRef]
29. Kersten, S.; Lichtenstein, L.; Steenbergen, E.; Mudde, K.; Hendriks, H.F.; Hesselink, M.K.; Schrauwen, P.; Muller, M. Caloric restriction and exercise increase plasma ANGPTL4 levels in humans via elevated free fatty acids. *Arter. Thromb. Vasc. Biol.* **2009**, *29*, 969–974. [CrossRef]
30. Catoire, M.; Alex, S.; Paraskevopulos, N.; Mattijssen, F.; Evers-van Gogh, I.; Schaart, G.; Jeppesen, J.; Kneppers, A.; Mensink, M.; Voshol, P.J.; et al. Fatty acid-inducible ANGPTL4 governs lipid metabolic response to exercise. *Proc. Natl. Acad. Sci. USA* **2014**, *111*, E1043–E1052. [CrossRef]
31. Fu, Z.; Yao, F.; Abou-Samra, A.B.; Zhang, R. Lipasin, thermoregulated in brown fat, is a novel but atypical member of the angiopoietin-like protein family. *Biochem. Biophys. Res. Commun.* **2013**, *430*, 1126–1131. [CrossRef]
32. Mattijssen, F.; Kersten, S. Regulation of triglyceride metabolism by Angiopoietin-like proteins. *Biochim. Biophys. Acta* **2012**, *1821*, 782–789. [CrossRef]
33. Conklin, D.; Gilbertson, D.; Taft, D.W.; Maurer, M.F.; Whitmore, T.E.; Smith, D.L.; Walker, K.M.; Chen, L.H.; Wattler, S.; Nehls, M.; et al. Identification of a mammalian angiopoietin-related protein expressed specifically in liver. *Genomics* **1999**, *62*, 477–482. [CrossRef]

34. Nidhina Haridas, P.A.; Soronen, J.; Sadevirta, S.; Mysore, R.; Quagliarini, F.; Pasternack, A.; Metso, J.; Perttila, J.; Leivonen, M.; Smas, C.M.; et al. Regulation of Angiopoietin-Like Proteins (ANGPTLs) 3 and 8 by Insulin. *J. Clin. Endocrinol. Metab.* **2015**, *100*, E1299–E1307. [CrossRef]
35. Wang, Y.; McNutt, M.C.; Banfi, S.; Levin, M.G.; Holland, W.L.; Gusarova, V.; Gromada, J.; Cohen, J.C.; Hobbs, H.H. Hepatic ANGPTL3 regulates adipose tissue energy homeostasis. *Proc. Natl. Acad. Sci. USA* **2015**, *112*, 11630–11635. [CrossRef]
36. Shimamura, M.; Matsuda, M.; Kobayashi, S.; Ando, Y.; Ono, M.; Koishi, R.; Furukawa, H.; Makishima, M.; Shimomura, I. Angiopoietin-like protein 3, a hepatic secretory factor, activates lipolysis in adipocytes. *Biochem. Biophys. Res. Commun.* **2003**, *301*, 604–609. [CrossRef]
37. Graham, M.J.; Lee, R.G.; Brandt, T.A.; Tai, L.J.; Fu, W.; Peralta, R.; Yu, R.; Hurh, E.; Paz, E.; McEvoy, B.W.; et al. Cardiovascular and Metabolic Effects of ANGPTL3 Antisense Oligonucleotides. *N. Engl. J. Med.* **2017**, *377*, 222–232. [CrossRef]
38. Zhang, Y.; Li, S.; Donelan, W.; Xie, C.; Wang, H.; Wu, Q.; Purich, D.L.; Reeves, W.H.; Tang, D.; Yang, L.J. Angiopoietin-like protein 8 (betatrophin) is a stress-response protein that down-regulates expression of adipocyte triglyceride lipase. *Biochim. Biophys. Acta* **2016**, *1861*, 130–137. [CrossRef]
39. Ren, G.; Kim, J.Y.; Smas, C.M. Identification of RIFL, a novel adipocyte-enriched insulin target gene with a role in lipid metabolism. *Am. J. Physiol. Endocrinol. Metab.* **2012**, *303*, E334–E351. [CrossRef]
40. Zhang, R. The ANGPTL3-4-8 model, a molecular mechanism for triglyceride trafficking. *Open Biol.* **2016**, *6*, 150272. [CrossRef]
41. Lee, J.; Hong, S.W.; Park, S.E.; Rhee, E.J.; Park, C.Y.; Oh, K.W.; Park, S.W.; Lee, W.Y. AMP-activated protein kinase suppresses the expression of LXR/SREBP-1 signaling-induced ANGPTL8 in HepG2 cells. *Mol. Cell. Endocrinol.* **2015**, *414*, 148–155. [CrossRef]
42. Kaplan, R.; Zhang, T.; Hernandez, M.; Gan, F.X.; Wright, S.D.; Waters, M.G.; Cai, T.Q. Regulation of the angiopoietin-like protein 3 gene by LXR. *J. Lipid Res.* **2003**, *44*, 136–143. [CrossRef]
43. Ge, H.; Cha, J.Y.; Gopal, H.; Harp, C.; Yu, X.; Repa, J.J.; Li, C. Differential regulation and properties of angiopoietin-like proteins 3 and 4. *J. Lipid Res.* **2005**, *46*, 1484–1490. [CrossRef]
44. Sorensen, L.P.; Andersen, I.R.; Sondergaard, E.; Gormsen, L.C.; Schmitz, O.; Christiansen, J.S.; Nielsen, S. Basal and insulin mediated VLDL-triglyceride kinetics in type 2 diabetic men. *Diabetes* **2011**, *60*, 88–96. [CrossRef]
45. Rong Guo, X.; Li Wang, X.; Chen, Y.; Hong Yuan, Y.; Mei Chen, Y.; Ding, Y.; Fang, J.; Jiao Bian, L.; Sheng Li, D. ANGPTL8/betatrophin alleviates insulin resistance via the Akt-GSK3beta or Akt-FoxO1 pathway in HepG2 cells. *Exp. Cell Res.* **2016**, *345*, 158–167. [CrossRef]
46. Heine, M.; Fischer, A.W.; Schlein, C.; Jung, C.; Straub, L.G.; Gottschling, K.; Mangels, N.; Yuan, Y.; Nilsson, S.K.; Liebscher, G.; et al. Lipolysis Triggers a Systemic Insulin Response Essential for Efficient Energy Replenishment of Activated Brown Adipose Tissue in Mice. *Cell Metab.* **2018**, *28*, 644–655. [CrossRef]
47. Blondin, D.P.; Labbe, S.M.; Noll, C.; Kunach, M.; Phoenix, S.; Guerin, B.; Turcotte, E.E.; Haman, F.; Richard, D.; Carpentier, A.C. Selective Impairment of Glucose but Not Fatty Acid or Oxidative Metabolism in Brown Adipose Tissue of Subjects with Type 2 Diabetes. *Diabetes* **2015**, *64*, 2388–2397. [CrossRef]
48. Karpe, F.; Dickmann, J.R.; Frayn, K.N. Fatty acids, obesity, and insulin resistance: Time for a reevaluation. *Diabetes* **2011**, *60*, 2441–2449. [CrossRef]
49. Arner, P. Human fat cell lipolysis: Biochemistry, regulation and clinical role. *Best Pract. Res. Clin. Endocrinol. Metab.* **2005**, *19*, 471–482. [CrossRef]

© 2019 by the authors. Licensee MDPI, Basel, Switzerland. This article is an open access article distributed under the terms and conditions of the Creative Commons Attribution (CC BY) license (http://creativecommons.org/licenses/by/4.0/).

Article

Low Serum Paraoxonase-1 Activity Associates with Incident Cardiovascular Disease Risk in Subjects with Concurrently High Levels of High-Density Lipoprotein Cholesterol and C-Reactive Protein

James P. Corsetti [1,*], Charles E. Sparks [1], Richard W. James [2], Stephan J. L. Bakker [3] and Robin P. F. Dullaart [4]

1. Department of Pathology and Laboratory Medicine, University of Rochester School of Medicine and Dentistry, Rochester, NY 1211, USA
2. Department of Medical Specialties-Endocrinology, Diabetology, Hypertension and Nutrition, University of Geneva, 1201 Geneva, Switzerland
3. Department of Nephrology, University of Groningen and University Medical Center Groningen, 9700 RB Groningen, The Netherlands
4. Department of Endocrinology, University of Groningen and University Medical Center Groningen, 9700 RB Groningen, The Netherlands
* Correspondence: james_corsetti@urmc.rochester.edu; Tel.: +1-585-275-4907; Fax: +1-585-273-3003

Received: 19 July 2019; Accepted: 27 August 2019; Published: 1 September 2019

Abstract: Paroxonase-1 (PON1) is a key enzyme that inhibits low-density lipoprotein oxidation and consequently atherogenesis. Here, we assessed whether low serum PON1 activity associates with incident cardiovascular disease (CVD) in subjects with high levels of high-density cholesterol (HDL-C) and C-reactive protein (CRP), a marker of low-grade systemic inflammation. Cox proportional-hazards modeling of incident CVD risk (11 years mean follow-up) adjusted for relevant clinical and biomarker covariates was performed on a population-based study (N = 7766) stratified into three groups: low CRP—(LR; event rate 4.9%); low HDL-C/high CRP—(HR1; event rate 14.4%); and high HDL-C/high CRP—(HR2; event rate 7.6%). Modeling results for PON1 activity in HR2 were significant and robust (hazard ratio/SD unit—0.68, 95% CI 0.55–0.83, p = 0.0003), but not so for LR and HR1. Analyses in HR2 of the interaction of PON1 with HDL-C, apoA-I, apoA-II, and apoE levels were significant only for PON1 with apoE (hazard ratio—1.77, 95% CI 1.29–2.41, p = 0.0003). Subsequent subgroup analysis revealed inverse risk dependence for apoE at low PON1 levels. In conclusion, in a population-based study of subjects with concurrently high HDL-C and CRP levels, low serum PON1 activity associates with incident CVD risk with risk accentuated at low apoE levels.

Keywords: cardiovascular disease; C-reactive protein; HDL; paraoxonase-1

1. Introduction

Paraoxonases make up a family of three enzymes (PON1, PON2, and PON3) with predominantly lactonase activity, each of which has been shown to manifest anti-atherogenic properties. This presumably relates to hydrolytic activity against lactone-like structures elaborated by oxidized polyunsaturated fatty acids on lipoproteins [1,2]. PON2 is an intracellular enzyme. After synthesis in the liver, PON1 and PON3 enter the circulation and become predominantly associated with high-density lipoprotein (HDL) particles. Anti-atherogenic action of PON1 has been demonstrated to derive in large part from the interaction of HDL-associated PON1 with low-density lipoprotein (LDL) particles. Such interaction leads to inhibition of oxidative modification of LDL through PON1-mediated hydrolysis of oxidized fatty acid species derived from oxidative stress-induced generation of hydroperoxides from phospholipids, cholesteryl esters, and triglycerides [2–6]. Likewise for HDL, studies have also demonstrated similar anti-oxidative activity by PON1 on lipid peroxidation products generated on HDL in oxidative stress environments [7–10]. The putative anti-atherogenic properties of PON1 stem from meta-analyses of human population studies that consistently show decreased PON1 activity to be associated with cardiovascular disease (CVD) risk [11–13]. It should also be noted that in addition to CVD, low PON1 activity has been reported in other disease states having a significant inflammatory component including diabetes mellitus, obesity, metabolic syndrome, cancer, and various rheumatic, renal, hepatic, and neurologic diseases [3,14].

Accumulating evidence demonstrates that the established anti-atherogenic actions of HDL become dysfunctional and potentially pro-atherogenic in chronic inflammatory and oxidative stress environments as a result of ensuing alterations in the HDL proteome and lipidome [15–17]. To elucidate the role of dysfunctional HDL in CVD risk, investigations of HDL-associated risk have been undertaken in populations with various inflammatory disorders. Notably, results from such studies demonstrate that high levels of HDL cholesterol (HDL-C) associate with increased CVD risk rather than with protective effects [18]. In our case, we have studied incident as well as recurrent coronary events in subjects with high HDL-C and concurrently high levels of C-reactive protein (CRP) as a marker of low-grade chronic inflammation. Compared to the controls, the results demonstrated significantly higher CVD risk in these patient groups [19,20]. Furthermore, within these populations, additional studies demonstrated risk associations for levels of apolipoprotein E (apoE) [21] and apolipoprotein A-II (apoA-II) [22]; and, in studies of single nucleotide polymorphisms, risk associations were found for genes associated with various aspects of HDL function including reverse cholesterol transport [19,23], thrombogenesis [24], oxidative stress [25], and fibrinolysis [26].

In light of the key role of PON1 in inhibiting oxidative changes in the HDL lipidome that acts to preserve anti-atherogenic functionality of HDL as described above, we hypothesized that low PON1 levels in subjects with high levels of HDL-C in the setting of chronic low-level inflammation would be at increased risk for CVD. To test this hypothesis, we studied risk of incident CVD in subjects having concurrently high HDL-C and CRP levels as an indicator of chronic low grade inflammation in comparison to two other subject groups, one with low HDL-C and high CRP levels and the other with "normal" HDL-C and low CRP levels. The present study was carried out in the frame of the prospective Prevention of Renal and Vascular End-stage Disease (PREVEND) cohort study.

2. Experimental Section

2.1. Study Population

PREVEND, a large general population-based prospective cohort study begun in 1997 that aimed to explore the role of albuminuria in cardiovascular and renal disease, served as the basis of the current study population [27]. Briefly for PREVEND recruitment, all inhabitants of the city of Groningen, the Netherlands, aged 28–75 years (N = 85,421), were sent a questionnaire requesting an early morning urine specimen. The response rate was 48% (N = 40,856). Of these, 9966 subjects had urine albumin levels ≥10 mg/L while 30,890 subjects had urine albumin <10 mg/L. Exclusions included pregnancy, type 1 diabetic individuals, and type 2 diabetic individuals using insulin. This resulted in 7768 subjects having urine albumin ≥10 mg/L. This group was combined with 3395 randomly chosen subjects with urine albumin <10 mg/L. Of this resultant group, 8592 individuals completed an additional screening program, thus establishing the PREVEND study cohort. For the present work, the only further exclusion from the PREVEND study cohort was for subjects with serum CRP levels ≥10 mg/L. This formed the study population of the current work (N = 7766). The PREVEND study was approved by the Medical Ethics Committee of the University of Groningen, the Netherlands. Informed written consent was obtained from all participants. The study adhered to the ethical principles set by the Declaration of Helsinki.

2.2. Clinical Parameters and Biomarkers

Questionnaires were used to ascertain cardiac history (hospitalization for myocardial infarction, revascularization procedures, or obstructive coronary artery disease). Diabetes was established either as fasting plasma glucose ≥7.0 mmol/L, self-report of physician diagnosis, or use of anti-diabetic medications. Medication use was retrieved via Groningen pharmacy dispensing data [28]. Smoking status was either current smoker or not; ethanol use was either one or less drinks per day or more than one drink per day (with one drink assumed to contain 10 g of alcohol). Body mass index (BMI) was calculated as weight divided by height squared (kg/m^2). Blood pressure was measured for 10 min at 1-min intervals using automatic instrumentation (Dinamap XL Model 9300; Johnson-Johnson Medical, Tampa, FL, USA) in the supine position; means of the last two recordings were the reported values [29].

The mean of two 24 h collections over two consecutive days at study entry was determined and used as the reported UAE value. Nephelometry (BNII, Dade Behring, Marburg, Germany) was used to measure urinary albumin concentration. Blood biomarker levels at study initiation were performed on overnight-fasted plasma and sera. Blood levels of creatinine, HDL-C, apoA-I, apoA-II, apoE, total cholesterol, triglycerides, apoB, glucose, and CRP were determined by standard techniques [20,21,30]. Estimated glomerular filtration rate (eGFR) was determined using the Chronic Kidney Disease Epidemiology Collaboration (CDK-EPI) equation [31]. NonHDL-C was calculated as the difference between total cholesterol and HDL-C. The Friedewald equation was used to estimate LDL-C levels. Serum PON1activity was measured in terms of arylesterase activity (rate of phenyl acetate hydrolysis to phenol) as described previously with inter-assay and intra-assay coefficients of variation of 8% and 6%, respectively [32].

2.3. Outcomes

CVD-related mortality and hospitalizations including acute MI (myocardial infarction), acute and subacute ischaemic heart disease, coronary artery bypass grafting, and percutaneous transluminal coronary angioplasty constituted incident cardiovascular events. Outcome events were followed over time (11 years of follow-up). Survival time was taken as date of initial urine collection (1997–1998) to date of first CVD event or to December 31, 2011. For subjects lost to follow-up (396 of the overall cohort of 8592 participants), censoring date was date of removal from the municipal registry. Record linkage with the Dutch Central Bureau of Statistics was used to acquire data on mortality and causes of death. PRISMANT, the Dutch national registry of hospital discharge diagnoses was used to obtain CVD morbidity data.

2.4. Data Analysis

Graphical and statistical analyses were performed using Statistica 13.3 (StatSoft, Inc., Tulsa, OK, USA). Graphical analyses portraying CVD risk over a bivariate parameter risk domain was performed using outcome event mapping as described previously [20]. Results are reported as means ± SD for normally distributed variables and as medians (interquartile range) for non-normally distributed variables. Differences among groups were assessed by ANOVA for continuous variables using Bonferroni correction for multiple comparisons and chi-square testing for categorical variables. Non-normal variables were log-transformed for analyses. Multivariable Cox proportional hazards modeling was used to follow cardiovascular outcomes over time. Continuous independent variables were standardized by transformation to distributions with means of zero and standard deviations of one for Cox analyses. Hazard ratios (HR) are per SD unit. Inclusion of multiplicative interaction terms into base models was used to assess interactions between risk variables. Two-sided p-values < 0.05 were considered statistically significant. The proportional hazards assumption was verified by correlation analysis of survival time with scaled Schoenfeld residuals.

3. Results

3.1. Study Population

During follow-up (11 years), a total of 643 events were recorded. The surface plot of Figure 1A demonstrates the estimated CVD outcome event rate as a function of HDL-C rank and CRP rank for the resultant total study population. The figure shows two peaks indicative of higher-risk populations: one at low HDL-C and high CRP rank and one at high HDL-C and high CRP rank. The corresponding contour plot of Figure 1B demonstrates the delineation of three populations based upon dichotomizations about median values of HDL-C and CRP ranks. The three resulting populations were designated as follows: low CRP—low risk (LR; N = 3889; 192 events; 4.9% event rate); low HDL-C/high CRP—high risk 1 (HR1; N = 2294; 331 events; event rate 14.4%); and high HDL-C/high CRP—high-risk 2 (HR2; N = 1583; 120 events; event rate 7.6%). Clinical and blood biomarker characterizations for the total, LR, HR1, and HR2 populations are given in Table 1. Results of chi square testing, unadjusted ANOVA, and ANOVA adjusted for gender and age revealed significant differences among the LR, HR1, and HR2 groups for all variables except ethanol use. Subsequent post-hoc testing revealed pair-wise significant differences between LR, HR1, and HR2 for all continuous variables except for: nonHDL-C (LR versus HR2, $p = 0.078$), triglycerides (LR versus HR2, $p = 0.90$), and UAE (LR versus HR2, $p = 0.55$).

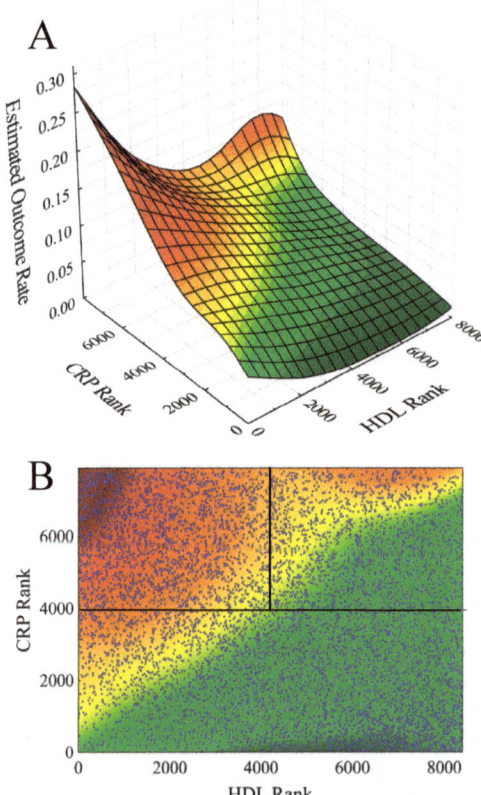

Figure 1. Cardiovascular event rates as a function of high-density lipoprotein (HDL) rank and C-reactive protein (CRP) rank given: (**A**) as a surface plot and (**B**) as a contour plot. The contour plot demarcates a low risk population (LR) characterized by lower CRP, a high-risk population (HR1) characterized by low HDL and high CRP, and another high-risk population (HR2) characterized by high HDL and high CRP.

Table 1. Clinical and biomarker parameters for the total study population and low risk (LR), high-risk 1 (HR1), and high-risk 2 (HR2) subpopulations. Biomarker levels are given as means ± standard deviations for normally distributed parameters and medians with interquartile ranges for non-normally distributed parameters.

Parameter	Total Population	Low Risk (86.1% of Subjects)	High-Risk 1	High-Risk 2	p^a	p^b
Subjects N (%)	7766	3889 (50.1)	2294 (29.5)	1583 (20.4)		
Outcome events N (%)	643 (8.3)	192 (4.9)	331 (14.4)	120 (7.6)	<0.001	
Age (years)	49.1 ± 12.6	46.2 ± 11.8	52.8 ± 12.5	50.9 ± 13.2	<0.001	
Females (%)	50.1	49.0	35.5	72.5	<0.001	
Cardiac history (%)	4.4	2.5	8.1	3.9	<0.001	
Diabetes (%)	3.2	1.4	7.0	2.1	<0.001	
Metabolic Syndrome (%)	24.4	13.4	51.3	9.7	<0.001	
Statins (%)	4.1	3.1	6.3	3.3	<0.001	
Anti-hypertensives (%)	13.6	8.4	20.8	15.7	<0.001	
Current Smoker (%)	33.3	27.7	42.7	33.6	<0.001	
Ethanol Use (>1 drink/day %)	25.3	25.9	24.1	25.7	0.28	
Pulse Rate (per minute)	69.0	68.7	70.4	70.3	<0.001	
Systolic BP (mmHg)	129 ± 20	124 ± 18	135 ± 21	131 ± 22	<0.001	
Diastolic BP (mmHg)	74 ± 10	72 ± 9	77 ± 10	74 ± 10	<0.001	
BMI (kg/m^2)	26.1 ± 4.2	24.6 ± 3.4	28.0 ± 4.3	26.2 ± 4.2	<0.001	
HDL-C (mM)	1.32 ± 0.40	1.40 ± 0.41	1.01 ± 0.18	1.62 ± 0.29	<0.001	
CRP (mg/L)	1.28 (0.56–2.98)	0.54 (0.30–0.82)	2.76 (1.79–4.62)	2.52 (1.71–4.16)	<0.001	
PON1 (U/L)	53.2 (43.2–65.1)	53.7 (43.7–65.6)	50.8 (40.9–62.0)	56.1 (46.6–68.3)	<0.001	
ApoA-I (μM)	47.8 ± 9.9	48.0 ± 10.0	42.8 ± 6.8	54.7 ± 9.3	<0.001	
ApoA-II (μM)	19.6 ± 3.8	19.5 ± 3.8	18.7 ± 3.2	21.4 ± 4.0	<0.001	
ApoA-I/HDL-C (μM/mM)	37.8 ± 8.2	35.7 ± 7.7	43.2 ± 7.7	34.1 ± 5.1	<0.001	
ApoA-II/HDL-C (μM/mM) (μmol/mmol)	15.9 ± 4.6	14.8 ± 4.4	19.0 ± 4.1	13.4 ± 2.7	<0.001	
Cholesterol (mM)	5.65 ± 1.13	5.46 ± 1.08	5.88 ± 1.18	5.75 ± 1.11	<0.001	
NonHDL-C (mM)	4.33 ± 1.21	4.05 ± 1.14	4.87 ± 1.19	4.13 ± 1.13	<0.001	
LDL-C (mM)	3.69 ± 1.05	3.51 ± 0.99	4.04 ± 1.04	3.58 ± 1.05	<0.001	
Triglycerides (mM)	1.16 (0.85–1.68)	1.00 (0.74–1.40)	1.59 (1.14–2.27)	1.09 (0.83–1.42)	<0.001	
ApoB (g/L)	1.04 ± 0.30	0.96 ± 0.28	1.17 ± 0.32	1.01 ± 0.27	<0.001	
Glucose (mM)	4.89 ± 1.19	4.67 ± 0.79	5.24 ± 1.58	4.76 ± 0.96	<0.001	
Creatinine (μM)	83.9 ± 19.5	83.1 ± 14.3	87.8 ± 26.3	80.7 ± 14.5	<0.001	
UAE (mg/24 h)	9.5 (6.3–17.8)	8.4 (6.0–13.6)	11.8 (7.2–27.1)	9.4 (6.1–17.4)	<0.001	
eGFR (mL/min/1.73 m^2)	84.0 ± 15.6	86.5 ± 14.8	81.3 ± 16.1	81.5 ± 15.5	<0.001	
ApoE (μM)	1.15 ± 0.47	1.09 ± 0.42	1.26 ± 0.57	1.14 ± 0.39	<0.001	
Paraoxonase/HDL-C (U/mM)	41.7 (32.3–53.8)	39.8 (30.7–50.7)	50.0 (40.5–63.8)	35.0 (28.6–43.5)	<0.001	<0.001

[a] For categorical variables, chi-square results showed significant differences (p < 0.0001) among the LR, HR1, and HR2 subgroups except for ethanol use (p = 0.28); for continuous variables, unadjusted ANOVA revealed statistically significant differences (p < 0.0001) among the subpopulations. Subsequent post-hoc testing revealed significant differences (p < 0.0001) between populations except for cholesterol (HR1 versus HR2, p = 0.0018), LDL-C (LR versus HR1, i = 0.035), glucose (LR versus HR1, p = 0.017), and apoE (LR versus HR1, p = 0.0012) and non-significant results for nonHDL-C (LR versus HR2, p = 0.078), triglycerides (LR versus HR2, p = 0.90), and UAE (LR versus HR2, p = 0.55). [b] ANOVA adjusted for age and gender also revealed statistically significant differences (p < 0.0001) among the subpopulations. BMI, body mass index; HDL-C, high-density lipoprotein cholesterol; CRP, C-reactive protein; PON1, Paraoxonase-1; ApoA-I, apolipoprotein A-I; ApoA-II, apolipoprotein A-II; NonHDL-C, non-HDL lipoprotein cholesterol; LDL-C, low-density lipoprotein cholesterol; ApoB, apolipoprotein B; ApoE, apolipoprotein E; UAE, urinary albumin excretion; eGFR, estimated glomerular filtration rate.

3.2. Correlation of PON1 Activity with Blood Biomarker Levels

To provide further characterization of the study populations focusing on PON1 activity, Table 2 gives Spearman correlation coefficients for PON1 activity levels with blood biomarkers for the total, LR, HR1, and HR2 populations. The table shows that although many of the correlations are statistically significant, magnitudes of the coefficients are generally small except for HDL-C, apoA-I, and apoA-II (highest at 0.28). Thus, PON1 activity did not appear to be highly correlated with any of the blood biomarkers in any of the populations which may be indicative potentially of a relatively independent role of PON1 activity in CVD risk.

Table 2. Spearman correlation coefficients of PON1 with biomarker levels for total, LR, HR1, and HR2 populations.

Parameter	Total Population	Low Risk	High-Risk 1	High-Risk 2
HDL-C	0.20 ***	0.19 ***	0.11 ***	0.15 ***
CRP	−0.05 ***	−0.02	−0.06 **	0.02
ApoA-I	0.19 ***	0.17 ***	0.10 ***	0.15 ***
ApoA-II	0.28 ***	0.26 ***	0.23 ***	0.26 ***
ApoA-I/HDL-C	−0.12 ***	−0.11 ***	−0.04	0.01
ApoA-II/HDL-C (µmol/mmol)	−0.04 ***	−0.03	0.08 ***	0.13 ***
Cholesterol	0.08 ***	0.13 ***	0.09 ***	0.01
NonHDL-C	0.01	0.05 **	0.07 **	−0.03
LDL-C	0.00	0.04 *	0.04	−0.06 *
Triglycerides	0.03 *	0.03 *	0.09 ***	0.13 ***
ApoB	0.00	0.03	0.05 *	−0.02
Glucose	−0.09 ***	−0.09 ***	−0.05 *	−0.07 **
Creatinine	−0.03 **	0.01	−0.04	−0.02
UAE	−0.02	0.02	−0.01	0.01
eGFR	0.02 *	−0.01	0.05 *	0.03
ApoE	0.00	0.04 *	0.04	−0.07 **

* $p < 0.05$; ** $p < 0.01$; *** $p < 0.001$.

3.3. PON1 Activity as a Marker of CVD Risk in LR, HR1, and HR2 Populations

To assess PON1 activity as a marker of CVD risk, Table 3 gives results of Cox multivariable proportional hazards modeling in the three populations. For the LR population, PON1 activity was significant (protective) in an unadjusted model, but lost significance upon adjustment for gender and age. For HR1, PON1 activity, even in an unadjusted model, did not achieve significance. For HR2, PON1 activity was highly significant (protective) in an unadjusted model ($p = 10^{-6}$); continued to be highly significant in a collection of models adjusted for gender, age, and a set of covariates added one-at time; and finally remained significant ($p = 0.0003$) in a model adjusted simultaneously for all covariates of the dataset (gender, age, UAE, past CVD, DM, apoB, use of statins, anti-hypertensives, smoking, ethanol use, and eGFR). Thus, PON1 activity appears to have a robust inverse relationship with CVD risk in the HR2 population.

Table 3. Cox multivariable proportional hazards modeling results for PON1 level giving hazard ratio (HR), 95% confidence interval (95% CI), and *p*-value. The first column lists the population examined (LR, HR1, or HR2) along with parameters adjusted in models. Hazard ratios are per SD unit. For regression calculations, PON1 levels were transformed to a distribution with mean of zero and SD of one.

Population	Model Adjustments	HR	95% CI (86.1% of Subjects)	*p*-Value
LR	unadjusted	0.86	0.75–0.98	0.024
LR	gender, age	0.95	0.82–1.09	0.46
HR1	unadjusted	0.93	0.83–1.04	0.22
HR1	gender, age	1.02	0.91–1.14	0.72
HR2	unadjusted	0.62	0.000001	0.000001
HR2	gender, age	0.72	0.59–0.87	0.0007
HR2	gender, age, UAE	0.70	0.58–0.86	0.0004
HR2	gender, age, UAE, apoB	0.70	0.57–0.85	0.0003
HR2	gender, age, UAE, past CVD	0.71	0.58–0.86	0.0005
HR2	gender, age, UAE, DM	0.71	0.58–0.86	0.0006
HR2	gender, age, UAE, statins	0.69	0.57–0.83	0.0002
HR2	gender, age, UAE, anti-hypertensives	0.68	0.56–0.83	0.0001
HR2	gender, age, UAE, SBP	0.70	0.58–0.85	0.0004
HR2	gender, age, UAE, DBP	0.70	0.58–0.85	0.0004
HR2	gender, age, UAE, smoking	0.70	0.58–0.85	0.0004
HR2	gender, age, UAE, ethanol use	0.69	0.57–0.84	0.0003
HR2	gender, age, UAE, eGFR	0.72	0.59–0.87	0.0009
HR2	gender, age, UAE, past CVD, DM, apoB, statins, anti-hypertensives smoking, ethanol use, eGFR, SBP, DBP	0.68	0.55–0.83	0.0003

3.4. HDL Particle Constituents in Addition to PON1 as Markers of CVD Risk in HR2

To assess HDL-associated markers of risk in HR2 beyond PON1, we determined proportional hazards modeling in HR2 as a function of PON1 activity along with additional single entry of HDL-associated parameters (apoA-I, apoA-II, apoA-I/HDL-C, apoA-II/HDL-C, and apoE levels) (Table 4). Models also included evaluation of the potential interaction of PON1 activity with each of the HDL-associated parameters. Again, all models were adjusted for gender, age, UAE, past CVD, DM, apoB, use of statins, anti-hypertensives, smoking, ethanol use, eGFR, SBP, and DBP. The results from Table 4 indicate that, in each case for models without interaction, an independent significant protective association continued for PON1. For HDL-associated parameters, an independent protective association was demonstrated for apoA-I/HDL-C. In models assessing interactions, only apoE levels were found to exhibit a statistically significant interaction with PON1 activity (HR–1.77, 95% CI 1.29–2.41, $p = 0.0003$).

To explore the nature of the PON1/apoE interaction in HR2, subgroup analysis using the proportional hazards model adjusted as before was performed on subjects of HR2 but this time with dichotomization about median PON1 levels giving low- and high-PON1 groups. Results indicated a trend in continued risk association with low PON1 levels both in the low PON1 group (HR—0.74, 95% CI 0.53–1.05, $p = 0.089$) and the high PON1 group (HR—0.60, 95% CI 0.26–1.40, $p = 0.24$). Regarding apoE and risk, while results for the high PON1 group demonstrated the generally expected trend in risk associated with high apoE levels (HR—1.41, 95% CI 0.91–2.18, $p = 0.13$), results for the low PON1 group revealed significant risk in association with low apoE levels (HR—0.69, 95% CI 0.50–0.96, $p = 0.029$). To illustrate the interaction, a contour plot of estimated CVD risk as a function of PON1 rank and apoE rank was generated (Figure 2). Consistent with the modeling results, the plot shows a high-risk population at concurrently low levels of both PON1 and apoE; while by contrast, at high PON1 levels, there is little indication of risk in association with low apoE levels. Hence in HR2, it would appear that the PON1/apoE interaction manifests as intensification of the usual CVD risk association for low PON1 activity but particularly in the setting of low apoE levels.

Table 4. Interaction of PON1 with HDL-related parameters for the HR2 population as assessed by Cox proportional hazards models of cardiovascular event occurrence adjusted for gender, age, UAE, past CVD history, diabetes, apoB, statin use, anti-hypertensive use, smoking, ethanol use, eGFR, SBP, and DBP. Hazard ratios are per SD unit. For regression calculations, continuous variables were transformed to distributions with means of zero and SDs of one.

Parameters	Models without Interaction			Models with Interaction		
	HR	95% CI	p	HR	95% CI	p
PON1	0.67	0.54–0.83	0.0003	0.76	0.57–1.01	0.059
HDL-C	1.04	0.75–1.45	0.82	1.01	0.72–1.40	0.98
interaction				0.81	0.057–1.16	0.25
PON1	0.69	0.56–0.86	0.0007	0.69	0.54–0.89	0.0036
ApoA-I	0.76	0.58–1.00	0.051	0.76	0.58–1.00	0.054
interaction				0.99	0.77–1.29	0.96
PON1	0.70	0.57–0.88	0.002	0.70	0.57–0.87	0.001
ApoA-II	0.78	0.60–1.02	0.070	0.79	0.61–1.02	0.070
interaction				1.09	0.90–1.31	0.39
PON1	0.66	0.54–0.82	0.0001	0.71	0.55–0.92	0.008
ApoA-I/HDL-C	0.56	0.37–0.85	0.0057	0.59	0.39–0.89	0.013
interaction				1.17	0.87–1.57	0.30
PON1	0.68	0.55–0.84	0.0004	0.81	0.60–1.08	0.15
ApoA-II/HDL-C	0.63	0.40–1.00	0.051	0.65	0.41–1.03	0.065
interaction				1.30	0.94–1.80	0.12
PON1	0.66	0.53–0.82	0.0002	0.64	0.51–0.80	0.0001
ApoE	0.82	0.60–1.12	0.21	0.85	0.62–1.16	0.30
interaction				1.77	1.29–2.41	0.0003

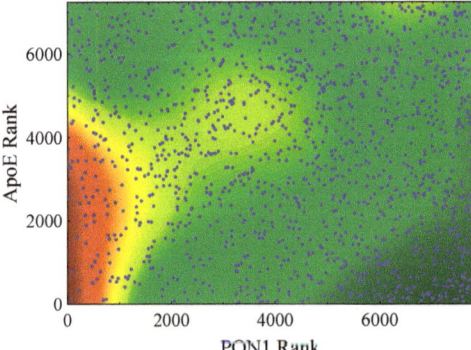

Figure 2. Contour plot of the estimated cardiac event rate as a function of PON1 rank and apoE rank in the high HDL/high CRP population (HR2).

4. Discussion

Results of the current study of incident CVD show a highly significant association of low serum PON1 activity with risk in subjects with high levels of HDL-C in the setting of chronic low-grade inflammation, as indicated by high CRP levels. Furthermore, interaction analysis demonstrated such risk to be accentuated in those subjects with low levels of serum apoE; no corresponding effect was observed for apoA-I, apoA-II, or HDL-C levels within the high HDL-C/high CRP group. These results from multivariable proportional hazards modeling were robust with regard to adjustment for relevant clinical and blood biomarkers including gender, age, past CVD event, diabetes, apoB, statin use, anti-hypertensive use, smoking, ethanol use, and eGFR. It should be noted that corresponding analyses performed both on subjects with low HDL-C and high CRP and on subjects with low CRP did not demonstrate in either case statistically significant associations for PON1 activity with CVD. Our results suggest that the putative anti-oxidative effects of serum PON1 are most evidently exhibited in subjects with high PON1 and high HDL-C levels in the setting of chronic inflammation that is signaled by higher CRP levels.

Results of a recent meta-analysis addressing the role of PON1 activity on incident CVD risk demonstrated an inverse relationship between PON1 activity and risk in general population settings [13]. It was also noted in this study that the inverse relationship became less robust when adjustments for established risk factors including HDL-C were included—a finding that could potentially be relevant to our results regarding PON1 activity in the high HDL-C/high CRP group. However, this does not appear to be the case as results of risk models including PON1 activity and HDL-C levels were not significant for HDL-C ($p = 0.82$); nor was interaction between the two significant ($p = 0.25$). It should also be noted that corresponding analyses of PON1 activity with apoA-I and apoA-II were also non-significant. Accordingly, although study results show a significant inverse relationship for PON1 activity with risk, this is not the case for other HDL parameters including levels of HDL-C, apoA-I, and apoA-II. These findings underscore the key role of low PON1 activity in establishment of CVD risk in this subject group, which likely derives from inhibition of oxidative modification of the HDL lipidome, thus helping to maintain HDL anti-atherogenic functionality. Consistent with our results regarding low PON1 activity in the high HDL-C/high CRP subgroup, as noted previously, many other disease states associated with low-grade chronic inflammation also manifest low PON1 activity [3,14]. In the case of CVD risk, in the current study we have identified a group of subjects where PON1 action is particularly relevant.

Further results from our study did not show association of PON1 activity with risk both in the low risk subject group (low CRP levels) and in a subject group with low HDL-C/high CRP levels. In the first case, we speculate that the lower CRP levels provide an environment not conducive to oxidative modification of HDL, thus preserving HDL functionality and consequently limiting the relevance of PON1 activity in this group. In the second case, we speculate that higher PON1 was not protective in HR1 stemming in one way or another from the inadequacy of PON1 to significantly inhibit oxidation of the presumably heavier load of small dense LDL particles very likely present in this population enriched with Metabolic Syndrome subjects (Table 1; HR1—51.3%, HR2—9.7%, $p < 0.001$). Inadequacy of PON1 in this setting is likely attributable to multiple factors. First, the load of LDL-C in HR1 was significantly higher than in HR2 (Table 1; 4.04 mM U/L versus 3.58 mM U/L, $p < 0.001$); and thus, the concentration of potentially atherogenic dense LDL particles would be higher in HR1 given their well-established smaller size. Second, from the very defining characteristics of HR1, the level of HDL-C was less than in HR2 (Table 1; 1.01 mM versus 1.62 mM, $p < 0.001$). Thus, the amount of PON1, known to be predominantly carried by HDL, should be and was observed to be smaller in HR1 (Table 1; 50.8 U/L versus 56.1 U/L, $p < 0.001$). Third, the carriage ability of HDL particles for PON1 might be less in HR1 versus HR2 subjects. Thus, using the apoA-I/HDL-C ratio as a crude measure of the number of apoA-I molecules per HDL particle, the observed higher apoA-I/HDL-C ratio in HR1 versus HR2 (Table 1, 43.2 µM/mM versus 34.1 µM/mM, $p < 0.001$) would tend to limit carriage of additional molecules like PON1 on HDL particles (the situation was similar for the apoA-II/HDL-C ratio).

Additional results in the high HDL-C/high CRP group of our study using interaction analysis showed accentuation of CVD risk at concurrently low levels of PON1 and apoE. This is consistent with in vitro studies of PON1 associations with reconstituted HDL particles carrying apoE that show such binding to stabilize and stimulate PON1 lactonase activity as well as to stimulate PON1 anti-atherogenic potential in a manner similar to apoA-I but with lower capacity [33]. Presumably then, lower apoE levels would work to decrease PON1 anti-atherogenic action. Furthermore, results of a clinical study of metabolic syndrome (MetS) showed that although subjects without MetS demonstrated a positive correlation of PON1 activity with serum apoE levels, suggestive of enhancement of PON1 anti-atherogenic action with increased apoE levels; the relationship was abrogated in subjects with MetS, a condition associated with inflammation and oxidative stress [34].

There were strengths and limitations in our study. Strengths included the availability of a large well-characterized population-based study cohort followed prospectively for CVD outcomes for an average of 11 years and availability of extensive corresponding clinical and laboratory parameters including serum apoA-I, apoA-II, and apoE levels. The study was limited in a number of ways. In terms of the study population, there were exclusions for pregnancy, and type 1 diabetes mellitus and for subjects with serum CRP ≥10 mg/L (to minimize confounding effects by acute inflammation, rheumatic or other disease conditions manifesting high levels of inflammation); while it should also be noted that the study cohort was almost exclusively comprised of subjects of white European origin thus potentially limiting relevance of study findings to other populations. Additionally, the study population was enriched with albuminuric subjects; however, risk modeling results included adjustments for UAE and eGFR to account for this. In terms of clinical and laboratory parameters, values were determined only at study initiation, thus limiting the effects of parameter value variation over time. Regarding PON1 action, arylesterase activity as a measure was chosen because of good assay precision and good correlation with PON1 mass assays [35]. Lastly, there was no experimental evidence to verify that the HDL lipidome was actually preserved from oxidative dysfunctional transformation by high PON1 activity as speculated in our study. Poor performance of the technically difficult characterization of lipoprotein lipidomics might determine the effect of *PON1* on the risk of functional genetic polymorphisms.

In summary, the results of our current work show that low serum PON1 activity is associated with increased CVD risk in a group of subjects from a population-based study having concurrently high levels of HDL-C and CRP indicative of a chronic low-grade inflammatory environment. We believe that high serum PON1 activity in this group serves to facilitate inhibition of oxidative modification of the HDL lipidome, thus tending to preserve anti-atherogenic functionality of HDL particles including the curtailing of LDL oxidation. As such, these findings serve to underscore the presumptive efficacy of PON1 activity in the inhibition of lipoprotein oxidative modification.

5. Conclusion

Individuals having high HDL-C levels in the setting of chronic systemic inflammation have increased risk for incident cardiovascular disease. Furthermore, in such individuals, low serum PON1 levels associate with risk with risk accentuated at low apoE levels."

Author Contributions: The authors' responsibilities were as follows: R.W.J. performed the PON-1 activity measurements. J.P.C. analyzed data/performed statistical analysis; J.P.C., C.E.S., R.W.J., S.J.L.B., and R.P.F.D. interpreted the data analysis; J.P.C. and R.P.F.D. wrote the paper; J.P.C. and R.P.F.D. had primary responsibility for final content. All authors read and approved the final manuscript.

Funding: The research was funded by the Dutch Kidney Foundation (Grant E.033) and the Dutch Heart Foundation (Grant 2001-005).

Acknowledgments: The PON-1 activity measurements were performed in the laboratory of James, University of Geneva, Switzerland.

Conflicts of Interest: The authors declare no conflict of interest.

References

1. Reddy, S.T.; Devarajan, A.; Bourquard, N.; Shih, D.; Fogelman, A.M. Is it just paraoxonase 1 or are other members of the paraoxonase gene family implicated in atherosclerosis? *Curr. Opin. Lipidol.* **2008**, *19*, 405–408. [CrossRef] [PubMed]
2. Chistiakov, D.A.; Melnichenko, A.A.; Orekhov, A.N.; Bobryshev, Y.V. Paraoxonase and atherosclerosis-related cardiovascular diseases. *Biochimie* **2017**, *132*, 19–27. [CrossRef] [PubMed]
3. Mackness, M.; Mackness, B. Human paraoxonase-1 (*PON1*); Gene structure and expression, promiscuous activities and multiple physiological roles. *Gene* **2015**, *567*, 12–21. [CrossRef] [PubMed]
4. Manolescu, B.N.; Busu, C.; Badita, D.; Stanculescu, R.; Berteanu, M. Paraoxoase 1–an update of the antioxidant properties of high-density lipoproteins. *Medica J. Clin. Med.* **2015**, *10*, 173–177.

5. Kowalska, K.; Socha, E.; Milerowicz, H. The role of paraoxonase in cardiovascular disease. *Ann. Clin. Lab. Sci.* **2015**, *45*, 226–233. [PubMed]
6. Shunmoogam, N.; Naidoo, P.; Chilton, R. Paraoxonase (PON1): A brief overview on genetics, structure, polymorphisms and clinical relevance. *Vasc. Health Risk Manag.* **2018**, *14*, 137–143. [CrossRef]
7. Aviram, M.; Rosenblat, M.; Bisgaier, C.L.; Newton, R.S.; Primo-Parmo, S.L.; La Du, B.N. Paraoxonase inhibits high-density lipoprotein oxidation and preserves its functions. *J. Clin. Investig.* **1998**, *101*, 1581–1590. [CrossRef]
8. Ahmed, Z.; Ravandi, A.; Maguire, G.F.; Emili, A.; Draganow, D.; La Du, B.N.; Kuksis, A.; Connelly, P.W. Apolipoprotein A-I promotes the formation of phosphatidylcholine core aldehydes that are hydrolyzed by paraoxonase (PON-1) during high density lipoprotein oxidation with a peroxynitrite donor. *J. Biol. Chem.* **2001**, *276*, 24473–24481. [CrossRef]
9. Mackness, M.; Durrington, P.; Mackness, B. Paraoxonase 1 activity, concentration and genotype in cardiovascular disease. *Curr. Opin. Lipidol.* **2004**, *15*, 399–404. [CrossRef]
10. Mastorikou, M.; Mackness, M.; Mackness, B. Defective metabolism of oxidized phospholipid by HDL from people with type 2 diabetes. *Diabetes* **2006**, *55*, 3099–3103. [CrossRef]
11. Zhao, Y.; Ma, Y.; Fang, Y.; Liu, L.; Wu, S.; Fu, D.; Wang, X. Association between PON1 activity and coronary heart disease risk: A meta-analysis based on 43 studies. *Mol. Genet. Metab.* **2012**, *105*, 141–148. [CrossRef] [PubMed]
12. Wang, M.; Lang, X.; Cui, S.; Zou, L.; Cao, J.; Wang, S.; Wu, X. Quantitative assessment of the influence of paraoxonase activity and coronary heart disease risk. *DNA Cell Biol.* **2012**, *31*, 975–982. [CrossRef] [PubMed]
13. Kunutsor, S.K.; Bakker, S.J.L.; James, R.W.; Dullaart, R.P.F. Serum paraoxonase-1 activity and risk of incident cardiovascular disease: The PREVEND study and meta-analysis of prospective population studies. *Atherosclerosis* **2016**, *245*, 143–154. [CrossRef] [PubMed]
14. Goswami, B.; Tayal, D.; Gupta, N.; Mallika, V. Paraoxonase: A multifaceted biomolecule. *Clin. Chem. Acta* **2009**, *410*, 1–12. [CrossRef] [PubMed]
15. Rosenson, R.S.; Brewer, H.B.; Ansell, B.J.; Barter, P.; Chapman, M.J.; Heinecke, J.W.; Kontush, A.; Tall, A.R.; Webb, N.R. Dysfunctional HDL and atherosclerotic cardiovascular disease. *Nat. Rev. Cardiol.* **2016**, *13*, 48–60. [CrossRef] [PubMed]
16. Annema, W.; von Eckardstein, A. Dysfunctional high-density lipoproteins in coronary heart disease: Implications for diagnostics and therapy. *Transl. Res.* **2016**, *173*, 30–57. [CrossRef] [PubMed]
17. Srivastava, R.A.K. Dysfunctional HDL in diabetes mellitus and its role in the pathogenesis of cardiovascular disease. *Mol. Cell Biochem.* **2018**, *440*, 167–187. [CrossRef]
18. Chang, T.I.; Streja, E.; Moradi, H. Could high-density lipoprotein cholesterol predict increased cardiovascular risk? *Curr. Opin. Endocrinol. Diabetes Obes.* **2017**, *24*, 140–147. [CrossRef]
19. Corsetti, J.P.; Ryan, D.; Rainwater, D.L.; Moss, A.J.; Zareba, W.; Sparks, C.E. Cholesteryl Ester Transfer Protein Polymorphism (TaqIB) Associates with Risk in Postinfarction Patients with High C-Reactive Protein and High-Density Lipoprotein Cholesterol Levels. *Atheroscler. Thromb. Vasc. Biol.* **2010**, *30*, 1657–1664. [CrossRef]
20. Corsetti, J.P.; Gansevoort, R.T.; Sparks, C.E.; Dullaart, R.P.F. Inflammation reduces HDL protection against primary cardiac risk. *Eur. J. Clin. Investig.* **2010**, *40*, 483–489. [CrossRef]
21. Corsetti, J.P.; Gansevoort, R.T.; Bakker, S.J.L.; Navis, G.J.; Sparks, C.E.; Dullaart, R.P.F. Apolipoprotein E Predicts Incident Cardiovascular Disease Risk in Women but not in Men with Concurrently High Levels of High-Density Lipoprotein Cholesterol and C-Reactive Protein. *Metab. Clin. Exp.* **2012**, *61*, 996–1002. [CrossRef] [PubMed]
22. Corsetti, J.P.; Bakker, S.J.L.; Sparks, C.E.; Dullaart, R.P.F. Apolipoprotein A-II Influences Apolipoprotein E-Linked Cardiovascular Disease Risk in Women with High Levels of HDL Cholesterol and C-Reactive Protein. *PLoS ONE* **2012**, *7*, e39110. [CrossRef] [PubMed]
23. Corsetti, J.P.; Gansevoort, R.T.; Navis, G.J.; Sparks, C.E.; Dullaart, R.P.F. *LPL* Polymorphism (D9N) Predicts Cardiovascular Disease Risk Directly and Through Interaction with *CETP* Polymorphism (TaqIB) in Women with High HDL Cholesterol and CRP. *Atherosclerosis* **2011**, *214*, 373–376. [CrossRef] [PubMed]
24. Corsetti, J.P.; Ryan, D.; Moss, A.J.; McCarthy, J.J.; Goldenberg, I.; Zareba, W.; Sparks, C.E. Thrombospondin-4 Polymorphism (A387P) Predicts Cardiovascular Risk in Postinfarction Patients with High HDL Cholesterol and C-Reactive Protein Levels. *Thromb. Haemost.* **2011**, *106*, 1170–1178. [PubMed]

25. Le, N.T.; Corsetti, J.P.; Dehoff-Sparks, J.L.; Sparks, C.E.; Fujiwara, K.; Abe, J.I. Reactive Oxygen Species (ROS), SUMOylation, and Endothelial Inflammation. *Int. J. Inflamm.* **2012**, *2012*, 678190. [CrossRef] [PubMed]
26. Corsetti, J.P.; Salzman, P.; Ryan, D.; Moss, A.J.; Zareba, W.; Sparks, C.E. Plasminogen Activator Inhibitor-2 Polymorphism Associates with Recurrent Coronary Event Risk in Patients with High HDL and C-Reactive Protein Levels. *PLoS ONE* **2013**, *8*, e68920. [CrossRef] [PubMed]
27. Pinto-Sietsma, S.J.; Janssen, W.M.; Hillege, H.L.; Navis, G.; De Zeeuw, D.; De Jong, P.E. Urinary albumin excretion is associated with renal functional abnormalities in a nondiabetic population. *J. Am. Soc. Nephrol.* **2000**, *11*, 1882–1888.
28. Kappelle, P.J.W.H.; Gansevoort, R.T.; Hillege, J.L.; Wolffenbuttel, B.H.R.; Dullaart, R.P.F. Apolipoprotein B/A-I and total cholesterol/high-density lipoprotein cholesterol ratios both predict cardiovascular events in the general population independently of nonlipid risk factors, albuminuria and C-reactive protein. *J. Intern. Med.* **2011**, *269*, 232–242. [CrossRef]
29. De Greeff, A.; Reggiori, F.; Shennan, A.H. Clinical assessment of the DINAMAP *ProCare* monitor in an adult population according to the British Hypertension Society Protocol. *Blood Press. Monit.* **2007**, *12*, 51–55. [CrossRef]
30. Borggreve, S.E.; Hillege, H.L.; Wolffenbuttel, B.H.R.; de Jong, P.E.; Zuurman, M.W.; van der Steege, G.; van Tol, A.; Dullaart, R.P.F.; PREVEND Study Group. An increased coronary risk is paradoxically associated with common cholesteryl ester transfer protein gene variations that relate to higher high-density lipoprotein cholesterol: A population-based study. *J. Clin. Endocrinol. Metab.* **2006**, *91*, 3382–3388. [CrossRef]
31. Levey, A.S.; Stevens, L.A.; Schmid, C.H.; Zhang, Y.L.; Castro, A.F.; Feldman, H.I.; Kusek, J.W.; Eggers, P.; Van Lente, F.; Greene, T.; et al. A new equation to estimate glomerular filtration rate. *Ann. Intern. Med.* **2009**, *150*, 604–612. [CrossRef] [PubMed]
32. Van den Berg, E.H.; Gruppen, E.G.; James, R.W.; Bakker, S.J.L.; Dullaart, R.P.F. Serum paraoxonase 1 activity is paradoxically maintained in nonalcoholic fatty liver disease despite low HDL cholesterol. *J. Lipid Res.* **2019**, *60*, 168–175. [CrossRef] [PubMed]
33. Gaidukov, L.; Viji, R.I.; Yacobson, S.; Rosenblat, M.; Aviram, M.; Tawfik, D.S. ApoE induces serum paraoxonase PON1 activity and stability similar to apoA-I. *Biochemistry* **2010**, *49*, 532–538. [CrossRef] [PubMed]
34. Dullaart, R.P.F.; Kwakernaak, A.J.; Dallinga-Thie, G.M. The positive relationship of serum paraoxonase-1 activity with apolipoprotein E is abrogated in metabolic syndrome. *Atherosclerosis* **2013**, *230*, 6–11. [CrossRef] [PubMed]
35. Dullaart, R.P.F.; Otvos, J.D.; James, R.W. Serum paraoxonase-1 activity is more closely related to HDL particle concentration and large HDL particles than to HDL cholesterol in Type 2 diabetic and non-diabetic subjects. *Clin. Biochem.* **2014**, *47*, 1022–1027. [CrossRef] [PubMed]

© 2019 by the authors. Licensee MDPI, Basel, Switzerland. This article is an open access article distributed under the terms and conditions of the Creative Commons Attribution (CC BY) license (http://creativecommons.org/licenses/by/4.0/).

Review

IL-1β and Statin Treatment in Patients with Myocardial Infarction and Diabetic Cardiomyopathy

Luca Liberale [1,2], Federico Carbone [2,3], Giovanni G. Camici [1,4,5] and Fabrizio Montecucco [3,6,*]

1. Center for Molecular Cardiology, University of Zürich, 8092 Schlieren, Switzerland; luca.liberale@uzh.ch (L.L.); giovanni.camici@uzh.ch (G.G.C.)
2. First Clinic of Internal Medicine, Department of Internal Medicine, University of Genoa, 16132 Genoa, Italy; federico.carbone@unige.it
3. IRCCS Ospedale Policlinico San Martino Genoa—Italian Cardiovascular Network, 16132 Genoa, Italy
4. University Heart Center, Department of Cardiology, University Hospital Zurich, 8001 Zurich, Switzerland
5. Department of Research and Education, University Hospital Zurich, 8001 Zurich, Switzerland
6. First Clinic of Internal Medicine, Department of Internal Medicine and Centre of Excellence for Biomedical Research (CEBR), University of Genoa, Genoa, University of Genoa, 16132 Genoa, Italy
* Correspondence: fabrizio.montecucco@unige.it; Tel.: +39-010-3351054

Received: 24 September 2019; Accepted: 21 October 2019; Published: 23 October 2019

Abstract: Statins are effective lipid-lowering drugs with a good safety profile that have become, over the years, the first-line therapy for patients with dyslipidemia and a real cornerstone of cardiovascular (CV) preventive therapy. Thanks to both cholesterol-related and "pleiotropic" effects, statins have a beneficial impact against CV diseases. In particular, by reducing lipids and inflammation statins, they can influence the pathogenesis of both myocardial infarction and diabetic cardiomyopathy. Among inflammatory mediators involved in these diseases, interleukin (IL)-1β is a pro-inflammatory cytokine that recently been shown to be an effective target in secondary prevention of CV events. Statins are largely prescribed to patients with myocardial infarction and diabetes, but their effects on IL-1β synthesis and release remain to be fully characterized. Of interest, preliminary studies even report IL-1β secretion to rise after treatment with statins, with a potential impact on the inflammatory microenvironment and glycemic control. Here, we will summarize evidence of the role of statins in the prevention and treatment of myocardial infarction and diabetic cardiomyopathy. In accordance with the dual lipid-lowering and anti-inflammatory effect of these drugs and in light of the important results achieved by IL-1β inhibition through canakinumab in CV secondary prevention, we will dissect the current evidence linking statins with IL-1β and outline the possible benefits of a potential double treatment with statins and canakinumab.

Keywords: cardiovascular disease; myocardial infarction; diabetic cardiomyopathy; cytokines; interleukin 1β; inflammation; CANTOS; canakinumab

1. Introduction

Statin discovery dates back to 1976, when mevastatin was isolated from cultures of *Penicillium citrinum* and proven to inhibit the production of cholesterol molecules [1]. Further experiments showed that statins occupy a portion of the rate-controlling enzyme of cholesterol synthesis 3-hydroxy-3-methylglutaryl-CoA (HMG-CoA) reductase (HMGR) by binding its active site with very high affinity, thus displacing the natural substrate, HMG-CoA, and inhibiting its function [2]. Furthermore, the statin-related reduction of circulatory lipoprotein induces the hepatic expression of low-density lipoprotein (LDL) receptor (LDLR) and LDL clearance from the bloodstream, thus accounting for a further decrease in circulating cholesterol levels [3]. Thanks to this dual mechanism of action and a good safety profile, both natural and synthetic statins became, over the years, the first-line therapy for dyslipidemia patients and a real cornerstone of cardiovascular (CV) preventive therapy. Soon after first trials with statins were published, evidence suggested that those compounds might have putative, non-lipid-related effects. Both Cholesterol and Recurrent Events (CARE) and Long-Term Intervention with Pravastatin in Ischaemic Disease (LIPID) trials showed that their overall cardiovascular benefit was disproportionate to the magnitude of lipid reduction [4,5]. In addition, the speed by which statins exercised their protective role was faster than that obtained with other lipid-lowering interventions such as ileal bypass [6]. These "pleiotropic" effects have been related to statins' inhibitory effect on the activation of different intracellular signaling mediators downstream the mevalonate pathways (i.e., Rho, Ras, and Rac proteins) alongside direct stimulatory effects on peroxisome proliferator-activated (PPAR)-α and -β [7].

Lipids and inflammation are closely interconnected and contribute to the pathogenesis of most CV disease [8,9]. Among those, myocardial infarction constantly rates among the most important causes of morbidity and mortality worldwide, while diabetic cardiomyopathy is an emerging disease whose incidence is set to rise in the next years following the increased prevalence of the diabetic population. Although the role of circulating lipoproteins in the determination of the individual CV risk have been appreciated since a long time ago, recently, clinical and experimental observations support a role for systemic inflammation [10]. Inflammatory cells and cytokines have been identified in human atherosclerotic vessels, and their dynamic regulation plays an important role in cardiac remodeling [11,12]. Observational studies reported a reduced CV risk in patients being treated with anti-inflammatory agents for immunological disease (e.g., rheumatoid arthritis), supporting the concept of inflammation as a valuable target for CV prevention [13]. However, not all anti-inflammatory drugs provided efficacy in reducing CV risk as different trials designed to test this hypothesis gave negative results (i.e., Cardiovascular Inflammation Reduction Trial [CIRT] testing methotrexate), and non-steroidal anti-inflammatory agents are even associated with an increased CV morbidity [14,15]. Of importance, in 2017, the Canakinumab Anti-Inflammatory Thrombosis Outcomes Study (CANTOS) trial showed the efficacy of IL-1β neutralization in patients with established coronary heart disease, highlighting this cytokine and its pathway as effective targets, as well as suggesting that specific interaction with inflammatory mediators might be a better strategy than providing anti-inflammation in a global fashion [16,17].

In this review article, we aim to summarize evidence of the role of statin treatment in myocardial infarction and prevention of myocardial remodeling in patients with diabetes mellitus. In accordance with the dual lipid-lowering and anti-inflammatory effect of these drugs and in light of the important results achieved by CANTOS in CV secondary prevention, we dissect the current evidence linking statins with IL-1β and outline possible benefits of double treatment with canakinumab.

2. Statins in Myocardial Infarction and Diabetic Cardiomyopathy

2.1. Myocardial Infarction

Coronary atherosclerotic heart disease is the major cause of CV events including stable and angina, non-ST-segment elevation myocardial infarction (NSTEMI), ST-segment elevation myocardial infarction (STEMI), and sudden coronary death [18]. Ample evidence demonstrated the key role of dyslipidemia and, in particular, of elevated LDL levels in the development of coronary heart disease and thus CV risk. As such, statins have become the first-line therapy for hyperlipidemia and reduction of CV risk in patients at high and very high risk [19]. With clinical guidelines becoming increasingly stringent with respect to cholesterol levels [20–22], the prescription and utilization of statins in the last 30 years has increased considerably, with most of the patients taking these drugs for primary prevention of CV events [23]. To date, several randomized clinical trials (RCTs) and systematic reviews have investigated the role of statins in this setting, reaching different conclusions [24–27]. These apparent discrepancies might be explained by different factors: (i) the population in primary prevention is highly heterogeneous, including patients with low CV risk and those with chronic kidney disease or diabetes mellitus with organ damage, who are usually considered as risk-equivalent to CV patients; (ii) most published systematic reviews, although focusing on primary prevention, included trials in which a proportion of patients had a history of CV disease. A very recent overview of systematic reviews tried to overcome these limitations by including exclusively primary prevention trials or individual patient data of trial participants using only data from patients without established CVD [28]. Here, the authors report a trend towards reduction of all-cause mortality in all systematic reviews, although this reached statistical significance only in one study out of three [28]. Furthermore, when patients where stratified for baseline risk, the effect of statin treatment lost statistical significance in almost all categories [28]. Similar inconclusive results were reported also when considering different outcomes such as vascular or non-vascular deaths or composite ones; here, again, stratification for baseline risk deeply impacted the magnitude of the results. The authors concluded that despite the high number of patients under statins treatment for primary CV prevention, the evidence for their prescription in this setting is very limited and should be substantiated by a careful individual assessment of baseline risk, absolute risk reduction, and potential harm [28].

On the other hand, the role of high-intensity statin treatment as a secondary prevention measure to reduce the recurrence of CV and cerebrovascular events is well established and highlighted by all international guidelines [20–22]. In patients with previous myocardial infarction and stroke, statins blunt the rate of recurrent CV events as well as the need for revascularization procedures. In addition, mortality is considerably reduced: In the five years after myocardial infarction, treatment of only 30 patients with statin is already able to prevent one cardiovascular death [29]. The pioneering Scandinavian Simvastatin Survival Study (4S) trial compared simvastatin treatment vs. placebo in $n = 4444$ patients with angina pectoris or previous myocardial infarction and found statin to greatly reduce the risk of death (both cardiovascular and non-cardiovascular ones) as well as that of undergoing revascularization procedures [29]. More recently, the Pravastatin or Atorvastatin Evaluation and Infection Therapy—Thrombolysis in Myocardial Infarction 22 (PROVE-IT TIMI 22) trial compared high-intensity (atorvastatin 80 mg/day) vs. moderate-intensity (pravastatin 40 mg/day) statin treatment early after ACS and found the strongest intervention to bring an additional 16% reduction of cardiovascular events as compared to pravastatin 40 mg/day, in 4162 patients [30]. Of interest, the benefit was already evident within 30 days and became statistical significant throughout the 2.5 years of follow-up [30]. After this, several other trials tested different statins at different dosages, and results have been summarized in numerous meta-analyses. Among them, in 2010 the Cholesterol Treatment Trialists (CTT) Collaboration analyzed five randomized trials comparing more intensive vs. less intensive statin regimens in $n = 39612$ patients with ACS or stable coronary disease [30]. As a result, the intensive statin treatment showed a greater reduction in major CV events compared to the less intensive one. Moreover, this research highlighted that statin benefit is maintained among patients with and without hypercholesterolemia, and no threshold was found under which LDL lowering was ineffective [30]. Recently, a specific analysis investigated specific population such as elderly people and the benefit of statin treatment in secondary prevention remained valid, although some warnings have been raised for specific high-dose regimens [31,32]. Importantly, the prognostic role of LDL reduction and the importance of an early start to high-dose statin treatment after ACS was consistently shown among the majority of clinical trials. This aspect has been taken further by the recent secondary prevention trials investigating the use of non-statin lipid-lowering agents in association with the maximally tolerated statin dose in ACS patients, confirming the concept of "the lower, the better" [33]. Accordingly, the very recent guidelines on dyslipidemia by the European Society of Cardiology (ESC)/European Atherosclerosis Society (EAS) have adopted a more aggressive approach with never-seen-before very low targets for LDL levels in high-risk categories (such as individuals with previous CV events) [20]. Indeed, the LDL target for patients at very high risk is now set at 1.4 mmol/L (<55 mg/dL), while in patients at very high-risk with multiple recent events, the target reaches 1.0 mmol/L (<40 mg/dL) [20].

2.2. Diabetic Cardiomyopathy

Diabetic patients are at increased risk of developing heart failure. The Framingham Heart Study clearly indicated that diabetes and heart failure are associated, independently of the presence of coronary artery disease and hypertension [34]. As such, hyperglycemia can cause cardiac insufficiency not only by increasing the risk of heart failure determinants but also by directly affecting cardiac structure and function. Diabetes is associated with cardiac oxidative stress, intracellular ion abnormalities, inflammation, and mitochondrial dysfunction, with metabolic turbulences directly causing the development of heart failure, and particularly, heart failure with preserved ejection fraction, by altering specific signaling pathways [35]. Although debated [36], diabetes is also thought to be associated with systolic heart failure, as previous work showed that indexes of systolic function may be slightly reduced in diabetic patients without overt coronary disease [37,38]. In this case, the chronic alteration of glucose levels may cause a reduction of myocardial flow reserve due to microvascular alterations and lead to subendocardial ischemia and systolic dysfunction [39]. Although the molecular signaling deranged by the chronic exposure to high glucose levels is very diverse and several pathways have been involved in the pathophysiology of diabetic cardiomyopathy, in general, they all converge towards the activation of the transcription factor NF-kB, which then leads to upregulation of cytokines, chemokines, and adhesion molecules [40]. Indeed, genetic or pharmacologic inhibition of this nuclear factor mitigates cardiac inflammation and oxidative stress in animal models of diabetes, thus preventing the development of diabetic cardiomyopathy [41,42]. Glycemia-oriented therapy does not effectively prevent cardiac complications of long-term type 2 diabetes mellitus [43], thus other drugs have been tested to reduce cardiac damage. Statins have been hypothesized to hold a protective role in the setting of diabetic cardiomyopathy thanks to their anti-inflammatory role. Furthermore, hyperlipidemia is associated with intracardiac accumulation of fatty acids and dysfunction due to lipotoxicity in the diabetic myocardium [44]. Pre-clinical evidence strongly supports this hypothesis; atorvastatin could improve left ventricular function by reducing cardiac intramyocardial inflammation and myocardial fibrosis in an experimental model of diabetic cardiomyopathy [45]. In addition, atorvastatin could reduce β-adrenergic dysfunction in rats with diabetic cardiomyopathy via increasing nitric oxide (NO) availability [46]. Rosuvastatin also exhibited protective properties in this setting by reducing NLRP3 inflammasome and IL-1β activation via suppression of MAPK pathways [47]. Finally, simvastatin could reduce cardiac dysfunction in streptozotocin-induced diabetic rats by attenuating hyperglycemia-induced cardiac oxidative stress, inflammation, and apoptosis. In the clinical setting, intensive lipid control with statins and other drugs is associated with an important decrease of cardiovascular risk in diabetic patients [48–50]. Accordingly, statins together with other lipid-modifying agents (i.e., peroxisome proliferator-activated receptor (PPAR) agonists) are suggested by diabetes guidelines for both primary and secondary CV prevention [51]. This being said, statins failed to effectively modify the course of diabetic cardiomyopathy, and they may even facilitate the onset of diabetes by impacting peripheral insulin sensitivity and β-cell function [52]. Nevertheless, discontinuing statin therapy in diabetic patients is not recommended [53].

3. Statins, Inflammation, and IL-1β

The effectiveness of statin anti-inflammatory properties in the CV setting has been definitively proven in clinical trials. The Justification for the Use of Statins in Prevention: an Intervention Trial Evaluating Rosuvastatin (JUPITER) trial enrolled apparently healthy persons without hyperlipidemia but with elevated hs-CRP and demonstrated that treatment with rosuvastatin significantly reduced the incidence of major CV events [54]. Similarly, the Pravastatin Or Atorvastatin Evaluation and Infection Therapy (PROVE-IT), Aggrastat to Zocor (AtoZ), and Improved Reduction of Outcomes: Vytorin Efficacy International Trial (IMPROVE-IT) trials also reported the clinical relevance of statin-related hs-CRP reduction [30,55,56]. Statins exert these anti-inflammatory effects by blunting the downstream synthesis of molecules in the mevalonate pathway through the inhibition of small GTPase prenylation and isoprenoid production [57–59]. Of note, small GTPases regulate different signaling pathways and thus cellular processes dependent on isoprenylation and involved in the development of CV diseases [60,61]. Rho and Rac cooperate in the surge of oxidative stress and inflammatory mediators that characterize different pathologic processes [62,63]. Furthermore, Ras is thought to play a central role in the regulation of cellular growth and proliferation [64]. The inhibition of those pathways is associated with a number of protective immunomodulatory effects in inflammatory cells and vascular and myocardial tissue [65,66]. In vascular cells, statins can enhance the availability of protective nitric oxide (NO) by increasing its synthesis and reducing its degradation [67,68]. This accompanies a sensible reduction of endothelial oxidative stress and modulation of redox-sensitive transcription factors such as NF-kβ and activator protein 1 (AP-1), key enzymes involved in the regulation of several pro-inflammatory genes [69]. As a result, treatment with statins is associated with blunted expression of pro-thrombotic factors as well as different adhesion molecules such as vascular cell adhesion molecule 1 (VCAM-1), platelet endothelial adhesion molecule 1 (PECAM-1), intercellular cell adhesion molecule-1 (ICAM-1), and P-selectin [70–72]. Moreover, statin treatment reduces monocyte, endothelial, and vascular smooth cell production of different pro-inflammatory cytokines including monocyte chemoattractant protein-1 (MCP-1), regulated on activation, normal T cell expressed and secreted (RANTES), and interleukin (IL)-6 and IL-8 [70,73,74]. Statins also exert their immunomodulatory effects by reducing monocyte expression of CD11β (an integrin with a key role in monocyte–endothelium interaction), suppressing the expression of major histocompatibility complex (MHC) class II protein, as well as reducing the proliferation and differentiation of activated T- and B-lymphocytes [75–77].

Dyslipidemia, altered glucose metabolism, and inflammation share several cardiac signaling pathways and are closely interconnected (Figure 1). Although the anti-inflammatory role of statins is widely accepted, different studies demonstrated that those molecules could paradoxically increase the production of IL-1β—among the most important pro-inflammatory cytokines—as a result of reduced protein prenylation in immune cells [78]. Mature, active IL-1β derives from the cleavage of its pro-form by the NOD-like receptor family, pyrin-domain-containing (NLRP) 3 inflammasome [79]. The NLRP3 inflammasome complex is formed by the sensor molecule NLRP3, the adaptor protein ASC, and pro-caspase-1. This multimeric protein complex regulates the release of cytokines IL-1β and IL-18, alongside initiating an inflammatory form of cell death known as pyroptosis [80]. Inflammasome activation is a two-step process that requires adequate priming of NLRP3 followed by a signal triggering the assembling [80]. The priming step occurs via different inflammatory stimuli such as TLR4 agonists, resulting in activation of NF-kB and transcription of NLRP3 and pro-IL-1β [81–83]. Furthermore, NLRP3 priming also associates with post-translational modifications of NLRP3 (such as phosphorylation and ubiquitination), further regulating its activation [84,85]. The second step is provided by the recognition of damage-associated molecular patterns (DAMPs) causing the perturbation of cellular metabolism with the production of reactive oxygen species, ion disturbances, and lysosomal disruption [86–89]. Due to its pro-inflammatory role, the NLRP3 inflammasome has progressively become an important molecular target to cope with different chronic diseases, including myocardial infarction and diabetes [90]. To date, five specific NLRP3 inhibitors have been validated in vivo or in vitro and entered clinical testing at different phases [91]. On the other hand, many drugs able to modify cellular metabolism and homeostasis have been shown to activate the inflammasome [92]. Although some controversies still exist due to possible differences between different molecules, statins are acknowledged among NLRP3 activators and thus IL-1β inducers. Of note, this characteristic is thought to account for their association with diabetes onset [78,93,94]. Various statins have been shown to increase IL-1β secretion from macrophages through NLRP3 activation, but none of them were able to act as a priming agent on the inflammasome as they all need bacterial liposaccharide to induce caspase-1-dependent cleavage of pro-IL1β into its active form [95,96]. In a report from 2014, Henriksbo and colleagues showed that long-term treatment of obese mice with fluvastatin promoted insulin resistance in adipose tissue and increased caspase-1 activity and IL-1β production in adipose tissue explants in the presence of LPS [95]. Of interest, this effect was not observed in NLRP3$^{-/-}$ explants and was reversed by glyburide, a known inflammasome inhibitor and antidiabetic drug [95].

Similarly, fluvastatin could increase the secretion of 1L-1β and IL-18 in peripheral blood mononuclear cells stimulated by *Mycobacterium tuberculosis* [97]. In additon, lovastatin increases reactive oxygen species (ROS) and synergizes with LPS to trigger IL-1β release in macrophages and monocytes [98]. Of interest, these pro-inflammatory effects have been shown to relate with statin-related disturbances on protein prenylation as addition of mevalonate or GGPP—an intermediate in the mevalonate pathway—could prevent IL-1β release [96,99]. Recently, this aspect has been further dissected by showing that, differently from LDL lowering, statin-related reduction of isoprenoids was required for NLRP3/caspase-1 inflammasome activation and IL-1β-dependent insulin resistance in adipose tissue [100]. Furthermore, supplementation of geranylgeranyl isoprenoids or caspase-1 inhibition could prevent statin-induced alteration of insulin signaling [100]. Moreover, IL-1β, but not IL-18, is necessary to induce insulin resistance in adipose tissue treated with atorvastatin [100]. Summarizing, inflammasome activation and IL-1β secretion likely link statins with impaired glucose metabolism. Thus, inflammasome might be an effective molecular target to reduce statin-related diabetes onset. On the other hand, targeting the inflammasome and IL-1β might reduce the effectiveness of statin treatment on CV prevention, as the importance of this interleukin has been recently highlighted in the CANTOS trial.

4. Perspective

Human coronary plaques are inflammatory lesions in which immune cells and inflammatory molecules are detectable at a high level and play pivotal roles [101–104]. Recently, the CANTOS trial confirmed the inflammatory theory of atherosclerosis and shed new light on the role of IL-1β in CV risk determination [16]. A total of 10,061 patients with a previous myocardial infarction and showing inflammatory residual risk (CRP > 2 mg/L) under optimal CV-protective therapy were enrolled in this randomized, double-blind trial to receive canakinumab, a IL-1β inhibitory monoclonal antibody, or a placebo every 3 months. Levels of lipids remained unaltered upon treatment with canakinumab, while a significant decrease in CRP levels was observed already after the first administration of the anti-inflammatory drug. The primary endpoint composed of cardiovascular death, non-fatal myocardial infarction, and non-fatal stroke was successfully met by the intermediate doses of the drugs (100 and 150 mg/administration) [16]. Of interest, those patients with CRP levels reduced to <2 mg/L after the first administration benefitted the most from the long-term treatment as this was associated with a 31% reduction in CV mortality, a 31% reduction in all-cause mortality, and a 25% reduction in major adverse CV events [105]. Conversely, in patients with on-treatment high-sensitivity CRP ≥ 2 mg/L, the treatment effects were non-significant [105]. Of interest, canakinumab was also effective in reducing rates of non-cardiovascular inflammatory disease, such as lung cancer, arthritis, and gout. As expected, patients treated with IL-1β inhibitory antibody had a higher rate of fatal infections as well as of neutropenia or thrombocytopenia [16]. The CANTOS trial not only provided solid proof of the effectiveness of IL-1β inhibition in secondary CV prevention; it also allowed a deepening of the complex relationship between lipids and inflammation. Indeed, cholesterol crystals can induce IL-1β activation via canonical (i.e., NLRP3-mediated) and non-canonical pathways, then IL-1β establishes a vicious circle that ends up in further pro-IL1β cleavage [106–108]. Then, blocking this pathway via canakinumab could reduce the effect of lipids on atherosclerotic inflammation. On the other hand, substantial evidence exists demonstrating a role for inflammation on the induction of dyslipidemia [109,110]. Differently from other anti-inflammatory drugs, in CANTOS, canakinumab did not affect cholesterol levels, while it slightly increased triglycerides [16]; thus, the cardiovascular preventive effects did not depend on any lipid effects related to IL-1β.

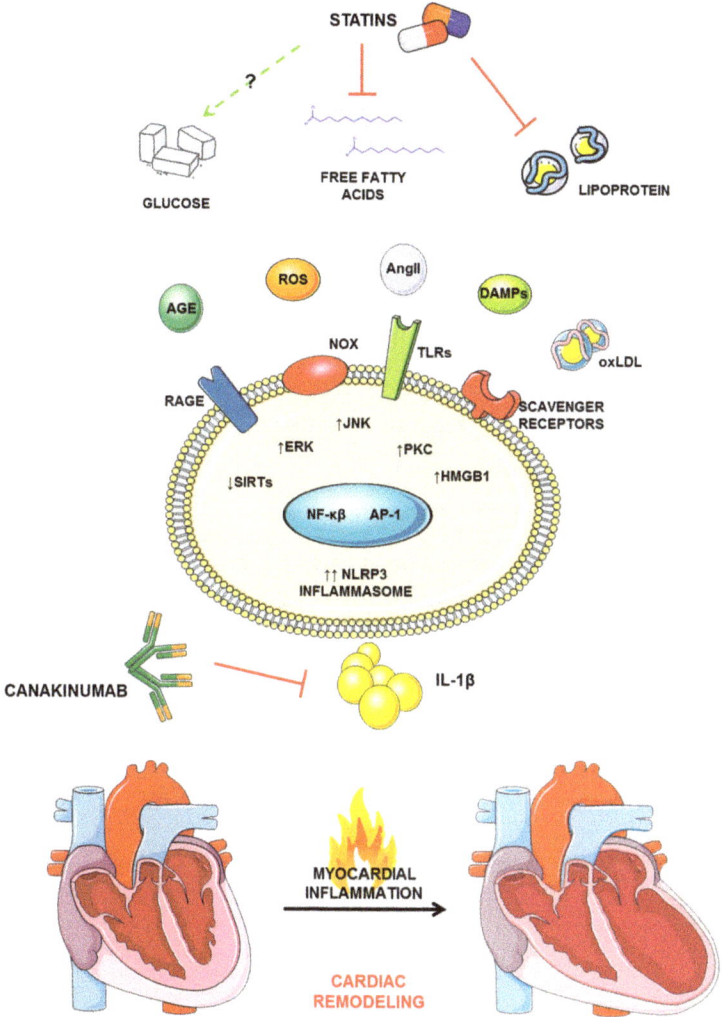

Figure 1. Altered lipid and glucose metabolisms share common molecular pathways in the pathophysiology of cardiac remodeling, relying on the release of pro-inflammatory interleukin (IL)-1β. The increased levels of different deleterious mediators (such as AGE, AngII, DAMP, and modified lipoproteins) are sensed by promiscuous receptors on cell surfaces and trigger secondary signaling pathways leading to activation of NF-κβ and AP-1, two transcription factors involved in the regulation of NLRP3 inflammasome activity. The activated inflammasome then leads to the activation and release of IL-1β, which fuel the sterile inflammation associated with cardiac remodeling. AGE: advanced glycation end-product; AngII: angiotensin II; AP-1: activator protein 1; DAMPs: damage-associated molecular patterns; ERK: extracellular signal-regulated kinases; HMGB1: high mobility group box 1; IL-1β: interleukin 1β; JNK: Janus kinases; NF-κβ: nuclear factor kappa-light-chain-enhancer of activated B cells; NLRP3: NACHT-, LRR-, and PYD-domain-containing protein 3; NOX: NADPH oxidase; PKC: protein kinase C; TLR: Toll-like receptor; RAGE: receptor of AGE; ROS: reactive oxygen species; oxLDL: oxidized low-density lipoproteins.

Canakinumab is effective and relatively safe for secondary prevention of CV events; whether this might also be the case for primary CV prevention or for the treatment of myocardial infarction sequelae such as cardiac remodeling remains to be investigated. Cardiac repair after myocardial infarction depends on the tight regulation of sterile inflammation, which serves to clear damaged cells and promote the formation of a functional scar; alterations of the inflammatory balance associate with deleterious myocardial remodeling, resulting in cardiac dysfunction and heart failure [111]. Being a key regulator of inflammation, IL-1β plays an important role in orchestrating the inflammatory response in an ischemic/reperfused myocardium [112]. In this setting, IL-1β can activate downstream mediators which further amplify inflammation via MAPK and NF-κB signaling. Furthermore, it allows for the spatial extension of inflammation by activating local parenchymal and infiltrating cells that express its receptor and facilitates leukocytes recruitment via increasing the expression of adhesion molecules and chemoattractant in the damaged myocardium [111]. In line with this evidence, previous experimental studies reported a role of IL-1 blockade in preventing adverse cardiac remodeling (Table 1).

Table 1. Experimental studies investigating the effect of IL-1 inhibition in preventing cardiac remodeling after myocardial infarction.

Author	Year	Drug(dose)	Schedule	Results
Abbate A. et al. [113]	2008	Anakinra (1 mg/kg)	Immediate or delayed (24 h after ischemia) and then daily for 6 days.	Anakinra-treated mice showed signs of more favorable ventricular remodeling.
Van Tassell et al. [114]	2010	IL-1 Trap (1, 5 or 30 mg/kg)	Every 48 h after surgery.	Mice treated with 5 or 30 mg/kg of IL-1 Trap had more favorable cardiac remodeling and echocardiographic assessment of infarct size at 7 days.
Toldo et al. [115]	2012	rhIL-1Ra	10 mg/kg given either 30 min or 4 h prior to surgery	Irrespective of dose, treated mice showed marked cardio-protection in terms of LVEF and the reduction of the infarct size.
Toldo et al. [116]	2013	Anti-IL-1β Ab	10 mg/kg immediately after surgery and then 1 week later.	When compared with control vehicle, anti-IL-1β Ab limit left ventricular enlargement and improve systolic dysfunction by inhibiting cardiomyocyte apoptosis.
Toldo et al. [117]	2014	Anti-IL-1β Ab	10 mg/kg 1 week after surgery and then weekly for 9 weeks.	After 10 weeks, anti-IL-1β Ab prevents reduction of LVEF, impairment in the myocardial performance index. and contractile reserve.
De Jesus et al. [118]	2017	Anakinra (10 mg/kg)	Daily, starting 24 h after surgery	Anakinra improved conduction velocity and reduced action potential duration dispersion, thus determining a reduction of spontaneous and inducible ventricular arrhythmias.
Mauro et al. [119]	2017	IL-1α-blocking antibody (15 µg/kg)	Single dose after reperfusion	At 24 h, IL-1α blockade significantly reduced inflammasome formation and infarct size, thus preserving LVFS.
Herouki et al. [120]	2017	Anti-IL-1β Ab	Single dose after reperfusion or 7 days after reperfusion	Immediate, but not delayed, administration of anti-IL-1β Ab reduces ischemia/reperfusion-related infarct size, left ventricular remodeling, and heart-failure-related coronary dysfunction.

IL: interleukin; rhIL-1Ra: recombinant human interleukin-1 receptor antagonist; LVEF: left ventricular ejection fraction; LVFS: left ventricular fractional shortening.

Unfortunately, the few randomized clinical trials and observational and cohort studies that have evaluated the effect of IL-1β inhibition in relation to the development of post-MI cardiac remodeling have provided conflicting results [121–124] (Table 2). Of interest, most of them were based on the unspecific blockage of IL-1 receptor, which recognizes both IL-1α and β isoforms; whether a more specific targeting of the IL-1β pathway via canakinumab might provide additional beneficial effects on top of statins in the context of post-myocardial remodeling remains to be fully determined. Recently, a sub-analysis of the previously mentioned CANTOS trial suggested post-MI treatment with canakinumab to dose-dependently reduce hospitalization for heart failure and the composite of hospitalization for heart failure or heart-failure-related mortality as compared to a placebo [125]. In this regard, it is important to take into consideration that the CV-protective role of canakinumab has been demonstrated only in patients with residual inflammatory risk, while that of statins is not restricted to this group. Furthermore, lipid-lowering therapies hold a very competitive risk/benefit balance even when very low LDL levels are reached [126], while this is not the case for IL-1β blockade, which is associated with a higher risk of sepsis and fatal infections.

Similarly, given the relevance of inflammatory mediators, and particularly IL-1β, in the pathophysiology of diabetes and diabetic cardiomyopathy [127], targeting the NLRP3/IL1β pathway could effectively reduce the burden of this disease. Given the high number of diabetic individuals under cardio-protective treatment with statins and the possible deleterious effect of these drugs on Il-1β activation and thus glycemic control, adding IL-1β inhibition on top of statin treatment might give additional benefit in terms of CV protection. In this sense, it will be very important to understand whether the inflammasome could be safely targeted without altering the general anti-inflammatory effect of statins. How statins could be generally anti-inflammatory and thus protective in the CV setting while increasing the risk of diabetes remains to be fully explained. In other words, on which pathophysiological aspect of the two diseases does the mechanism of action of statins differ? Different investigators have previously tried to address this question, with different hypotheses being made [78]. Statins might have different effects on different cells with different roles in the diseases. In this sense, the effect of statins on endothelial cells should drive the protective CV effects, while their roles on adipocytes, pancreatic islet cells, or myocytes could be of more relevance in diabetes onset and diabetic cardiomyopathy [78]. In addition, the cholesterol-lowering effect might play a more important role in the prevention of CV disease as compared to that played in diabetes development. The connection between statins, NLRP3/IL-1β, and insulin resistance remains to be characterized in depth, and many mechanistic questions are still unsolved; understanding these aspects might pave the way for new therapeutic strategies, including a combination of statins and IL-1β inhibition.

Table 2. Clinical studies investigating the effect of IL-1 inhibition in preventing cardiac remodeling after myocardial infarction.

Author	Year	Drug	Treatment	Disease (cohort)	Results
Abbate et al. VCU-ART [128]	2010	Anakinra	100 mg/daily sc for 14 days	STEMI (n = 10)	In this pilot double blind RCT, treatment with anakinra showed to be safe and to reduce left ventricular remodeling (assessed by both echocardiography and cardiac magnetic resonance) after STEMI as compared to placebo.
Morton et al. MRC-ILA-HEART [129]	2015	Anakinra	100 mg/daily sc for 14 days	NSTEMI (n = 182)	In this proof-of-principle double blind RCT, patients treated with anakinra showed reduced levels of hsCRP and IL-6 as compared to those receiving a placebo.

Table 2. Cont.

Author	Year	Drug	Treatment	Disease (cohort)	Results
Abbate et al. VCU-ART2 [121]	2013	Anakinra	100 mg/daily sc for 14 days	STEMI (n = 30)	In this pilot double blind RCT, treatment with anakinra could reduce hsCRP levels as compared to a placebo. Anakinra-treated patients also showed a numerically lower incidence of heart failure, although this was not statistically significant.
Ridker et al. CANTOS [125]	2019	Canakinumab	50, 100 or 150 mg/daily sc every 3 months	STEMI (n = 10'061)	In this double blind RCT, treatment with canakinumab after STEMI was shown to dose-dependently reduce hospitalization for heart failure and the composite of hospitalization for heart failure or heart-failure-related mortality as compared to a placebo.
Van Tassell et al. VCU-ART3 [130]	2019	Anakinra	100 mg once or twice/daily for 14 days	STEMI (n = 99)	Preliminary results of this double blind RCT were presented at the 2019 Congress of the European Society of Cardiology. Patients treated with anakinra showed significant improvement in cardiac systolic function after STEMI, as compared to a placebo.

CANTOS: Canakinumab Anti-Inflammatory Thrombosis Outcomes Study; hsCRP: high-sensitivity C-reactive protein; IL-6: interleukin-6; NSTEMI: non-ST-elevation myocardial infarction; RCT: randomized clinical trial; STEMI: ST-elevation myocardial infarction; VCU-ART: Virginia Commonwealth University Anakinra Remodeling Trial.

5. Conclusions

Firstly introduced to reduce circulating LDL, statins soon became pillars of prevention and treatment of CV diseases. Aside from their lipid-lowering actions, statins hold different pleiotropic effects that are thought to deeply contribute to their CV-protective effect and involve the modulation of the inflammatory response. Statins are recommended for prevention of myocardial infarction in patients with dyslipidemia, high, or very high cardiovascular risk. In diabetic subjects, statins have been hypothesized to reduce the development of diabetic cardiomyopathy thanks to their anti-inflammatory effect. Despite the encouraging results of the pre-clinical tests, statins failed to effectively modify the course of heart failure in diabetic patients and may even facilitate the onset of diabetes in patients without previous glucose disturbances. Thanks to the CANTOS trial involving the IL-1β inhibitory antibody canakinumab, this pro-inflammatory cytokine has recently emerged as an effective and relatively safe target for secondary CV prevention in patients with residual inflammatory risk. Although generally seen as anti-inflammatory drugs, statins may have different effects on IL-1β synthesis in different cells, with some studies even demonstrating a paradoxical increase. In consideration of the detrimental role of IL-1β in the pathophysiology of myocardial infarction and diabetic cardiomyopathy, adding canakinumab on top of statins in these patients might then provide a stronger inhibition of the IL-1β-mediated inflammatory response, with additional beneficial effects on the pathophysiology of these diseases. In addition, in patients treated with statins, canakinumab might even be able to reduce statin-induced insulin resistance as this is thought to depend on the activation of the NLRP3 inflammasome/IL-1β pathway. Further specific investigations will be needed to test this hypothesis in order to reduce the very high global burden of myocardial infarction and diabetic cardiomyopathy.

Author Contributions: Conceptualization, L.L. and F.M.; writing—original draft preparation, L.L. and F.C.; writing—review and editing, F.C., G.G.C., and F.M..; fund acquisition, G.G.C. and F.M.

Funding: This research was funded by a grant from the Italian Ministry of Health to the Italian Cardiovascular Network Grant number 2754291 (to F.M.). Furthermore, the present work was supported by the Swiss National Science Foundation (to G.G.C.) [310030_175546], the Alfred and Annemarie von Sick Grants for Translational and Clinical Research Cardiology and Oncology (to G.G.C.), and the Foundation for Cardiovascular Research–Zurich Heart House. G.G.C. is a recipient of an H.H. Sheikh Khalifa bin Hamad Al Thani Foundation Assistant Professorship at the Faculty of Medicine of the University of Zurich.

Acknowledgments: Figure 1 was designed using Servier Medical Art by Servier under a Creative Commons Attribution 3.0 Unported License.

Conflicts of Interest: The authors declare no conflict of interest.

References

1. Endo, A.; Kuroda, M. Citrinin, an inhibitor of cholesterol synthesis. *J. Antibiot. (Tokyo)* **1976**, *29*, 841–843. [CrossRef] [PubMed]
2. Istvan, E.S.; Deisenhofer, J. Structural mechanism for statin inhibition of hmg-coa reductase. *Science* **2001**, *292*, 1160–1164. [CrossRef] [PubMed]
3. Goldstein, J.L.; Brown, M.S. Regulation of the mevalonate pathway. *Nature* **1990**, *343*, 425–430. [CrossRef] [PubMed]
4. Sacks, F.M.; Pfeffer, M.A.; Moye, L.A.; Rouleau, J.L.; Rutherford, J.D.; Cole, T.G.; Brown, L.; Warnica, J.W.; Arnold, J.M.; Wun, C.C.; et al. The effect of pravastatin on coronary events after myocardial infarction in patients with average cholesterol levels. Cholesterol and recurrent events trial investigators. *N. Engl. J. Med.* **1996**, *335*, 1001–1009. [CrossRef] [PubMed]
5. Long-Term Intervention with Pravastatin in Ischaemic Disease Study Group. Prevention of cardiovascular events and death with pravastatin in patients with coronary heart disease and a broad range of initial cholesterol levels. *N. Engl. J. Med.* **1998**, *339*, 1349–1357. [CrossRef] [PubMed]
6. Schonbeck, U.; Libby, P. Inflammation, immunity, and hmg-coa reductase inhibitors: Statins as antiinflammatory agents? *Circulation* **2004**, *109*, 18–26. [CrossRef]
7. Liao, J.K.; Laufs, U. Pleiotropic effects of statins. *Annu. Rev. Pharmacol. Toxicol.* **2005**, *45*, 89–118. [CrossRef]
8. Liberale, L.; Montecucco, F.; Camici, G.G.; Dallegri, F.; Vecchie, A.; Carbone, F.; Bonaventura, A. Treatment with proprotein convertase subtilisin/kexin type 9 (pcsk9) inhibitors to reduce cardiovascular inflammation and outcomes. *Curr. Med. Chem.* **2017**, *24*, 1403–1416. [CrossRef]
9. Carbone, F.; Liberale, L.; Bonaventura, A.; Cea, M.; Montecucco, F. Targeting inflammation in primary cardiovascular prevention. *Curr. Pharm. Des.* **2016**, *22*, 5662–5675. [CrossRef]
10. Montecucco, F.; Liberale, L.; Bonaventura, A.; Vecchie, A.; Dallegri, F.; Carbone, F. The role of inflammation in cardiovascular outcome. *Curr. Atheroscler. Rep.* **2017**, *19*, 11. [CrossRef]
11. Bonaventura, A.; Montecucco, F.; Dallegri, F.; Carbone, F.; Luscher, T.F.; Camici, G.G.; Liberale, L. Novel findings in neutrophil biology and their impact on cardiovascular disease. *Cardiovasc. Res.* **2019**, *115*, 1266–1285. [CrossRef] [PubMed]
12. Liberale, L.; Camici, G.G. The role of vascular aging in atherosclerotic plaque development and vulnerability. *Curr. Pharm. Des.* **2019**. [CrossRef] [PubMed]
13. Carbone, F.; Bonaventura, A.; Liberale, L.; Paolino, S.; Torre, F.; Dallegri, F.; Montecucco, F.; Cutolo, M. Atherosclerosis in rheumatoid arthritis: Promoters and opponents. *Clin. Rev. Allergy Immunol.* **2019**, 1–14. [CrossRef] [PubMed]
14. Ridker, P.M.; Everett, B.M.; Pradhan, A.; MacFadyen, J.G.; Solomon, D.H.; Zaharris, E.; Mam, V.; Hasan, A.; Rosenberg, Y.; Iturriaga, E.; et al. Low-dose methotrexate for the prevention of atherosclerotic events. *N. Engl. J. Med.* **2019**, *380*, 752–762. [CrossRef] [PubMed]
15. Bally, M.; Dendukuri, N.; Rich, B.; Nadeau, L.; Helin-Salmivaara, A.; Garbe, E.; Brophy, J.M. Risk of acute myocardial infarction with nsaids in real world use: Bayesian meta-analysis of individual patient data. *BMJ* **2017**, *357*, 1909. [CrossRef]
16. Ridker, P.M.; Everett, B.M.; Thuren, T.; MacFadyen, J.G.; Chang, W.H.; Ballantyne, C.; Fonseca, F.; Nicolau, J.; Koenig, W.; Anker, S.D.; et al. Antiinflammatory therapy with canakinumab for atherosclerotic disease. *N. Engl. J. Med.* **2017**, *377*, 1119–1131. [CrossRef]

17. Baylis, R.A.; Gomez, D.; Mallat, Z.; Pasterkamp, G.; Owens, G.K. The cantos trial: One important step for clinical cardiology but a giant leap for vascular biology. *Arterioscler. Thromb. Vasc. Biol.* **2017**, *37*, 174–177. [CrossRef]
18. Thygesen, K.; Alpert, J.S.; Jaffe, A.S.; Chaitman, B.R.; Bax, J.J.; Morrow, D.A.; White, H.D.; Group E.S.C.S.D. Fourth universal definition of myocardial infarction (2018). *Eur. Heart J.* **2019**, *40*, 237–269. [CrossRef]
19. Adhyaru, B.B.; Jacobson, T.A. Safety and efficacy of statin therapy. *Nat. Rev. Cardiol.* **2018**, *15*, 757–769. [CrossRef]
20. Mach, F.; Baigent, C.; Catapano, A.L.; Koskinas, K.C.; Casula, M.; Badimon, L.; Chapman, M.J.; De Backer, G.G.; Delgado, V.; Ference, B.A.; et al. 2019 esc/eas guidelines for the management of dyslipidaemias: Lipid modification to reduce cardiovascular risk. *Eur. Heart J.* **2019**. [CrossRef]
21. Piepoli, M.F.; Hoes, A.W.; Agewall, S.; Albus, C.; Brotons, C.; Catapano, A.L.; Cooney, M.T.; Corra, U.; Cosyns, B.; Deaton, C.; et al. 2016 european guidelines on cardiovascular disease prevention in clinical practice: The sixth joint task force of the european society of cardiology and other societies on cardiovascular disease prevention in clinical practice (constituted by representatives of 10 societies and by invited experts)developed with the special contribution of the european association for cardiovascular prevention & rehabilitation (eacpr). *Eur. Heart J.* **2016**, *37*, 2315–2381. [PubMed]
22. Grundy, S.M.; Stone, N.J.; Bailey, A.L.; Beam, C.; Birtcher, K.K.; Blumenthal, R.S.; Braun, L.T.; de Ferranti, S.; Faiella-Tommasino, J.; Forman, D.E.; et al. 2018 aha/acc/aacvpr/aapa/abc/acpm/ada/ags/apha/aspc/nla/pcna guideline on the management of blood cholesterol: A report of the american college of cardiology/american heart association task force on clinical practice guidelines. *J. Am. Coll. Cardiol.* **2019**, *73*, 285–350. [CrossRef] [PubMed]
23. Byrne, P.; Cullinan, J.; Murphy, C.; Smith, S.M. Cross-sectional analysis of the prevalence and predictors of statin utilisation in ireland with a focus on primary prevention of cardiovascular disease. *BMJ Open* **2018**, *8*, 18524. [CrossRef] [PubMed]
24. Jang, T.L.; Bekelman, J.E.; Liu, Y.; Bach, P.B.; Basch, E.M.; Elkin, E.B.; Zelefsky, M.J.; Scardino, P.T.; Begg, C.B.; Schrag, D. Physician visits prior to treatment for clinically localized prostate cancer. *Arch. Intern. Med.* **2010**, *170*, 440–450. [CrossRef] [PubMed]
25. Petretta, M.; Costanzo, P.; Perrone-Filardi, P.; Chiariello, M. Impact of gender in primary prevention of coronary heart disease with statin therapy: A meta-analysis. *Int. J. Cardiol.* **2010**, *138*, 25–31. [CrossRef]
26. Brugts, J.J.; Yetgin, T.; Hoeks, S.E.; Gotto, A.M.; Shepherd, J.; Westendorp, R.G.; de Craen, A.J.; Knopp, R.H.; Nakamura, H.; Ridker, P.; et al. The benefits of statins in people without established cardiovascular disease but with cardiovascular risk factors: Meta-analysis of randomised controlled trials. *BMJ* **2009**, *338*, 2376. [CrossRef]
27. de Vries, F.M.; Denig, P.; Pouwels, K.B.; Postma, M.J.; Hak, E. Primary prevention of major cardiovascular and cerebrovascular events with statins in diabetic patients: A meta-analysis. *Drugs* **2012**, *72*, 2365–2373. [CrossRef]
28. Byrne, P.; Cullinan, J.; Smith, A.; Smith, S.M. Statins for the primary prevention of cardiovascular disease: An overview of systematic reviews. *BMJ Open* **2019**, *9*, 23085. [CrossRef]
29. Scandinavian Simvastatin Survival Study Group. Randomised trial of cholesterol lowering in 4444 patients with coronary heart disease: The scandinavian simvastatin survival study (4s) *Lancet* **1994**, *344*, 1383–1389.
30. Cannon, C.P.; Braunwald, E.; McCabe, C.H.; Rader, D.J.; Rouleau, J.L.; Belder, R.; Joyal, S.V.; Hill, K.A.; Pfeffer, M.A.; Skene, A.M.; et al. Intensive versus moderate lipid lowering with statins after acute coronary syndromes. *N. Engl. J. Med.* **2004**, *350*, 1495–1504. [CrossRef]
31. Rodriguez, F.; Maron, D.J.; Knowles, J.W.; Virani, S.S.; Lin, S.; Heidenreich, P.A. Association between intensity of statin therapy and mortality in patients with atherosclerotic cardiovascular disease. *JAMA Cardiol.* **2017**, *2*, 47–54. [CrossRef] [PubMed]
32. Armitage, J.; Bowman, L.; Wallendszus, K.; Bulbulia, R.; Rahimi, K.; Haynes, R.; Parish, S.; Peto, R.; Collins, R. Intensive lowering of ldl cholesterol with 80 mg versus 20 mg simvastatin daily in 12,064 survivors of myocardial infarction: A double-blind randomised trial. *Lancet* **2010**, *376*, 1658–1669. [PubMed]
33. Ference, B.A.; Majeed, F.; Penumetcha, R.; Flack, J.M.; Brook, R.D. Effect of naturally random allocation to lower low-density lipoprotein cholesterol on the risk of coronary heart disease mediated by polymorphisms in npc1l1, hmgcr, or both: A 2 × 2 factorial mendelian randomization study. *J. Am. Coll. Cardiol.* **2015**, *65*, 1552–1561. [CrossRef] [PubMed]

34. Kannel, W.B.; McGee, D.L. Diabetes and cardiovascular disease. The framingham study. *JAMA* **1979**, *241*, 2035–2038. [CrossRef] [PubMed]
35. Jia, G.; Hill, M.A.; Sowers, J.R. Diabetic cardiomyopathy: An update of mechanisms contributing to this clinical entity. *Circ. Res.* **2018**, *122*, 624–638. [CrossRef]
36. Holscher, M.E.; Bode, C.; Bugger, H. Diabetic cardiomyopathy: Does the type of diabetes matter? *Int. J. Mol. Sci.* **2016**, *17*, 2136. [CrossRef]
37. Zarich, S.W.; Arbuckle, B.E.; Cohen, L.R.; Roberts, M.; Nesto, R.W. Diastolic abnormalities in young asymptomatic diabetic patients assessed by pulsed doppler echocardiography. *J. Am. Coll. Cardiol.* **1988**, *12*, 114–120. [CrossRef]
38. Palmieri, V.; Bella, J.N.; Arnett, D.K.; Liu, J.E.; Oberman, A.; Schuck, M.Y.; Kitzman, D.W.; Hopkins, P.N.; Morgan, D.; Rao, D.C.; et al. Effect of type 2 diabetes mellitus on left ventricular geometry and systolic function in hypertensive subjects: Hypertension genetic epidemiology network (hypergen) study. *Circulation* **2001**, *103*, 102–107. [CrossRef]
39. Boudina, S.; Abel, E.D. Diabetic cardiomyopathy revisited. *Circulation* **2007**, *115*, 3213–3223. [CrossRef]
40. Frati, G.; Schirone, L.; Chimenti, I.; Yee, D.; Biondi-Zoccai, G.; Volpe, M.; Sciarretta, S. An overview of the inflammatory signalling mechanisms in the myocardium underlying the development of diabetic cardiomyopathy. *Cardiovasc. Res.* **2017**, *113*, 378–388. [CrossRef]
41. Thomas, C.M.; Yong, Q.C.; Rosa, R.M.; Seqqat, R.; Gopal, S.; Casarini, D.E.; Jones, W.K.; Gupta, S.; Baker, K.M.; Kumar, R. Cardiac-specific suppression of nf-kappab signaling prevents diabetic cardiomyopathy via inhibition of the renin-angiotensin system. *Am. J. Physiol. Heart Circ. Physiol.* **2014**, *307*, 1036–1045. [CrossRef] [PubMed]
42. Mariappan, N.; Elks, C.M.; Sriramula, S.; Guggilam, A.; Liu, Z.; Borkhsenious, O.; Francis, J. Nf-kappab-induced oxidative stress contributes to mitochondrial and cardiac dysfunction in type ii diabetes. *Cardiovasc. Res.* **2010**, *85*, 473–483. [CrossRef] [PubMed]
43. Fuentes-Antras, J.; Picatoste, B.; Ramirez, E.; Egido, J.; Tunon, J.; Lorenzo, O. Targeting metabolic disturbance in the diabetic heart. *Cardiovasc. Diabetol.* **2015**, *14*, 17. [CrossRef] [PubMed]
44. Costantino, S.; Akhmedov, A.; Melina, G.; Mohammed, S.A.; Othman, A.; Ambrosini, S.; Wijnen, W.J.; Sada, L.; Ciavarella, G.M.; Liberale, L.; et al. Obesity-induced activation of jund promotes myocardial lipid accumulation and metabolic cardiomyopathy. *Eur. Heart J.* **2019**, *40*, 997–1008. [CrossRef]
45. Van Linthout, S.; Riad, A.; Dhayat, N.; Spillmann, F.; Du, J.; Dhayat, S.; Westermann, D.; Hilfiker-Kleiner, D.; Noutsias, M.; Laufs, U.; et al. Anti-inflammatory effects of atorvastatin improve left ventricular function in experimental diabetic cardiomyopathy. *Diabetologia* **2007**, *50*, 1977–1986. [CrossRef]
46. Carillion, A.; Feldman, S.; Na, N.; Biais, M.; Carpentier, W.; Birenbaum, A.; Cagnard, N.; Loyer, X.; Bonnefont-Rousselot, D.; Hatem, S.; et al. Atorvastatin reduces beta-adrenergic dysfunction in rats with diabetic cardiomyopathy. *PLoS ONE* **2017**, *12*, 180103. [CrossRef]
47. Luo, B.; Li, B.; Wang, W.; Liu, X.; Liu, X.; Xia, Y.; Zhang, C.; Zhang, Y.; Zhang, M.; An, F. Rosuvastatin alleviates diabetic cardiomyopathy by inhibiting nlrp3 inflammasome and mapk pathways in a type 2 diabetes rat model. *Cardiovasc. Drugs Ther.* **2014**, *28*, 33–43. [CrossRef]
48. Heart Protection Study Collaborative Group. Mrc/bhf heart protection study of cholesterol lowering with simvastatin in 20,536 high-risk individuals: A randomised placebo-controlled trial. *Lancet* **2002**, *360*, 7–22. [CrossRef]
49. Colhoun, H.M.; Betteridge, D.J.; Durrington, P.N.; Hitman, G.A.; Neil, H.A.; Livingstone, S.J.; Thomason, M.J.; Mackness, M.I.; Charlton-Menys, V.; Fuller, J.H.; et al. Primary prevention of cardiovascular disease with atorvastatin in type 2 diabetes in the collaborative atorvastatin diabetes study (cards): Multicentre randomised placebo-controlled trial. *Lancet* **2004**, *364*, 685–696. [CrossRef]
50. Gaede, P.; Lund-Andersen, H.; Parving, H.H.; Pedersen, O. Effect of a multifactorial intervention on mortality in type 2 diabetes. *N. Engl. J. Med.* **2008**, *358*, 580–591. [CrossRef]
51. Cosentino, F.; Grant, P.J.; Aboyans, V.; Bailey, C.J.; Ceriello, A.; Delgado, V.; Federici, M.; Filippatos, G.; Grobbee, D.E.; Hansen, T.B.; et al. 2019 esc guidelines on diabetes, pre-diabetes, and cardiovascular diseases developed in collaboration with the easd. *Eur. Heart J.* **2019**. [CrossRef] [PubMed]
52. Paseban, M.; Butler, A.E.; Sahebkar, A. Mechanisms of statin-induced new-onset diabetes. *J. Cell. Physiol.* **2019**, *234*, 12551–12561. [CrossRef] [PubMed]

53. Barylski, M.; Nikolic, D.; Banach, M.; Toth, P.P.; Montalto, G.; Rizzo, M. Statins and new-onset diabetes. *Curr. Pharm. Des.* **2014**, *20*, 3657–3664. [CrossRef] [PubMed]
54. Ridker, P.M.; Danielson, E.; Fonseca, F.A.; Genest, J.; Gotto, A.M., Jr.; Kastelein, J.J.; Koenig, W.; Libby, P.; Lorenzatti, A.J.; MacFadyen, J.G.; et al. Rosuvastatin to prevent vascular events in men and women with elevated c-reactive protein. *N. Engl. J. Med.* **2008**, *359*, 2195–2207. [CrossRef]
55. de Lemos, J.A.; Blazing, M.A.; Wiviott, S.D.; Lewis, E.F.; Fox, K.A.; White, H.D.; Rouleau, J.L.; Pedersen, T.R.; Gardner, L.H.; Mukherjee, R.; et al. Early intensive vs. a delayed conservative simvastatin strategy in patients with acute coronary syndromes: Phase z of the a to z trial. *JAMA* **2004**, *292*, 1307–1316. [CrossRef]
56. Cannon, C.P.; Blazing, M.A.; Giugliano, R.P.; McCagg, A.; White, J.A.; Theroux, P.; Darius, H.; Lewis, B.S.; Ophuis, T.O.; Jukema, J.W.; et al. Ezetimibe added to statin therapy after acute coronary syndromes. *N. Engl. J. Med.* **2015**, *372*, 2387–2397. [CrossRef]
57. Liu, L.; Moesner, P.; Kovach, N.L.; Bailey, R.; Hamilton, A.D.; Sebti, S.M.; Harlan, J.M. Integrin-dependent leukocyte adhesion involves geranylgeranylated protein(s). *J. Biol. Chem.* **1999**, *274*, 33334–33340. [CrossRef]
58. Li, X.; Liu, L.; Tupper, J.C.; Bannerman, D.D.; Winn, R.K.; Sebti, S.M.; Hamilton, A.D.; Harlan, J.M. Inhibition of protein geranylgeranylation and rhoa/rhoa kinase pathway induces apoptosis in human endothelial cells. *J. Biol. Chem.* **2002**, *277*, 15309–15316. [CrossRef]
59. Rasmussen, L.M.; Hansen, P.R.; Nabipour, M.T.; Olesen, P.; Kristiansen, M.T.; Ledet, T. Diverse effects of inhibition of 3-hydroxy-3-methylglutaryl-coa reductase on the expression of vcam-1 and e-selectin in endothelial cells. *Biochem. J.* **2001**, *360*, 363–370. [CrossRef]
60. Hodge, R.G.; Ridley, A.J. Regulating rho gtpases and their regulators. *Nat. Rev. Mol. Cell Biol.* **2016**, *17*, 496–510. [CrossRef]
61. Simanshu, D.K.; Nissley, D.V.; McCormick, F. Ras proteins and their regulators in human disease. *Cell* **2017**, *170*, 17–33. [CrossRef] [PubMed]
62. Nimnual, A.S.; Taylor, L.J.; Bar-Sagi, D. Redox-dependent downregulation of rho by rac. *Nat. Cell Biol.* **2003**, *5*, 236–241. [CrossRef] [PubMed]
63. Satoh, M.; Ogita, H.; Takeshita, K.; Mukai, Y.; Kwiatkowski, D.J.; Liao, J.K. Requirement of rac1 in the development of cardiac hypertrophy. *Proc. Natl. Acad. Sci. USA* **2006**, *103*, 7432–7437. [CrossRef] [PubMed]
64. Stout, M.C.; Asiimwe, E.; Birkenstamm, J.R.; Kim, S.Y.; Campbell, P.M. Analyzing ras-associated cell proliferation signaling. *Methods Mol. Biol.* **2014**, *1170*, 393–409. [PubMed]
65. Treasure, C.B.; Klein, J.L.; Weintraub, W.S.; Talley, J.D.; Stillabower, M.E.; Kosinski, A.S.; Zhang, J.; Boccuzzi, S.J.; Cedarholm, J.C.; Alexander, R.W. Beneficial effects of cholesterol-lowering therapy on the coronary endothelium in patients with coronary artery disease. *N. Engl. J. Med.* **1995**, *332*, 481–487. [CrossRef] [PubMed]
66. Anderson, T.J.; Meredith, I.T.; Yeung, A.C.; Frei, B.; Selwyn, A.P.; Ganz, P. The effect of cholesterol-lowering and antioxidant therapy on endothelium-dependent coronary vasomotion. *N. Engl. J. Med.* **1995**, *332*, 488–493. [CrossRef]
67. Yamakuchi, M.; Greer, J.J.; Cameron, S.J.; Matsushita, K.; Morrell, C.N.; Talbot-Fox, K.; Baldwin, W.M., 3rd; Lefer, D.J.; Lowenstein, C.J. Hmg-coa reductase inhibitors inhibit endothelial exocytosis and decrease myocardial infarct size. *Circ. Res.* **2005**, *96*, 1185–1192. [CrossRef]
68. Meda, C.; Plank, C.; Mykhaylyk, O.; Schmidt, K.; Mayer, B. Effects of statins on nitric oxide/cgmp signaling in human umbilical vein endothelial cells. *Pharmacol. Rep.* **2010**, *62*, 100–112. [CrossRef]
69. Dichtl, W.; Dulak, J.; Frick, M.; Alber, H.F.; Schwarzacher, S.P.; Ares, M.P.; Nilsson, J.; Pachinger, O.; Weidinger, F. Hmg-coa reductase inhibitors regulate inflammatory transcription factors in human endothelial and vascular smooth muscle cells. *Arterioscler. Thromb. Vasc. Biol.* **2003**, *23*, 58–63. [CrossRef]
70. Crisby, M.; Nordin-Fredriksson, G.; Shah, P.K.; Yano, J.; Zhu, J.; Nilsson, J. Pravastatin treatment increases collagen content and decreases lipid content, inflammation, metalloproteinases, and cell death in human carotid plaques: Implications for plaque stabilization. *Circulation* **2001**, *103*, 926–933. [CrossRef]
71. Aikawa, M.; Rabkin, E.; Sugiyama, S.; Voglic, S.J.; Fukumoto, Y.; Furukawa, Y.; Shiomi, M.; Schoen, F.J.; Libby, P. An hmg-coa reductase inhibitor, cerivastatin, suppresses growth of macrophages expressing matrix metalloproteinases and tissue factor in vivo and in vitro. *Circulation* **2001**, *103*, 276–283. [CrossRef] [PubMed]
72. Xenos, E.S.; Stevens, S.L.; Freeman, M.B.; Cassada, D.C.; Goldman, M.H. Nitric oxide mediates the effect of fluvastatin on intercellular adhesion molecule-1 and platelet endothelial cell adhesion molecule-1 expression on human endothelial cells. *Ann. Vasc. Surg.* **2005**, *19*, 386–392. [CrossRef] [PubMed]

73. Ito, T.; Ikeda, U.; Yamamoto, K.; Shimada, K. Regulation of interleukin-8 expression by hmg-coa reductase inhibitors in human vascular smooth muscle cells. *Atherosclerosis* **2002**, *165*, 51–55. [CrossRef]
74. Weitz-Schmidt, G.; Welzenbach, K.; Brinkmann, V.; Kamata, T.; Kallen, J.; Bruns, C.; Cottens, S.; Takada, Y.; Hommel, U. Statins selectively inhibit leukocyte function antigen-1 by binding to a novel regulatory integrin site. *Nat. Med.* **2001**, *7*, 687–692. [CrossRef]
75. Simon, D.I.; Dhen, Z.; Seifert, P.; Edelman, E.R.; Ballantyne, C.M.; Rogers, C. Decreased neointimal formation in mac-1(-/-) mice reveals a role for inflammation in vascular repair after angioplasty. *J. Clin. Investig.* **2000**, *105*, 293–300. [CrossRef]
76. Kwak, B.; Mulhaupt, F.; Myit, S.; Mach, F. Statins as a newly recognized type of immunomodulator. *Nat. Med.* **2000**, *6*, 1399–1402. [CrossRef]
77. Hillyard, D.Z.; Cameron, A.J.; McDonald, K.J.; Thomson, J.; MacIntyre, A.; Shiels, P.G.; Panarelli, M.; Jardine, A.G. Simvastatin inhibits lymphocyte function in normal subjects and patients with cardiovascular disease. *Atherosclerosis* **2004**, *175*, 305–313. [CrossRef]
78. Henriksbo, B.D.; Schertzer, J.D. Is immunity a mechanism contributing to statin-induced diabetes? *Adipocyte* **2015**, *4*, 232–238. [CrossRef]
79. Libby, P. Interleukin-1 beta as a target for atherosclerosis therapy: Biological basis of cantos and beyond. *J. Am. Coll. Cardiol.* **2017**, *70*, 2278–2289. [CrossRef]
80. Swanson, K.V.; Deng, M.; Ting, J.P. The nlrp3 inflammasome: Molecular activation and regulation to therapeutics. *Nat. Rev. Immunol.* **2019**, *19*, 477–489. [CrossRef]
81. Bauernfeind, F.G.; Horvath, G.; Stutz, A.; Alnemri, E.S.; MacDonald, K.; Speert, D.; Fernandes-Alnemri, T.; Wu, J.; Monks, B.G.; Fitzgerald, K.A.; et al. Cutting edge: Nf-kappab activating pattern recognition and cytokine receptors license nlrp3 inflammasome activation by regulating nlrp3 expression. *J. Immunol.* **2009**, *183*, 787–791. [CrossRef] [PubMed]
82. Franchi, L.; Eigenbrod, T.; Nunez, G. Cutting edge: Tnf-alpha mediates sensitization to atp and silica via the nlrp3 inflammasome in the absence of microbial stimulation. *J. Immunol.* **2009**, *183*, 792–796. [CrossRef] [PubMed]
83. Xing, Y.; Yao, X.; Li, H.; Xue, G.; Guo, Q.; Yang, G.; An, L.; Zhang, Y.; Meng, G. Cutting edge: Traf6 mediates tlr/il-1r signaling-induced nontranscriptional priming of the nlrp3 inflammasome. *J. Immunol.* **2017**, *199*, 1561–1566. [CrossRef] [PubMed]
84. Song, N.; Liu, Z.S.; Xue, W.; Bai, Z.F.; Wang, Q.Y.; Dai, J.; Liu, X.; Huang, Y.J.; Cai, H.; Zhan, X.Y.; et al. Nlrp3 phosphorylation is an essential priming event for inflammasome activation. *Mol. Cell* **2017**, *68*, 185–197. [CrossRef]
85. Juliana, C.; Fernandes-Alnemri, T.; Kang, S.; Farias, A.; Qin, F.; Alnemri, E.S. Non-transcriptional priming and deubiquitination regulate nlrp3 inflammasome activation. *J. Biol. Chem.* **2012**, *287*, 36617–36622. [CrossRef]
86. Murakami, T.; Ockinger, J.; Yu, J.; Byles, V.; McColl, A.; Hofer, A.M.; Horng, T. Critical role for calcium mobilization in activation of the nlrp3 inflammasome. *Proc. Natl. Acad. Sci. USA* **2012**, *109*, 11282–11287. [CrossRef]
87. Tang, T.; Lang, X.; Xu, C.; Wang, X.; Gong, T.; Yang, Y.; Cui, J.; Bai, L.; Wang, J.; Jiang, W.; et al. Clics-dependent chloride efflux is an essential and proximal upstream event for nlrp3 inflammasome activation. *Nat. Commun.* **2017**, *8*, 202. [CrossRef]
88. Hornung, V.; Bauernfeind, F.; Halle, A.; Samstad, E.O.; Kono, H.; Rock, K.L.; Fitzgerald, K.A.; Latz, E. Silica crystals and aluminum salts activate the nalp3 inflammasome through phagosomal destabilization. *Nat. Immunol.* **2008**, *9*, 847–856. [CrossRef]
89. Munoz-Planillo, R.; Kuffa, P.; Martinez-Colon, G.; Smith, B.L.; Rajendiran, T.M.; Nunez, G. K(+) efflux is the common trigger of nlrp3 inflammasome activation by bacterial toxins and particulate matter. *Immunity* **2013**, *38*, 1142–1153. [CrossRef]
90. Wang, Z.; Hu, W.; Lu, C.; Ma, Z.; Jiang, S.; Gu, C.; Acuna-Castroviejo, D.; Yang, Y. Targeting nlrp3 (nucleotide-binding domain, leucine-rich-containing family, pyrin domain-containing-3) inflammasome in cardiovascular disorders. *Arterioscler. Thromb. Vasc. Biol.* **2018**, *38*, 2765–2779. [CrossRef]
91. Yang, Y.; Wang, H.; Kouadir, M.; Song, H.; Shi, F. Recent advances in the mechanisms of nlrp3 inflammasome activation and its inhibitors. *Cell Death Dis.* **2019**, *10*, 128. [CrossRef] [PubMed]
92. Mauro, A.G.; Bonaventura, A.; Abbate, A. Drugs to inhibit the nlrp3 inflammasome: Not always on target. *J. Cardiovasc. Pharmacol.* **2019**, *74*, 225–227. [CrossRef] [PubMed]

93. Mitchell, P.; Marette, A. Statin-induced insulin resistance through inflammasome activation: Sailing between scylla and charybdis. *Diabetes* **2014**, *63*, 3569–3571. [CrossRef] [PubMed]
94. Banach, M.; Malodobra-Mazur, M.; Gluba, A.; Katsiki, N.; Rysz, J.; Dobrzyn, A. Statin therapy and new-onset diabetes: Molecular mechanisms and clinical relevance. *Curr. Pharm. Des.* **2013**, *19*, 4904–4912. [CrossRef] [PubMed]
95. Henriksbo, B.D.; Lau, T.C.; Cavallari, J.F.; Denou, E.; Chi, W.; Lally, J.S.; Crane, J.D.; Duggan, B.M.; Foley, K.P.; Fullerton, M.D.; et al. Fluvastatin causes nlrp3 inflammasome-mediated adipose insulin resistance. *Diabetes* **2014**, *63*, 3742–3747. [CrossRef]
96. Massonnet, B.; Normand, S.; Moschitz, R.; Delwail, A.; Favot, L.; Garcia, M.; Bourmeyster, N.; Cuisset, L.; Grateau, G.; Morel, F.; et al. Pharmacological inhibitors of the mevalonate pathway activate pro-il-1 processing and il-1 release by human monocytes. *Eur. Cytokine Netw.* **2009**, *20*, 112–120. [CrossRef]
97. Montero, M.T.; Hernandez, O.; Suarez, Y.; Matilla, J.; Ferruelo, A.J.; Martinez-Botas, J.; Gomez-Coronado, D.; Lasuncion, M.A. Hydroxymethylglutaryl-coenzyme a reductase inhibition stimulates caspase-1 activity and th1-cytokine release in peripheral blood mononuclear cells. *Atherosclerosis* **2000**, *153*, 303–313. [CrossRef]
98. Liao, Y.H.; Lin, Y.C.; Tsao, S.T.; Lin, Y.C.; Yang, A.J.; Huang, C.T.; Huang, K.C.; Lin, W.W. Hmg-coa reductase inhibitors activate caspase-1 in human monocytes depending on atp release and p2 × 7 activation. *J. Leukoc. Biol.* **2013**, *93*, 289–299. [CrossRef]
99. Frenkel, J.; Rijkers, G.T.; Mandey, S.H.; Buurman, S.W.; Houten, S.M.; Wanders, R.J.; Waterham, H.R.; Kuis, W. Lack of isoprenoid products raises ex vivo interleukin-1beta secretion in hyperimmunoglobulinemia d and periodic fever syndrome. *Arthritis Rheum.* **2002**, *46*, 2794–2803. [CrossRef]
100. Henriksbo, B.D.; Tamrakar, A.K.; Xu, J.; Duggan, B.M.; Cavallari, J.F.; Phulka, J.; Stampfli, M.R.; Ashkar, A.A.; Schertzer, J.D. Statins promote interleukin-1beta-dependent adipocyte insulin resistance through lower prenylation, not cholesterol. *Diabetes* **2019**, *68*, 1441–1448. [CrossRef]
101. Bonaventura, A.; Liberale, L.; Carbone, F.; Vecchie, A.; Diaz-Canestro, C.; Camici, G.G.; Montecucco, F.; Dallegri, F. The pathophysiological role of neutrophil extracellular traps in inflammatory diseases. *Thromb. Haemost.* **2018**, *118*, 6–27. [CrossRef] [PubMed]
102. Liberale, L.; Dallegri, F.; Montecucco, F.; Carbone, F. Pathophysiological relevance of macrophage subsets in atherogenesis. *Thromb. Haemost.* **2017**, *117*, 7–18. [CrossRef] [PubMed]
103. Liberale, L.; Bertolotto, M.; Carbone, F.; Contini, P.; Wust, P.; Spinella, G.; Pane, B.; Palombo, D.; Bonaventura, A.; Pende, A.; et al. Resistin exerts a beneficial role in atherosclerotic plaque inflammation by inhibiting neutrophil migration. *Int. J. Cardiol.* **2018**, *272*, 13–19. [CrossRef] [PubMed]
104. Carbone, F.; Rigamonti, F.; Burger, F.; Roth, A.; Bertolotto, M.; Spinella, G.; Pane, B.; Palombo, D.; Pende, A.; Bonaventura, A.; et al. Serum levels of osteopontin predict major adverse cardiovascular events in patients with severe carotid artery stenosis. *Int. J. Cardiol.* **2018**, *255*, 195–199. [CrossRef]
105. Ridker, P.M.; MacFadyen, J.G.; Everett, B.M.; Libby, P.; Thuren, T.; Glynn, R.J.; Group, C.T. Relationship of c-reactive protein reduction to cardiovascular event reduction following treatment with canakinumab: A secondary analysis from the cantos randomised controlled trial. *Lancet* **2018**, *391*, 319–328. [CrossRef]
106. Duewell, P.; Kono, H.; Rayner, K.J.; Sirois, C.M.; Vladimer, G.; Bauernfeind, F.G.; Abela, G.S.; Franchi, L.; Nunez, G.; Schnurr, M.; et al. Nlrp3 inflammasomes are required for atherogenesis and activated by cholesterol crystals. *Nature* **2010**, *464*, 1357–1361. [CrossRef]
107. Rajamaki, K.; Lappalainen, J.; Oorni, K.; Valimaki, E.; Matikainen, S.; Kovanen, P.T.; Eklund, K.K. Cholesterol crystals activate the nlrp3 inflammasome in human macrophages: A novel link between cholesterol metabolism and inflammation. *PLoS ONE* **2010**, *5*, 11765. [CrossRef]
108. Warner, S.J.; Auger, K.R.; Libby, P. Interleukin 1 induces interleukin 1. Ii. Recombinant human interleukin 1 induces interleukin 1 production by adult human vascular endothelial cells. *J. Immunol.* **1987**, *139*, 1911–1917.
109. Liberale, L.; Bonaventura, A.; Vecchie, A.; Casula, M.; Dallegri, F.; Montecucco, F.; Carbone, F. The role of adipocytokines in coronary atherosclerosis. *Curr. Atheroscler. Rep.* **2017**, *19*, 10. [CrossRef]
110. Feingold, K.R.; Grunfeld, C. The Effect of Inflammation and Infection on Lipids and Lipoproteins. In *Endotext*; Feingold, K.R., Anawalt, B., Boyce, A., Chrousos, G., Dungan, K., Grossman, A., Hershman, J.M., Kaltsas, G., Koch, C., Kopp, P., et al., Eds.; MDText.com, Inc.: South Dartmouth, MA, USA, 2000.
111. Frangogiannis, N.G. The inflammatory response in myocardial injury, repair, and remodelling. *Nat. Rev. Cardiol.* **2014**, *11*, 255–265. [CrossRef]

112. Hartman, M.H.T.; Groot, H.E.; Leach, I.M.; Karper, J.C.; van der Harst, P. Translational overview of cytokine inhibition in acute myocardial infarction and chronic heart failure. *Trends Cardiovasc. Med.* **2018**, *28*, 369–379. [CrossRef] [PubMed]
113. Abbate, A.; Salloum, F.N.; Vecile, E.; Das, A.; Hoke, N.N.; Straino, S.; Biondi-Zoccai, G.G.; Houser, J.E.; Qureshi, I.Z.; Ownby, E.D.; et al. Anakinra, a recombinant human interleukin-1 receptor antagonist, inhibits apoptosis in experimental acute myocardial infarction. *Circulation* **2008**, *117*, 2670–2683. [CrossRef] [PubMed]
114. Van Tassell, B.W.; Varma, A.; Salloum, F.N.; Das, A.; Seropian, I.M.; Toldo, S.; Smithson, L.; Hoke, N.N.; Chau, V.Q.; Robati, R.; et al. Interleukin-1 trap attenuates cardiac remodeling after experimental acute myocardial infarction in mice. *J. Cardiovasc. Pharmacol.* **2010**, *55*, 117–122. [CrossRef] [PubMed]
115. Toldo, S.; Schatz, A.M.; Mezzaroma, E.; Chawla, R.; Stallard, T.W.; Stallard, W.C.; Jahangiri, A.; Van Tassell, B.W.; Abbate, A. Recombinant human interleukin-1 receptor antagonist provides cardioprotection during myocardial ischemia reperfusion in the mouse. *Cardiovasc. Drugs Ther.* **2012**, *26*, 273–276. [CrossRef]
116. Toldo, S.; Mezzaroma, E.; Van Tassell, B.W.; Farkas, D.; Marchetti, C.; Voelkel, N.F.; Abbate, A. Interleukin-1beta blockade improves cardiac remodelling after myocardial infarction without interrupting the inflammasome in the mouse. *Exp. Physiol.* **2013**, *98*, 734–745. [CrossRef]
117. Toldo, S.; Mezzaroma, E.; Bressi, E.; Marchetti, C.; Carbone, S.; Sonnino, C.; Van Tassell, B.W.; Abbate, A. Interleukin-1beta blockade improves left ventricular systolic/diastolic function and restores contractility reserve in severe ischemic cardiomyopathy in the mouse. *J. Cardiovasc. Pharmacol.* **2014**, *64*, 1–6. [CrossRef]
118. De Jesus, N.M.; Wang, L.; Lai, J.; Rigor, R.R.; Francis Stuart, S.D.; Bers, D.M.; Lindsey, M.L.; Ripplinger, C.M. Antiarrhythmic effects of interleukin 1 inhibition after myocardial infarction. *Heart Rhythm* **2017**, *14*, 727–736. [CrossRef]
119. Mauro, A.G.; Mezzaroma, E.; Torrado, J.; Kundur, P.; Joshi, P.; Stroud, K.; Quaini, F.; Lagrasta, C.A.; Abbate, A.; Toldo, S. Reduction of myocardial ischemia-reperfusion injury by inhibiting interleukin-1 alpha. *J. Cardiovasc. Pharmacol.* **2017**, *69*, 156–160. [CrossRef]
120. Harouki, N.; Nicol, L.; Remy-Jouet, I.; Henry, J.P.; Dumesnil, A.; Lejeune, A.; Renet, S.; Golding, F.; Djerada, Z.; Wecker, D.; et al. The il-1beta antibody gevokizumab limits cardiac remodeling and coronary dysfunction in rats with heart failure. *JACC Basic Transl. Sci.* **2017**, *2*, 418–430. [CrossRef]
121. Abbate, A.; Van Tassell, B.W.; Biondi-Zoccai, G.; Kontos, M.C.; Grizzard, J.D.; Spillman, D.W.; Oddi, C.; Roberts, C.S.; Melchior, R.D.; Mueller, G.H.; et al. Effects of interleukin-1 blockade with anakinra on adverse cardiac remodeling and heart failure after acute myocardial infarction [from the virginia commonwealth university-anakinra remodeling trial (2) (vcu-art2) pilot study]. *Am. J. Cardiol.* **2013**, *111*, 1394–1400. [CrossRef]
122. Van Tassell, B.W.; Canada, J.; Carbone, S.; Trankle, C.; Buckley, L.; Oddi Erdle, C.; Abouzaki, N.A.; Dixon, D.; Kadariya, D.; Christopher, S.; et al. Interleukin-1 blockade in recently decompensated systolic heart failure: Results from redhart (recently decompensated heart failure anakinra response trial). *Circ. Heart Fail.* **2017**, *10*, 4373. [CrossRef] [PubMed]
123. Van Tassell, B.W.; Arena, R.; Biondi-Zoccai, G.; Canada, J.M.; Oddi, C.; Abouzaki, N.A.; Jahangiri, A.; Falcao, R.A.; Kontos, M.C.; Shah, K.B.; et al. Effects of interleukin-1 blockade with anakinra on aerobic exercise capacity in patients with heart failure and preserved ejection fraction (from the d-hart pilot study). *Am. J. Cardiol.* **2014**, *113*, 321–327. [CrossRef] [PubMed]
124. Abbate, A.; Kontos, M.C.; Abouzaki, N.A.; Melchior, R.D.; Thomas, C.; Van Tassell, B.W.; Oddi, C.; Carbone, S.; Trankle, C.R.; Roberts, C.S.; et al. Comparative safety of interleukin-1 blockade with anakinra in patients with st-segment elevation acute myocardial infarction (from the vcu-art and vcu-art2 pilot studies). *Am. J. Cardiol.* **2015**, *115*, 288–292. [CrossRef] [PubMed]
125. Everett, B.M.; Cornel, J.H.; Lainscak, M.; Anker, S.D.; Abbate, A.; Thuren, T.; Libby, P.; Glynn, R.J.; Ridker, P.M. Anti-inflammatory therapy with canakinumab for the prevention of hospitalization for heart failure. *Circulation* **2019**, *139*, 1289–1299. [CrossRef]
126. Giugliano, R.P.; Pedersen, T.R.; Park, J.G.; De Ferrari, G.M.; Gaciong, Z.A.; Ceska, R.; Toth, K.; Gouni-Berthold, I.; Lopez-Miranda, J.; Schiele, F.; et al. Clinical efficacy and safety of achieving very low ldl-cholesterol concentrations with the pcsk9 inhibitor evolocumab: A prespecified secondary analysis of the fourier trial. *Lancet* **2017**, *390*, 1962–1971. [CrossRef]
127. Peiro, C.; Lorenzo, O.; Carraro, R.; Sanchez-Ferrer, C.F. Il-1beta inhibition in cardiovascular complications associated to diabetes mellitus. *Front. Pharmacol.* **2017**, *8*, 363. [CrossRef]

128. Abbate, A.; Kontos, M.C.; Grizzard, J.D.; Biondi-Zoccai, G.G.; Van Tassell, B.W.; Robati, R.; Roach, L.M.; Arena, R.A.; Roberts, C.S.; Varma, A.; et al. Interleukin-1 blockade with anakinra to prevent adverse cardiac remodeling after acute myocardial infarction (virginia commonwealth university anakinra remodeling trial [vcu-art] pilot study). *Am. J. Cardiol.* **2010**, *105*, 1371–1377. [CrossRef]
129. Morton, A.C.; Rothman, A.M.; Greenwood, J.P.; Gunn, J.; Chase, A.; Clarke, B.; Hall, A.S.; Fox, K.; Foley, C.; Banya, W.; et al. The effect of interleukin-1 receptor antagonist therapy on markers of inflammation in non-st elevation acute coronary syndromes: The mrc-ila heart study. *Eur. Heart J.* **2015**, *36*, 377–384. [CrossRef]
130. Van Tassell, B.W.; Lipinski, M.J.; Appleton, D.; Roberts, C.S.; Kontos, M.C.; Abouzaki, N.; Melchior, R.; Mueller, G.; Garnett, J.; Canada, J.; et al. Rationale and design of the virginia commonwealth university-anakinra remodeling trial-3 (vcu-art3): A randomized, placebo-controlled, double-blinded, multicenter study. *Clin. Cardiol.* **2018**, *41*, 1004–1008. [CrossRef]

© 2019 by the authors. Licensee MDPI, Basel, Switzerland. This article is an open access article distributed under the terms and conditions of the Creative Commons Attribution (CC BY) license (http://creativecommons.org/licenses/by/4.0/).

Article

Plasma Levels of Retinol Binding Protein 4 Relate to Large VLDL and Small LDL Particles in Subjects with and without Type 2 Diabetes

Hanna Wessel [1,2], Ali Saeed [2,3], Janette Heegsma [2,4], Margery A. Connelly [5], Klaas Nico Faber [2,4] and Robin P. F. Dullaart [1,*]

[1] Department of Endocrinology, University of Groningen and University Medical Center Groningen, 9700 RB Groningen, The Netherlands; wessel.hanna@gmail.com
[2] Department of Gastroenterology and Hepatology, University of Groningen and University Medical Center Groningen, 9700 RB Groningen, The Netherlands; a.saeed@umcg.nl (A.S.); j.heegsma@umcg.nl (J.H.); k.n.faber@umcg.nl (K.N.F.)
[3] Institute of Molecular Biology and Biotechnology, Bahauddin Zakariya University, Multan 66000, Pakistan
[4] Department of Laboratory Medicine, University of Groningen and University Medical Center Groningen, 9700 RB Groningen, The Netherlands
[5] Laboratory Corporation of America Holdings (LabCorp), Morrisville, NC 27560, USA; connem5@labcorp.com
* Correspondence: r.p.f.dullaart@umcg.nl

Received: 18 September 2019; Accepted: 22 October 2019; Published: 25 October 2019

Abstract: Background: Retinol binding protein 4 (RBP4) carries retinol in plasma, but is also considered an adipokine, as it is implicated in insulin resistance in mice. Plasma RBP4 correlates with total cholesterol, low density lipoprotein (LDL)-cholesterol and triglycerides, and may confer increased cardiovascular risk. However, controversy exists about circulating RPB4 levels in type 2 diabetes mellitus (T2DM) and obesity. Here, we analyzed the relationships of RBP4 and retinol with lipoprotein subfractions in subjects with and without T2DM. Methods: Fasting plasma RBP4 (enzyme-linked immunosorbent assay) and retinol (high performance liquid chromatography) were assayed in 41 T2DM subjects and 37 non-diabetic subjects. Lipoprotein subfractions (NMR spectroscopy) were measured in 36 T2DM subjects and 27 non-diabetic subjects. Physical interaction of RBP4 with lipoproteins was assessed by fast protein liquid chromatography (FPLC). Results: Plasma RBP4 and retinol were strongly correlated ($r = 0.881$, $p < 0.001$). RBP4, retinol and the RBP4/retinol ratio were not different between T2DM and non-diabetic subjects (all $p > 0.12$), and were unrelated to body mass index. Notably, RBP4 and retinol were elevated in subjects with metabolic syndrome ($p < 0.05$), which was attributable to an association with elevated triglycerides ($p = 0.013$). Large VLDL, total LDL and small LDL were increased in T2DM subjects ($p = 0.035$ to 0.003). Taking all subjects together, RBP4 correlated with total cholesterol, non-HDL cholesterol, LDL cholesterol, triglycerides and apolipoprotein B in univariate analysis ($p < 0.001$ for each). Age-, sex- and diabetes status-adjusted multivariable linear regression analysis revealed that RBP4 was independently associated with large VLDL ($\beta = 0.444$, $p = 0.005$) and small LDL particles ($\beta = 0.539$, $p < 0.001$). Its relationship with large VLDL remained after further adjustment for retinol. RBP4 did not co-elute with VLDL nor LDL particles in FPLC analyses. Conclusions: Plasma RBP4 levels are related to but do not physically interact with large VLDL and small LDL particles. Elevated RBP4 may contribute to a proatherogenic plasma lipoprotein profile.

Keywords: retinol binding protein 4; retinol; lipoprotein subfractions; large VLDL; small LDL; Type 2 diabetes mellitus; metabolic syndrome; nuclear magnetic resonance spectroscopy

1. Introduction

The incidence of the metabolic syndrome (MetS) and Type 2 diabetes mellitus (T2DM) has rapidly increased in the past few years [1]. Both MetS and T2DM are associated with an increased risk of cardiovascular disease and reduced life span [2]. Impaired insulin action is a main determinant responsible for the development of T2DM [3]. Hence, continued attention is being paid to the pathogenic mechanisms underlying insulin resistance.

Retinol-binding protein 4 (RBP4) is produced by the liver and adipose tissue and transports retinol (vitamin A) via the circulation to peripheral tissues. However, RBP4 is also considered to be an adipokine, as it has been implicated in the pathogenesis of insulin resistance [4]. In rodent studies, Rbp4 overexpression was convincingly shown to enhance insulin resistance, impair insulin signaling in muscle and induce pathways involved in hepatic gluconeogenesis [4]. Conversely, genetic deletion of Rbp4 in mice was found to enhance insulin sensitivity [4]. Nonetheless, studies in humans with obesity, impaired glucose tolerance and T2DM have shown inconsistent results regarding the associations of these pathologies with circulating RBP4. Initial studies documented that RBP4 levels were elevated in patients with obesity, impaired glucose tolerance and T2DM, and were correlated with HbA1c, fasting glucose and insulin [5]. Enhanced RBP4 levels were also observed in obese children, and (pronounced) weight loss was shown to reduce RBP4 levels in some but not in all studies [6–8]. Likewise, studies reporting potential associations between RBP4 and T2DM were also inconsistent. Initial studies observed elevated RBP4 levels in T2DM patients [9–11], whereas more recent reports either detected no abnormalities in RBP4 [12] or even found a negative correlation of RBP4 with diabetes status [13]. RBP4 is the major carrier of retinol in plasma although not all RBP4 may be "loaded" with retinol. The RBP4/retinol ratio was found to be elevated in T2DM despite lower RBP4 levels as such [13], and similar observations were made in obese children [14]. The RBP4/retinol ratio has therefore been advocated to better reflect alterations in obesity and T2DM [13,14].

In contrast to the equivocally reported relationships of circulating RBP4 with obesity, dysglycemia and insulin resistance, positive relationships of RBP4 with total cholesterol, low density lipoprotein (LDL) cholesterol and triglycerides have repeatedly and almost invariably been demonstrated [6,7,12,13,15–17]. Quantification of specific lipoprotein subfractions is increasingly used to better delineate dyslipidemia in pathological conditions, such as obesity and (pre)diabetes [18–21]. Interestingly, RBP4 associates with small dense LDL particles and oxidized LDL [17,22]. No data are currently available regarding the relationship of RBP4 with nuclear magnetic resonance (NMR)-quantified lipoprotein subfractions in adult subjects with and without T2DM, the latter condition being expected to be characterized by elevated large VLDL and small-sized LDL particles [18,20]. Given that an increasing number of human studies suggest that elevated plasma RBP4 levels may confer (subclinical) cardiovascular disease (CVD) risk [15,23–26], it is clinically relevant to more precisely delineate the association of RBP4 with various lipoprotein subfractions.

Therefore, the present study aimed to determine whether plasma levels of RBP4 and/or retinol relate to VLDL and LDL subfractions in adults with and without T2DM.

2. Experimental Section

2.1. Subjects

The study protocol was approved by the medical ethics committee of the University Medical Center Groningen. All participants provided written informed consent and were aged > 18 years. They were recruited by advertisement in local newspapers. T2DM was diagnosed by primary care physicians based on fasting plasma glucose ≥ 7.0 mmol/L and/or a non-fasting plasma glucose ≥ 11.1 mmol/L. Diabetic patients using metformin, sulfonylurea and antihypertensives were included. Insulin use was an exclusion criterion. A positive history of cardiovascular disease, chronic kidney disease (estimated glomerular filtration rate < 60 mL/min/1.73 m^2 and/or proteinuria), abnormal liver function tests (transaminases > 3 times the upper reference limit) or thyroid dysfunction (thyroid stimulating hormone > 10 or < 0.40 mU/L or use of thyroid function influencing medication), as well as current smoking and use of lipid lowering drugs was also an exclusion criterion. Blood pressure was measured after 15 min of rest in the left arm in sitting position using a sphygmomanometer. Body mass index (BMI in kg/m^2) was calculated as weight divided by height squared. Waist circumference was measured on bare skin between the 10th rib and the iliac crest. The participants were studied after an overnight fast. Metabolic syndrome (MetS) was defined according to NCEP-ATP III criteria [27]. Three or more of the following criteria were required for categorization of subjects with MetS: waist circumference > 102 cm for men and > 88 cm for women; hypertension (blood pressure ≥ 130/85 mmHg or use of antihypertensive drugs); fasting plasma triglycerides ≥ 1.70 mmol/L; HDL cholesterol <1.00 mmol/L for men and <1.30 mmol/L for women; fasting glucose ≥ 5.60 mmol/l. Insulin sensitivity was estimated by homeostasis model assessment of insulin resistance (HOMA-IR) applying the following equation: fasting plasma insulin (mU/L) × glucose (mmol/L)/22.5.

2.2. Laboratory Analysis

Serum and EDTA-anticoagulated plasma samples were stored at −80 °C until analysis. Plasma glucose was measured shortly after blood collection with an APEC glucose analyzer (APEC Inc., Danvers, MA, USA). Plasma total cholesterol and triglycerides were measured by routine enzymatic methods (Roche/Hitachi cat. nos 11,875,540 and 1,187,602, respectively; Roche Diagnostics GmBH, Mannheim, Germany). HDL cholesterol was assayed by a homogeneous enzymatic colorimetric test (Roche/Hitachi, cat.no 04713214). Non-HDL cholesterol was calculated as the difference between total cholesterol and HDL cholesterol. LDL cholesterol was calculated by the Friedewald formula if triglycerides were <4.5 mmol/L. Apolipoprotein A-I (apoA-I) and apolipoprotein B (apoB) were measured by immunoturbidimetry (Roche/Cobas Integra Tinaquant cat no. 03032566, Roche Diagnostics). HbA1c was measured by high-performance liquid chromatography (HPLC; Bio-Rad, Veenendaal, the Netherlands; normal range: 27–43 mmol/mol).

Plasma RBP4 was assayed by enzyme-linked immunosorbent assay (ELISA; R&D Systems; Catalog Number DRB400). In brief, 96-well plates precoated with anti-human RBP4 monoclonal antibody were stored at 4 °C. After adding 200 μL of 2 vials (11 mL/vial) of a buffered protein solution to each well, 20 μL of sample material was added to each well and left for 1 h at room temperature on a horizontal orbital microplate shaker (0.12" orbit) set at 500 rpm. This procedure was followed by aspiration and washing of each well with 400 μL buffered surfactant, repeating the process three times for a total of four washes. 200 μL of monoclonal antibody specific for human RBP4 conjugated to horseradish peroxidase were added to each well and incubated for 1 h at room temperature under gentle agitation. After repeating aspiration/wash as described, 200 μL of Substrate Solution consisting of stabilized hydrogen peroxide and stabilized chromogen (tetramethylbenzidine) were added to each well and incubated for 30 min at room temperature under protection from light. The reaction was then stopped by adding 50 μL of 2 N sulfuric acid; the absorbance was measured at 450 nm, 540 nm and 570 nm.

Plasma retinol levels were analyzed by reverse phase high performance liquid chromatography (HPLC) as described [28]. Briefly, 50 μL plasma samples were added in antioxidant solutions (2.8 mL) and vortexed thoroughly for 1 min. Retinol was extracted and deproteinized twice with n-hexane in the presence of retinol acetate (100 μL, concentration 4 μmol/L) as external standard to assess the level of recovery after the extraction procedure. Standard curves created from a range of concentrations of retinol were used to determine absolute serum retinol concentrations. Additionally, two negative controls (only containing internal standard) and two positive controls (low and high concentrations of retinol plus internal standard) were included in each series of extractions. Samples were evaporated under N2 and diluted in 300 μL 100% ultrapure ethanol. Then, 50 μL were injected in HPLC (Waters 2795 Alliance HT Separations Module, Connecticut, USA) for phase separation on a C18 column (Waters Symmetry C18, dimension 150 × 3.0 mm, particle size 5 μm, Waters Corporation, Milford, MA, USA) (UV-VIS, dual wavelength, UV-4075 Jasco, Tokyo, Japan). Retinoids in samples were identified by applying exact retention time of known standards in ultraviolet absorption at 325 nm by HPLC. Finally, retinol concentrations were calculated and normalized to final volume. The intra-assay coefficients of variation (CV) of RBP4 and retinol amount to 5.7–8.1% and 3.6%, respectively.

Lipoprotein separation was performed by fast protein liquid chromatography (FPLC) as described before [29]. One mL of plasma of four control individuals were subjected to size exclusion gel FPLC filtration using a Superose 6 column (GE Healthcare, Uppsala, Sweden) at a flow rate of 0.5 mL/min. Fractions of 500 µL each were collected. Total protein (Bradford), total cholesterol, triglycerides and RBP4 were determined in the respective lipoprotein fractions by ELISA as detailed above. One representative FPLC profile is presented.

Lipoprotein particle profiles were determined by nuclear magnetic resonance (NMR) spectroscopy with the LP3 algorithm (LipoScience, now LabCorp, Morrisville, NC, USA) [30,31]. Very low density lipoprotein (VLDL), low density lipoprotein (LDL) and high density lipoprotein (HDL) particle classes and subfractions were quantified from the amplitudes of their spectroscopically distinct lipid methyl group NMR signals. Diameter range estimates were for VLDL (including chylomicrons if present): >60 nm to 29 nm, for LDL: 29 nm to 18 nm, and for HDL: 14 nm to 7.3 nm. The VLDL, LDL and HDL particle concentrations (VLDL-P, LDL-P and HDL-P, respectively) were calculated as the weighted average of the respective lipoprotein subclasses. The intra-assay CVs for the lipoprotein parameters are: VLDL-P (11.0%), LDL-P (4.1%), HDL-P (2.0%) and amount to 6.6–27.9%. for the various VLDL, LDL and HDL subfractions.

2.3. Statistical Analysis

IBM SPSS software (SPSS, version 24.0, SPSS Inc. Chicago, IL, USA) was used for data analysis. Continuous variables were expressed in means ± SD or medians (interquartile ranges) in case of not-normally distributed data. HOMA-IR, triglycerides and lipoprotein subfractions were $\log_e$ transformed to achieve approximately normal distributions. Between group differences in variables were determined by T-tests for unpaired data and Chi-square tests where appropriate. Univariate correlations were determined by Pearson correlation coefficients. Multivariable linear regression analyses were carried out to disclose the independent relationships of RBP4, retinol and the RBP4/retinol ratio with either T2DM or MetS. Additionally, multivariable linear regression analyses were performed to disclose the independent associations with RBP4 and retinol with VLDL, LDL and HDL subfraction characteristics, each in separate models. Two-sided p-values < 0.05 were considered statistically significant.

3. Results

Forty-one (41) subjects with T2DM and 37 subjects without T2DM were included. Clinical characteristics and biochemical data are presented in Table 1. Twenty-nine (29) of the diabetic patients used glucose-lowering drugs (metformin and/or sulfonylurea). Antihypertensive drugs, particularly angiotensin-converting enzyme (ACE) inhibitors, angiotensin II receptor antagonists and diuretics, alone or in combination were used by 19 T2DM subjects; none of the non-diabetic subjects were using antihypertensive drugs. Three non-diabetic women were using estrogens. Other medications were not taken.

As expected, T2DM subjects were classified with MetS more frequently, had higher blood pressure, were more obese and had higher glucose and HbA1c levels and had higher HOMA-IR when compared to non-diabetic controls (Table 1). In general, metabolic control was adequate in the diabetic patients given an average HbA1c level of 50 mmol/mol. Total cholesterol, non-HDL cholesterol, LDL cholesterol and apoB levels were not significantly different between diabetic and non-diabetic individuals. Triglycerides were higher, whereas HDL cholesterol was lower in T2DM subjects (Table 1). Plasma levels of RBP4, retinol and the RBP4/retinol ratio were not different between diabetic and non-diabetic subjects (Table 1 and Figure 1A), also not after adjustment for age and sex (RBP4, $p = 0.63$; retinol, $p = 0.25$; RBP4/retinol ratio, $p = 0{,}19$, respectively; data not shown). In contrast, RBP4 and retinol levels were higher in subjects with MetS vs. subjects without MetS (Figure 1B; RBP4:

41.38 ± 7.98 mg/L in subjects with MetS and 37.48 ± 8.48 mg/L in subjects without MetS, $p = 0.041$; retinol: 2.37 ± 0.50 µmol/L in subjects with MetS vs. 2.14 ± 0.46 µmol/L in subjects without MetS, $p = 0.038$). As a consequence, the RBP4/retinol ratio was not different between subjects with and without MetS (17.61 ± 1.73 mg/µmol vs. 17.57 ± 1.91 mg/µmol, respectively, $p = 0.78$). The associations of RBP4 and retinol with the presence of MetS remained close to significance after adjustment for age and sex (retinol: $\beta = 0.223$, $p = 0.052$ and RBP4: $\beta = 0.224$, $p = 0.051$; data not shown). Additionally, in age- and sex-adjusted multivariable linear regression analysis, RBP4 was independently associated with elevated triglycerides ($\beta = 0.348$, $p = 0.013$), but not with the other individual MetS components ($p \geq 0.75$ for each). Serum retinol levels were not associated with any of the MetS components ($p > 0.40$ for each; data not shown). In univariate analysis, RBP4 and retinol were not significantly associated with HOMA-IR ($r = 0.088$, $p = 0.441$ and $r = 0.149$, $p = 0.19$, respectively). Furthermore, RBP4, retinol and the RBP4/retinol ratio were not significantly different between men and women (RBP4: 39.01 ± 8.40 mg/L in men and 39.43 ± 8.53 mg/L in women, $p = 0.83$; retinol: 2.29 ± 0.46 µmol/L in men and 2.22 ± 0.51 µmol/L in women, $p = 0.57$; RBP4/retinol ratio: 17.10 ± 1.77 mg/µmol in men and 17.90 ± 1.80 mg/µmol in women, $p = 0.071$).

Table 1. Clinical and laboratory characteristics in 41 subjects with Type 2 diabetes mellitus (T2DM) and 37 subjects without T2DM.

	T2DM Subjects (n = 41)	Non-Diabetic Subjects (n = 37)	p-Value
Age (years)	60 ± 10	52 ± 9	<0.001
Gender (men/women)	19/22	10/27	0.127
Metabolic syndrome (yes/no)	29/12	7/30	<0.001
Systolic blood pressure (mm Hg)	145 ± 20	129 ± 20	0.001
Diastolic blood pressure (mm Hg)	87 ± 9	81 ± 12	0.025
BMI (kg/m^2)	29.0 ± 4.9	25.5 ± 4.1	0.001
Waist (cm)	100 ± 14	84 ± 13	<0.001
Glucose (mmol/L)	8.9 ± 2.3	5.6 ± 0.7	<0.001
HbA1c (mmol/mol)	50 ± 9	33 ± 3	<0.001
HOMA-IR (mU mmol/L^2/22.5)	4.01 (2.94–6.99)	1.56 (1.13–2.03)	<0.001
Total cholesterol (mmol/L)	5.53 ± 0.97	5.65 ± 0.98	0.578
Non-HDL cholesterol (mmol/L)	4.21 ± 1.05	4.06 ± 1.10	0.547
LDL cholesterol (mmol/L)	3.39 ± 0.88	3.46 ± 0.97	0.73
HDL cholesterol (mmol/L)	1.31 ± 0.39	1.59 ± 0.35	0.002
Triglycerides (mmol/L)	1.90 ± 1.60	1.34 ± 0.53	0.021
ApoB (g/L)	0.97 ± 0.24	0.91 ± 0.26	0.324
Apo A-1 (g/L)	1.37 ± 0.26	1.46 ± 0.21	0.080
RBP4 (mg/L)	40.13 ± 8.71	38.33 ± 8.11	0.349
Retinol (µmol/L)	2.33 ± 0.53	2.16 ± 0.43	0.129
RBP4/retinol ratio (mg/µmol)	17.38 ± 1.70	17.82 ± 1.94	0.290

Data are expressed in means ± SD, medians (interquartile range) or numbers. Abbreviations: Apo: apolipoprotein; BMI: body mass index; HOMA-IR: homeostasis model assessment of insulin resistance; HbA1c: glycated hemoglobin; HDL: high density lipoproteins; LDL: low density lipoproteins; RBP4: retinol binding protein 4. Differences between subjects with and without T2DM were determined by T-tests (using log$_e$ transformed values in case of not-normally distributed data) and Chi-square tests where appropriate. LDL cholesterol was calculated in 39 T2DM subjects and in 36 non-diabetic subjects.

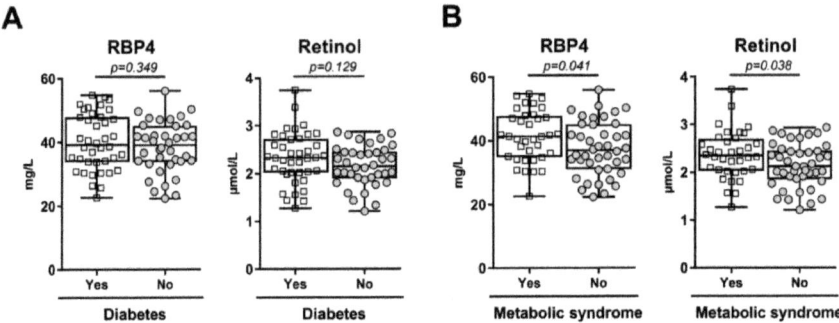

Figure 1. (**A**) Plasma retinol binding protein 4 (RBP4) and retinol in 41 subjects with and 37 subjects without Type 2 diabetes mellitus; (**B**) Plasma retinol binding protein 4 (RBP4) and retinol in 36 subjects with and 42 without the metabolic syndrome (MetS). Data are expressed in box and whiskers plots with mean and minimum to maximum values. All data points are shown.

Lipoprotein subfractions were quantified in 36 T2DM subjects and 27 non-diabetic subjects (Table 2). T2DM patients had more large-sized VLDL particles, more LDL particles, more small-sized LDL particles and more small-sized HDL particles, but less large- and medium-sized HDL particles.

Table 2. Lipoprotein subfraction characteristics in 36 subjects with Type 2 diabetes mellitus (T2DM) and 27 subjects without T2DM.

	T2DM Subjects ($n = 36$)	Non-Diabetic Subjects ($n = 27$)	p-Value
Total VLDL (nmol/L)	68.9 (48.6–82.1)	58.7 (51.4–86.7)	0.56
Large VLDL (nmol/L)	6.8 (2.8–9.9)	2.9 (2.3–4.7)	0.035
Medium VLDL (nmol/L)	23.0 (15.8–40.6)	24.8 (12.5–38.2)	0.45
Small VLDL (nmol/L)	29.9 (20.2–42.9)	32.2 (21.1–44.3)	0.51
Total LDL (nmol/L)	1257 (1022–1540)	981 (856–1284)	0.004
IDL (nmol/L)	170 (118–234)	188 (140–257)	0.14
Large LDL (nmol/L)	509 (345–612)	469 (435–597)	0.57
Small LDL (nmol/L)	586 (400–850)	338 (149–442)	0.003
Total HDL (μmol/L)	32.9 (29.3–37.9)	33.8 (32.0–36.1)	0.75
Large HDL (μmol/L)	5.0 (2.6–6.6)	6.7 (4.6–9.9)	0.021
Medium HDL (μmol/L)	10.1 (8.3–14.4)	12.8 (11.5–16.1)	0.007
Small HDL (μmol/L)	17.7 (14.6–20.9)	14.3 (10.1–16.30	0.006

Data are expressed in medians (interquartile ranges). Abbreviations: VLDL: very low density lipoproteins; LDL low density lipoproteins; IDL: intermediate density lipoproteins; HDL: high density lipoproteins. Differences between subjects with and without T2DM were determined by T-tests using $\log_e$ transformed values.

In univariate regression analysis, RBP4 was strongly correlated with retinol in all subjects combined, and this relationship did not vary according to diabetes status (Table 3A–C). Plasma retinol and RBP4 levels strongly correlated in all subjects combined ($r = 0.881$, $p < 0.001$), as well as in T2DM subjects ($r = 0.900$, $p < 0.001$) and non-diabetic subjects separately ($r = 0.859$, $p < 0.001$) (Figure 2). In all subjects combined, as well as in T2DM subjects and non-diabetic subjects separately, RBP4 and retinol were unrelated to age, blood pressure, obesity measures and glycemic control, except for a positive correlation of RBP4 with age in non-diabetic subjects. Notably, both RBP4 and retinol were positively correlated with total cholesterol, non-HDL cholesterol, LDL cholesterol, triglycerides and apoB levels with essentially similar relationships in T2DM subjects and non-diabetic subjects separately (Table 3A–C).

Table 3. Univariate relationships of retinol binding protein 4 (RBP4) and retinol with clinical and laboratory variables in 78 subjects (**A**), 41 subjects with Type 2 diabetes mellitus (T2DM) (**B**) and 37 subjects without T2DM (**C**).

All Subjects (A)	RBP4	Retinol
RBP4		0.881 ***
Retinol	0.881 ***	
Age	0.153	0.117
Systolic blood pressure	0.151	0.133
Diastolic blood pressure	0.099	0.139
BMI	0.021	0.005
Waist	0.099	0.167
Glucose	0.104	0.196
HbA1c	0.176	0.171
Total cholesterol	0.435 ***	0.399 ***
Non-HDL cholesterol	0.463 ***	0.421 ***
LDL cholesterol	0.384 ***	0.368 ***
HDL cholesterol	−0.186	−0.163
Triglycerides	0.391 ***	0.313 **
ApoB	0.445 ***	0.372 **
ApoA-I	−0.007	−0.039
T2DM Subjects (B)		
RBP4		0.859 ***
Retinol	0.859 ***	
Age	−0.082	−0.008
Systolic blood pressure	0.022	0.026
Diastolic blood pressure	0.030	0.065
BMI	−0.117	−0.091
Waist	−0.039	0.030
Glucose	0.082	0.161
HbA1c	0.184	0.089
Total cholesterol	0.376 *	0.321 *
Non-HDL cholesterol	0.425 **	0.340 *
LDL cholesterol	0.315 *	0.271
HDL cholesterol	−0.212	−0.122
Triglycerides	0.415 **	0.293
ApoB	0.394 *	0.262
ApoA-I	−0.150	−0.091

Table 3. Cont.

All Subjects (A)	RBP4	Retinol
Non-Diabetic Subjects (C)		
RBP4		0.900 ***
Retinol	0.900 ***	
Age	0.381 *	0.258
Systolic blood pressure	0.235	0.140
Diastolic blood pressure	0.115	0.144
BMI	−0.117	−0.018
Waist	0.170	0.186
Glucose	−0.100	−0.099
HbA1c	0.051	0.041
Total cholesterol	0.528 ***	0.555 ***
Non-HDL cholesterol	0.500 ***	0.526 ***
LDL cholesterol	0.514 ***	0.463 **
HDL cholesterol	−0.090	−0.096
Triglycerides	0.416 *	0.332 *
ApoB	0.489 ***	0.494 ***
ApoA-I	0.184	0.146

Pearson correlation coefficients are shown. Triglycerides are log$_e$ transformed. Abbreviations: BMI: body mass index; HbA1c: glycated hemoglobin; LDL: low density lipoproteins; HDL: high density lipoproteins. * $p < 0.05$; ** $p \leq 0.01$; *** $p \leq 0.001$.

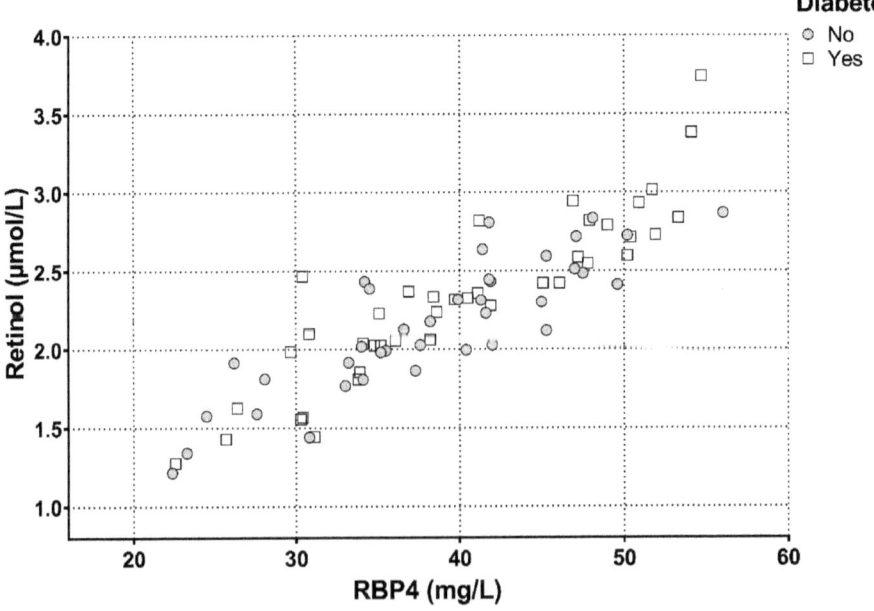

Figure 2. Correlation of total plasma retinol with retinol binding protein 4 (RBP4) in 41 subjects with and 37 without type 2 diabetes mellitus. All subjects: r = 0.881, p < 0.001; diabetic subjects: r = 0.900, p < 0.001; non-diabetic subjects: r = 0.859, p < 0.001.

With respect to lipoprotein subfractions, we observed positive univariate correlations of RBP4 and retinol with VLDL-P, large and medium VLDL, LDL-P, small LDL and small HDL, but inverse correlations with large HDL (Table 4A–C). Again, these relationships were mostly similar in T2DM subjects and non-diabetic subjects separately, although for HDL subfractions these relationships appeared to be stronger in non-diabetic subjects. In addition, large VLDL ($r = 0.537$, $p < 0.001$), small LDL ($r = 0.476$, $p < 0.001$) and large HDL ($r = -0.403$, $p = 0.001$) were correlated with HOMA-IR in all subjects combined. Comparable relationships were found in T2DM and non-diabetic subjects separately (data not shown).

Table 4. Univariate relationships of retinol binding protein 4 (RBP4) and retinol with lipoprotein subfractions in 63 subjects; (**A**): all subjects; (**B**): 36 subjects with Type 2 diabetes mellitus (T2DM) and (**C**): 27 subjects without T2DM.

All Subjects (A) (n = 63)	RBP4	Retinol
Total VLDL	0.452 ***	0.381 **
Large VLDL	0.433 ***	0.341 **
Medium VLDL	0.293 *	0.248 *
Small VLDL	0.178	0.187
Total LDL	0.398 ***	0.340 **
IDL	−0.053	−0.098
Large LDL	−0.007	0.118
Small LDL	0.423 ***	0.353 **
Total HDL	0.136	0.126
Large HDL	−0.243	−0.258 *
Medium HDL	−0.056	−0.010
Small HDL	0.317 *	0.290 *
T2DM Subjects (B) (n = 36)	**RBP4**	**Retinol**
Total VLDP	0.423 *	0.322
Large VLDL	0.401 *	0.262
Medium VLDL	0.238	0.168
Small VLDL	0.160	0.176
Total LDL	0.255	0.191
IDL	−0.227	−0.233
Large LDL	−0.079	0.082
Small LDL	0.351 *	0.266
Total HDL	0.016	0.010
Large HDL	−0.170	−0.190
Medium HDL	−0.004	0.064
Small HDL	0.019	0.002
Non-Diabetic Subjects (C) (n = 27)	**RBP4**	**Retinol**
Total VLDP	0.487 **	0.472 *
Large VLDL	0.433 *	0.385 *
Medium VLDL	0.344	0.333
Small VLDL	0.270	0.293
Total LDL	0.529 **	0.442 *

Table 4. Cont.

All Subjects (A) (n = 63)	RBP4	Retinol
IDL	0.309	0.257
Large LDL	0.223	0.301
Small LDL	0.467 *	0.366
Total HDL	0.478 *	0.502 **
Large HDL	−0.287	−0.254
Medium HDL	−0.027	0.051
Small HDL	0.523 **	0.476 *

Pearson correlation coefficients are shown. All lipoprotein subfractions are $\log_e$ transformed. Abbreviations: VLDL: very low density lipoproteins; LDL: low density lipoproteins; IDL: intermediate density lipoproteins; HDL: high density lipoproteins. * $p < 0.05$; ** $p \leq 0.01$; *** $p \leq 0.001$.

We next performed multivariable linear regression analysis to disclose the extent to which RBP4 and retinol were independently related to VLDL, LDL and HDL subfractions (Table 5). In age-, sex- and diabetes status-adjusted analysis, RBP4 and retinol were independently related to large VLDL and small LDL (Table 5, models A and B). RBP4 and retinol were also independently and inversely associated with large HDL (Table 5, models A and B). When these analyses were additionally adjusted for the use of metformin, sulfonylurea and antihypertensive medication, the positive relationships of large VLDL with RBP4 ($\beta = 0.494$, $p < 0.001$) and retinol ($\beta = 0.373$, $p = 0.003$), of small LDL with RBP4 ($\beta = 0.511$, $p < 0.001$) and with retinol ($\beta = 0.376$, $p = 0.009$), and the inverse relationship of large HDL with RBP4 ($\beta = -0.416$, $p = 0.011$) and retinol ($\beta = -0.343$, $p = 0.039$) remained statistically significant. Remarkably, when retinol was added as an independent variable to the model of RBP4 with VLDL subfractions (Table 5, model A), the relationship of RBP4 with large VLDL remained significant ($\beta = 0.173$, $p = 0.022$). Conversely, when RBP4 was added as independent variable in the model of retinol with VLDL subfractions (Table 5, model B), the relation of retinol with large VLDL was lost ($\beta = -0.091$, $p = 0.28$).

Given the correlation of plasma RBP4 levels with certain subfractions, we also assessed whether RBP4 may actually physically interact with such lipoprotein particles. Lipoprotein particles, as well as soluble proteins, were separated by FPLC from freshly-obtained (non-frozen) plasma of non-diabetic individuals and fractions were analyzed for cholesterol, triglycerides, protein and RBP4 concentrations. A representative FPLC profile is shown in Figure 3. RBP4 remained undetectable in the VLDL enriched samples (B1–B9), while very high levels were detected in the fractions with soluble proteins (C2–C10). Some overlap of RBP4 was detected with HDL-containing samples (B10–C2), but RBP4 was absent in the peak fraction of HDL (C1), indicating that the overlap is due to co-elution rather than a physical interaction between RBP4 and HDL. From these experiments, it is concluded that while plasma levels of RBP4 strongly correlate with certain lipoproteins particles; this occurs in the absence of a physical interaction.

Table 5. Multivariable regression analyses showing independent relationships of retinol binding protein 4 (RBP4) and retinol with very low density lipoprotein (VLDL) subfractions (A), low density lipoprotein (LDL) subfractions (B) and high density lipoprotein (HDL) subfractions (C) in 63 subjects (36 subjects with and 27 subjects without Type 2 diabetes mellitus).

A VLDL Subfractions	Model A		Model B	
	β	p-Value	β	p-Value
Age	0.305	0.027	0.226	0.12
Sex (men vs. women)	−0.072	0.57	0.018	0.89
T2DM	−0.079	0.60	0.017	0.92

Table 5. Cont.

B LDL Subfractions	Model A		Model B	
	β	p-Value	β	p-Value
Large VLDL	0.444	0.005	0.324	0.046
Medium VLDL	0.044	0.770	0.032	0.84
Small VLDL	0.183	0.13	0.206	0.105
Age	0.306	0.035	0.195	0.19
Sex (men vs. women)	−0.180	0.19	−0.099	0.48
T2DM	0.157	0.33	−0.030	0.86
IDL	−0.150	0.24	−0.138	0.30
Large LDL	0.106	0.39	0.221	0.090
Small LDL	0.539	<0.001	0.440	0.003
B HDL Subfractions	Model A		Model B	
	β	p-value	β	p-value
Age	0.309	0.044	0.257	0.095
Sex (men vs. women)	−0.180	0.26	0.006	0.97
T2DM	−0.073	0.65	−0.008	0.96
Large HDL	−0.336	0.034	−0.323	0.043
Medium HDL	0.145	0.38	0.302	0.073
Small HDL	0.262	0.076	0.254	0.086

β: standardized regression coefficient. All lipoprotein subfractions are loge transformed. IDL: intermediate density lipoproteins; **Models A**: RBP4 as dependent variable; **Models B**: retinol as dependent variable.

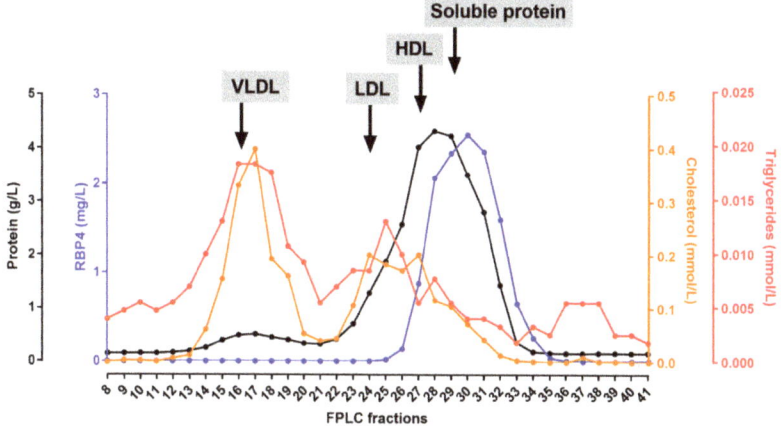

Figure 3. Total protein, retinol binding protein 4 (RBP4), cholesterol and triglyceride concentrations in fast protein liquid chromatography (FPLC) fractions from plasma of a representative control subject. The location of peak fractions of very low density lipoprotein (VLDL), low density lipoprotein (LDL) and high density lipoprotein (HDL) particles, as well as of the soluble proteins are indicated.

4. Discussion

This study confirms and extends previous findings [6,7,12,13,16,17], showing that plasma RBP4 is positively correlated with total cholesterol, non-HDL cholesterol, LDL cholesterol, triglycerides and apoB in subjects with and without T2DM. The main novel finding is that the plasma concentration of RBP4 is positively correlated with total VLDL and LDL particles, which appeared to be due to relationships with large VLDL and small LDL particles. The univariate correlations did not differ in diabetic and non-diabetic subjects. In multivariable linear regression analysis adjusted for age, sex, diabetes status and the use of glucose lowering and antihypertensive medication, it was demonstrated that RBP4 was independently related to large VLDL particles, as well as to small LDL particles. The current study thus suggests that the positive relationship of RBP4 with (triglyceride-rich) apoB-containing lipoproteins is attributable to associations with large VLDL and small LDL particles. Notably, no physical interaction appeared to exist between RBP4 and lipoprotein particles.

In our study, plasma RBP4 levels were not elevated in T2DM subjects, in agreement with earlier studies [12,13], but in contrast to other reports [5,9–11]. In addition, we did not observe a significant correlation with insulin resistance as determined by HOMA-IR. However, we did find that RBP4 was elevated in subjects classifying with MetS, which was attributable to its association with triglycerides. In comparison, Ingelsson et al. also observed RBP4 elevations in MetS and demonstrated incrementally higher RBP4 levels with an increasing number of MetS components [15]. Of note, while retinol was found to be elevated in the subjects categorized with MetS as well, there was no significant relationship with individual MetS components, and we observed no abnormalities in the RBP4/retinol ratio in T2DM or MetS. In fact, there was a close correlation of retinol with RBP4 as anticipated [14], which appeared to be linear over the full range of RBP4 concentrations observed. Taken together, these findings support the notion that elevated levels of circulating RBP4, rather than retinol, represent the primary abnormality in MetS.

RBP4 levels were predominantly associated with large VLDL particles, as well as with small LDL particles, the latter finding being consistent with earlier observations using conventional methods for lipoprotein subfraction quantification [17]. Given the product-precursor relationships of small LDL particles with large VLDL [18,32,33], it is conceivable that the relationships of RBP4 with these specific VLDL and LDL subfractions are interdependent. We also observed an independent and inverse relationship of RBP4 with large HDL particles. It is well established that HDL cholesterol is inversely associated with large VLDL, consequent to the process of cholesteryl ester transfer [19]. It was recently shown that large VLDL particles determine an increased cholesteryl ester transfer in T2DM [20], but it is unclear whether RBP4 could interfere with this process. Of further interest, RBP4 was associated with large VLDL independent of retinol, but retinol was not associated with large VLDL independent of RBP4. This suggests that circulating RBP4 rather than retinol is primarily related to large VLDL particles. In addition, large VLDL and small LDL particle concentrations were expectedly elevated in T2DM [18,20], again consistent with the concept that large VLDL and small LDL particles are metabolically interrelated [33]. Since our study neither showed an effect of the diabetic state on plasma RBP4 nor on retinol, we surmise that circulating levels of RBP4 and retinol do not to a considerable extent explain the predominance of large VLDL and small LDL particles in this condition. Furthermore, large VLDL and small LDL particles were correlated with HOMA-IR whereas RBP4 and retinol were not. Circulating RBP4 and VLDL particles largely originate from the liver, but their hepatic production and secretion machineries appear not interconnected. Still, the correlation between the circulating levels led us to determine whether a possible physical interaction between RBP4 and VLDL and (small) LDL particles exists. FPLC was used to separate lipoprotein particles from soluble proteins as a mild approach to retain protein-protein and protein-lipid interactions. Still, no co-elution of RBP4 with VLDL was observed, providing no evidence that such physical interaction actually exists. We, therefore, hypothesize that RBP4 could directly interfere with the metabolism of triglyceride-rich apoB-containing lipoproteins. As a potential mechanism, it is of interest that RBP4 transcriptionally enhances CD36 expression [26], a protein that impairs VLDL secretion from mouse liver in vivo [34].

While the current study was primarily focused on RBP4, our findings with respect to retinol levels are also relevant. Retinol is the circulating form of vitamin A that is locally (in tissues) converted to retinoic acids (such as all-trans retinoic acid and 9-cis retinoic acid) that activate transcription factors (Retinoic Acid Receptor (RAR) and Retinoic X Receptor (RXR)) [35,36]. Retinol and retinoic acids simulate hepatic RBP4 secretion and enhance de novo lipogenesis in the liver [34]. In line, retinoic acids also stimulate VLDL secretion in rats [37]. More recently, the enzyme 17-beta hydroxysteroid dehydrogenase 13 (HSD17b13) with retinol dehydrogenase activity, was implicated in the pathogenesis of non-alcoholic fatty liver disease [38]. Retinol may, therefore, play multiple roles in hepatic fat accumulation and triglyceride metabolism, both directly and indirectly, by interacting with RBP4.

In view of the accumulating though sometimes controversial epidemiological evidence indicating that RBP4 is associated with incident CVD [23,24,26,39,40], the present findings with regard to the association of RBP4 with small LDL particles are likely to be clinically relevant. Small dense LDL particles are prone to oxidative modification [17] and may predict incident CVD beyond conventional lipid measures [41–43]. Moreover, smaller-sized LDL particles may be more electronegatively charged and upregulate the lectin-like oxidized LDL receptor in vitro, a pro-atherogenic phenomenon [44]. Such particles are associated with the extent and severity of coronary artery atherosclerosis [45]. We, therefore, propose that the relationship of RBP4 with small LDL particles could contribute to its alleged pro-atherogenic potential.

Several methodological issues of the present study need consideration. Diabetic subjects using insulin and lipid lowering were not allowed to participate, in order to obviate confounding due to effects of insulin and lipid-lowering drugs on glucose and lipid metabolism. As a consequence, T2DM subjects with rather mild hyperglycemia and dyslipidemia were preferentially included, which could to some extent mask relationships of RBP4 with dysglycemia and could underestimate relationships of RBP4 with (apo)lipoproteins and lipoprotein subfractions. Furthermore, we excluded subjects with impaired glomerular filtration rate. This was done to minimize confounding with respect to circulating RBP4 due to chronic kidney disease in view of the inverse relationship of RBP4 with glomerular filtration rate [15]. Obviously, the cross-sectional design of our study precludes to establish cause-effect relationships with certainty. Finally, we consider the present findings as preliminary in view of the rather low numbers of diabetic and non-diabetic participants included.

5. Conclusions

Serum RBP4 levels are related to large VLDL and small LDL particles, but in the absence of a physical interaction between RBP4 and the lipoprotein particles. Higher RBP4 may be part of a proatherogenic plasma lipoprotein profile. However, it seems unlikely that alterations in circulating RBP4 per se contribute significantly to the predominance of large VLDL and small LDL in the diabetic state.

Author Contributions: H.W. and R.P.F.D. analyzed the data. H.W., K.N.F. and R.P.F.D. wrote the manuscript. All authors discussed the results and contributed to the final manuscript. All authors approved the final version of this article.

Funding: This research received no external funding.

Acknowledgments: Plasma lipid and apolipoprotein measurement was carried out in the laboratory of L.D. Dikkeschei, Department of Clinical Chemistry, Isala Clinics Zwolle, The Netherlands.

Conflicts of Interest: M.A.C. is an employee of LabCorp. The rest of the authors declare that they have no competing interests.

Abbreviations

atRA receptor (RAR); apolipoprotein A-I (apoA-I); apolipoprotein B (apoB); body mass index (BMI); cardiovascular disease (CVD); enzyme linked immunosorbent assay (ELISA); fast protein liquid chromatography (FPLC); high density lipoproteins (HDL); high density lipoprotein particle concentration (HDL-P); high performance liquid chromatography (HPLC); homeostasis model assessment of insulin resistance (HOMA-IR); intermediate density lipoproteins (IDL); low density lipoproteins (LDL); low density lipoprotein particle concentration (LDL-P); nuclear magnetic resonance spectrometry (NMR); retinol binding protein 4 (RBP4); very low density lipoproteins (VLDL); very low density lipoprotein particle concentration (VLDL-P)

References

1. Sullivan, P.W.; Morrato, E.H.; Ghushchyan, V.; Wyatt, H.R.; Hill, J.O. Obesity, inactivity, and the prevalence of diabetes and diabetes-related cardiovascular comorbidities in the, U.S.; 2000–2002. *Diabetes Care* **2005**, *28*, 1599–1603. [CrossRef] [PubMed]
2. Kendall, D.M.; Harmel, A.P. The metabolic syndrome, type 2 diabetes, and cardiovascular disease: Understanding the role of insulin resistance. *Am. J. Manag. Care* **2002**, *8*, 635–653.
3. Kahn, S.E.; Hull, R.L.; Utzschneider, K.M. Mechanisms linking obesity to insulin resistance and type 2 diabetes. *Nature* **2006**, *444*, 840–846. [CrossRef] [PubMed]
4. Yang, Q.; Graham, T.E.; Mody, N.; Preitner, F.; Peroni, O.D.; Zabolotny, J.M.; Kotani, K.; Quadro, L.; Kahn, B.B. Serum retinol binding protein 4 contributes to insulin resistance in obesity and type 2 diabetes. *Nature* **2005**, *436*, 356–362. [CrossRef]
5. Graham, T.E.; Yang, Q.; Blüher, M.; Hammarstedt, A.; Ciaraldi, T.P.; Henry, R.R.; Wason, C.J.; Andreas Oberbach Andreas Oberbach, B.S.; Jansson, P.-A.; Smith, U.; et al. Retinol-binding protein 4 and insulin resistance in lean, obese, and diabetic subjects. *N. Engl. J. Med.* **2006**, *354*, 2552–2563. [CrossRef]
6. Broch, M.; Gómez, J.M.; Auguet, M.T.; Vilarrasa, N.; Pastor, R.; Elio, I.; Olona, M.; García-España, A.; Richart, C. Association of retinol-binding protein-4 (RBP4) with lipid parameters in obese women. *Obes. Surg.* **2010**, *20*, 1258–1264. [CrossRef]
7. Haider, D.G.; Schindler, K.; Prager, G.; Bohdjalian, A.; Luger, A.; Wolzt, M.; Ludvik, B. Serum retinol-binding protein 4 is reduced after weight loss in morbidly obese subjects. *J. Clin. Endocrinol. Metab.* **2007**, *92*, 1168–1171. [CrossRef]
8. Gómez-Ambrosi, J.; Rodríguez, A.; Catalán, V.; Ramírez, B.; Silva, C.; Rotellar, F.; Gil, M.J.; Salvador, J.; Frühbeck, G. Serum retinol-binding protein 4 is not increased in obesity or obesity-associated type 2 diabetes mellitus, but is reduced after relevant reductions in body fat following gastric bypass. *Clin. Endocrinol. (Oxf).* **2008**, *69*, 208–215. [CrossRef]
9. Abahusain, M.A.; Wright, J.; Dickerson, J.W.; de Vol, E.B. Retinol, alpha-tocopherol and carotenoids in diabetes. *Eur. J. Clin. Nutr.* **1999**, *53*, 630–635. [CrossRef]
10. Basualdo, C.G.; Wein, E.E.; Basu, T.K. Vitamin A (retinol) status of first nation adults with non-insulin-dependent diabetes mellitus. *J. Am. Coll. Nutr.* **1997**, *16*, 39–45. [CrossRef]
11. Cho, Y.M.; Youn, B.S.; Lee, H.; Lee, N.; Min, S.S.; Kwak, S.H.; Lee, H.K.; Park, K.S. Plasma retinol-binding protein-4 concentrations are elevated in human subjects with impaired glucose tolerance and type 2 diabetes. *Diabetes Care* **2006**, *29*, 2457–2461. [CrossRef] [PubMed]
12. von Eynatten, M.; Lepper, P.M.; Liu, D.; Lang, K.; Baumann, M.; Nawroth, P.P.; Bierhaus, A.; Dugi, K.A.; Heemann, U.; Allolio, B.; et al. Retinol-binding protein 4 is associated with components of the metabolic syndrome, but not with insulin resistance, in men with type 2 diabetes or coronary artery disease. *Diabetologia* **2007**, *50*, 1930–1937. [CrossRef] [PubMed]
13. Erikstrup, C.; Mortensen, O.H.; Nielsen, A.R.; Fischer, C.P.; Plomgaard, P.; Petersen, A.M.; Krogh-Madsen, R.; Lindegaard, B.; Erhardt, J.G.; Ullum, H.; et al. RBP-to-retinol ratio, but not total RBP, is elevated in patients with type 2 diabetes. *Diabetes Obes. Metab.* **2009**, *11*, 204–212. [CrossRef] [PubMed]
14. Aeberli, I.; Biebinger, R.; Lehmann, R.; l'Allemand, D.; Spinas, G.A.; Zimmermann, M.B. Serum retinol-binding protein 4 concentration and its ratio to serum retinol are associated with obesity and metabolic syndrome components in children. *J. Clin. Endocrinol. Metab.* **2007**, *92*, 4359–4365. [CrossRef]
15. Ingelsson, E.; Lind, L. Circulating retinol-binding protein 4 and subclinical cardiovascular disease in the elderly. *Diabetes Care* **2009**, *32*, 733–735. [CrossRef]
16. Comerford, K.B.; Buchan, W.; Karakas, S.E. The effects of weight loss on FABP4 and RBP4 in obese women with metabolic syndrome. *Horm. Metab. Res.* **2014**, *46*, 224–231. [CrossRef]
17. Wu, J.; Shi, Y.H.; Niu, D.M.; Li, H.Q.; Zhang, C.N.; Wang, J.J. Association among retinol-binding protein 4, small dense LDL cholesterol and oxidized LDL levels in dyslipidemia subjects. *Clin. Biochem.* **2012**, *45*, 619–622. [CrossRef]
18. Taskinen, M.R.; Borén, J. New insights into the pathophysiology of dyslipidemia in type 2 diabetes. *Atherosclerosis* **2015**, *239*, 483–495. [CrossRef]

19. Dallinga-Thie, G.M.; Dullaart, R.P.; van Tol, A. Derangements of intravascular remodeling of lipoproteins in type 2 diabetes mellitus: Consequences for atherosclerosis development. *Curr. Diabete Rep.* **2008**, *8*, 65–70. [CrossRef]
20. Dullaart, R.P.; de Vries, R.; Kwakernaak, A.J.; Perton, F.; Dallinga-Thie, G.M. Increased large VLDL particles confer elevated cholesteryl ester transfer in diabetes. *Eur. J. Clin. Investig.* **2015**, *45*, 36–44. [CrossRef]
21. Shah, A.S.; Davidson, W.S.; Gao, Z.; Dolan, L.M.; Kimball, T.R.; Urbina, E.M. Superiority of lipoprotein particle number to detect associations with arterial thickness and stiffness in obese youth with and without prediabetes. *J. Clin. Lipidol.* **2016**, *10*, 610–618. [CrossRef] [PubMed]
22. Usui, S.; Ichimura, M.; Ikeda, S.; Okamoto, M. Association between serum retinol-binding protein 4 and small dense low-density lipoprotein cholesterol levels in young adult women. *Clin. Chim. Acta* **2009**, *399*, 45–48. [CrossRef] [PubMed]
23. Mallat, Z.; Simon, T.; Benessiano, J.; Clément, K.; Taleb, S.; Wareham, N.J.; Luben, R.; Khaw, K.-T.; Tedgui, A.; Boekholdt, S.M. Retinol-binding protein 4 and prediction of incident coronary events in healthy men and women. *J. Clin. Endocrinol. Metab.* **2009**, *94*, 255–260. [CrossRef] [PubMed]
24. Sun, Q.; Kiernan, U.A.; Shi, L.; Phillips, D.A.; Kahn, B.B.; Hu, F.B.; Manson, J.E.; Albert, C.M.; Rexrode, K.M. Plasma retinol-binding protein 4 (RBP4) levels and risk of coronary heart disease: A prospective analysis among women in the nurses' health study. *Circulation* **2013**, *127*, 1938–1947. [CrossRef]
25. Liu, Y.; Wang, D.; Chen, H.; Xia, M. Circulating retinol binding protein 4 is associated with coronary lesion severity of patients with coronary artery disease. *Atherosclerosis* **2015**, *238*, 45–51. [CrossRef]
26. Liu, Y.; Zhong, Y.; Chen, H.; Wang, D.; Wang, M.; Ou, J.S.; Xia, M. Retinol-Binding Protein-Dependent Cholesterol Uptake Regulates Macrophage Foam Cell Formation and Promotes Atherosclerosis. *Circulation* **2017**, *135*, 1339–1354. [CrossRef]
27. Grundy, S.M.; Cleeman, J.I.; Daniels, S.R.; Donato, K.A.; Eckel, R.H.; Franklin, B.A.; Gordon, D.J.; Krauss, R.M.; Savage, P.J.; Smith, S.C.; et al. Diagnosis and management of the metabolic syndrome: An American Heart Association/National Heart, Lung, and Blood Institute Scientific Statement. *Circulation* **2005**, *112*, 2735–2752. [CrossRef]
28. Zaman, Z.; Fielden, P.; Frost, P.G. Simultaneous determination of vitamins A and E and carotenoids in plasma by reversed-phase HPLC in elderly and younger subjects. *Clin. Chem.* **1993**, *39*, 2229–2234.
29. Nijstad, N.; Wiersma, H.; Gautier, T.; van der Giet, M.; Maugeais, C.; Tietge, U.J. Scavenger receptor BI-mediated selective uptake is required for the remodeling of high density lipoprotein by endothelial lipase. *J. Biol. Chem.* **2009**, *284*, 6093–6100. [CrossRef]
30. Jeyarajah, E.J.; Cromwell, W.C.; Otvos, J.D. Lipoprotein particle analysis by nuclear magnetic resonance spectroscopy. *Clin. Lab. Med.* **2006**, *26*, 847–870. [CrossRef]
31. Matyus, S.P.; Braun, P.J.; Wolak-Dinsmore, J.; Jeyarajah, E.J.; Shalaurova, I.; Xu, Y.; Warner, S.M.; Clement, T.S.; Connelly, M.A.; Fischer, T.J. NMR measurement of LDL particle number using the Vantera Clinical Analyzer. *Clin. Biochem.* **2014**, *47*, 203–210. [CrossRef] [PubMed]
32. Packard, C.J.; Shepherd, J. Lipoprotein heterogeneity and apolipoprotein B metabolism. *Arterioscler. Thromb. Vasc. Biol.* **1997**, *17*, 3542–3556. [CrossRef] [PubMed]
33. Adiels, M.; Olofsson, S.O.; Taskinen, M.R.; Borén, J. Overproduction of very low-density lipoproteins is the hallmark of the dyslipidemia in the metabolic syndrome. *Arterioscler. Thromb. Vasc. Biol.* **2008**, *28*, 1225–1236. [CrossRef] [PubMed]
34. Nassir, F.; Adewole, O.L.; Brunt, E.M.; Abumrad, N.A. CD36 deletion reduces VLDL secretion, modulates liver prostaglandins, and exacerbates hepatic steatosis in ob/ob mice. *J. Lipid Res.* **2013**, *54*, 2988–2997. [CrossRef]
35. Saeed, A.; Dullaart, R.P.F.; Schreuder, T.C.M.A.; Blokzijl, H.; Faber, K.N. Disturbed Vitamin A Metabolism in Non-Alcoholic Fatty Liver Disease (NAFLD). *Nutrients* **2017**, *10*, 29. [CrossRef]
36. Flajollet, S.; Staels, B.; Lefebvre, P. Retinoids and nuclear retinoid receptors in white and brown adipose tissues: Physiopathologic aspects. *Horm. Mol. Biol. Clin. Investig.* **2013**, *14*, 75–86. [CrossRef]
37. Gerber, L.E.; Erdman, J.W. Retinoic acid and hypertriglyceridemia. *Ann. N. Y. Acad. Sci.* **1981**, *359*, 391–392. [CrossRef]
38. Ma, Y.; Belyaeva, O.V.; Brown, P.M.; Fujita, K.; Valles, K.; Karki, S.; de Boer, Y.S.; Koh, C.; Chen, Y.; Du, X.; et al. HSD17B13 is a Hepatic Retinol Dehydrogenase Associated with Histological Features of Non-Alcoholic Fatty Liver Disease. *Hepatology* **2019**, *69*, 1504–1509. [CrossRef]

39. Misra, S.; Kumar, A.; Kumar, P.; Yadav, A.K.; Mohania, D.; Pandit, A.K.; Prasad, K.; Vibha, D. Blood-based protein biomarkers for stroke differentiation: A systematic review. *Proteom. Clin. Appl.* **2017**, *11*, 1700007. [CrossRef]
40. Rist, P.M.; Jiménez, M.C.; Tworoger, S.S.; Hu, F.B.; Manson, J.E.; Sun, Q.; Rexrode, K.M. Plasma Retinol-Binding Protein 4 Levels and the Risk of Ischemic Stroke among Women. *J. Stroke Cerebrovasc. Dis.* **2018**, *27*, 68–75. [CrossRef]
41. Kuller, L.; Arnold, A.; Tracy, R.; Otvos, J.; Burke, G.; Psaty, B.; Siscovick, D.; Freedman, D.S.; Kronmal, R. Nuclear magnetic resonance spectroscopy of lipoproteins and risk of coronary heart disease in the cardiovascular health study. *Arterioscler. Thromb. Vasc. Biol.* **2002**, *22*, 1175–1180. [CrossRef] [PubMed]
42. Tsai, M.Y.; Steffen, B.T.; Guan, W.; McClelland, R.L.; Warnick, R.; McConnell, J.; Hoefner, D.M.; Remaley, A.T. New automated assay of small dense low-density lipoprotein cholesterol identifies risk of coronary heart disease: The Multi-ethnic Study of Atherosclerosis. *Arterioscler. Thromb. Vasc. Biol.* **2014**, *34*, 196–201. [CrossRef] [PubMed]
43. Hoogeveen, R.C.; Gaubatz, J.W.; Sun, W.; Dodge, R.C.; Crosby, J.R.; Jiang, J.; Couper, D.; Virani, S.S.; Kathiresan, S.; Boerwinkle, E.; et al. Small dense low-density lipoprotein-cholesterol concentrations predict risk for coronary heart disease: The Atherosclerosis Risk In Communities (ARIC) study. *Arterioscler. Thromb. Vasc. Biol.* **2014**, *34*, 1069–1077. [CrossRef] [PubMed]
44. Chang, C.Y.; Chen, C.H.; Chen, Y.M.; Hsieh, T.Y.; Li, J.P.; Shen, M.Y.; Lan, J.L.; Chen, D.Y. Association between Negatively Charged Low-Density Lipoprotein L5 and Subclinical Atherosclerosis in Rheumatoid Arthritis Patients. *J. Clin. Med.* **2019**, *8*, 177. [CrossRef] [PubMed]
45. Niccoli, G.; Bacà, M.; De Spirito, M.; Parasassi, T.; Cosentino, N.; Greco, G.; Conte, M.; Montone, R.A.; Arcovito, G.; Crea, F. Impact of electronegative low- density lipoprotein on angiographic coronary atherosclerotic burden. *Atherosclerosis* **2012**, *223*, 166–170. [CrossRef]

© 2019 by the authors. Licensee MDPI, Basel, Switzerland. This article is an open access article distributed under the terms and conditions of the Creative Commons Attribution (CC BY) license (http://creativecommons.org/licenses/by/4.0/).

Article

High Betaine, a Trimethylamine N-Oxide Related Metabolite, Is Prospectively Associated with Low Future Risk of Type 2 Diabetes Mellitus in the PREVEND Study

Erwin Garcia [1,*], Maryse C. J. Osté [2], Dennis W. Bennett [1], Elias J. Jeyarajah [1], Irina Shalaurova [1], Eke G. Gruppen [2,3], Stanley L. Hazen [4,5], James D. Otvos [1], Stephan J. L. Bakker [2], Robin P.F. Dullaart [3] and Margery A. Connelly [1]

[1] Laboratory Corporation of America Holdings (LabCorp), Morrisville, NC 27560, USA; dwbent@uwm.edu (D.W.B.); eliasjey@gmail.com (E.J.J.); shalaui@labcorp.com (I.S.); otvosj@labcorp.com (J.D.O.); connem5@labcorp.com (M.A.C.)
[2] Department of Nephrology, University of Groningen and University Medical Center Groningen, 9700 RB Groningen, The Netherlands; m.c.j.oste@umcg.nl (M.C.J.O.); e.g.gruppen@umcg.nl (E.G.G.); s.j.l.bakker@umcg.nl (S.J.L.B.)
[3] Department of Endocrinology, University of Groningen and University Medical Center Groningen, 9700 RB Groningen, The Netherlands; r.p.f.dullaart@umcg.nl
[4] Department of Cardiovascular and Metabolic Sciences, Cleveland Clinic, Cleveland, OH 44195, USA; HAZENS@ccf.org
[5] Department of Cardiovascular Medicine, Cleveland Clinic, Cleveland, OH 44195, USA
* Correspondence: garce14@labcorp.com; Tel.: +(919)-388-5551

Received: 20 August 2019; Accepted: 29 October 2019; Published: 1 November 2019

Abstract: Background: Gut microbiota-related metabolites, trimethylamine-N-oxide (TMAO), choline, and betaine, have been shown to be associated with cardiovascular disease (CVD) risk. Moreover, lower plasma betaine concentrations have been reported in subjects with type 2 diabetes mellitus (T2DM). However, few studies have explored the association of betaine with incident T2DM, especially in the general population. The goals of this study were to evaluate the performance of a newly developed betaine assay and to prospectively explore the potential clinical associations of betaine and future risk of T2DM in a large population-based cohort. Methods: We developed a high-throughput, nuclear magnetic resonance (NMR) spectroscopy procedure for acquiring spectra that allow for the accurate quantification of plasma/serum betaine and TMAO. Assay performance for betaine quantification was assessed and Cox proportional hazards regression was employed to evaluate the association of betaine with incident T2DM in 4336 participants in the Prevention of Renal and Vascular End-Stage Disease (PREVEND) study. Results: Betaine assay results were linear (y = 1.02X − 3.75) over a wide range of concentrations (26.0–1135 µM). The limit of blank (LOB), limit of detection (LOD) and limit of quantitation (LOQ) were 6.4, 8.9, and 13.2 µM, respectively. Coefficients of variation for intra- and inter-assay precision ranged from 1.5–4.3% and 2.5–5.5%, respectively. Deming regression analysis of results produced by NMR and liquid chromatography coupled to tandem mass spectrometry(LC-MS/MS) revealed an R^2 value of 0.94 (Y = 1.08x − 1.89) and a small bias for higher values by NMR. The reference interval, in a cohort of apparently healthy adult participants ($n = 501$), was determined to be 23.8 to 74.7 µM (mean of 42.9 ± 12.6 µM). In the PREVEND study ($n = 4336$, excluding subjects with T2DM at baseline), higher betaine was associated with older age and lower body mass index, total cholesterol, triglycerides, and hsCRP. During a median follow-up of 7.3 (interquartile range (IQR), 5.9–7.7) years, 224 new T2DM cases were ascertained. Cox proportional hazards regression models revealed that the highest tertile of betaine was associated with a lower incidence of T2DM. Hazard ratio (HR) for the crude model was 0.61 (95% CI: 0.44–0.85, $p = 0.004$). The association remained significant even after adjusting for multiple clinical covariates and T2DM risk factors, including fasting glucose. HR for the fully-adjusted model was 0.50 (95% CI: 0.32–0.80, $p = 0.003$). Conclusions: The newly developed NMR-based betaine assay exhibits performance characteristics that are consistent with usage in the clinical laboratory. Betaine levels may be useful for assessing the risk of future T2DM.

Keywords: betaine; trimethylamine N-oxide related metabolites; nuclear magnetic resonance spectroscopy; type 2 diabetes mellitus

---

1. Introduction

Betaine (N,N,N-trimethylglycine) is an osmoprotectant as well as a methyl donor in one-carbon metabolism [1,2]. Betaine is required for remethylation of homocysteine to methionine, a precursor of the universal methyl donor S-adenosylmethionine (SAM) [1,2]. By decreasing SAM availability, betaine deficiency may decrease phosphatidylcholine synthesis, promote hepatic steatosis and modify very low-density lipoprotein (VLDL) synthesis and secretion [3]. Betaine was shown to be inversely associated with triglycerides (TG) and phospholipid transfer protein (PLTP) activity, further supporting the notion that low betaine levels may alter liver fat accumulation and lipid/lipoprotein metabolism [4]. Lower plasma betaine concentrations have been reported in subjects with metabolic syndrome, type 2 diabetes mellitus (T2DM), non-alcoholic fatty liver disease (NAFLD) and/or non-alcoholic steatohepatitis (NASH) [3,5–8]. Moreover, betaine levels were associated with disease severity in subjects with NAFLD, betaine levels being significantly lower in subjects with NASH compared to subjects with hepatic steatosis [3]. Plasma betaine levels arise from both dietary intake and endogenous synthesis from dietary choline [1,2,9]. Having a high-throughput assay to measure plasma/serum betaine levels may be a way of gaining a better understanding of the role of betaine in hepatic and metabolic diseases.

The gut microbe-derived metabolite TMAO has been shown to be an independent marker of cardiovascular disease (CVD) and mortality with higher levels of TMAO correlating with higher risk [10–22]. Betaine has also been reported to be associated with CVD. However, in contrast to TMAO, low plasma betaine levels were associated with increased CVD risk in most studies [14,23–26]. In one study, however, Lever et al. reported that low betaine levels were associated with CVD events in patients without T2DM, whereas in contrast, elevated betaine levels were associated with CVD in patients with T2DM [27]. In addition, it has been shown that the association of betaine with CVD risk was attenuated by TMAO, suggesting that it is TMAO that drives the risk of future CVD [14]. In the same cohort [14], it was also found that a modest but statistically significant correlation exists between TMAO and betaine.

Besides the cross-sectional associations of betaine with metabolic diseases, as well as prospective relationships of betaine with CVD events, prospective studies have shown that lower betaine levels were associated with a higher risk of incident T2DM. In a study of Norwegian subjects with suspected angina pectoris, plasma betaine levels were inversely associated with incident T2DM even after adjusting for multiple clinical characteristics and risk factors [7]. In subjects enrolled in the Prevención con Dieta Mediterránea (PREDIMED) trial, higher betaine levels were associated with a reduced risk of future T2DM [28]. In the Diabetes Prevention Program (DPP), a program designed to study progression to T2DM in subjects with elevated fasting plasma glucose (FPG) and impaired glucose tolerance, low betaine levels were associated with T2DM onset [6]. Moreover, this association remained significant even after adjusting for covariates such as age, sex, body mass index (BMI), hypertension, ethnicity, and FPG [6]. Additionally, the DPP lifestyle intervention led to an increase in betaine levels, suggesting that betaine may have utility in monitoring the effects of interventions designed to prevent progression to T2DM [6]. However, no studies to date have explored the association of betaine with incident T2DM in a general population-based cohort.

The goals of this study were to evaluate the laboratory performance of the newly developed nuclear magnetic resonance spectroscopy (NMR)-based betaine assay and to explore potential clinical associations of betaine and incident T2DM in a large prospective population-based cohort.

2. Methods

2.1. NMR Data Acquisition and Betaine Quantification by Peak Deconvolution

Serum/plasma specimens were diluted with citrate/phosphate buffer (3:1 v/v) to lower the pH to 5.3 in order to separate the betaine and TMAO signals which overlap at physiological pH (Figure 1). One-dimensional (1D) 1H-NMR spectra were recorded as previously described [4,29]. Briefly, the Carr–Purcell–Meiboom–Gill (CPMG) acquisition technique was utilized to suppress resonances from macromolecules (e.g., proteins and lipoproteins) along with Water suppression Enhanced through T_1 effects (WET) gradient sequence to attenuate the water signal. Spectra were acquired at 47 °C (spectral width = 4496.4 Hz, relaxation delay between scans = 5 s, direct detection time = 1.2 s, number of scans = 48) on a Vantera NMR clinical analyzer (LabCorp, Morrisville, NC) equipped with 400 MHz (9.4 T) Agilent spectrometer [29,30]. Betaine was quantified from the acquired spectra using a proprietary deconvolution algorithm that resolved the betaine region into its spectral components (Figure 1). The α-anomeric glucose signal is used as a reference to locate the betaine peak. The betaine peak at 3.22 ppm, as well as the glucose peaks on either side, were mathematically modeled using lognormal and Lorentzian peak shapes. A linear function was incorporated into the algorithm to adequately model the baseline around these peaks. After subtracting the baseline, the lineshape deconvolution was achieved by a bound-constrained non-linear least-squares fitting algorithm. The derived betaine signal amplitudes were converted to μM units using a factor that was empirically determined by plotting data from dialyzed serum samples spiked with known amounts of betaine and relating the betaine signal area to the expected concentrations.

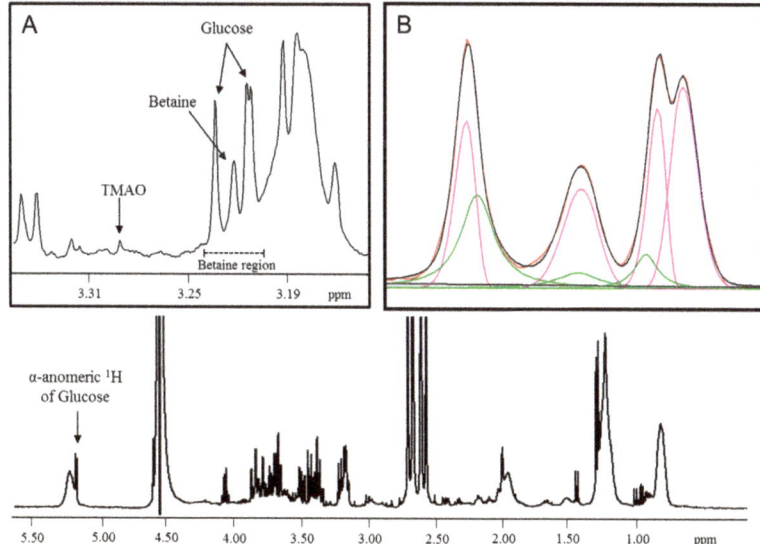

Figure 1. 1D 1H CPMG NMR spectrum of serum used for modeling the betaine peak. (**A**) Expansion of the spectrum where the betaine region is extracted for lineshape deconvolution. (**B**) The experimental (red) spectrum overlaid with a mathematical fit (black) from a composite of lognormal (pink) and Lorentzian (green) lineshapes after subtraction of the baseline in the betaine region. ppm = parts per million; CPMG, Carr–Purcell–Meiboom–Gill; NMR, nuclear magnetic resonance.

2.2. Assay Performance Testing

All assay performance studies were conducted according to Clinical and Laboratory Standards Institute (CLSI) guidelines. Dialyzed serum pools devoid of betaine were analyzed as blanks (five pools, four replicates, three days). Serum pools containing low concentrations of betaine were tested to determine the limits of detection (LOD) (five pools, four replicates, three days) and quantitation (LOQ) (eight pools, four replicates, three days). The LOB, LOD, and LOQ were calculated as previously described [30]. Within-run ($n = 20$ replicates) and within-laboratory ($n = 80$ replicates) imprecision were determined using serum pools with low, intermediate and high betaine values. These values were targeted around the 50th and 100th percentile of the reference interval, as well a high value that had been observed in clinical samples. Mean concentrations and coefficients of variation (%CV) were calculated for each pool. For the tube comparison, specimens were drawn from 21 donors into three blood collection tube types: Serum LipoTubes (Greiner Bio-One, Monroe, NC, USA), K_2EDTA plasma tubes and plain red top serum tubes without gel barriers (BD Diagnostics, Durham, NC, USA). Specimen tubes were processed as per the manufacturer's recommendations. Results for EDTA plasma and plain serum tubes were compared to results for the LipoTube by calculating the mean % bias for all 21 samples. Linearity was evaluated by comparing known spiked concentrations of betaine with expected concentrations tested in quadruplicates across the biological range from 25 to 1200 μM ($n = 13$).

A method comparison study was performed to compare betaine quantification by NMR versus mass spectrometry as per CLSI guidelines [31]. Serum specimens were obtained from multiple donors and aliquots were frozen at −80°C until the time of analysis. The same frozen serum samples ($n = 24$) were analyzed via NMR at LabCorp and in parallel by stable isotope dilution, high-performance liquid chromatography with online electrospray ionization tandem mass spectrometry (LC-MS/MS) at the Cleveland Clinic. Samples analyzed via the LC-MS/MS method were injected at a flow rate of 0.8 ml/min to a 4.6 × 250 mm, 5 μm Luna silica column interfaced with an AB SCIEX 5000 triple quadrupole mass spectrometer using d9-(trimethyl)-labeled internal standards [32]. Details of the method for the LC-MS/MS assay have been previously described [10,32,33]. Samples for the method comparison study spanned the reported reference interval for betaine [1].

To confirm the previously reported reference interval for betaine [1], two studies were evaluated. For the first study, normal, apparently healthy ($n = 501$) men and women aged 18 to 80 were recruited at LipoScience (now LabCorp, Morrisville, NC, USA). Informed consent was obtained from all donors whose samples were analyzed and the study protocol was approved by a local Institutional Review Board. Non-fasting serum specimens collected in LipoTubes and tested by NMR as described above. Detailed descriptions of this study have been previously described [29,30]. The second study entailed reanalyzing the digitally stored NMR spectra, from previously tested fasting EDTA plasma samples from the Prevention of Renal and Vascular End-Stage Disease (PREVEND) Study ($n = 5621$), using the newly developed proprietary betaine assay software (LabCorp, Morrisville, NC, USA) [29]. Quantiles for sample results from both studies were determined, and the reference intervals were estimated at the 2.5th and 97.5th percentiles. Pearson's correlation coefficients were determined, and no statistically significant correlation was found between betaine and TMAO in either the normal healthy population study that was used for the reference interval determination ($r = 0.065$, $p = 0.149$) or the PREVEND study ($r = -0.01$, $p = 0.508$).

2.3. Cross-sectional and Prospective Analyses in Participants in the Prevention of Renal and Vascular End-Stage Disease (PREVEND) Study

Details of the PREVEND study design and recruitment have been described before [34]. Briefly, the PREVEND study is a Dutch cohort of predominantly Caucasian men and women drawn from the general population of the city of Groningen in the northern part of the Netherlands. After exclusion of subjects with insulin-treated diabetes and pregnant women, subjects with a urinary albumin concentration ≥10 mg/L were invited to participate ($n = 7768$), 6000 accepted. In addition, a random sample of 2592 individuals with a urinary albumin concentration <10 mg/L was included. During the years 1997–1998, 8592 subjects aged 28–75 years completed the baseline survey. The baseline for the current study was the second screening which took place between 2001 and 2003 ($n = 6892$). Individuals who were diagnosed with T2DM or were missing data for diabetes at baseline or follow-up, as well as those missing NMR at baseline or follow-up, were excluded, leaving 4336 subjects for the present cross-sectional and prospective analyses. The follow-up period (period between baseline and date of T2DM diagnosis) was determined to be 7.3 (IQR, 5.9–7.7) years for this analysis. T2DM was ascertained if one or more of the following criteria were met: 1) FPG ≥7.0 mmol/L (126 mg/dL), 2) random sample plasma glucose ≥11.1 mmol/L (200 mg/dL), 3) self-report of a physician diagnosis of T2DM, and 4) initiation of glucose-lowering medication use, retrieved from a central pharmacy registry [35]. These criteria are consistent with current guidelines for the diagnosis of T2DM [36]. Calculations of BMI, as well as determinations of blood pressure, smoking status, alcohol intake, hypertension and estimated glomerular filtration rate (eGFR), have been described previously [37]. At the second screening, venous blood was obtained after an overnight fast. EDTA plasma samples were prepared and stored at −80 °C until testing occurred. Total cholesterol was measured on a Beckman Coulter®AU680 analyzer, and high-density lipoprotein cholesterol (HDL-C) and triglycerides (TG) were measured on an Olympus AU400 analyzer (Beckman Coulter, Brea, CA, USA) [38]. FPG was measured by dry chemistry (Eastman Kodak, Rochester, NY, USA) and C-reactive protein (CRP) was measured by nephelometry with a threshold of 0.18 mg/L (BNII, Dade Behring). Betaine results were obtained as described above. This study was carried out in accordance with The Code of Ethics of the World Medical Association (Declaration of Helsinki), cleared by an Institutional Review Board and all donors signed consent forms.

2.4. Statistical Analyses

Statistical analyses were performed using SAS v9.4 (SAS Institute, Cary, NC, USA), Analyze-it v3.90.1 (Analyze-it Software, Ltd., Leeds, UK) or IBM Statistics SPSS v23.0 (IBM Inc., Chicago, IL, USA). For the analytical validation studies, linear regression analyses were performed. Deming regression analysis and Bland–Altman plots were used to evaluate the correlation between the results obtained on the two platforms for the method comparison study. Reference intervals were compared by assessing the means using the Wilcoxon Rank Sum and Mann–Whitney tests and the distributions using the Kruskal-Wallis test. For the epidemiological studies, data are expressed in mean ± SD (or SEM for figures) when normally distributed, median (interquartile range) for skewed distribution and number and percentages in case of categorical data. Differences between the tertiles of betaine were tested by ANOVA or Kruskal Wallis for continuous variables and with χ^2- test for categorical variables. Pearson's correlation coefficients were used to evaluate any potential relation between betaine and TMAO.

Crude and multivariable Cox proportional hazards regression analyses were performed to estimate the effect of betaine on T2DM in a crude analysis as well as analyses adjusted for age and sex (model 1). Further adjustments were then made for eGFR (model 2), BMI and smoking status (model 3), ethnicity, FPG, total cholesterol, HDL-C, TG, CRP, and use of lipid-lowering drugs (model 4). In continuous Cox proportional hazards regression models, betaine was log-base 2 transformed to allow for the expression of the hazard ratios (HRs) per doubling of betaine. Additionally, betaine was used as a categorical variable for analyses by tertiles. Data were presented as HRs and 95% CI confidence intervals (CIs). Potential effect modification was evaluated by age, sex, and eGFR. Potential effect modifiers were tested by entering both main effects and the cross-product term in Cox regression analyses. When effect modification was observed, we proceeded with stratified analyses. For the sensitivity analyses, we performed crude and multivariable Cox proportional hazards regression analyses excluding subjects with previous CVD, use of lipid-lowering drugs, microalbuminuria and eGFR <60 mL/min/1.73 m^2 at baseline. Two-sided p-values <0.05 were considered statistically significant for all reported analyses.

2.5. Ethic Approval and Consent to Participate

This study was carried out in accordance with The Code of Ethics of the World Medical Association (Declaration of Helsinki), cleared by an Institutional Review Board and all donors signed consent forms.

3. Results

The NMR signals from betaine and TMAO overlap in the Carr-Purcell-Meiboom-Gill (CPMG) collected spectrum of serum at physiological pH [29]. Hence, an assay was developed that reduces the pH of the sample to 5.3. The betaine peak is not affected by the change in pH, however, the TMAO peak is shifted downfield in the NMR spectrum which allows for accurate quantification of both betaine and TMAO separately [29] (Figure 1). In order to assess its clinical usefulness, the analytical performance of the NMR-based betaine assay was evaluated. The LOB, LOD, and LOQ were determined to be 6.4, 8.9, and 13.2 µM, respectively, with the LOQ being below the previously reported normal reference interval for betaine [1]. The coefficients of variation for intra-assay and inter-assay precision were 1.5–4.3% and 2.5–5.5%, respectively (Table 1). A tube comparison study revealed no significant differences between results from serum collected in LipoTubes vs. results from serum collected in plain red top tubes (mean % bias = −1.9) or EDTA plasma (mean % bias = 4.1). Linearity was demonstrated between 26.0 and 1135 µM, well above the upper limit of the normal reference interval [1], with a correlation coefficient (R^2) of 1.00 and a linear equation of $Y = 1.02x - 3.75$ (Figure 2A). Based on these data, the reportable range for the NMR-based betaine assay was determined to be 13.2–1135 µM.

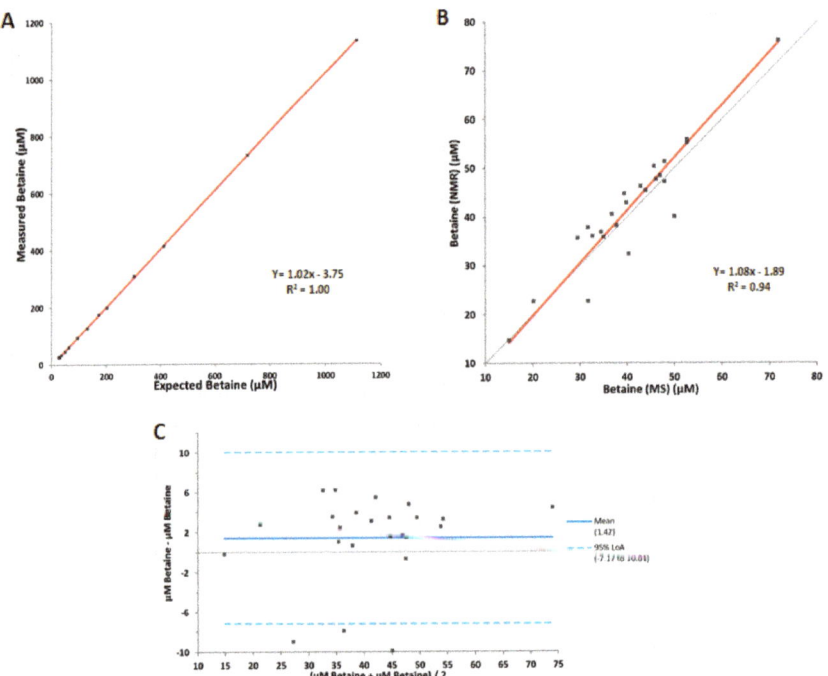

Figure 2. Linearity and method comparison data for the betaine assay. (**A**) Results of linearity study ($n = 13$), (**B**) Deming regression comparison between LC-MS/MS and NMR measured betaine in serum samples ($n = 24$), (**C**) Bland–Altman plot ($n = 24$). The limits of agreement (LoA) are depicted as dotted blue lines, and the 0% bias is a solid black line.

Table 1. Within-laboratory (inter-assay) and within-run (intra-assay) imprecision.

Imprecision	Betaine (µM)		
	Low	Medium	High
Within-lab [a]			
Mean	45.2	97.2	205.9
SD	2.5	4.0	5.2
CV (%)	5.5	4.1	2.5
Within-run [b]			
Mean	44.1	95.6	205.6
SD	1.9	3.1	3.1
CV (%)	4.3	3.2	1.5

Abbreviations: CV, coefficients of variation; SD, standard deviation; [a] Based on CLSI EP5-A2 tested using three controls, two runs per day in duplicate for 20 days (total n = 80); [b] Based on one run of 20 tests.

A method comparison study was performed to compare the quantification of betaine by NMR versus LC-MS/MS. Deming regression analysis of betaine results from both platforms produced an R^2 value of 0.94 (Y = 1.08x − 1.89) (Figure 2B). The Bland–Altman plot revealed that there was a small but systematic bias for higher values produced by the NMR assay compared to results from the LC-MS/MS assay (Figure 2C). The normal reference interval for betaine was confirmed in two populations, a cohort of apparently healthy volunteers and the full set of PREVEND cohort participants (Table 2). The distributions were similar between the two study populations, and the reference intervals were similar to those previously reported [1]. The distributions for men and women in the cohort of apparently healthy volunteers were not statistically different, and neither were the means. Therefore, there were no significant between-gender differences in the distribution of results or the reference intervals, although the results tended to be somewhat lower in women compared to men, as previously noted [1].

Table 2. Distribution and population means for betaine (µM) in generally healthy adults.

Percentile	Normal Healthy Adult Volunteers (n = 501)	Normal Healthy Adult Volunteers Female (n = 292)	Normal Healthy Adult Volunteers Male (n = 209)	PREVEND Study Participants (n = 5621)
0th	<13.2 [a]	<13.2 [a]	24.6	13.3
2.5th	23.8	22.0	28.5	21.0
25.0th	34.1	32.4	37.3	30.8
50.0th	41.0	39.1	44.6	36.8
75.0th	49.5	46.9	52.4	43.8
97.5th	74.7	74.4	75.3	63.0
100th	104.1	101.6	104.1	190.7
Mean (SD)	42.9 (12.6)	40.7 (12.4)	46.0 (12.2)	38.1 (11.2)

Abbreviations: SD, standard deviation. [a] Lower bound of the reportable range is limited by the LOQ of this assay.

Of the 6892 PREVEND participants that completed the second round of screening, 4336 subjects were included in this study. Subjects were excluded if they were missing NMR data, if they were missing information regarding T2DM diagnosis at baseline or follow-up, or if they were diagnosed with T2DM at baseline. Baseline clinical and laboratory characteristics for the entire cohort as well as according to tertiles of betaine are shown in Table 3. Subjects with higher betaine levels were older, more likely to be men and less likely to be Caucasian. Subjects with higher betaine levels also had lower BMI, were more likely to be non-smokers and less likely to have high diastolic blood pressure. Subjects in the highest tertile of betaine had lower, total cholesterol, TG and C-reactive protein (CRP). They were more likely to be on lipid-lowering medications.

Table 3. Baseline characteristics of the 4336 subjects of the Prevention of Renal and Vascular End-Stage Disease (PREVEND) study of the overall population and according to tertiles of betaine.

	Overall	Tertiles of betaine			p-Value
		T1	T2	T3	
Participants, n	4336	1445	1446	1445	
Betaine, µM	36.9 (31.0–44.0)	28.7 (25.3–31.0)	36.9 (35.1–38.8)	47.3 (44.0–52.6)	
Participants, n	4236	1398	1418	1420	
TMAO, µM	4.6 ± 5.9	4.6 ± 5.1	4.7 ± 6.3	4.6 ± 6.4	
General characteristics					
Age, years	52.6 ± 11.5	51.3 ± 10.9	52.8 ± 11.4	53.7 ± 12.1	<0.001
Male sex, n (%)	2159 (49.8)	456 (31.6)	750 (51.9)	953 (66.0)	<0.001
Ethnicity, Caucasian, n (%)	4164 (96.0)	1403 (97.1)	1400 (96.8)	1361 (94.2)	0.001
BMI, kg/m^2	26.4 ± 4.2	26.7 ± 4.3	26.6 ± 4.2	26.0 ± 4.1	<0.001
Smoking status, n (%)					0.008
Never	1286 (29.7)	420 (29.1)	424 (29.3)	442 (30.6)	
Former	1828 (42.2)	574 (39.7)	639 (44.2)	615 (42.6)	
Current	1165 (26.9)	438 (30.3)	358 (24.8)	369 (25.5)	
Alcohol consumption, never, n (%)	1004 (23.2)	329 (22.8)	351 (24.3)	324 (22.4)	0.26
eGFR, mL/min/1.73m^2	93.5 ± 16.3	94.5 ± 15.9	93.1 ± 16.7	93.0 ± 16.3	0.02
Hypertension, n (%)	1280 (29.5)	406 (28.1)	443 (30.6)	431 (29.8)	0.33
Hypercholesterolemia, n (%)	1235 (28.5)	468 (32.4)	390 (27.0)	377 (26.1)	<0.001
Parental history of CKD, n (%)	20 (0.5)	8 (0.6)	7 (0.5)	5 (0.3)	0.70
Circulation					
SBP, mmHg	124.8 ± 18.1	123.9 ± 17.8	125.4 ± 18.2	125.1 ± 18.3	0.06
DBP, mmHg	73.0 ± 9.0	72.5 ± 9.1	73.4 ± 8.8	72.3 ± 9.0	0.02
Laboratory parameters					
Total cholesterol, mmol/L	5.4 ± 1.0	5.6 ± 1.1	5.4 ± 1.0	5.2 ± 1.0	<0.001
HDL cholesterol, mmol/L	1.3 ± 0.3	1.2 ± 0.3	1.3 ± 0.3	1.2 ± 0.3	<0.001
Triglycerides, mmol/L	1.1 (0.8–1.6)	1.2 (0.8–1.7)	1.1 (0.8–1.6)	1.0 (0.8–1.5)	0.005
Fasting glucose, mmol/L	4.7 (4.4–5.2)	4.7 (4.4–5.2)	4.7 (4.4–5.2)	4.7 (4.4–5.2)	0.34
C-reactive protein, mg/L	1.2 (0.6–2.8)	1.4 (0.6–3.0)	1.2 (0.5–2.6)	1.1 (0.5–2.8)	0.005
Medication					
Antihypertensive drugs, n (%)	716 (16.5)	221 (15.3)	239 (16.5)	256 (17.7)	0.12
Lipid lowering drug use, n (%)	304 (7.0)	62 (4.3)	94 (6.5)	148 (10.2)	<0.001

Data are represented as mean ± SD, median (interquartile range) or n (%). Differences were tested by ANOVA or Kruskal Wallis for continuous variables and with χ^2- test for categorical variables. The eGFR is based on the creatinine–cystatin C equation. Abbreviations: BMI, body mass index; eGFR, estimated glomerular filtration rate; CKD, chronic kidney disease; SBP, systolic blood pressure; DBP, diastolic blood pressure; HDL, high-density lipoproteins.

After a median (interquartile range) follow up period of 7.3 (IQR, 5.9–7.7) years, 224 new T2DM cases were ascertained. Cox proportional hazards regression models revealed that the highest tertile of betaine was associated with a lower incidence of T2DM. Hazard ratio (HR) for the crude model was 0.61 (95% CI: 0.44–0.85, $p = 0.004$) (Table 4). The association remained largely unchanged after adjustment for multiple clinical covariates and T2DM risk factors, including age, sex, eGFR, BMI, smoking status, ethnicity, FPG, total cholesterol, HDL-C, TG, CRP, and use of lipid-lowering drugs. In the fully adjusted model, the HR for the upper vs the lower tertile was 0.50 (95% CI: 0.32–0.80, $p = 0.004$). When the association of betaine with T2DM development was assessed as a continuous variable, the same trend was observed albeit the association of betaine with incident T2DM did not reach statistical significance in the fully adjusted model.

We observed that the association of betaine with T2DM was modified by sex ($P_{interaction} = 0.01$), but not by age ($P_{interaction} = 0.93$) or eGFR ($P_{interaction} = 0.38$). Betaine as continuous variable was inversely associated with incident T2DM in men (HR: 0.44, 95% CI: 0.24–0.80, $p = 0.007$), but not in women (HR: 0.97, 95% CI: 0.55–1.71, $p = 0.93$), after adjustment for age, eGFR, BMI, smoking status, ethnicity, FPG, total cholesterol, HDL-C, TG, CRP, and use of lipid-lowering drugs (Table 5).

Table 4. Association of betaine as a continuous variable and according to tertiles with the development of T2DM.

	Betaine as Continuous Variable (^{2}log)		Tertiles of Betaine					
			T1		T2		T3	
	HR (95% CI)	p			HR (95% CI)	p	HR (95% CI)	p
Diabetes, no. of events (%)	224 (5.2%)		93 (6.4%)		74 (5.1%)		57 (3.9%)	
Crude	0.79 (0.59–1.05)	0.10	1.00 (ref)		0.78 (0.57–1.05)	0.10	0.61 (0.44–0.85)	0.004
Model 1	0.60 (0.46–0.79)	<0.001	1.00 (ref)		0.64 (0.47–0.88)	0.005	0.45 (0.32–0.64)	<0.001
Model 2	0.59 (0.45–0.78)	<0.001	1.00 (ref)		0.61 (0.45–0.84)	0.002	0.42 (0.29–0.59)	<0.001
Model 3	0.63 (0.47–0.85)	0.002	1.00 (ref)		0.68 (0.50–0.94)	0.02	0.47 (0.33–0.66)	<0.001
Model 4	0.69 (0.46–1.02)	0.06	1.00 (ref)		0.65 (0.43–0.96)	0.03	0.50 (0.32–0.80)	0.004

Abbreviations: eGFR, estimated glomerular filtration rate; T2DM, type 2 diabetes mellitus. Association between betaine and development of diabetes in 4336 (224 cases) subjects of the Prevention of Renal and Vascular End-Stage Disease (PREVEND) study as a continuous variable (^{2}log-transformed) and according to tertiles (T1-T3). Hazard ratios and 95% confidence intervals were derived from Cox proportional hazards regression models. The eGFR is based on creatinine–cystatin C equation; Model 1: Adjustment for age and sex; Model 2: Model 1+ adjustment for eGFR; Model 3: Model 2 + adjustment for body mass index and smoking; Model 4: Model 3 + adjustment for ethnicity, fasting glucose, total cholesterol, high-density lipoprotein cholesterol, triglycerides, C-reactive protein, and use of lipid-lowering drugs.

Table 5. Sex-stratified analyses of the association of betaine as a continuous variable and according to tertiles with the development of T2DM.

	Betaine as Continuous Variable (^{2}log)		Tertiles of Betaine					
			T1		T2		T3	
	HR (95% CI)	p			HR (95% CI)	p	HR (95% CI)	p
				Men				
Diabetes, no. of events (%)	139 (6.4%)		54 (2.5%)		46 (2.1%)		39 (1.8%)	
Crude	0.41 (0.26–0.64)	<0.001	1.00 (ref)		0.49 (0.33–0.73)	<0.001	0.34 (0.22–0.51)	<0.001
Model 1	0.40 (0.26–0.62)	<0.001	1.00 (ref)		0.49 (0.33–0.72)	<0.001	0.32 (0.21–0.49)	<0.001
Model 2	0.41 (0.26–0.65)	<0.001	1.00 (ref)		0.46 (0.31–0.69)	<0.001	0.32 (0.21–0.48)	<0.001
Model 3	0.56 (0.35–0.90)	0.02	1.00 (ref)		0.54 (0.36–0.82)	0.003	0.43 (0.28–0.65)	<0.001
Model 4	0.44 (0.24–0.80)	0.007	1.00 (ref)		0.43 (0.25–0.74)	0.003	0.36 (0.20–0.65)	0.001
				Women				
Diabetes, no. of events (%)	85 (3.9%)		39 (1.8%)		28 (1.3%)		18 (0.8%)	
Crude	0.96 (0.59–1.53)	0.85	1.00 (ref)		1.01 (0.62–1.63)	0.98	0.95 (0.54–1.66)	0.86
Model 1	0.81 (0.52–1.27)	0.35	1.00 (ref)		0.89 (0.55–1.45)	0.65	0.83 (0.47–1.45)	0.51
Model 2	0.74 (0.48–1.15)	0.18	1.00 (ref)		0.86 (0.53–1.41)	0.56	0.67 (0.37–1.22)	0.19
Model 3	0.73 (0.47–1.15)	0.18	1.00 (ref)		0.96 (0.58–1.56)	0.85	0.62 (0.33–1.15)	0.13
Model 4	0.97 (0.55–1.71)	0.93	1.00 (ref)		1.02 (0.56–1.89)	0.94	0.74 (0.35–1.60)	0.45

Abbreviations: eGFR, estimated glomerular filtration rate; T2DM, type 2 diabetes mellitus. Association between betaine and development of diabetes in 2159 men (139 cases) and 2177 women (85 cases) of the Prevention of Renal and Vascular End-Stage Disease (PREVEND) study as a continuous variable (^{2}log-transformed) and according to tertiles (T1-T3). Hazard ratios and 95% confidence intervals were derived from Cox proportional hazards regression models. The eGFR is based on creatinine–cystatin C equation; Model 1: Adjustment for age; Model 2: Model 1+ adjustment for eGFR; Model 3: Model 2 + adjustment for body mass index and smoking; Model 4: Model 3 + adjustment for ethnicity, fasting glucose, total cholesterol, high-density lipoprotein cholesterol, triglycerides, C-reactive protein, and use of lipid-lowering drugs.

To ensure the results would be similar if the analyses were conducted in subjects at a lower risk of CVD, a sensitivity analysis was performed excluding subjects with previous CVD, use of lipid-lowering drugs, microalbuminuria and eGFR <60 mL/min/1.73 m^2 at baseline ($n = 2810$ subjects with 112 cases of T2DM ascertained) (Table 6). Cox proportional hazards regression models revealed that the highest tertile of betaine was associated with a lower incidence of T2DM. HR for the crude model was 0.58 (95% CI: 0.36–0.93, $p = 0.03$). The association remained significant after adjustment for multiple clinical covariates and T2DM risk factors, including age, sex, eGFR, BMI, and smoking status. Only the fully-adjusted model, with additional cumulative adjustment for ethnicity, FPG, total cholesterol, HDL-C, TG, and CRP, lost statistical significance (0.57 95% CI: 0.32–1.04, $p = 0.07$).

Table 6. Sensitivity analysis of the association of betaine as a continuous variable and according to tertiles with development of T2DM, excluding subjects with previous CVD history, microalbuminuria, eGFR < 60 mL/min/1.73 m^2, and use of lipid-lowering drugs.

	Betaine as Continuous Variable (^{2}log)		Tertiles of Betaine					
			T1	T2		T3	
	HR (95% CI)	p		HR (95% CI)	p	HR (95% CI)	p
Diabetes, no. of events (%)	112 (4.0%)		48 (4.9%)	39 (4.2%)		25 (2.8%)	
Crude	0.75 (0.51–1.10)	0.14	1.00 (ref)	0.84 (0.55–1.28)	0.41	0.58 (0.36–0.93)	0.03
Model 1	0.57 (0.41–0.81)	0.002	1.00 (ref)	0.69 (0.45–1.05)	0.09	0.41 (0.25–0.67)	<0.001
Model 2	0.59 (0.40–0.80)	0.001	1.00 (ref)	0.69 (0.45–1.05)	0.08	0.41 (0.25–0.67)	<0.001
Model 3	0.62 (0.42–0.93)	0.02	1.00 (ref)	0.75 (0.49–1.15)	0.19	0.50 (0.31–0.82)	0.006
Model 4	0.69 (0.43–1.10)	0.12	1.00 (ref)	0.89 (0.55–1.46)	0.65	0.57 (0.32–1.04)	0.07

Abbreviations: CVD, cardiovascular disease; eGFR, estimated glomerular filtration rate; T2DM, type 2 diabetes mellitus. Association between betaine and development of diabetes in 2810 (112 cases) subjects of the Prevention of Renal and Vascular End-Stage Disease (PREVEND) study as a continuous variable (^{2}log-transformed) and according to tertiles (T1-T3). Hazard ratios and 95% confidence intervals were derived from Cox proportional hazards regression models. The eGFR is based on creatinine–cystatin C equation; Model 1: Adjustment for age and sex; Model 2: Model 1+ adjustment for Egfr; Model 3: Model 2 + adjustment for body mass index and smoking; Model 4: Model 3 + adjustment for ethnicity, fasting glucose, total cholesterol, high-density lipoprotein cholesterol, triglycerides, and C-reactive protein.

4. Discussion

Several chromatographic, chemical, and mass spectrometry (MS)-based techniques for quantification of betaine have been reported [10,32,33,39,40]. With the advent of the Vantera clinical analyzer, an automated high-throughput NMR spectrometer, NMR-based assays are now available in the diagnostic laboratory for clinical use [29,30,41–44]. Hence, an NMR-based assay that quantifies plasma/serum TMAO and betaine was developed for the diagnostic laboratory [4,21,29]. While the NMR signals for TMAO and betaine align in NMR spectra acquired at physiological pH, this assay uses a buffer with a lower pH that separates the two metabolites, allowing for their simultaneous quantification [29]. Analytical validation of the NMR-based betaine assay revealed that it has performance characteristics that are robust enough for use as a diagnostic or prognostic test in the clinical laboratory.

The major finding of this study, however, is that lower circulating levels of betaine are associated with future development of T2DM in a large population-based cohort. Lower betaine levels were associated with increased risk of developing T2DM even after adjusting for clinical characteristics as well as typical risk factors for T2DM such as BMI, lipids, and FPG. Previously, lower betaine levels were reported to be associated with incident T2DM development in subjects across a spectrum of risks for CVD, such as in Norwegian subjects with suspected angina pectoris, as well as in primary prevention individuals enrolled in the PREDIMED dietary intervention trial [7,28]. In addition, betaine was associated with future diabetes in the DPP, a program designed to study progression to, and possibly prevention of, T2DM in subjects at high risk of T2DM [6]. Moreover, the prospective association of low betaine levels with future T2DM in DPP remained significant after adjusting for risk factors such as age, sex, BMI, hypertension, ethnicity and FPG [6]. Furthermore, the DPP lifestyle intervention led to an increase in betaine, raising the hypothesis that betaine may have utility in monitoring the effects of interventions designed to prevent progression to T2DM [6].

Betaine is required for remethylation of homocysteine to methionine, a precursor of the universal methyl donor SAM [1,2]. By decreasing SAM availability, betaine deficiency may decrease phosphatidylcholine synthesis, promote hepatic steatosis, and modify VLDL synthesis and secretion [3]. Betaine was shown to be inversely associated with TG and PLTP activity, further supporting the notion that low betaine levels may alter liver fat accumulation and lipid/lipoprotein metabolism [4]. Furthermore, lower plasma betaine concentrations have been reported in subjects with metabolic syndrome, T2DM, NAFLD, and/or NASH, supporting the association of low levels of betaine with diseases that are associated with liver fat accumulation [3,5–8]. Moreover, betaine levels were associated with disease severity in subjects with NAFLD, betaine levels being significantly lower in subjects with NASH compared to subjects with hepatic steatosis [3]. These observations support the findings of this study, that lower betaine levels may be a good predictor of future T2DM.

Given that betaine has been described as being a precursor to TMAO, it seems counterintuitive that lower betaine levels would coincide with higher TMAO levels and vice versa. However, we did not find a statistically significant correlation between TMAO and betaine in the normal healthy population we used for our reference interval study or in the PREVEND study. This may be attributed to the fact that there are other precursors for TMAO such as carnitine [11], choline, phosphatidylcholine [10], and trimethyllysine [45] that can give rise to TMAO via trimethylamine. In addition, choline is oxidized to betaine [27] in humans, and betaine can be shunted for multiple metabolic functions [1,2] Therefore, there may not be a clear relationship between TMAO and betaine, especially in normal, healthy populations such as the ones analyzed in this study.

Subjects with cardiometabolic diseases such as T2DM and non-alcoholic fatty liver disease (NAFLD) exhibit altered gut microbiome, and have higher levels of TMAO and lower levels of betaine compared to healthy subjects [3,5–8,20,25,46,47]. Betaine, the gut-microbiome generated metabolite TMAO, and dietary metabolites such as choline and L-carnitine, may help determine if a subject's gut microbiome is altered (gut dysbiosis), and identify individuals who may benefit from intensive dietary or lifestyle intervention in order to reduce the progression to more severe metabolic disease such as T2DM. Further studies are needed to confirm this to be the case.

There are several strengths and limitations in the current study. A strength of this study is that it includes a large number of participants in the general population, whereas previously published studies reported associations of betaine with incident T2DM in subjects who may have been at risk of CVD or T2DM. In addition, the PREVEND study has a large number of participants with a large age range, which could be considered a strength. Notably, the PREVEND study was designed to study the impact of albuminuria on renal and CVD outcomes. Hence, the PREVEND subjects were preferentially recruited based on an initially elevated urinary albumin concentration. However, a sensitivity analysis excluding subjects with microalbuminuria and compromised renal function showed an essentially similar association of betaine with future risk of T2DM. Furthermore, the majority of the PREVEND participants were of Caucasian descent, and study results may not be applicable to subjects of non-white ethnicities.

5. Conclusions

The newly developed NMR-based betaine assay exhibits excellent performance characteristics. We found that lower circulating betaine was associated with increased risk of developing T2DM in a large population-based cohort even after adjusting for clinical covariates and T2DM risk factors. Potential clinical applications may include assessing the risk of future T2DM.

Author Contributions: E.G., D.W.B., E.J.J., I.S. and J.D.O. designed the NMR-based betaine assay; E.G. collected performance data; E.G. and M.A.C. analyzed the analytical performance data; S.L.H. collected the mass spectrometry data; M.C.J.O., E.G.G., S.J.L.B. and R.P.F.D. designed and analyzed the clinical data; R.P.F.D., M.A.C. and E.G. interpreted the data and wrote the manuscript; All authors contributed to critical review of the manuscript.

Funding: SLH reports support by a grant from the National Institutes of Health (NIH) and Office of Dietary Supplements (R01HL103866, P01HL147823).

Conflicts of Interest: E.G., D.W.B., E.J., I.S., J.D.O. and M.A.C. are currently, or were at the time of this study, employees of LabCorp. M.C.J.O., E.G.G., X.S.L., S.J.L.B. and R.P.F.D. have no conflicts of interest to reveal. S.L.H. reports being listed as a co-inventor on pending and issued patents held by the Cleveland Clinic relating to cardiovascular diagnostics and therapeutics, having been paid as a consultant for Proctor & Gamble, and having received research funds from LipoScience, Pfizer, Inc., and Proctor & Gamble.

References

1. Lever, M.; Slow, S. The clinical significance of betaine, an osmolyte with a key role in methyl group metabolism. *Clin. Biochem.* **2010**, *43*, 732–744. [CrossRef] [PubMed]

2. Day, C.R.; Kempson, S.A. Betaine chemistry, roles, and potential use in liver disease. *Biochim. Biophys. Acta* **2016**, *1860*, 1098–1106. [CrossRef] [PubMed]
3. Sookoian, S.; Puri, P.; Castano, G.O.; Scian, R.; Mirshahi, F.; Sanyal, A.J.; Pirola, C.J. Nonalcoholic steatohepatitis is associated with a state of betaine-insufficiency. *Liver Int.* **2017**, *37*, 611–619. [CrossRef] [PubMed]
4. Dullaart, R.P.; Garcia, E.; Jeyarajah, E.; Gruppen, E.G.; Connelly, M.A. Plasma phospholipid transfer protein activity is inversely associated with betaine in diabetic and non-diabetic subjects. *Lipids Health Dis.* **2016**, *15*, 143. [CrossRef] [PubMed]
5. Konstantinova, S.V.; Tell, G.S.; Vollset, S.E.; Nygard, O.; Bleie, O.; Ueland, P.M. Divergent associations of plasma choline and betaine with components of metabolic syndrome in middle age and elderly men and women. *J. Nutr.* **2008**, *138*, 914–920. [CrossRef]
6. Walford, G.A.; Ma, Y.; Clish, C.; Florez, J.C.; Wang, T.J.; Gerszten, R.E.; Diabetes Prevention Program Research Group. Metabolite Profiles of Diabetes Incidence and Intervention Response in the Diabetes Prevention Program. *Diabetes* **2016**, *65*, 1424–1433. [CrossRef]
7. Svingen, G.F.; Schartum-Hansen, H.; Pedersen, E.R.; Ueland, P.M.; Tell, G.S.; Mellgren, G.; Njolstad, P.R.; Seifert, R.; Strand, E.; Karlsson, T.; et al. Prospective Associations of Systemic and Urinary Choline Metabolites with Incident Type 2 Diabetes. *Clin. Chem.* **2016**, *62*, 755–765. [CrossRef]
8. Chen, Y.M.; Liu, Y.; Zhou, R.F.; Chen, X.L.; Wang, C.; Tan, X.Y.; Wang, L.J.; Zheng, R.D.; Zhang, H.W.; Ling, W.H.; et al. Associations of gut-flora-dependent metabolite trimethylamine-N-oxide, betaine and choline with non-alcoholic fatty liver disease in adults. *Sci. Rep.* **2016**, *6*, 19076. [CrossRef]
9. Konstantinova, S.V.; Tell, G.S.; Vollset, S.E.; Ulvik, A.; Drevon, C.A.; Ueland, P.M. Dietary patterns, food groups, and nutrients as predictors of plasma choline and betaine in middle-aged and elderly men and women. *Am. J. Clin. Nutr.* **2008**, *88*, 1663–1669. [CrossRef]
10. Wang, Z.; Klipfell, E.; Bennett, B.J.; Koeth, R.; Levison, B.S.; Dugar, B.; Feldstein, A.E.; Britt, E.B.; Fu, X.; Chung, Y.M.; et al. Gut flora metabolism of phosphatidylcholine promotes cardiovascular disease. *Nature* **2011**, *472*, 57–63. [CrossRef]
11. Koeth, R.A.; Wang, Z.; Levison, B.S.; Buffa, J.A.; Org, E.; Sheehy, B.T.; Britt, E.B.; Fu, X.; Wu, Y.; Li, L.; et al. Intestinal microbiota metabolism of l-carnitine, a nutrient in red meat, promotes atherosclerosis. *Nat. Med.* **2013**. [CrossRef] [PubMed]
12. Tang, W.H.; Wang, Z.; Levison, B.S.; Koeth, R.A.; Britt, E.B.; Fu, X.; Wu, Y.; Hazen, S.L. Intestinal microbial metabolism of phosphatidylcholine and cardiovascular risk. *N. Engl. J. Med.* **2013**, *368*, 1575–1584. [CrossRef] [PubMed]
13. Bennett, B.J.; de Aguiar Vallim, T.Q.; Wang, Z.; Shih, D.M.; Meng, Y.; Gregory, J.; Allayee, H.; Lee, R.; Graham, M.; Crooke, R.; et al. Trimethylamine-N-oxide, a metabolite associated with atherosclerosis, exhibits complex genetic and dietary regulation. *Cell Metab.* **2013**, *17*, 49–60. [CrossRef] [PubMed]
14. Wang, Z.; Tang, W.H.; Buffa, J.A.; Fu, X.; Britt, E.B.; Koeth, R.A.; Levison, B.S.; Fan, Y.; Wu, Y.; Hazen, S.L. Prognostic value of choline and betaine depends on intestinal microbiota-generated metabolite trimethylamine-N-oxide. *Eur. Heart J.* **2014**, *35*, 904–910. [CrossRef]
15. Zhu, W.; Gregory, J.C.; Org, E.; Buffa, J.A.; Gupta, N.; Wang, Z.; Li, L.; Fu, X.; Wu, Y.; Mehrabian, M.; et al. Gut Microbial Metabolite TMAO Enhances Platelet Hyperreactivity and Thrombosis Risk. *Cell* **2016**, *165*, 111–124. [CrossRef]
16. Senthong, V.; Li, X.S.; Hudec, T.; Coughlin, J.; Wu, Y.; Levison, B.; Wang, Z.; Hazen, S.L.; Tang, W.H. Plasma Trimethylamine N-Oxide, a Gut Microbe-Generated Phosphatidylcholine Metabolite, Is Associated With Atherosclerotic Burden. *J. Am. Coll. Cardiol.* **2016**, *67*, 2620–2628. [CrossRef]
17. Senthong, V.; Wang, Z.; Fan, Y.; Wu, Y.; Hazen, S.L.; Tang, W.H. Trimethylamine N-Oxide and Mortality Risk in Patients With Peripheral Artery Disease. *J. Am. Heart Assoc.* **2016**, *5*. [CrossRef]
18. Senthong, V.; Wang, Z.; Li, X.S.; Fan, Y.; Wu, Y.; Tang, W.H.; Hazen, S.L. Intestinal Microbiota-Generated Metabolite Trimethylamine-N-Oxide and 5-Year Mortality Risk in Stable Coronary Artery Disease: The Contributory Role of Intestinal Microbiota in a COURAGE-Like Patient Cohort. *J. Am. Heart Assoc.* **2016**, *5*. [CrossRef]
19. Suzuki, T.; Heaney, L.M.; Bhandari, S.S.; Jones, D.J.; Ng, L.L. Trimethylamine N-oxide and prognosis in acute heart failure. *Heart* **2016**, *102*, 841–848. [CrossRef]

20. Tang, W.H.; Wang, Z.; Li, X.S.; Fan, Y.; Li, D.S.; Wu, Y.; Hazen, S.L. Increased Trimethylamine N-Oxide Portends High Mortality Risk Independent of Glycemic Control in Patients with Type 2 Diabetes Mellitus. *Clin. Chem.* **2017**, *63*, 297–306. [CrossRef]
21. Gruppen, E.G.; Garcia, E.; Connelly, M.A.; Jeyarajah, E.J.; Otvos, J.D.; Bakker, S.J.L.; Dullaart, R.P.F. TMAO is Associated with Mortality: Impact of Modestly Impaired Renal Function. *Sci. Rep.* **2017**, *7*, 13781. [CrossRef] [PubMed]
22. Tang, W.H.W.; Backhed, F.; Landmesser, U.; Hazen, S.L. Intestinal Microbiota in Cardiovascular Health and Disease: JACC State-of-the-Art Review. *J. Am. Coll. Cardiol.* **2019**, *73*, 2089–2105. [CrossRef] [PubMed]
23. Lever, M.; George, P.M.; Elmslie, J.L.; Atkinson, W.; Slow, S.; Molyneux, S.L.; Troughton, R.W.; Richards, A.M.; Frampton, C.M.; Chambers, S.T. Betaine and secondary events in an acute coronary syndrome cohort. *PLoS ONE* **2012**, *7*, e37883. [CrossRef] [PubMed]
24. Tang, W.H.; Hazen, S.L. The contributory role of gut microbiota in cardiovascular disease. *J. Clin. Investig.* **2014**, *124*, 4204–4211. [CrossRef]
25. Lynch, S.V.; Pedersen, O. The Human Intestinal Microbiome in Health and Disease. *N. Engl. J. Med.* **2016**, *375*, 2369–2379. [CrossRef]
26. Wang, Z.; Zhao, Y. Gut microbiota derived metabolites in cardiovascular health and disease. *Protein Cell* **2018**, *9*, 416–431. [CrossRef]
27. Lever, M.; George, P.M.; Slow, S.; Bellamy, D.; Young, J.M.; Ho, M.; McEntyre, C.J.; Elmslie, J.L.; Atkinson, W.; Molyneux, S.L.; et al. Betaine and Trimethylamine-N-Oxide as Predictors of Cardiovascular Outcomes Show Different Patterns in Diabetes Mellitus: An Observational Study. *PLoS ONE* **2014**, *9*, e114969. [CrossRef]
28. Papandreou, C.; Bullo, M.; Zheng, Y.; Ruiz-Canela, M.; Yu, E.; Guasch-Ferre, M.; Toledo, E.; Clish, C.; Corella, D.; Estruch, R.; et al. Plasma trimethylamine-N-oxide and related metabolites are associated with type 2 diabetes risk in the Prevencion con Dieta Mediterranea (PREDIMED) trial. *Am. J. Clin. Nutr.* **2018**, *108*, 163–173. [CrossRef]
29. Garcia, E.; Wolak-Dinsmore, J.; Wang, Z.; Li, X.S.; Bennett, D.W.; Connelly, M.A.; Otvos, J.D.; Hazen, S.L.; Jeyarajah, E.J. NMR quantification of trimethylamine-N-oxide in human serum and plasma in the clinical laboratory setting. *Clin. Biochem.* **2017**, *50*, 947–955. [CrossRef]
30. Matyus, S.P.; Braun, P.J.; Wolak-Dinsmore, J.; Jeyarajah, E.J.; Shalaurova, I.; Xu, Y.; Warner, S.M.; Clement, T.S.; Connelly, M.A.; Fischer, T.J. NMR measurement of LDL particle number using the Vantera Clinical Analyzer. *Clin. Biochem.* **2014**, *47*, 203–210. [CrossRef]
31. Clinical and Laboratory Standards Institute. *CLSI Document EP9-A2: Method Comaprison and Bias Estimation Using Patient Samples; Approved Guideline*, 2nd ed.; Clinical and Laboratory Standards Institute: Wayne, PA, USA, 2002.
32. Wang, Z.; Levison, B.S.; Hazen, J.E.; Donahue, L.; Li, X.M.; Hazen, S.L. Measurement of trimethylamine-N-oxide by stable isotope dilution liquid chromatography tandem mass spectrometry. *Anal. Biochem.* **2014**, *455C*, 35–40. [CrossRef] [PubMed]
33. Koeth, R.A.; Lam-Galvez, B.R.; Kirsop, J.; Wang, Z.; Levison, B.S.; Gu, X.; Copeland, M.F.; Bartlett, D.; Cody, D.B.; Dai, H.J.; et al. l-Carnitine in omnivorous diets induces an atherogenic gut microbial pathway in humans. *J. Clin. Investig.* **2019**, *129*, 373–387. [CrossRef] [PubMed]
34. Lambers Heerspink, H.J.; Brantsma, A.H.; de Zeeuw, D.; Bakker, S.J.; de Jong, P.E.; Gansevoort, R.T.; Group, P.S. Albuminuria assessed from first-morning-void urine samples versus 24-hour urine collections as a predictor of cardiovascular morbidity and mortality. *Am. J. Epidemiol.* **2008**, *168*, 897–905. [CrossRef] [PubMed]
35. Abbasi, A.; Corpeleijn, E.; Postmus, D.; Gansevoort, R.T.; de Jong, P.E.; Gans, R.O.; Struck, J.; Hillege, H.L.; Stolk, R.P.; Navis, G.; et al. Plasma procalcitonin and risk of type 2 diabetes in the general population. *Diabetologia* **2011**, *54*, 2463–2465. [CrossRef] [PubMed]
36. American Diabetes Association. 2. Classification and Diagnosis of Diabetes: Standards of Medical Care in Diabetes-2019. *Diabetes Care* **2019**, *42*, S13–S28. [CrossRef]
37. Gruppen, E.G.; Connelly, M.A.; Sluiter, W.J.; Bakker, S.J.L.; Dullaart, R.P.F. Higher plasma GlycA, a novel pro-inflammatory glycoprotein biomarker, is associated with reduced life expectancy: The PREVEND study. *Clin. Chim. Acta* **2019**, *488*, 7–12. [CrossRef]

38. Corsetti, J.P.; Bakker, S.J.; Sparks, C.E.; Dullaart, R.P. Apolipoprotein A-II influences apolipoprotein E-linked cardiovascular disease risk in women with high levels of HDL cholesterol and C-reactive protein. *PLoS ONE* **2012**, *7*, e39110. [CrossRef]
39. Laryea, M.D.; Steinhagen, F.; Pawliczek, S.; Wendel, U. Simple method for the routine determination of betaine and N,N-dimethylglycine in blood and urine. *Clin. Chem.* **1998**, *44*, 1937–1941.
40. Awwad, H.M.; Kirsch, S.H.; Geisel, J.; Obeid, R. Measurement of concentrations of whole blood levels of choline, betaine, and dimethylglycine and their relations to plasma levels. *J. Chromatogr. B Anal. Technol. Biomed. Life Sci.* **2014**, *957*, 41–45. [CrossRef]
41. Shalaurova, I.; Connelly, M.A.; Garvey, W.T.; Otvos, J.D. Lipoprotein insulin resistance index: A lipoprotein particle-derived measure of insulin resistance. *Metab. Syndr. Relat. Disord.* **2014**, *12*, 422–429. [CrossRef]
42. Matyus, S.P.; Braun, P.J.; Wolak-Dinsmore, J.; Saenger, A.K.; Jeyarajah, E.J.; Shalaurova, I.; Warner, S.M.; Fischer, T.J.; Connelly, M.A. HDL particle number measured on the Vantera(R), the first clinical NMR analyzer. *Clin. Biochem.* **2015**, *48*, 148–155. [CrossRef]
43. Otvos, J.D.; Shalaurova, I.; Wolak-Dinsmore, J.; Connelly, M.A.; Mackey, R.H.; Stein, J.H.; Tracy, R.P. GlycA: A Composite Nuclear Magnetic Resonance Biomarker of Systemic Inflammation. *Clin. Chem.* **2015**, *61*, 714–723. [CrossRef] [PubMed]
44. Wolak-Dinsmore, J.; Gruppen, E.G.; Shalaurova, I.; Matyus, S.P.; Grant, R.P.; Gegen, R.; Bakker, S.J.L.; Otvos, J.D.; Connelly, M.A.; Dullaart, R.P.F. A novel NMR-based assay to measure circulating concentrations of branched-chain amino acids: Elevation in subjects with type 2 diabetes mellitus and association with carotid intima media thickness. *Clin. Biochem.* **2018**, *54*, 92–99. [CrossRef] [PubMed]
45. Li, X.S.; Wang, Z.; Cajka, T.; Buffa, J.A.; Nemet, I.; Hurd, A.G.; Gu, X.; Skye, S.M.; Roberts, A.B.; Wu, Y.; et al. Untargeted metabolomics identifies trimethyllysine, a TMAO-producing nutrient precursor, as a predictor of incident cardiovascular disease risk. *J. Clin. Investig. Insight* **2018**, *3*, e99096. [CrossRef] [PubMed]
46. Tang, W.H.; Hazen, S.L. Microbiome, trimethylamine N-oxide, and cardiometabolic disease. *Transl. Res.* **2017**, *179*, 108–115. [CrossRef]
47. Org, E.; Blum, Y.; Kasela, S.; Mehrabian, M.; Kuusisto, J.; Kangas, A.J.; Soininen, P.; Wang, Z.; Ala-Korpela, M.; Hazen, S.L.; et al. Relationships between gut microbiota, plasma metabolites, and metabolic syndrome traits in the METSIM cohort. *Genome Biol.* **2017**, *18*, 70. [CrossRef]

© 2019 by the authors. Licensee MDPI, Basel, Switzerland. This article is an open access article distributed under the terms and conditions of the Creative Commons Attribution (CC BY) license (http://creativecommons.org/licenses/by/4.0/).

Review

Cross-Talk between Lipoproteins and Inflammation: The Role of Microvesicles

Gemma Chiva-Blanch [1] and Lina Badimon [1,2,*]

1. Cardiovascular Program ICCC, Institut de Recerca Hospital Santa Creu i Sant Pau—IIB Sant Pau, Sant Antoni Maria Claret, 167, 08025 Barcelona, Spain; gchiva@santpau.cat
2. CIBER Enfermedades Cardiovasculares (CIBERCV), Instituto de Salud Carlos III (ISCIII), 28029 Madrid, Spain
* Correspondence: lbadimon@santpau.cat; Tel.: +34-9355-65882

Received: 25 October 2019; Accepted: 20 November 2019; Published: 22 November 2019

Abstract: Atherothrombosis is the principal underlying cause of cardiovascular disease (CVD). Microvesicles (MV) are small blebs originated by an outward budding at the cell plasma membranes, which are released in normal conditions. However, MV release is increased in pathophysiologic conditions such as CVD. Low density lipoprotein (LDL) and MV contribute to atherothrombosis onset and progression by promoting inflammation and leukocyte recruitment to injured endothelium, as well as by increasing thrombosis and plaque vulnerability. Moreover, (oxidized)LDL induces MV release and vice-versa, perpetuating endothelium injury leading to CVD progression. Therefore, MV and lipoproteins exhibit common features, which should be considered in the interpretation of their respective roles in the pathophysiology of CVD. Understanding the pathways implicated in this process will aid in developing novel therapeutic approaches against atherothrombosis.

Keywords: microvesicles; inflammation; lipoproteins; LDL cholesterol; microparticles; cardiovascular disease; platelets; endothelial cells; leukocytes; atherothrombosis

---

1. Introduction

Atherothrombosis is the principal underlying cause of cardiovascular disease (CVD). Atherosclerosis is caused by lipid accumulation associated with endothelial dysfunction and chronic low-grade inflammation and oxidative stress. Innate immunity cells and vascular smooth muscle cells respond to these perturbations by initiating interactions and gene programs that contribute to vascular dysfunction and atherosclerotic plaque formation. Advanced atherosclerotic plaques, with a lipid-rich atheroma, show inward remodeling encroaching in the arterial lumen, which decrease blood flow, thus leading to tissue ischemia. Eventually, atherosclerotic plaques can rupture and provoke thrombus formation that may occlude the lumen interrupting oxygen supply [1].

High plasma concentrations of low density lipoprotein (LDL) cholesterol induce atherosclerosis, while decreasing LDL cholesterol levels associates with a reduced incidence of major CV events [2]. In fact, the life-long exposure of an artery to LDL cholesterol remains a principal determinant of atherosclerotic progression [3]. Recently, enhanced perivascular adipose tissue mass and local inflammation has been associated with increased atherosclerotic plaque burden [4], and immunity cells and inflammation have a causal role in atherosclerosis progression by modulating the resident cells in the artery wall [3].

Microvesicles (MV) derived from blood and vascular cells seem also being able to participate in the initiation, progression and complications of atherothrombosis by a direct and a paracrine regulation of target cells. Therefore, this review is aimed at summarizing the crosstalk between lipoproteins, inflammation, and microvesicles in the pathophysiology of atherothrombosis.

2. Atherosclerosis: Lipids and Inflammation

Atherosclerosis is considered a lipid-initiated, chronic and progressive inflammatory systemic disease of large and medium arteries and characterized by inflammatory and immune responses contributing to the onset and evolution of the disease and to plaque instability and rupture. It is triggered by the presence of elevated levels of cholesterol in the vessel wall. On the other hand, high density lipoprotein (HDL) particles, when functionally capable, can exert the opposite effect by removing cholesterol from the circulation through the induction of reverse cholesterol transport, by protecting LDL and other proteins from oxidative damage [5], and by inhibiting monocyte production of inflammatory cytokines [6,7] and monocyte differentiation [8]. However, alterations in the HDL structure and function can render dysfunctional HDL particles that may exert deleterious effects in the CV system [9,10]. Inflammation plays a pivotal role in atherosclerotic plaque formation and instability by promoting endothelial cell activation, endothelial dysfunction, loss of integrity and lipid deposition in the intima and by impairing endothelial-repairing capacity [11].

The involvement of inflammation besides LDL in atherosclerosis has been recently proven in the proof-of-concept CANTOS (Canakinumab Anti-Inflammatory Thrombosis Outcome Study) trial, showing that interleukin 1β inhibition resulted in a 15% reduction in cardiovascular events [12]. Supporting these findings, it has been recently observed that interleukin 6 trans-signaling increases the risk of cardiovascular events [13]. Interleukin 1β/ interleukin 6 signaling is activated by the nucleotide-binding leucine-rich repeat-containing pyrine receptor 3 (NLRP3) inflammasome platform [14]. However, (minimally modified) LDL activates the inflammasome [15–17], in its turn activating the Interleukin 1β/ interleukin 6 signaling pathway, and thus hampering the dissection of the separate roles of lipids and inflammation in atherosclerosis progression.

Atherosclerosis initiates with increases in endothelial permeability to circulating LDL and its accumulation in the intima. Activated endothelial cells then release cytokines, chemokines, and adhesion molecules attracting circulating leukocytes, and more specifically monocytes, into the atherosclerotic lesion, inducing the differentiation of monocytes into proinflammatory macrophages and finally foam cells [18]. Although the primary origin of foam cells in atherosclerotic lesions are leukocytes [19], smooth muscle cells have been shown to differentiate to foam cells [20]. The rupture of the atherosclerotic plaque is the most common trigger of thrombosis, leading to an infarction.

Therefore, atherosclerosis is driven by LDL and inflammation mediated by immune cells, leading to thrombosis and a major CV event. In this process, MV act as a paracrine complementary system ensuing atherothrombosis as will be further discussed.

3. Microvesicles

MV belong to the family of extracellular vesicles. Extracellular vesicles are literally defined by the International Society for Extracellular Vesicles as "the generic term for particles naturally released from the cell that are delimited by a lipid bilayer and cannot replicate, i.e. do not contain a functional nucleus" [21]. MV are medium/large extracellular vesicles (0.1–1 μm) originated by an outward budding at the plasma membrane from almost all cell types [21–24], and their release is increased when activated, injured or undergoing apoptosis. The biogenesis of MV requires a structured rearrangement of the plasma membrane which provokes physical bending of the membrane and restructuring of the underlying actin cytoskeleton, inducing membrane budding, exposure of phosphatidylserine (PS) from the inner leaflet to the cell surface and formation of membrane blebs. However, MV biogenesis also occurs even if membrane lipid asymmetry is maintained [25]. Given their biogenesis, they contain bioactive molecules from their parental cells, and lipids are essential components of MV. Cholesterol is a structural component of the plasma membrane, which regulates cell functionality and subcellular compartmentalization, and is a precursor or cofactor for several signaling molecules/pathways. In fact, MV are enriched in cholesterol compared to plasma membranes [26,27], and cholesterol mediates their formation and release, ensures membrane stability, and it is necessary for their uptake by target cells. Although MV are known to carry biomolecules such as proteins, RNAs and even DNA, their complete molecular fingerprint is largely unexplored [28], probably because the molecular composition of MV is largely influenced by the stimulus originating them [29,30].

Once released from the cell, MV reach target cells and transfer their molecular cargo, triggering functional responses and phenotypic changes affecting cell functionality, both at the local and systemic environments. The net effect of MV in target cells involves signaling at multiple levels including their interaction with cell surface receptors, and transference of genetic material, metabolites and cytosolic enzymes.

Although the current review is focused in the lipid-mediated role of MV in the pathogenesis of atherothrombosis, it is worth mentioning that there are smaller extracellular vesicles sized ≤150 nm called exosomes, formed by the fusion of intracellular multivesicular bodies with the plasma membrane [23]. Exosomes are rich in biomolecules such as protein, mRNA and non-coding RNA such as miRNA, and have also emerged as both regulators and biomarkers of CVD progression [31]. For further information please refer to reviews [32–37].

4. Role of Microvesicles in Atherosclerosis

As previously mentioned, MV are released from all cell types, including platelets, endothelial cells, smooth muscle cells, leukocytes and erythrocytes. Therefore, MV are released in the bloodstream affecting local and distal vulnerable zones.

MV are involved in CVD progression by supporting cellular cross-talk leading to vascular inflammation and dysfunction, leukocyte adhesion and tissue remodeling, thus creating an inflammatory and prothrombotic milieu. In parallel to LDL, MV accumulate and promote the progression of atherosclerotic plaques [38]. Given the externalization of PS during their biogenesis, and, in some cases, tissue factor on their surfaces, MV are potent inducers of coagulation [39], and within the atherosclerotic plaque, retained MV account for the procoagulant activity of the lipid core. MV are the main reservoir of tissue factor activity, thus MV increase plaque vulnerability by promoting coagulation after erosion or rupture of the atherosclerotic plaque. In fact, MV exposing PS in their surface enhance clot propagation, cause apoptosis of nearby cells, and modulate gene expression via a variety of pathways [25], for instance, by transferring monomeric C-reactive protein (CRP) to the cell surface and generating pro-inflammatory signals [40]. Endothelial release of MV is associated to increased cytokine production, and in its turn, increased cytokine production increases MV release. In fact, it has been observed that pharmacological attenuation of inflammation induces a decreased shedding of endothelial-derived MV [41]. Monocyte-derived MV increase endothelial thrombogenicity and apoptosis in vitro by increasing endothelial tissue factor expression, also by the adherence of tissue factor derived from MV to the endothelial cell membrane, by decreasing levels of tissue factor pathway inhibitor (TFPI) and thrombomodulin, and by disrupting endothelial integrity [42]. In addition, MV can transfer cluster of differentiation 40 ligand (CD40L) which binds to monocyte CD40 triggering the activation of tumor necrosis factor (TNF) receptor-associated factor 6 (TRAF6) downstream signaling, activating in its turn nuclear factor (NF)-kB, thus inducing monocyte release of inflammatory mediators [43]. Besides their prothrombotic activity [44], platelet-derived MV enhance the expression of adhesion molecules and trigger the production of interleukins and TNF-α [45]. In addition, platelet MV recruit activated platelets to the endothelial injury area and can also activate platelets; and platelet-derived MV also contribute to atherogeneis by inducing smooth muscle cell proliferation [46]. Finally, it has been shown that erythrocytes can induce the vulnerability and rupture of plaques in a dose-dependant manner [47]. This effect has been proposed to be indirectly mediated by erythrocyte MV by activating the endothelium, attracting leukocytes and platelets, and enhancing the whole inflammatory pathway [46], although scarce research has been done in this direction.

Therefore, MV contribute to atherothrombosis progression by indirectly elevating the thrombotic risk through the induction of an inflammatory response and directly activating platelets and coagulation. Although up to 200-fold elevated concentrations of MV have been found at the local area of atherosclerotic plaques compared to plasma circulating levels [48], circulating MV from CVD patients [49] and healthy subjects [44] have shown significant prothrombotic activity per se, a role that should not be underestimated in atherothrombotic progression.

Crosstalk between Lipoproteins and MV

Recent findings pinpoint some similarities between MV and lipoproteins: both MV and HDL and LDL contain and transfer miRNA to target cells [50,51]; MV extracts have been shown to carry ApoE, ApoB, ApoC-II, among others; and the size and density of small- and middle-sized MV closely overlap with lipoproteins, as can be observed in Figure 1. As a matter of fact, some isolation techniques result in co-purification of MV and lipoprotein particles [22,23,52].

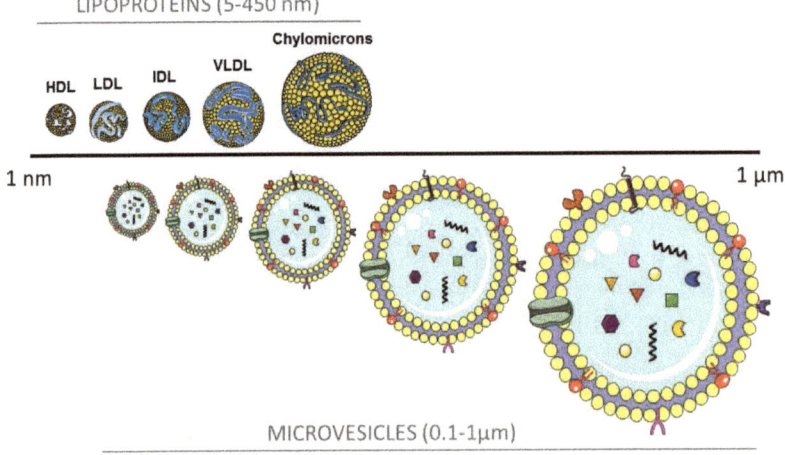

Figure 1. Particle size of lipoproteins and microvesicles. HDL, high density lipoprotein; LDL, low density lipoprotein; IDL, intermediate density lipoprotein; VLDL, very low density lipoprotein; and MV, microvesicles.

Given this size overlap, it is of extreme importance that blood extraction for MV isolation and characterization is being performed in fasting conditions. Depletion of lipoproteins for MV analyses and vice-versa should be carefully considered in biomarker and functional studies. However, the half-life of lipoproteins and MV differs significantly. In physiological conditions, remnant chylomicrons, LDL and HDL, are readily metabolized and stored as intracellular lipid droplets in the liver and adipose tissue. Oppositely, MV rapidly accumulate in resident macrophages of the liver, lungs, and spleen [53]. During atherosclerosis progression, oxidized lipoproteins accumulate in macrophages and other subendothelial cells of the vascular wall provoking an inflammatory response and the progression of the plaque formation. The contribution of circulating (not local) MV in atherothrombosis needs further research.

As previously mentioned and summarized in Figure 2, monocyte-derived macrophages accumulate cholesterol modified within the arterial wall during atherogenesis [54]. In its turn, cholesterol accumulation in macrophages and foam cells enhance AV[+] and tissue factor[+] MV release [55], potentially contributing to the prothrombotic state in hypercholesterolemia [56], and to the prothrombotic core of the lipid-rich vulnerable plaque [57].

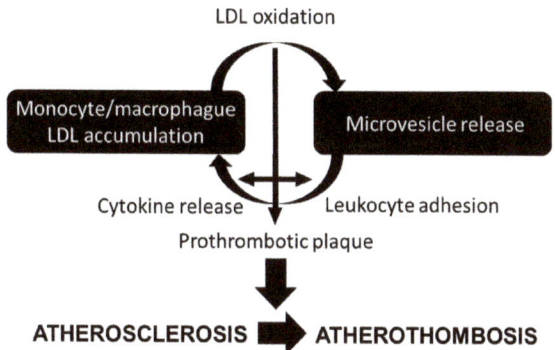

Figure 2. Lipid and microvesicle crosstalk contributing to atherosclerosis progression.

It is known that oxidized low-density lipoprotein (oxLDL) activates platelets [58–60], thus inducing a prothrombotic state [39]. In addition, oxLDL induces the cellular release of MV [33,39]. In fact, total, AV^+ and $CD41a^+$ MV release was shown increased after challenging platelets with oxLDL (but not native LDL) [61], and this effect was comparable to that of ADP. However, Nielsen et al. [62] observed that *in vitro* incubation of platelets with oxLDL (or native LDL as well) did not significantly stimulate $CD41^+$ and $CD41^+/CD36^+$ MV release, suggesting that MV release induced by oxLDL may not be mediated by the interaction with CD36 on platelets.

It has been shown that oxLDL, in a time- and dose-dependent manner, promotes the *in vitro* release of endothelial MV rich in intercellular adhesion molecule 1 (ICAM-1) [63], which can be transferred from MV to endothelial cells increasing monocyte adhesion to endothelial cells [64], further propagating the atheroprone effects of oxLDL even in its absence. Moreover, enrichment of THP-1 monocytic cells with unsterilized cholesterol resulted in increased MV production [65], and these MV induced extensive leukocyte rolling and adherence to the endothelium. In the presence of oxLDL, high shear stress-induced platelet-derived MV were able to activate THP-1 monocytes and induce them to generate tissue factor-rich MV in vitro [66].

Elevated concentrations of circulating autoantibodies for oxLDL, a surrogate biomarker for LDL oxidation in vivo, have been associated with increased levels of platelet- and monocyte-derived circulating MV in acute coronary syndrome patients [66]. As previously stated, oxLDL induces the cellular release of tissue factor-exposing MV, promoting coagulation and thrombosis, and also disseminating the inflammatory response [39]. Although surface molecules of parental cells are transferred to MV, the interactions between of oxLDL and MV are not entirely elucidated and the consequent effects still remain unknown.

In addition to oxidized LDL, aggregated LDL but not native LDL increases tissue factor-loaded MV from smooth muscle cells [67]. In their turn, oxidized MV have been shown to stimulate monocyte adhesion to endothelial cells through oxidized membrane phospholipids, thus also contributing to atherosclerosis progression [65,68,69].

Given the heterogenic composition of MV, they may have cardioprotective functions as well. As recently reviewed, some MV carry antioxidant enzymes, conferring antioxidant activity at MV under specific stimuli [70]. In its turn, MV have also been shown to both stimulate or inhibit angiogenesis by several mechanisms of action, again depending on the cellular origin and molecular composition derived from the trigger or condition originating their release [71].

5. Role of Microvesicles in Dyslipidemia

The pathophysiological link between postprandial hypertriglyceridemia, inflammation and endothelial injury may be provoked by an excessive retention of lipoproteins in the extracellular matrix and increased uptake by macrophages, thus initiating the atherogenic process. A large body of evidence indicates a direct relationship between postprandial hypertriglyceridemia and CVD risk [72,73]. Postprandial dyslipidemia, independently of the caloric intake or the postprandial state itself, is associated with increased endothelial-derived CD31$^+$/CD42$^-$ [74], and total circulating MV levels in healthy subjects [75], and to increased platelet-derived MV in men with different CV risk burden [76], in parallel to increased markers of oxidative stress such as oxLDL and impaired flow-mediated dilation [75]. The effects of postprandial hypertriglyceridemia on MV release in subjects with metabolic dysregulation are quite unexplored and deserve further research, as the reported results are controversial. Type 2 diabetic patients, who show exacerbated postprandial dyslipidemia, show around 3.5 fold increased concentration of endothelial-derived CD144 circulating MV after a meal [77]. However, patients with carotid atherosclerosis show a similar postprandial elevation of circulating platelet-derived MV, despite having higher postprandial hypertriglyceridemia than control subjects free of atherosclerosis [78].

Patients with hypercholesterolemia show higher levels of monocyte- and platelet-derived MV than healthy subjects [79–81]. Familial hypercholesterolemia (FH) is an autosomal dominant genetic disorder associated with elevated LDL cholesterol levels and deposition in tendons (xanthomas), and premature heart disease [82]. MV have been associated with atherosclerosis progression and with a higher risk of atherothrombosis in FH patients. In fact, elevated concentrations of total, endothelial cell-derived, erythrocyte-derived, monocyte-derived, tissue factor-loaded MV [83], and platelet-derived MV [84], have been found in FH patients compared to healthy controls [85], and identify subclinical atherosclerosis [80]. In addition, circulating CD36$^+$ MV derived from endothelial cells and monocytes were significantly higher in FH patients compared to healthy controls [85], and monocyte-derived circulating MV in FH patients directly correlated with oxLDL plasma concentrations [79]. Moreover, these patients showed increased concentrations of circulating MV derived from leukocytes, and lymphocyte-derived CD3$^+$/CD45$^+$ circulating MV have been shown elevated in FH patients with lipid-rich atherosclerotic plaques [86]. In addition, platelet- (CD41a$^+$/AV$^+$, CD31$^+$/AV$^+$, CD41a$^+$/CD31$^+$/AV$^+$), granulocyte- (CD66$^+$/AV$^+$), neutrophil- (CD11b$^+$/CD66$^+$/AV$^+$), and endothelial cell-derived (CD62E$^+$/AV$^{+/-}$) circulating MV discriminate and map coronary atherosclerotic plaque and calcification [84], and leukocyte- and activated platelet-derived circulating MV predict a major cardiovascular event in these patients three years before it takes place [87].

In FH patients, plasma apheresis reduces circulating MV concentration, mainly platelet-derived MV AV$^+$ [88].

Pharmacological Treatment of Dyslipidemia and Circulating MV

Given the crosstalk between lipids and MV, as discussed in a previous section, it appears plausible that lipid lowering therapy may elicit a pleiotropic effect by (partially) inhibiting MV release, contributing to explain their beneficial effects in atherothrombotic complications. We have observed that statins reduce MV shedding from platelets, endothelial cells and leukocytes carrying markers of cell activation [89], and this effect was found to be accumulative throughout years of treatment. Other researchers have also observed that statins decrease MV release from several cell origins [90–92], and modify their molecular fingerprint by decreasing their cargo of cell activation markers [93,94]. In stroke patients with hyperlipidemia, simvastatin treatment for 6 months reduced the percentage of CD61$^+$ platelet-derived MV to similar levels to those of age- and sex-matched controls [95].

However, some authors did not observe such effect of statins [96,97], and, as discussed in the 2019 ESC/EAS Guidelines for the management of dyslipidemias, the clinical relevance of the pleiotropic effects of statins remains clinically unproven [98], and thus may be object of future research.

6. Conclusions and Future Perspectives

The studies summarized in this review evidence that MV are settled in the tandem lipids inflammation, contributing to the progression of atherothrombosis leading to a major CV event. Despite a huge body of research towards the understanding of the exact contribution of MV in this process, several questions remain unanswered, such as the specific interaction of MV with native and modified LDL, the molecular fingerprint of MV, how do they qualitatively (by which pathways) and quantitatively (in what amounts) contribute in the presentation of major CV events, and if they can be considered real candidates as drug delivery vectors in the clinical setting. Given the central role of MV in systemic inflammation and endothelial injury within the pathophysiology of atherothrombosis, their link with lipids and lipoproteins are of particular interest. Understanding the pathways implicated in this process will aid in developing novel therapeutic approaches against early atherosclerosis. In this setting, MV and lipoproteins exhibit common features, which should be considered in the interpretation of their respective roles in the pathophysiology of CVD.

Author Contributions: Conceptualization, G.C.-B. and L.B.; writing—original draft preparation, G.C.-B. and L.B.; writing—review and editing, G.C.-B. and L.B.; funding acquisition, L.B.

Funding: This research was funded by the Spanish Ministry of Economy and Competitiveness of Science, SAF2016-76819-R; the Institute of Health Carlos III (ISCIII), TERCEL—RD16/0011/0018 and CIBERCV CB16/11/00411; FEDER "Una Manera de Hacer Europa", and by the Agency for Management of University and Research Grants, 2017SGR1480. The APC was funded by SAF2016-76819-R.

Acknowledgments: We thank the Fundació d'Investigació Cardiovascular (FIC)-Fundación Jesús Serra, Barcelona, Spain, for their continuous support. Figures have been created with templates of Servier Medical Art.

Conflicts of Interest: The authors declare no conflict of interest.

References

1. Badimon, L.; Vilahur, G. Thrombosis formation on atherosclerotic lesions and plaque rupture. *J. Intern. Med.* **2014**, *276*, 618–632. [CrossRef]
2. Silverman, M.G.; Ference, B.A.; Im, K.; Wiviott, S.D.; Giugliano, R.P.; Grundy, S.M.; Braunwald, E.; Sabatine, M.S. Association Between Lowering LDL-C and Cardiovascular Risk Reduction Among Different Therapeutic Interventions: A Systematic Review and Meta-analysis. *JAMA* **2016**, *316*, 1289–1297. [CrossRef]
3. Libby, P.; Buring, J.E.; Badimon, L.; Hansson, G.K.; Deanfield, J.; Bittencourt, M.S.; Tokgözoğlu, L.; Lewis, E.F. Atherosclerosis. *Nat. Rev. Dis. Prim.* **2019**, *5*, 56. [CrossRef]
4. Vilahur, G.; Ben-Aicha, S.; Badimon, L. New insights into the role of adipose tissue in thrombosis. *Cardiovasc. Res.* **2017**, *113*, 1046–1054. [CrossRef]
5. Zerrad-Saadi, A.; Therond, P.; Chantepie, S.; Couturier, M.; Rye, K.-A.; Chapman, M.J.; Kontush, A. HDL3-Mediated Inactivation of LDL-Associated Phospholipid Hydroperoxides Is Determined by the Redox Status of Apolipoprotein A-I and HDL Particle Surface Lipid Rigidity. *Arterioscler. Thromb. Vasc. Biol.* **2009**, *29*, 2169–2175. [CrossRef] [PubMed]
6. Baker, P.W.; Rye, K.A.; Gamble, J.R.; Vadas, M.A.; Barter, P.J. Ability of reconstituted high density lipoproteins to inhibit cytokine-induced expression of vascular cell adhesion molecule-1 in human umbilical vein endothelial cells. *J. Lipid Res.* **1999**, *40*, 345–353. [PubMed]
7. Jonas, K.; Kopeć, G. HDL Cholesterol as a Marker of Disease Severity and Prognosis in Patients with Pulmonary Arterial Hypertension. *Int. J. Mol. Sci.* **2019**, *20*, 3514. [CrossRef] [PubMed]
8. Grün, J.L.; Manjarrez-Reyna, A.N.; Gómez-Arauz, A.Y.; Leon-Cabrera, S.; Rückert, F.; Fragoso, J.M.; Bueno-Hernández, N.; Islas-Andrade, S.; Meléndez-Mier, G.; Escobedo, G. High-Density Lipoprotein Reduction Differentially Modulates to Classical and Nonclassical Monocyte Subpopulations in Metabolic Syndrome Patients and in LPS-Stimulated Primary Human Monocytes In Vitro. *J. Immunol. Res.* **2018**, *2018*, 2737040. [CrossRef]
9. Padró, T.; Cubedo, J.; Camino, S.; Béjar, M.T.; Ben-Aicha, S.; Mendieta, G.; Escolà-Gil, J.C.; Escate, R.; Gutiérrez, M.; Casani, L.; et al. Detrimental Effect of Hypercholesterolemia on High-Density Lipoprotein Particle Remodeling in Pigs. *J. Am. Coll. Cardiol.* **2017**, *70*, 165–178. [CrossRef]
10. Ben-Aicha, S.; Escate, R.; Casaní, L.; Padró, T.; Peña, E.; Arderiu, G.; Mendieta, G.; Badimón, L.; Vilahur, G. High-density lipoprotein remodelled in hypercholesterolaemic blood induce epigenetically driven down-regulation of endothelial HIF-1α expression in a preclinical animal model. *Cardiovasc. Res.* **2019**. [CrossRef]

11. Marchio, P.; Guerra-Ojeda, S.; Vila, J.M.; Aldasoro, M.; Victor, V.M.; Mauricio, M.D. Targeting Early Atherosclerosis: A Focus on Oxidative Stress and Inflammation. *Oxid. Med. Cell. Longev.* **2019**, *2019*, 8563845. [CrossRef] [PubMed]
12. Ridker, P.M.; Everett, B.M.; Thuren, T.; MacFadyen, J.G.; Chang, W.H.; Ballantyne, C.; Fonseca, F.; Nicolau, J.; Koenig, W.; Anker, S.D.; et al. Antiinflammatory Therapy with Canakinumab for Atherosclerotic Disease. *N. Engl. J. Med.* **2017**, *377*, 1119–1131. [CrossRef] [PubMed]
13. Ziegler, L.; Frumento, P.; Wallén, H.; de Faire, U.; Gigante, B. The predictive role of interleukin 6 trans-signalling in middle-aged men and women at low-intermediate risk of cardiovascular events. *Eur. J. Prev. Cardiol.* **2019**. [CrossRef] [PubMed]
14. Von Moltke, J.; Trinidad, N.J.; Moayeri, M.; Kintzer, A.F.; Wang, S.B.; Van Rooijen, N.; Brown, C.R.; Krantz, B.A.; Leppla, S.H.; Gronert, K.; et al. Rapid induction of inflammatory lipid mediators by the inflammasome in vivo. *Nature* **2012**, *490*, 107–111. [CrossRef] [PubMed]
15. Jukema, R.A.; Ahmed, T.A.N.; Tardif, J.-C. Does low-density lipoprotein cholesterol induce inflammation? If so, does it matter? Current insights and future perspectives for novel therapies. *BMC Med.* **2019**, *17*, 197. [CrossRef] [PubMed]
16. Duewell, P.; Kono, H.; Rayner, K.J.; Sirois, C.M.; Vladimer, G.; Bauernfeind, F.G.; Abela, G.S.; Franchi, L.; Núñez, G.; Schnurr, M.; et al. NLRP3 inflammasomes are required for atherogenesis and activated by cholesterol crystals. *Nature* **2010**, *464*, 1357–1361. [CrossRef]
17. Varghese, J.F.; Patel, R.; Yadav, U.C.S. Sterol regulatory element binding protein (SREBP) -1 mediates oxidized low-density lipoprotein (oxLDL) induced macrophage foam cell formation through NLRP3 inflammasome activation. *Cell. Signal.* **2019**, *53*, 316–326. [CrossRef]
18. Libby, P.; Ridker, P.M.; Maseri, A. Inflammation and atherosclerosis. *Circulation* **2002**, *105*, 1135–1143. [CrossRef]
19. Obama, T.; Ohinata, H.; Takaki, T.; Iwamoto, S.; Sawada, N.; Aiuchi, T.; Kato, R.; Itabe, H. Cooperative Action of Oxidized Low-Density Lipoproteins and Neutrophils on Endothelial Inflammatory Responses Through Neutrophil Extracellular Trap Formation. *Front. Immunol.* **2019**, *10*, 1899. [CrossRef]
20. Llorente-Cortés, V.; Royo, T.; Juan-Babot, O.; Badimon, L. Adipocyte differentiation-related protein is induced by LRP1-mediated aggregated LDL internalization in human vascular smooth muscle cells and macrophages. *J. Lipid Res.* **2007**, *48*, 2133–2140. [CrossRef]
21. Théry, C.; Witwer, K.W.; Aikawa, E.; Alcaraz, M.J.; Anderson, J.D.; Andriantsitohaina, R.; Antoniou, A.; Arab, T.; Archer, F.; Atkin-Smith, G.K.; et al. Minimal information for studies of extracellular vesicles 2018 (MISEV2018): A position statement of the International Society for Extracellular Vesicles and update of the MISEV2014 guidelines. *J. Extracell. Vesicles* **2018**, *7*, 1535750. [CrossRef] [PubMed]
22. Jeppesen, D.K.; Fenix, A.M.; Franklin, J.L.; Higginbotham, J.N.; Zhang, Q.; Zimmerman, L.J.; Liebler, D.C.; Ping, J.; Liu, Q.; Evans, R.; et al. Reassessment of Exosome Composition. *Cell* **2019**, *177*, 428–445.e18. [CrossRef] [PubMed]
23. Mathieu, M.; Martin-Jaular, L.; Lavieu, G.; Théry, C. Specificities of secretion and uptake of exosomes and other extracellular vesicles for cell-to-cell communication. *Nat. Cell Biol.* **2019**, *21*, 9–17. [CrossRef] [PubMed]
24. Badimon, L.; Suades, R.; Arderiu, G.; Peña, E.; Chiva-Blanch, G.; Padró, T. Microvesicles in Atherosclerosis and Angiogenesis. From Bench to Bedside and Reverse. *Front. Cardiovasc. Med.* **2017**, *4*, 77. [CrossRef]
25. Del Conde, I.; Shrimpton, C.N.; Thiagarajan, P.; López, J.A. Tissue-factor–bearing microvesicles arise from lipid rafts and fuse with activated platelets to initiate coagulation. *Blood* **2005**, *106*, 1604–1611. [CrossRef]
26. Pfrieger, F.W.; Vitale, N. Cholesterol and the journey of extracellular vesicles. *J. Lipid Res.* **2018**, *59*, 2255–2261. [CrossRef]
27. BIRO, E.; Akkerman, J.W.N.; Hoek, F.J.; Gorter, G.; Pronk, L.M.; Sturk, A.; Nieuwland, R. The phospholipid composition and cholesterol content of platelet-derived microparticles: A comparison with platelet membrane fractions. *J. Thromb. Haemost.* **2005**, *3*, 2754–2763. [CrossRef]
28. Gudbergsson, J.M.; Jønsson, K.; Simonsen, J.B.; Johnsen, K.B. Systematic review of targeted extracellular vesicles for drug delivery—Considerations on methodological and biological heterogeneity. *J. Control. Release* **2019**, *306*, 108–120. [CrossRef]
29. Williams, C.; Palviainen, M.; Reichardt, N.-C.; Siljander, P.R.-M.; Falcón-Pérez, J.M. Metabolomics Applied to the Study of Extracellular Vesicles. *Metabolites* **2019**, *9*, 276. [CrossRef]

30. Turchinovich, A.; Drapkina, O.; Tonevitsky, A. Transcriptome of extracellular vesicles: State-of-the-art. *Front. Immunol.* **2019**, *10*, 202. [CrossRef]
31. De Giusti, C.J.; Santalla, M.; Das, S. Exosomal non-coding RNAs (Exo-ncRNAs) in cardiovascular health. *J. Mol. Cell. Cardiol.* **2019**, *137*, 143–151. [CrossRef] [PubMed]
32. Jiang, W.; Wang, M. New insights into the immunomodulatory role of exosomes in cardiovascular disease. *Rev. Cardiovasc. Med.* **2019**, *20*, 153–160. [PubMed]
33. Yang, W.; Zou, B.; Hou, Y.; Yan, W.; Chen, T.; Qu, S. Extracellular vesicles in vascular calcification. *Clin. Chim. Acta* **2019**, *499*, 118–122. [CrossRef] [PubMed]
34. Kita, S.; Maeda, N.; Shimomura, I. Interorgan communication by exosomes, adipose tissue, and adiponectin in metabolic syndrome. *J. Clin. Investig.* **2019**, *129*, 4041–4049. [CrossRef]
35. Yu, H.; Wang, Z. Cardiomyocyte-Derived Exosomes: Biological Functions and Potential Therapeutic Implications. *Front. Physiol.* **2019**, *10*, 1049. [CrossRef]
36. Jafarzadeh-Esfehani, R.; Soudyab, M.; Parizadeh, S.M.; Jaripoor, M.E.; Nejad, P.S.; Shariati, M.; Nabavi, A.S. Circulating exosomes and their role in stroke. *Curr. Drug Targets* **2019**, *20*. [CrossRef]
37. Aghabozorgi, A.S.; Ahangari, N.; Eftekhaari, T.E.; Torbati, P.N.; Bahiraee, A.; Ebrahimi, R.; Pasdar, A. Circulating exosomal miRNAs in cardiovascular disease pathogenesis: New emerging hopes. *J. Cell. Physiol.* **2019**, *234*, 21796–21809. [CrossRef]
38. Owens, A.P.; Mackman, N. Microparticles in Hemostasis and Thrombosis. *Circ. Res.* **2011**, *108*, 1284–1297. [CrossRef]
39. Obermayer, G.; Afonyushkin, T.; Binder, C.J. Oxidized low-density lipoprotein in inflammation-driven thrombosis. *J. Thromb. Haemost.* **2018**, *16*, 418–428. [CrossRef]
40. Habersberger, J.; Strang, F.; Scheichl, A.; Htun, N.; Bassler, N.; Merivirta, R.-M.; Diehl, P.; Krippner, G.; Meikle, P.; Eisenhardt, S.U.; et al. Circulating microparticles generate and transport monomeric C-reactive protein in patients with myocardial infarction. *Cardiovasc. Res.* **2012**, *96*, 64–72. [CrossRef]
41. Pirro, M.; Bianconi, V.; Paciullo, F.; Mannarino, M.R.; Bagaglia, F.; Sahebkar, A. Lipoprotein(a) and inflammation: A dangerous duet leading to endothelial loss of integrity. *Pharmacol. Res.* **2017**, *119*, 178–187. [CrossRef] [PubMed]
42. Tsimerman, G.; Roguin, A.; Bachar, A.; Melamed, E.; Brenner, B.; Aharon, A. Involvement of microparticles in diabetic vascular complications. *Thromb. Haemost.* **2011**, *106*, 310–321. [CrossRef] [PubMed]
43. Bei, J.-J.; Liu, C.; Peng, S.; Liu, C.-H.; Zhao, W.-B.; Qu, X.-L.; Chen, Q.; Zhou, Z.; Yu, Z.-P.; Peter, K.; et al. Staphylococcal SSL5-induced platelet microparticles provoke proinflammatory responses via the CD40/TRAF6/NFKB signalling pathway in monocytes. *Thromb. Haemost.* **2016**, *115*, 632–645. [CrossRef] [PubMed]
44. Suades, R.; Padró, T.; Vilahur, G.; Badimon, L. Circulating and platelet-derived microparticles in human blood enhance thrombosis on atherosclerotic plaques. *Thromb. Haemost.* **2012**, *108*, 1208–1219. [CrossRef] [PubMed]
45. Hugel, B.; Martínez, M.C.; Kunzelmann, C.; Freyssinet, J.-M. Membrane Microparticles: Two Sides of the Coin. *Physiology* **2005**, *20*, 22–27. [CrossRef]
46. Blum, A. The possible role of red blood cell microvesicles in atherosclerosis. *Eur. J. Intern. Med.* **2009**, *20*, 101–105. [CrossRef]
47. Lin, H.-L.; Xu, X.-S.; Lu, H.-X.; Zhang, L.; Li, C.-J.; Tang, M.-X.; Sun, H.-W.; Liu, Y.; Zhang, Y. Pathological mechanisms and dose dependency of erythrocyte-induced vulnerability of atherosclerotic plaques. *J. Mol. Cell. Cardiol.* **2007**, *43*, 272–280. [CrossRef]
48. Leroyer, A.S.; Isobe, H.; Lesèche, G.; Castier, Y.; Wassef, M.; Mallat, Z.; Binder, B.R.; Tedgui, A.; Boulanger, C.M. Cellular Origins and Thrombogenic Activity of Microparticles Isolated from Human Atherosclerotic Plaques. *J. Am. Coll. Cardiol.* **2007**, *49*, 772–777. [CrossRef]
49. Mallat, Z.; Benamer, H.; Hugel, B.; Benessiano, J.; Steg, P.G.; Freyssinet, J.M.; Tedgui, A. Elevated levels of shed membrane microparticles with procoagulant potential in the peripheral circulating blood of patients with acute coronary syndromes. *Circulation* **2000**, *101*, 841–843. [CrossRef]
50. Lee, H.; Zhang, D.; Zhu, Z.; Dela Cruz, C.S.; Jin, Y. Epithelial cell-derived microvesicles activate macrophages and promote inflammation via microvesicle-containing microRNAs. *Sci. Rep.* **2016**, *6*, 35250. [CrossRef]
51. Shu, Z.; Tan, J.; Miao, Y.; Zhang, Q. The role of microvesicles containing microRNAs in vascular endothelial dysfunction. *J. Cell. Mol. Med.* **2019**. [CrossRef] [PubMed]

52. Mahmood Hussain, M. A proposed model for the assembly of chylomicrons. *Atherosclerosis* **2000**, *148*, 1–15. [CrossRef]
53. Wiklander, O.P.B.; Nordin, J.Z.; O'Loughlin, A.; Gustafsson, Y.; Corso, G.; Mäger, I.; Vader, P.; Lee, Y.; Sork, H.; Seow, Y.; et al. Extracellular vesicle in vivo biodistribution is determined by cell source, route of administration and targeting. *J. Extracell. Vesicles* **2015**, *4*, 26316. [CrossRef] [PubMed]
54. Skålén, K.; Gustafsson, M.; Rydberg, E.K.; Hultén, L.M.; Wiklund, O.; Innerarity, T.L.; Borén, J. Subendothelial retention of atherogenic lipoproteins in early atherosclerosis. *Nature* **2002**, *417*, 750–754. [CrossRef]
55. Liu, M.-L.; Reilly, M.P.; Casasanto, P.; McKenzie, S.E.; Williams, K.J. Cholesterol enrichment of human monocyte/macrophages induces surface exposure of phosphatidylserine and the release of biologically-active tissue factor-positive microvesicles. *Arterioscler. Thromb. Vasc. Biol.* **2007**, *27*, 430–435. [CrossRef]
56. Cipollone, F.; Mezzetti, A.; Porreca, E.; Di Febbo, C.; Nutini, M.; Fazia, M.; Falco, A.; Cuccurullo, F.; Davì, G. Association Between Enhanced Soluble CD40L and Prothrombotic State in Hypercholesterolemia. *Circulation* **2002**, *106*, 399–402. [CrossRef]
57. Williams, K.J.; Tabas, I. Lipoprotein retention—And clues for atheroma regression. *Arterioscler. Thromb. Vasc. Biol.* **2005**, *25*, 1536–1540. [CrossRef]
58. Podrez, E.A.; Byzova, T.V.; Febbraio, M.; Salomon, R.G.; Ma, Y.; Valiyaveettil, M.; Poliakov, E.; Sun, M.; Finton, P.J.; Curtis, B.R.; et al. Platelet CD36 links hyperlipidemia, oxidant stress and a prothrombotic phenotype. *Nat. Med.* **2007**, *13*, 1086–1095. [CrossRef]
59. Chen, K.; Febbraio, M.; Li, W.; Silverstein, R.L. A Specific CD36-Dependent Signaling Pathway Is Required for Platelet Activation by Oxidized Low-Density Lipoprotein. *Circ. Res.* **2008**, *102*, 1512–1519. [CrossRef]
60. Wraith, K.S.; Magwenzi, S.; Aburima, A.; Wen, Y.; Leake, D.; Naseem, K.M. Oxidized low-density lipoproteins induce rapid platelet activation and shape change through tyrosine kinase and Rho kinase-signaling pathways. *Blood* **2013**, *122*, 580–589. [CrossRef]
61. Wang, H.; Wang, Z.-H.; Kong, J.; Yang, M.-Y.; Jiang, G.-H.; Wang, X.-P.; Zhong, M.; Zhang, Y.; Deng, J.-T.; Zhang, W. Oxidized Low-Density Lipoprotein-Dependent Platelet-Derived Microvesicles Trigger Procoagulant Effects and Amplify Oxidative Stress. *Mol. Med.* **2012**, *18*, 159–166. [CrossRef] [PubMed]
62. Nielsen, T.B.; Nielsen, M.H.; Handberg, A. In vitro Incubation of Platelets with oxLDL Does Not Induce Microvesicle Release When Measured by Sensitive Flow Cytometry. *Front. Cardiovasc. Med.* **2015**, *2*, 37. [CrossRef] [PubMed]
63. Fu, Z.; Zhou, E.; Wang, X.; Tian, M.; Kong, J.; Li, J.; Ji, L.; Niu, C.; Shen, H.; Dong, S.; et al. Oxidized low-density lipoprotein-induced microparticles promote endothelial monocyte adhesion via intercellular adhesion molecule 1. *Am. J. Physiol. Physiol.* **2017**, *313*, C567–C574. [CrossRef] [PubMed]
64. Rautou, P.-E.; Leroyer, A.S.; Ramkhelawon, B.; Devue, C.; Duflaut, D.; Vion, A.-C.; Nalbone, G.; Castier, Y.; Leseche, G.; Lehoux, S.; et al. Microparticles From Human Atherosclerotic Plaques Promote Endothelial ICAM-1–Dependent Monocyte Adhesion and Transendothelial Migration. *Circ. Res.* **2011**, *108*, 335–343. [CrossRef] [PubMed]
65. Liu, M.L.; Scalia, R.; Mehta, J.L.; Williams, K.J. Cholesterol-induced membrane microvesicles as novel carriers of damage-associated molecular patterns: Mechanisms of formation, action, and detoxification. *Arterioscler. Thromb. Vasc. Biol.* **2012**, *32*, 2113–2121. [CrossRef] [PubMed]
66. Matsumoto, N.; Nomura, S.; Kamihata, H.; Kimura, Y.; Iwasaka, T. Increased level of oxidized LDL-dependent monocytederived microparticles in acute coronary syndrome. *Thromb. Haemost.* **2004**, *91*, 146–154. [CrossRef] [PubMed]
67. Llorente-Cortés, V.; Otero-Viñas, M.; Camino-López, S.; Llampayas, O.; Badimon, L. Aggregated low-density lipoprotein uptake induces membrane tissue factor procoagulant activity and microparticle release in human vascular smooth muscle cells. *Circulation* **2004**, *110*, 452–459. [CrossRef]
68. Huber, J.; Vales, A.; Mitulovic, G.; Blumer, M.; Schmid, R.; Witztum, J.L.; Binder, B.R.; Leitinger, N. Oxidized membrane vesicles and blebs from apoptotic cells contain biologically active oxidized phospholipids that induce monocyte-endothelial interactions. *Arterioscler. Thromb. Vasc. Biol.* **2002**, *22*, 101–107. [CrossRef]
69. Papac-Milicevic, N.; Busch, C.J.L.; Binder, C.J. Malondialdehyde Epitopes as Targets of Immunity and the Implications for Atherosclerosis. In *Advances in Immunology*; Academic Press Inc.: Cambridge, MA, USA, 2016; Volume 131, pp. 1–59.

70. Bodega, G.; Alique, M.; Puebla, L.; Carracedo, J.; Ramírez, R.M. Microvesicles: ROS scavengers and ROS producers. *J. Extracell. Vesicles* **2019**, *8*, 1626654. [CrossRef]
71. Todorova, D.; Simoncini, S.; Lacroix, R.; Sabatier, F.; Dignat-George, F. Extracellular vesicles in angiogenesis. *Circ. Res.* **2017**, *120*, 1658–1673. [CrossRef]
72. Karpe, F. Postprandial lipoprotein metabolism and atherosclerosis. *J. Intern. Med.* **1999**, *246*, 341–355. [CrossRef] [PubMed]
73. Nappo, F.; Esposito, K.; Cioffi, M.; Giugliano, G.; Molinari, A.M.; Paolisso, G.; Marfella, R.; Giugliano, D. Postprandial endothelial activation in healthy subjects and in type 2 diabetic patients: Role of fat and carbohydrate meals. *J. Am. Coll. Cardiol.* **2002**, *39*, 1145–1150. [CrossRef]
74. Ferreira, A.C.; Peter, A.A.; Mendez, A.J.; Jimenez, J.J.; Mauro, L.M.; Chirinos, J.A.; Ghany, R.; Virani, S.; Garcia, S.; Horstman, L.L.; et al. Postprandial Hypertriglyceridemia Increases Circulating Levels of Endothelial Cell Microparticles. *Circulation* **2004**, *110*, 3599–3603. [CrossRef] [PubMed]
75. Tushuizen, M.E.; Nieuwland, R.; Scheffer, P.G.; Sturk, A.; Heine, R.J.; Diamant, M. Two consecutive high-fat meals affect endothelial-dependent vasodilation, oxidative stress and cellular microparticles in healthy men. *J. Thromb. Haemost.* **2006**, *4*, 1003–1010. [CrossRef] [PubMed]
76. Tamburrelli, C.; Gianfagna, F.; D'imperio, M.; De Curtis, A.; Rotilio, D.; Iacoviello, L.; De Gaetano, G.; Donati, M.B.; Cerletti, C. Postprandial cell inflammatory response to a standardised fatty meal in subjects at different degree of cardiovascular risk. *Thromb. Haemost.* **2012**, *107*, 530–537. [CrossRef] [PubMed]
77. Tushuizen, M.E.; Nieuwland, R.; Rustemeijer, C.; Hensgens, B.E.; Sturk, A.; Heine, R.J.; Diamant, M. Elevated endothelial microparticles following consecutive meals are associated with vascular endothelial dysfunction in type 2 diabetes. *Diabetes Care* **2007**, *30*, 728–730. [CrossRef]
78. Elevated Levels of Platelet Microparticles in Carotid Atherosclerosis and during the Postprandial State-ClinicalKey. Available online: https://www.clinicalkey.es/#!/content/playContent/1-s2.0-S0049384808005082?scrollTo=%23hl0000494 (accessed on 14 November 2019).
79. Hjuler Nielsen, M.; Irvine, H.; Vedel, S.; Raungaard, B.; Beck-Nielsen, H.; Handberg, A. Elevated Atherosclerosis-Related Gene Expression, Monocyte Activation and Microparticle-Release Are Related to Increased Lipoprotein-Associated Oxidative Stress in Familial Hypercholesterolemia. *PLoS ONE* **2015**, *10*, e0121516. [CrossRef]
80. Suades, R.; Padró, T.; Alonso, R.; Mata, P.; Badimon, L. High levels of TSP1+/CD142+ platelet-derived microparticles characterise young patients with high cardiovascular risk and subclinical atherosclerosis. *Thromb. Haemost.* **2015**, *114*, 1310–1321. [CrossRef]
81. Pirro, M.; Schillaci, G.; Paltriccia, R.; Bagaglia, F.; Menecali, C.; Mannarino, M.R.; Capanni, M.; Velardi, A.; Mannarino, E. Increased ratio of CD31+/CD42− microparticles to endothelial progenitors as a novel marker of atherosclerosis in hypercholesterolemia. *Arterioscler. Thromb. Vasc. Biol.* **2006**, *26*, 2530–2535. [CrossRef]
82. Badimon, L.; Chiva-Blanch, G. Lipid Metabolism in Dyslipidemia and Familial Hypercholesterolemia. *Mol. Nutr. Fats* **2019**, 307–322.
83. Owens, A.P.; Passam, F.H.; Antoniak, S.; Marshall, S.M.; McDaniel, A.L.; Rudel, L.; Williams, J.C.; Hubbard, B.K.; Dutton, J.-A.; Wang, J.; et al. Monocyte tissue factor-dependent activation of coagulation in hypercholesterolemic mice and monkeys is inhibited by simvastatin. *J. Clin. Investig.* **2012**, *122*, 558–568. [CrossRef] [PubMed]
84. Chiva-Blanch, G.; Padró, T.; Alonso, R.; Crespo, J.; Perez de Isla, L.; Mata, P.; Badimon, L. Liquid biopsy of extracellular microvesicles maps coronary calcification and atherosclerotic plaque in asymptomatic patients with familial hypercholesterolemia. *Arterioscler. Thromb. Vasc. Biol.* **2019**, *39*, 945–955. [CrossRef] [PubMed]
85. Nielsen, M.H.; Irvine, H.; Vedel, S.; Raungaard, B.; Beck-Nielsen, H.; Handberg, A. The Impact of Lipoprotein-Associated Oxidative Stress on Cell-Specific Microvesicle Release in Patients with Familial Hypercholesterolemia. *Oxid. Med. Cell. Longev.* **2016**, *2016*, 2492858. [CrossRef] [PubMed]
86. Suades, R.; Padró, T.; Alonso, R.; López-Miranda, J.; Mata, P.; Badimon, L. Circulating CD45+/CD3+ lymphocyte-derived microparticles map lipid-rich atherosclerotic plaques in familial hypercholesterolaemia patients. *Thromb. Haemost.* **2013**, *111*, 111–121. [CrossRef] [PubMed]
87. Suades, R.; Padró, T.; Crespo, J.; Sionis, A.; Alonso, R.; Mata, P.; Badimon, L. Liquid Biopsy of Extracellular Microvesicles Predicts Future Major Ischemic Events in Genetically Characterized Familial Hypercholesterolemia Patients. *Arterioscler. Thromb. Vasc. Biol.* **2019**, *39*, 1172–1181. [CrossRef]

88. Connolly, K.D.; Willis, G.R.; Datta, D.B.N.; Ellins, E.A.; Ladell, K.; Price, D.A.; Guschina, I.A.; Rees, D.A.; James, P.E. Lipoprotein-apheresis reduces circulating microparticles in individuals with familial hypercholesterolemia. *J. Lipid Res.* **2014**, *55*, 2064–2072. [CrossRef]
89. Suades, R.; Padró, T.; Alonso, R.; Mata, P.; Badimon, L. Lipid-lowering therapy with statins reduces microparticle shedding from endothelium, platelets and inflammatory cells. *Thromb. Haemost.* **2013**, *110*, 366–377. [CrossRef]
90. Nomura, S.; Inami, N.; Shouzu, A.; Omoto, S.; Kimura, Y.; Takahashi, N.; Tanaka, A.; Urase, F.; Maeda, Y.; Ohtani, H.; et al. The effects of pitavastatin, eicosapentaenoic acid and combined therapy on platelet-derived microparticles and adiponectin in hyperlipidemic, diabetic patients. *Platelets* **2009**, *20*, 16–22. [CrossRef]
91. Nomura, S.; Shouzu, A.; Omoto, S.; Inami, N.; Ueba, T.; Urase, F.; Maeda, Y. Effects of eicosapentaenoic acid on endothelial cell-derived microparticles, angiopoietins and adiponectin in patients with type 2 diabetes. *J. Atheroscler. Thromb.* **2009**, *16*, 83–90. [CrossRef]
92. Tramontano, A.F.; O'Leary, J.; Black, A.D.; Muniyappa, R.; Cutaia, M.V.; El-Sherif, N. Statin decreases endothelial microparticle release from human coronary artery endothelial cells: Implication for the Rho-kinase pathway. *Biochem. Biophys. Res. Commun.* **2004**, *320*, 34–38. [CrossRef]
93. Sommeijer, D.W.; Joop, K.; Leyte, A.; Reitsma, P.H.; Cate, H.T. Pravastatin reduces fibrinogen receptor gpIIIa on platelet-derived microparticles in patients with type 2 diabetes. *J. Thromb. Haemost.* **2005**, *3*, 1168–1171. [CrossRef] [PubMed]
94. Mobarrez, F.; He, S.; Bröijersen, A.; Wiklund, B.; Antovic, A.; Antovic, J.; Egberg, N.; Jörneskog, G.; Wallén, H. Atorvastatin reduces thrombin generation and expression of tissue factor, P-selectin and GPIIIa on platelet-derived microparticles in patients with peripheral arterial occlusive disease. *Thromb. Haemost.* **2011**, *106*, 344–352. [CrossRef] [PubMed]
95. Pawelczyk, M.; Chmielewski, H.; Kaczorowska, B.; Przybyła, M.; Baj, Z. The influence of statin therapy on platelet activity markers in hyperlipidemic patients after ischemic stroke. *Arch. Med. Sci.* **2015**, *1*, 115–121. [CrossRef] [PubMed]
96. Camargo, L.M.; França, C.N.; Izar, M.C.; Bianco, H.T.; Lins, L.S.; Barbosa, S.P.; Pinheiro, L.F.; Fonseca, F.A.H. Effects of simvastatin/ezetimibe on microparticles, endothelial progenitor cells and platelet aggregation in subjects with coronary heart disease under antiplatelet therapy. *Braz. J. Med. Biol. Res.* **2014**, *47*, 432–437. [CrossRef]
97. Lins, L.C.A.; França, C.N.; Fonseca, F.A.H.; Barbosa, S.P.M.; Matos, L.N.; Aguirre, A.C.; Bianco, H.T.; do Amaral, J.B.; Izar, M.C. Effects of Ezetimibe on Endothelial Progenitor Cells and Microparticles in High-Risk Patients. *Cell Biochem. Biophys.* **2014**, *70*, 687–696. [CrossRef]
98. Mach, F.; Baigent, C.; Catapano, A.L.; Koskinas, K.C.; Casula, M.; Badimon, L.; Chapman, M.J.; De Backer, G.G.; Delgado, V.; Ference, B.A.; et al. 2019 ESC/EAS Guidelines for the management of dyslipidaemias: Lipid modification to reduce cardiovascular risk. *Eur. Heart J.* **2019**. [CrossRef]

© 2019 by the authors. Licensee MDPI, Basel, Switzerland. This article is an open access article distributed under the terms and conditions of the Creative Commons Attribution (CC BY) license (http://creativecommons.org/licenses/by/4.0/).

Article

Impaired HDL Metabolism Links GlycA, A Novel Inflammatory Marker, with Incident Cardiovascular Events

Kayla A. Riggs [1], Parag H. Joshi [2], Amit Khera [2], Kavisha Singh [2], Oludamilola Akinmolayemi [1], Colby R. Ayers [2] and Anand Rohatgi [2,*]

[1] Department of Internal Medicine, The University of Texas Southwestern Medical Center, Dallas, TX 75390, USA; kayla.riggs@phhs.org (K.A.R.); Dami.Akinmolayemi@UTSouthwestern.edu (O.A.)
[2] Department of Internal Medicine, Division of Cardiology, The University of Texas Southwestern Medical Center, Dallas, TX 75390, USA; parag.joshi@utsouthwestern.edu (P.H.J.); amit.khera@utsouthwestern.edu (A.K.); kavisha.singh@phhs.org (K.S.); colby.ayers@utsouthwestern.edu (C.R.A.)
* Correspondence: anand.rohatgi@utsouthwestern.edu; Tel.: +1-214-645-7500

Received: 5 November 2019; Accepted: 27 November 2019; Published: 3 December 2019

Abstract: High-density lipoproteins (HDL) exert anti-atherosclerotic effects via reverse cholesterol transport, yet this salutary property is impaired in the setting of inflammation. GlycA, a novel integrated glycosylation marker of five acute phase reactants, is linked to cardiovascular (CV) events. We assessed the hypothesis that GlycA is associated with measures of impaired HDL function and that dysfunctional HDL may contribute to the association between GlycA and incident CV events. Baseline measurements of HDL cholesterol (HDL-C), HDL particle concentration (HDL-P), apolipoprotein A1 (Apo A1), cholesterol efflux capacity, GlycA and high-sensitivity C-reactive protein (hs-CRP) were obtained from the Dallas Heart Study, a multi-ethnic cohort of 2643 adults (median 43 years old; 56% women, 50% black) without cardiovascular disease (CVD). GlycA was derived from nuclear magnetic resonance imaging. Participants were followed for first nonfatal MI, nonfatal stroke, coronary revascularization, or CV death over a median of 12.4 years (n = 197). The correlation between GlycA and hs-CRP was 0.58 (p < 0.0001). In multivariate models with HDL-C, GlycA was directly associated with HDL-P and Apo A1 and inversely associated with cholesterol efflux (standardized beta estimates: 0.08, 0.29, -0.06, respectively; all p ≤ 0.0004) GlycA was directly associated with incident CV events (adjusted hazard ratio (HR) for Q4 vs. Q1: 3.33, 95% confidence interval (CI) 1.99, 5.57). Adjustment for cholesterol efflux mildly attenuated this association (HR for Q4 vs. Q1: 3.00, 95% CI 1.75 to 5.13). In a multi-ethnic cohort, worsening inflammation, as reflected by higher GlycA levels, is associated with higher HDL-P and lower cholesterol efflux. Impaired cholesterol efflux likely explains some of the association between GlycA and incident CV events. Further studies are warranted to investigate the impact of inflammation on HDL function and CV disease.

Keywords: HDL; lipids; inflammation; atherosclerotic cardiovascular disease (ASCVD); cardiovascular events; GlycA

1. Introduction

Although low high-density lipoprotein cholesterol (HDL-C) is considered an atherosclerotic cardiovascular disease (ASCVD) risk factor, contemporary epidemiological studies suggest that HDL particle concentration and HDL function better reflect HDL metabolism, and are better predictors of ASCVD [1–4]. In particular, cholesterol efflux capacity (CEC) as a measure of reverse cholesterol transport has been shown to inversely associate with prevalent and incident ASCVD in both low-risk and high-risk cohorts, independent of HDL-C and HDL particle concentration (HDL-P) [5]. Similarly, HDL particle concentration is also inversely associated with ASCVD, independent of HDL-C [2].

Chronic inflammation accelerates atherosclerosis, presumably in part by modifying HDL and its ability to promote reverse cholesterol transport [6–9]. GlycA is a circulating biomarker that reflects inflammation via glycosylated acute phase reactants and is directly associated with incident ASCVD [10–15]. GlycA predicts cardiovascular mortality, but the association was attenuated after adjustment for cardiovascular risk factors. Further, a recent meta-analysis also reported that GlycA is significantly associated with all-cause mortality [16].

GlycA is measured through nuclear magnetic resonance assay reflecting the mobile glycan residues from N-acetyl methyl group attachments added in the inflammatory setting. GlycA is modestly correlated with high-sensitivity C-reactive protein (hs-CRP) [12,17]. The most prominent contributions to the GlycA signal are alpha1-antitrypsin, haptoglobin, alpha1-antichymotrypsin, alpha1-acid glycoprotein, and transferrin [17]. These proteins are all associated with HDL in multiple studies, except transferrin [18–20]. In the setting of inflammation, protein glycan structures undergo modification, which can result in association with different receptors and changes in function [21,22]. For example, the protein composition of HDL differs in those with cardiovascular disease compared to those without cardiovascular disease [23]. hs-CRP is associated with impaired HDL cholesterol efflux [24]. Whether dysfunctional HDL is associated with this aggregated inflammatory biomarker, GlycA, is unknown. Furthermore, though inflammation and dysfunctional HDL have been linked in preclinical investigations, it remains unknown whether dysfunctional HDL explains part of the link between inflammation and ASCVD in humans.

We assessed the hypothesis that GlycA is associated with impaired HDL function measures and that the association between GlycA and incident cardiovascular (CV) events is partially explained by dysfunctional HDL. We tested this hypothesis in the multiethnic, population-based Dallas Heart Study (DHS) by investigating cross-sectional associations between GlycA with multiple HDL parameters and longitudinal associations with incident ASCVD events.

2. Experimental Section

2.1. Study Design

The Dallas Heart Study is a multiethnic, population-based cohort of Dallas County residents aged 30 to 65 years [25]. There was intentional oversampling of black persons in the DHS, to comprise 50% of the cohort. Persons with a history of cardiovascular disease (self-reported history of myocardial infarction, stroke, heart failure, arterial revascularization, or arrhythmia), chronic kidney disease stage V or self-reported dialysis were excluded from the study, as well as individuals without values of GlycA. The final study sample included 2643 individuals. This research was approved by the UT Southwestern Medical Center Institutional Review, and all participants provided informed written consent.

2.2. Assessment of HDL Parameters and GlycA

Participant blood collected in a fasting state at baseline through venipuncture was placed in ethylenediamine tetraacetic acid (EDTA) tubes and stored at four degrees Celsius for less than 4 h. Then, blood samples were centrifuged, and plasma was removed and stored at −70 degrees Celsius.

Traditional lipids were measured through previously described methods [25]. Cholesterol efflux capacity was measured through the efflux of fluorescence-labeled cholesterol (boron dipyrromethene difluoride reagent, BODIPY) from J774 macrophages to apolipoprotein B depleted plasma [5]. Apolipoprotein A1 (Apo A1), HDL particle concentration and size, and GlycA were measured by nuclear magnetic resonance (NMR) spectroscopy (LabCorp, Inc.; formerly LipoScience, Raleigh, SC, USA). GlycA is measured through detection of glycosylation of N-acetyl methyl groups of specific serum proteins (predominately alpha1-acid glycoprotein, haptoglobin, alpha1-antitrypsin, alpha1-antichymotrypsin and transferrin) by NMR to calculate the concentration of GlycA in micromol/L [11].

2.3. Clinical End Points

The primary end point of the study was incident atherosclerotic cardiovascular disease (ASCVD), including first nonfatal myocardial infarction, nonfatal stroke, coronary revascularization, or CV death over a median of 12.4 years (n = 197). A secondary end point, total cardiovascular disease (CVD), includes ASCVD and hospitalization for congestive heart failure, atrial fibrillation, or peripheral artery disease. Death from cardiovascular causes was defined by codes I00 to I99 in the International Classification of Diseases, 10th Revision, from the National Death Index. Nonfatal cardiovascular events data were obtained from the annual detailed surveying of subjects and hospitalization admissions were tracked quarterly through the Dallas-Fort Worth Hospital Council Data Initiative database. All end points were adjudicated by two cardiologists who were blinded to all exposure variables. The vital status of participants was reviewed through 31 December 2013 [26].

2.4. Statistical Analysis

GlycA was the main exposure variable and was assessed for relationships with demographics and risk factors as both a continuous variable (Spearman correlation coefficients) and as a categorical variable across increasing quartiles (Jonckheere–Terpstra Test). Multivariable linear regression models adjusted for age, sex, race/ethnicity, current smoking, hypertension, body mass index (BMI), waist circumference, diabetes, homeostatic model assessment of insulin resistance (HOMA-IR), non-HDL cholesterol, triglycerides, anti-hypertensive medications, and hs-CRP were used to assess the cross-sectional association between HDL parameters and GlycA levels.

The hazard associated with baseline GlycA and incident ASCVD (the primary end point) and total CVD (secondary end point) was estimated using Cox proportional hazards models. Multivariable models included age, sex, race/ethnicity, smoking status, systolic blood pressure, BMI, waist circumference, diabetes, HOMA-IR, non-HDL-C, triglycerides and statins. The variance inflation factor between BMI and waist circumference is four, which does not suggest collinearity. hs-CRP and HDL parameters (HDL-C, HDL-P, ApoA1, and cholesterol efflux) were serially added to these models.

Two-sided P values of 0.05 or less were considered statistically significant. The statistical analysis was performed with SAS software version 9.4 (SAS, Raleigh, NC, USA).

3. Results

3.1. Demographics

The mean age of study participants was 44, with 56% women and 50% blacks. GlycA was normally distributed within the DHS (Figure 1).

The median value of GlycA in DHS was 327 µmol/L (interquartile range 291 µmol/L to 369 µmol/L), with higher levels in women vs. men ($p < 0.0001$; Table 1) and significant differences by race/ethnicity ($p < 0.0001$, Table 1).

Table 1. GlycA levels by gender and race/ethnicity in the Dallas Heart Study.

Category	Mean +/− Standard Deviation
Overall (n = 2643)	332 +/− 62
Men (n = 1152)	318 +/− 60
Women (n = 1491)	344 +/− 60
White (n = 1305)	337 +/− 64
Black (n = 865)	327 +/− 60
Hispanic (n = 417)	332 +/− 59

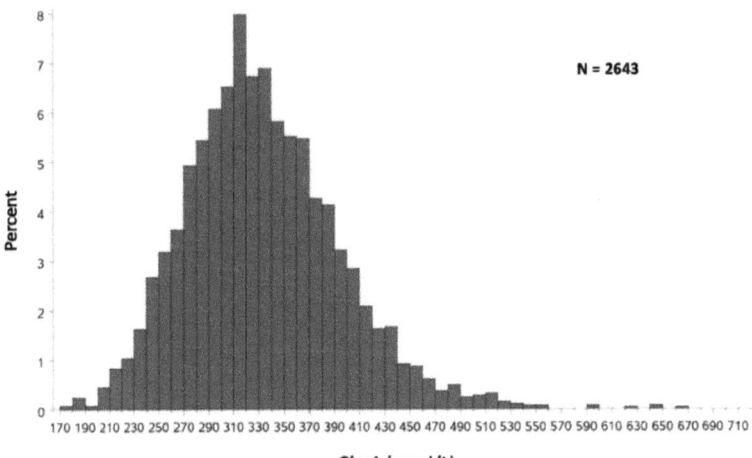

Figure 1. GlycA Distribution in DHS.

3.2. Univariate Correlations with GlycA

Increasing quartiles of GlycA were directly associated with age, systolic blood pressure, BMI, waist circumference, diabetes, non-HDL-C, triglycerides, and hs-CRP (Table 2). GlycA was not associated with HDL-C but was associated with increasing HDL-P. There was an inverse trend but no statistical association with cholesterol efflux ($p = 0.06$).

Table 2. Cardiovascular Risk Factors and high density lipoprotein (HDL) Parameters by Quartiles of GlycA (µmol/L).

	Q1 170–291 µmol/L n = 600 Mean (SD)	Q2 292–327 µmol/L n = 603 Mean (SD)	Q3 328–369 µmol/L n = 597 Mean (SD)	Q4 370–660 µmol/L n = 592 Mean (SD)	p-Value
Age (years)	42 (9.4)	44 (9.7)	44 (9.6)	45 (9.8)	<0.0001
Systolic Blood Pressure (mmHg)	120 (16.2)	122 (17.4)	125 (18.2)	128 (19.7)	<0.0001
BMI (kg/m^2)	27 (5.5)	29 (5.7)	30 (7.1)	32 (8.0)	<0.0001
Waist Circumference (cm)	93 (15.0)	98 (14.8)	100 (17.0)	104 (16.6)	<0.0001
HOMA-IR	3.2 (4.1)	3.6 (3.1)	4.6 (4.8)	5.1 (4.5)	<0.0001
Non-HDL (mg/dL)	127 (39.1)	130 (39.0)	132 (39.9)	136 (41.8)	0.001
Triglycerides (mg/dL)	84 (59, 123)	91 (69, 136)	104 (72, 153)	111 (78, 162)	<0.0001
hs-CRP (mg/L)	1.9 (2.0)	3.4 (3.8)	5.2 (4.8)	9.5 (6.7)	<0.0001
HDL-C (mg/dL)	49.9 (14.6)	49.6 (14.3)	50.3 (13.9)	50.2 (15.9)	0.7600
HDL-P (µmol/L)	32.5 (5.5)	33.0 (6.1)	33.4 (5.8)	34.3 (7.5)	<0.0001
Cholesterol Efflux (normalized to pool)	1.06 (0.33)	1.01 (0.30)	1.03 (0.33)	1.02 (0.30)	0.06
Apo A1 (µmol/L)	122.1 (27.0)	125.4 (28.2)	128.2 (28.0)	132.3 (32.4)	<0.0001

BMI = body mass index, HOMA-IR = homeostatic model assessment of insulin resistance, non-HDL = non-high-density lipoprotein, hs-CRP = high-sensitivity C-reactive protein, HDL-C = high-density lipoprotein cholesterol, HDL-P = high-density lipoprotein particle concentration, Apo A1 = apolipoprotein A1. Jonckheere–Terpstra test reported for two-sided p-value. All values reported as mean (standard deviation), except for triglycerides, which are reported as median (interquartile range).

Analyzed continuously in unadjusted univariate correlation analyses, GlycA was moderately correlated with hs-CRP in those (Spearman r = 0.58; $p < 0.0001$). These correlations were qualitatively higher in women vs. men (Spearman r = 0.59 for women, $p < 0.0001$; Spearman r = 0.50 for men, $p < 0.0001$) and highest in whites (Spearman r = 0.61; $p < 0.0001$).

Unadjusted associations between continuous GlycA and HDL parameters varied across parameters and by sex and ethnicity (Supplemental Table S1). GlycA correlated inversely with HDL-C (Spearman r = −0.11, $p < 0.001$) and directly with HDL-P (Spearman r = 0.14, $p < 0.001$) in women but not in men. GlycA did not correlate with efflux within any sex or ethnicity (Spearman r = −0.05–0.02, p = 0.07–0.74).

3.3. Understanding Variation in GlycA

GlycA was modeled as an outcome in adjusted linear regression models that included age, sex, race/ethnicity, current smoking, systolic blood pressure, BMI, waist circumference, diabetes, HOMA-IR, non-HDL-C, triglyceride, and antihypertensive medication. hs-CRP was added serially to this model to assess its additional contribution to variation in GlycA. Factors directly associated with GlycA included age, women, Hispanic race/ethnicity, current smoking, systolic blood pressure, BMI, waist circumference, triglycerides, and hs-CRP (all $p < 0.05$). The cardiovascular risk factor model explained 17% of the variation in GlycA, increasing to 36% of the variation when hs-CRP was included.

HDL parameters were then individually entered into separate multivariable models, and HDL-C was included in all models. GlycA was directly associated with HDL-P and Apo A1 and inversely associated with cholesterol efflux (Table 3). Analysis of HDL subfractions revealed significant associations with medium and large HDL particles but not small HDL particles.

Table 3. Association of GlycA with HDL Parameters.

Variable	Standardized Estimate	*p*-Value
HDL-P	0.08	<0.0001
Apo A1	0.29	<0.0001
Cholesterol Efflux	−0.06	0.0004
HDL Medium	0.10	<0.0001
HDL Large	−0.07	0.02

Each row is a separate linear regression model for GlycA adjusted for age, sex, race/ethnicity, current smoking, systolic blood pressure, body mass index, waist circumference, diabetes, homeostatic model assessment of insulin resistance, high-density lipoprotein cholesterol (HDL-C), non-HDL-C, antihypertensive medication, and high-sensitivity C-reactive protein. The association displayed is for the independent addition of a single HDL parameter to this model. Only statistically significant HDL parameters are included. HDL-P = high-density lipoprotein particle concentration; Apo A1 = apolipoprotein A1.

There was a significant interaction between HDL-P and gender with respect to GlycA (p for interaction < 0.001). HDL-P was associated with GlycA among women (standardized estimate 0.18, $p < 0.0001$) but not among men (p = ns) (Supplemental Table S2). Medium and small HDL-P were both associated with GlycA in men and women (all $p < 0.01$).

3.4. GlycA, HDL Parameters, and Incident Cardiovascular Events

Over a median of 12.4 years, there were 197 first ASCVD events. Increasing GlycA concentration was directly associated with ASCVD events in adjusted models (hazard ratio (HR) Q4 vs. Q1: 3.33, 95% confidence interval (CI) 1.99, 5.57). Further adjustment for cholesterol efflux mildly attenuated this association (HR for Q4 vs. Q1: 3.00, 95% CI 1.75 to 5.13, Figure 2). Similar attenuation was seen when analyzing GlycA and cholesterol efflux as continuous variables (Supplemental Table S3).

Increasing quartiles of GlycA remained directly associated with ASCVD in models adjusted for HDL-P (HR for GlycA, Q4 vs. Q1: 3.46, 95% CI 2.06 to 5.84) or HDL-C (HR for GlycA, Q4 vs. Q1: 3.31, 95% CI 1.98 to 5.55). In contrast to the modest attenuation with addition of efflux, there was no attenuation with adjustment for HDL-P (HR for HDL-P, Q4 vs. Q1: 0.61, 95% CI 0.41 to 0.90).

Over a median of 12.4 years, there were 239 total CVD events. In models adjusted for risk factors, GlycA remained directly associated with total CV events (HR for Q4 vs. Q1: 2.70, 95% CI 1.76, 4.14). Again, further adjustment for cholesterol efflux mildly attenuated this association (HR for Q4 vs. Q1: 2.37, 95% CI 1.52, 3.70); however, cholesterol efflux remained inversely associated with total CVD when adjusted for risk factors and GlycA (HR for Q4 vs. Q1: 0.69, 95% CI 0.48, 0.98).

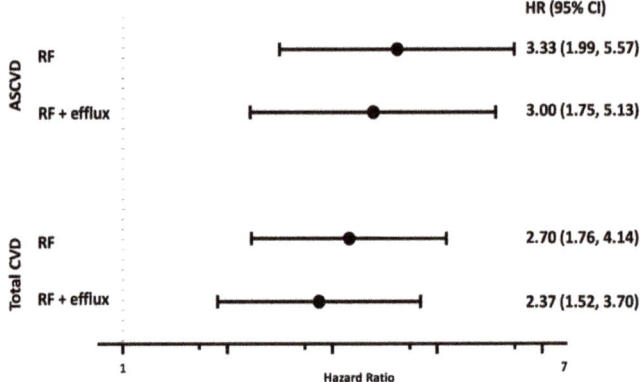

Figure 2. Hazard Ratios of Quartiles of GlycA (Q4 vs. Q1) derived from Cox Proportional Hazards Models adjusted for age, sex, race/ethnicity, diabetes, current smoking, systolic blood pressure, antihypertensive medications, non-high-density lipoprotein, body mass index, statin. Cholesterol efflux was serially added to the models.

4. Discussion

In a large, low-risk, multi-ethnic cohort, we found that GlycA as an integrated measure of inflammation associates with several HDL parameters, and impaired cholesterol efflux explains some of the relationship of GlycA with CV events. We noted gender-specific associations between GlycA and metabolic and lipid parameters and a modest correlation with hs-CRP.

Unlike most circulating markers, which reflect single enzymes, proteins, lipids, etc., GlycA is a unique cardiometabolic marker as it directly reflects concentrations of five glycosylated acute phase reactants—alpha1-antitrypsin, haptoglobin, alpha1-antichymotrypsin, alpha1-acid glycoprotein, and transferrin [17]. Prior reports suggest that this NMR-derived marker of inflammation robustly correlates with a variety of CV phenotypes (cardiovascular disease, coronary artery disease, cardiovascular mortality, and total death), including the risk of incident CV events [10–15]. However, the relationships with cardiometabolic processes underlying these risk associations have not been fully explored.

Inflammation has been linked to disordered lipid metabolism, especially dysfunctional HDL. In particular, both acute and chronic inflammation can alter the HDL proteome and impair cholesterol transport, anti-oxidant, and other atheroprotective functions [6–9]. However, most of these observations have occurred at the preclinical level or on a limited basis among humans. Moreover, although CRP is an established generalized marker of inflammation, GlycA is distinct from CRP in that it represents multiple glycosylated proteins in response to cardiometabolic stress and perhaps is a unique reflection of the intersection between inflammation and altered protein structure and function. This is supported by the moderate but not high correlation between hs-CRP and GlycA seen in our study and previous studies. Thus, we were keenly interested in the relationship between GlycA and parameters of HDL particle composition and function.

Beyond HDL's cholesterol load (HDL-C), total HDL particle number is a robust marker of ASCVD and outperforms HDL-C in predicting ASCVD risk [1–4]. In addition, Apo AI is the most abundant protein on HDL and associated with the majority of HDL's atheroprotective functions. Unlike Apolipoprotein B-containing particles, which have a 1:1 ratio of Apolipoprotein B to particle number, the number of Apo AIs varies per HDL particle (2–4), resulting in only a moderate correlation between HDL particle number and Apo AI concentration [27]. Therefore, both measures of HDL particles offer complementary information on cardiometabolic risk. Lastly, the ability of HDL to promote reverse cholesterol transport is considered its most important atheroprotective function, and cholesterol efflux capacity has been the most studied aspect of this process in human epidemiologic studies, demonstrating links to prevalent and incident ASCVD [5].

The Dallas Heart Study is a population-based low risk cohort balanced by gender and enriched for African Americans, offering the opportunity to explore both gender- and race/ethnicity-specific correlates of GlycA. We discovered that several cardiometabolic risk factors including adiposity, blood pressure, and dyslipidemia are associated with higher GlycA levels. Intriguingly, several of these associations, in particular measures of adiposity, were enhanced and most prevalent among women, and either weak or non-existent among men. With respect to markers of HDL metabolism, GlycA was most strongly and directly associated with particle-related parameters (HDL-P and Apo AI) rather than cholesterol content. These relationships were most evident in non-African Americans but also varied by gender. Lastly, GlycA appeared to be inversely related to large HDL, mostly among women, mirroring the associations with adiposity.

Thus, GlycA appears to be associated with altered lipid metabolism, especially among women, and is most strongly linked to increased HDL particle number and Apo AI levels, with a trend toward smaller HDL particles. Perhaps the glycosylation of acute phase reactants reflects a milieu in which HDL particles are also modified with increased shedding of apolipoproteins and lipid-poor particles. Accounting for these relationships with HDL particle number and Apo AI levels, there emerged a modest inverse association between GlycA and cholesterol efflux, suggesting that on a per particle basis, increased glycosylation and inflammation is linked to impaired HDL function at a population level.

Having demonstrated a modest but significant link between inflammation, as reflected by GlycA and HDL metabolism, we then determined to what extent this link may explain the association between GlycA and CV events. We found a modest effect of cholesterol efflux on GlycA's relationship with incident CV events. These findings at a population level suggest that dysfunctional HDL likely contributes to the pathways linking inflammation and CVD.

In summary, despite the independent associations of GlycA, HDL particle number, and cholesterol efflux with CV risk, markers of HDL metabolism were only modestly associated with GlycA, and cholesterol efflux was the only marker to mildly explain the link between GlycA and CV risk. Future studies may help to clarify the role of inflammation and dysfunctional HDL in CV risk by assessing higher risk cohorts with chronic CV disease and/or those undergoing acute illnesses such as acute coronary syndromes.

Supplementary Materials: The following are available online at http://www.mdpi.com/2077-0383/8/12/2137/s1, Table S1: GlycA Correlations with Variables in DHS, Table S2: Associations between HDL parameters and GlycA stratified by gender, Table S3: Cox Proportional Hazards Models of GlycA per 1 SD Association with Cardiovascular Events.

Author Contributions: Data curation, O.A.; Formal analysis, C.R.A.; Supervision, A.R.; Writing—original draft, K.A.R.; Writing—review and editing, P.H.J., A.K., K.S., O.A. and A.R.

Funding: This research was funded by National Center for Advancing Translational Sciences of the National Institutes of Health under award Number UL1TR001105. The content is solely the responsibility of the authors and does not necessarily represent the official views of the NIH. Rohatgi is supported by NIH/NHLBI R01HL136724, NIH/NHLBI K24HL146838, and AHA 17UNPG33840006.

Acknowledgments: Research reported in this publication was supported by the National Center for Advancing Translational Sciences of the National Institutes of Health under award Number UL1TR001105. The content is solely the responsibility of the authors and does not necessarily represent the official views of the NIH. Rohatgi is supported by NIH/NHLBI R01HL136724, NIH/NHLBI K24HL146838, and AHA 17UNPG33840006.

Conflicts of Interest: K.A.R.—none; P.H.J.—Grant support (sig): AHA, NovoNordisk, AstraZeneca, GSK, Personal Fees: Bayer, Stock (sig): Global Genomics Group; A.K.—none; K.S.—none; O.A.—none; C.R.A.—none; A.R.—Merck, research grant, significant; Merck, consultant, modest; CSL Limited, consultant, modest; HDL Diagnostics, Advisory Board, modest.

References

1. Tehrani, D.M.; Zhao, Y.; Blaha, M.J.; Mora, S.; Mackey, R.H.; Michos, E.D.; Wong, N.D. Discordance of Low-Density Lipoprotein and High-Density Lipoprotein Cholesterol Particle Versus Cholesterol Concentration for the Prediction of Cardiovascular Disease in Patients with Metabolic Syndrome and

Diabetes Mellitus (from the Multi-Ethnic Study of Atherosclerosis [MESA]). *Am. J. Cardiol.* **2016**, *117*, 1921–1927. [CrossRef]
2. Mora, S.; Glynn, R.J.; Ridker, P.M. High-density lipoprotein cholesterol, size, particle number, and residual vascular risk after potent statin therapy. *Circulation* **2013**, *128*, 1189–1197. [CrossRef]
3. Mackey, R.H.; Greenland, P.; Goff, D.C., Jr.; Lloyd-Jones, D.; Sibley, C.T.; Mora, S. High-density lipoprotein cholesterol and particle concentrations, carotid atherosclerosis, and coronary events: MESA (multi-ethnic study of atherosclerosis). *J. Am. Coll. Cardiol.* **2012**, *60*, 508–516. [CrossRef]
4. Chandra, A.; Neeland, I.J.; Das, S.R.; Khera, A.; Turer, A.T.; Ayers, C.R.; Rohatgi, A. Relation of black race between high density lipoprotein cholesterol content, high density lipoprotein particles and coronary events (from the Dallas Heart Study). *Am. J. Cardiol.* **2015**, *115*, 890–894. [CrossRef]
5. Rohatgi, A.; Khera, A.; Berry, J.D.; Givens, E.G.; Ayers, C.R.; Wedin, K.E.; Shaul, P.W. HDL cholesterol efflux capacity and incident cardiovascular events. *N. Engl. J. Med.* **2014**, *371*, 2383–2393. [CrossRef]
6. Vaisar, T.; Tang, C.; Babenko, I.; Hutchins, P.; Wimberger, J.; Suffredini, A.F.; Heinecke, J.W. Inflammatory remodeling of the HDL proteome impairs cholesterol efflux capacity. *J. Lipid Res.* **2015**, *56*, 1519–1530. [CrossRef]
7. McGillicuddy, F.C.; de la Llera Moya, M.; Hinkle, C.C.; Joshi, M.R.; Chiquoine, E.H.; Billheimer, J.T.; Reilly, M.P. Inflammation impairs reverse cholesterol transport in vivo. *Circulation* **2009**, *119*, 1135–1145. [CrossRef]
8. He, L.; Qin, S.; Dang, L.; Song, G.; Yao, S.; Yang, N.; Li, Y. Psoriasis decreases the anti-oxidation and anti-inflammation properties of high-density lipoprotein. *Biochim. Biophys. Acta* **2014**, *1841*, 1709–1715. [CrossRef]
9. De la Llera Moya, M.; McGillicuddy, F.C.; Hinkle, C.C.; Byrne, M.; Joshi, M.R.; Nguyen, V.; Mehta, N.N. Inflammation modulates human HDL composition and function in vivo. *Atherosclerosis* **2012**, *222*, 390–394. [CrossRef]
10. Akinkuolie, A.O.; Buring, J.E.; Ridker, P.M.; Mora, S. A novel protein glycan biomarker and future cardiovascular disease events. *J. Am. Heart Assoc.* **2014**, *3*, e001221. [CrossRef]
11. Akinkuolie, A.O.; Glynn, R.J.; Padmanabhan, L.; Ridker, P.M.; Mora, S. Circulating N-Linked Glycoprotein Side-Chain Biomarker, Rosuvastatin Therapy, and Incident Cardiovascular Disease: An Analysis from the JUPITER Trial. *J. Am. Heart Assoc.* **2016**, *5*. [CrossRef] [PubMed]
12. Gruppen, E.G.; Riphagen, I.J.; Connelly, M.A.; Otvos, J.D.; Bakker, S.J.; Dullaart, R.P. GlycA, a Pro-Inflammatory Glycoprotein Biomarker, and Incident Cardiovascular Disease: Relationship with C-Reactive Protein and Renal Function. *PLoS ONE* **2015**, *10*, e0139057. [CrossRef] [PubMed]
13. Duprez, D.A.; Otvos, J.; Sanchez, O.A.; Mackey, R.H.; Tracy, R.; Jacobs, D.R., Jr. Comparison of the Predictive Value of GlycA and Other Biomarkers of Inflammation for Total Death, Incident Cardiovascular Events, Noncardiovascular and Noncancer Inflammatory-Related Events, and Total Cancer Events. *Clin. Chem.* **2016**, *62*, 1020–1031. [CrossRef] [PubMed]
14. McGarrah, R.W.; Kelly, J.P.; Craig, D.M.; Haynes, C.; Jessee, R.C.; Huffman, K.M.; Shah, S.H. A Novel Protein Glycan-Derived Inflammation Biomarker Independently Predicts Cardiovascular Disease and Modifies the Association of HDL Subclasses with Mortality. *Clin. Chem.* **2017**, *63*, 288–296. [CrossRef] [PubMed]
15. Otvos, J.D.; Guyton, J.R.; Connelly, M.A.; Akapame, S.; Bittner, V.; Kopecky, S.L.; Boden, W.E. Relations of GlycA and lipoprotein particle subspecies with cardiovascular events and mortality: A post hoc analysis of the AIM-HIGH trial. *J. Clin. Lipidol.* **2018**. [CrossRef]
16. Gruppen, E.G.; Kunutsor, S.K.; Kieneker, L.M.; van der Vegt, B.; Connelly, M.A.; de Bock, G.H.; Dullaart, R.P. GlycA, a novel pro-inflammatory glycoprotein biomarker is associated with mortality: Results from the PREVEND study and meta-analysis. *J. Intern. Med.* **2019**, *286*, 596–609. [CrossRef]
17. Otvos, J.D.; Shalaurova, I.; Wolak-Dinsmore, J.; Connelly, M.A.; Mackey, R.H.; Stein, J.H.; Tracy, R.P. GlycA: A Composite Nuclear Magnetic Resonance Biomarker of Systemic Inflammation. *Clin. Chem.* **2015**, *61*, 714–723. [CrossRef]
18. The Davidson/Shah Lab. HDL Proteome Watch. Available online: http://homepages.uc.edu/~{}davidswm/HDLproteome.html (accessed on 19 November 2019).
19. Holzer, M.; Wolf, P.; Curcic, S.; Birner-Gruenberger, R.; Weger, W.; Inzinger, M.; Marsche, G. Psoriasis alters HDL composition and cholesterol efflux capacity. *J. Lipid Res.* **2012**, *53*, 1618–1624. [CrossRef]

20. Gordon, S.M.; Deng, J.; Tomann, A.B.; Shah, A.S.; Lu, L.J.; Davidson, W.S. Multi-dimensional co-separation analysis reveals protein-protein interactions defining plasma lipoprotein subspecies. *Mol. Cell Proteom.* **2013**, *12*, 3123–3134. [CrossRef]
21. Gornik, O.; Lauc, G. Glycosylation of serum proteins in inflammatory diseases. *Dis. Markers* **2008**, *25*, 267–278. [CrossRef]
22. Ceciliani, F.; Pocacqua, V. The acute phase protein alpha1-acid glycoprotein: A model for altered glycosylation during diseases. *Curr. Protein Pept. Sci.* **2007**, *8*, 91–108. [CrossRef] [PubMed]
23. Vaisar, T.; Pennathur, S.; Green, P.S.; Gharib, S.A.; Hoofnagle, A.N.; Cheung, M.C.; Chea, H. Shotgun proteomics implicates protease inhibition and complement activation in the antiinflammatory properties of HDL. *J. Clin. Investig.* **2007**, *117*, 746–756. [CrossRef] [PubMed]
24. Annema, W.; Dikkers, A.; De Boer, J.F.; Van Greevenbroek, M.M.; Van Der Kallen, C.J.; Schalkwijk, C.G.; Tietge, U.J. Impaired HDL cholesterol efflux in metabolic syndrome is unrelated to glucose tolerance status: The CODAM study. *Sci. Rep.* **2016**, *6*, 27367. [CrossRef] [PubMed]
25. Victor, R.G.; Haley, R.W.; Willett, D.L.; Peshock, R.M.; Vaeth, P.C.; Leonard, D.; Staab, J.M. The Dallas Heart Study: A population-based probability sample for the multidisciplinary study of ethnic differences in cardiovascular health. *Am. J. Cardiol.* **2004**, *93*, 1473–1480. [CrossRef]
26. Maroules, C.D.; Rosero, E.; Ayers, C.; Peshock, R.M.; Khera, A. Abdominal aortic atherosclerosis at MR imaging is associated with cardiovascular events: The Dallas heart study. *Radiology* **2013**, *269*, 84–91. [CrossRef]
27. Rosenson, R.S.; Brewer, H.B.; Chapman, M.J.; Fazio, S.; Hussain, M.M.; Kontush, A.; Schaefer, E.J. HDL measures, particle heterogeneity, proposed nomenclature, and relation to atherosclerotic cardiovascular events. *Clin. Chem.* **2011**, *57*, 392–410. [CrossRef]

© 2019 by the authors. Licensee MDPI, Basel, Switzerland. This article is an open access article distributed under the terms and conditions of the Creative Commons Attribution (CC BY) license (http://creativecommons.org/licenses/by/4.0/).

Article

Higher Sodium Intake Assessed by 24 Hour Urinary Sodium Excretion Is Associated with Non-Alcoholic Fatty Liver Disease: The PREVEND Cohort Study

Eline H. van den Berg [1,2,*], Eke G. Gruppen [1,3], Hans Blokzijl [2], Stephan J.L. Bakker [3] and Robin P.F. Dullaart [1]

1. Department of Endocrinology, University of Groningen, University Medical Center Groningen, 9700RB Groningen, The Netherlands; e.g.gruppen@umcg.nl (E.G.G.); r.p.f.dullaart@umcg.nl (R.P.F.D.)
2. Department of Gastroenterology and Hepatology, University of Groningen, University Medical Center Groningen, 9700RB Groningen, The Netherlands; h.blokzijl@umcg.nl
3. Department of Nephrology, University of Groningen, University Medical Center Groningen, 9700RB Groningen, The Netherlands; s.j.l.bakker@umcg.nl
* Correspondence: e.h.van.den.berg@umcg.nl; Tel.: +31-50-3616161

Received: 29 October 2019; Accepted: 4 December 2019; Published: 6 December 2019

Abstract: A higher sodium intake is conceivably associated with insulin resistant conditions like obesity, but associations of non-alcoholic fatty liver disease (NAFLD) with a higher sodium intake determined by 24 hours (24 h) urine collections are still unclear. Dietary sodium intake was measured by sodium excretion in two complete consecutive 24 h urine collections in 6132 participants of the Prevention of Renal and Vascular End-Stage Disease (PREVEND) cohort. Fatty Liver Index (FLI) ≥60 and Hepatic Steatosis Index (HSI) >36 were used as proxies of suspected NAFLD. 1936 (31.6%) participants had an FLI ≥60, coinciding with the increased prevalence of type 2 diabetes (T2D), metabolic syndrome, hypertension and history of cardiovascular disease. Sodium intake was higher in participants with an FLI ≥60 (163.63 ± 61.81 mmol/24 h vs. 136.76 ± 50.90 mmol/24 h, $p < 0.001$), with increasing incidence in ascending quartile categories of sodium intake ($p < 0.001$). Multivariably, an FLI ≥60 was positively associated with a higher sodium intake when taking account for T2D, a positive cardiovascular history, hypertension, alcohol intake, smoking and medication use (odds ratio (OR) 1.54, 95% confidence interval (CI) 1.44–1.64, $p < 0.001$). Additional adjustment for the Homeostasis Model Assessment of Insulin Resistance (HOMA-IR) diminished this association (OR 1.30, 95% CI 1.21–1.41, $p < 0.001$). HSI >36 showed similar results. Associations remained essentially unaltered after adjustment for body surface area or waist/hip ratio. In conclusion, suspected NAFLD is a feature of higher sodium intake. Insulin resistance-related processes may contribute to the association of NAFLD with sodium intake.

Keywords: non-alcoholic fatty liver; sodium intake; insulin resistance; fatty liver index; hepatic steatosis index; HOMA-IR

1. Introduction

Non-alcoholic fatty liver disease (NAFLD) is characterized by hepatic steatosis in the absence of excessive alcohol use, and is emerging as the most common cause of chronic liver disease [1,2]. The spectrum of NAFLD ranges from simple steatosis to non-alcoholic steatohepatitis (NASH), fibrosis and eventually cirrhosis [1,3]. NAFLD coincides with obesity and insulin resistance, and is seen as the liver manifestation of the metabolic syndrome (MetS) [4]. NAFLD may in itself also increase the risk for the development of MetS and type 2 diabetes mellitus (T2D) [5,6]. NAFLD predisposes to plasma lipoprotein abnormalities, including elevations in apolipoprotein B-containing lipoproteins, as evidenced by elevations in very low-density lipoproteins (VLDL) and consequently in higher triglycerides, increased levels of low density lipoprotein (LDL) cholesterol, as well as in decreased levels of high density lipoprotein (HDL) cholesterol [2,7,8], which predisposes to atherosclerotic cardiovascular disease (CVD) [9]. The pathophysiological mechanisms underlying the development of NAFLD are not fully clarified, but multifactorial contributors including environmental factors (diet), central obesity, insulin resistance, alterations in gut microbiota and genetic factors are likely to play an important role [10].

In recent years, individual sodium intake is presumably increased, with the majority of extra sodium coming from diets filled-up with highly processed foods, particularly convenience foods [11]. High sodium intake is related to various metabolic disorders, such as obesity, insulin resistance, T2D, MetS, hypertension and CVD [11–14]. In turn, high sodium intake may also deleteriously influence metabolic diseases and is associated with expanding body fat and a diminished fat free mass [13,15]. Given increasingly documented associations of sodium intake with obesity [13,16,17], it is plausible to postulate that high sodium intake coincides with NAFLD and could even play a role in its pathogenesis. So far such a relationship has only been explored in two studies [14,18]. In these Asian studies, sodium intake was either estimated by dietary recall questionnaires [18] or by estimation from spot urine specimens [14]. Both studies found a positive relationship of NAFLD with estimated sodium intake [14,18]. However, the preferred method for assessing dietary salt intake is by measurement of sodium excretion in multiple 24 hour (24 h) urine collections, with other estimates being considered to be less accurate [19]. Thus, large-scale studies on the association of NAFLD with sodium intake measured by using multiple 24 h urine collections are still lacking.

Therefore, we initiated the present study to interrogate the impact of high sodium intake, assessed by using multiple 24 h urine collections, on NAFLD. We carried out a cross-sectional analysis among 6132 participants of in the Prevention of REnal and Vascular ENd-stage Disease (PREVEND) cohort study, comprising a large and well-characterized population from the North of The Netherlands.

2. Materials and Methods

2.1. Study Design and Population

This study was performed among participants of the Prevention of REnal and Vascular ENd-stage Disease (PREVEND) cohort study, a large prospective general population-based study that started in 1997 [20,21]. The PREVEND study was approved by the Medical Ethics Committee of the University Medical Center Groningen and is performed in accordance with Declaration of Helsinki guidelines [20,21]. Written informed consent was obtained from all participants. PREVEND was initiated to investigate cardiovascular and renal disease with a focus on albuminuria. All inhabitants (28 to 75 years old) of Groningen, the Netherlands were sent a questionnaire on demographics and cardiovascular morbidity, and were asked to supply urine specimens. Pregnant women, type 1 diabetic subjects and T2D subjects using insulin were excluded from participation. Participants with an urinary albumin concentration ≥10 mg/L were invited to our clinic together with randomly selected subjects with an urinary albumin concentration <10 mg/L. The initial study population of the PREVEND study comprised 8592 subjects who completed the total study screening program.

For the present study, we used data of participants who completed the second screening round in the PREVEND study (2001–2003; n = 6893), in which 24 h urine collections are available [22,23]. We excluded all subjects with missing values on urinary sodium excretion and subjects in which clinical and biochemical variables required to calculate the Fatty Liver Index (FLI), a proxy of NAFLD, were not available, leaving 6132 participants (Figure 1).

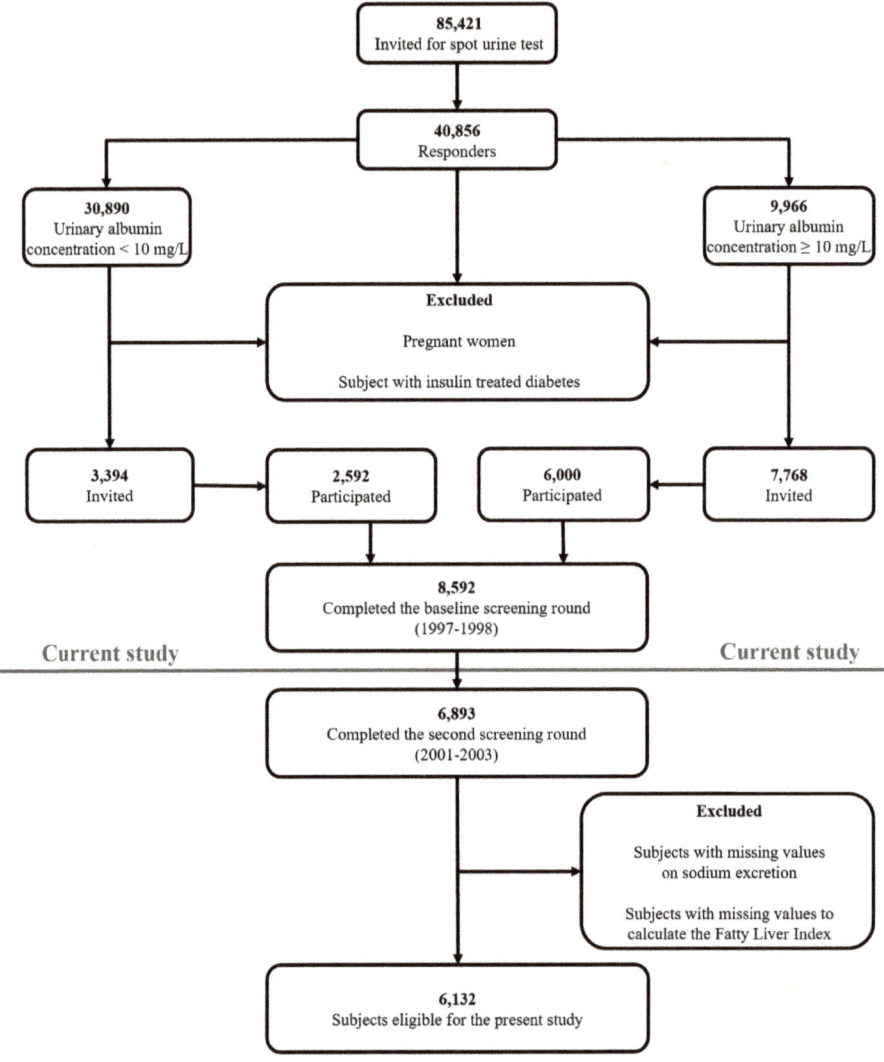

Figure 1. Flowchart of the study population.

2.2. Measurements and Definitions

Procedures at examination, measurements and definitions in the PREVEND study are reported in detail previously [20–24]. Body mass index (BMI) was calculated as weight (kg) divided by height squared (meter). Body surface area (BSA) was calculated as the square root of height (cm) multiplied by weight (kg) divided by 3600. Waist circumference was measured as the smallest girth between rib cage and iliac crest [20]. The waist/hip ratio was determined as the waist circumference divided by the largest girth between waist and thigh. Blood pressure was assessed in a standardized fashion on the right arm in supine position, every minute for 10 min, with an automatic Dinamap XL Model 9300 series device. The mean of the last 2 recordings was used [23]. Lipid lowering drugs and other medications were not stopped prior to the clinical evaluation and blood sample collection. T2D was defined as a fasting glucose $\geq$7.0 mmol/L, a random glucose $\geq$11.1 mmol/L, self-report of a physician diagnosis or the use of glucose lowering drugs. Daily alcohol consumption was recorded with one alcoholic drink being assumed to contain 10 grams of alcohol. Smoking was categorized into current and never/former smokers. A positive cardiovascular history included: hospitalization for myocardial ischemia, obstructive coronary artery disease or revascularization procedures. Information on medication use was combined with information from a pharmacy-dispensing registry, which has complete information on drug usage of >95% of subjects in the PREVEND study [20,21].

Participants collected two consecutive 24 h urine specimens after thorough oral and written instruction. They were asked to avoid heavy exercise during the urine collection and instructed to postpone the urine collection in case of a urinary tract infection, menstruation or fever [22,23]. Urinary sodium concentration (in mmol/L) was multiplied by the urine volume in liters per 24 h to obtain a value in mmol/24 h. Results of two consecutive 24 h urine collections were averaged for analyses. Estimated glomerular filtration rate (eGFR) was calculated applying the combined creatinine cystatin C-based Chronic Kidney Disease Epidemiology Collaboration equation [25].

For the diagnosis of suspected NAFLD, we used the algorithm of the Fatty Liver Index (FLI) [26]. The FLI was calculated according to the following formula [26]:

$$[e^{(0.953 \times \log_e (\text{triglycerides}) + 0.139 \times \text{BMI} + 0.718 \times \log_e (\text{GGT}) + 0.053 \times \text{waist circumference} - 15.745)}/[1 + e^{(0.953 \times \log_e (\text{triglycerides}) + 0.139 \times \text{BMI} + 0.718 \times \log_e (\text{GGT}) + 0.053 \times \text{waist circumference} - 15.745)}] \times 100 \quad (1)$$

where GGT is gamma-glutamyltransferase.

The optimal cut-off value for the FLI is documented to be 60 with an accuracy of 84%, a sensitivity of 61% and a specificity of 86% for detecting suspected NAFLD as determined by ultrasonography [26]. Therefore, FLI ≥60 was used as proxy of NAFLD. The FLI is currently considered as one of the best-validated steatosis scores for larger scale screening studies [27]. Alternatively, we used the Hepatic Steatosis Index (HSI) [28]. The HSI is defined as follows:

$$\text{HSI} = 8 \times \text{ALT/AST ratio} + \text{BMI} (+2, \text{if diabetes}; +2, \text{if female}) \quad (2)$$

where ALT is alanine aminotransferase and AST is aspartate aminotransferase.

The cut-off value of the HSI for detecting suspected NAFLD is 36 [28]. In these equations, BMI is expressed in kg/m^2, triglycerides are expressed in mmol/L, and GGT, ALT and AST are expressed in U/L.

The MetS was defined according to the revised National Cholesterol Education Program Adult Treatment Panel (NCEP-ATP) III criteria [29]. Participants were categorized with MetS when at least three out of five of the following criteria were present: waist circumference >102 cm for men and >88 cm for women; plasma triglycerides ≥1.7 mmol/L; HDL cholesterol <1.0 mmol/L for men and <1.3 mmol/L for women; hypertension (blood pressure ≥130/85 mm Hg or the use of antihypertensive medication); hyperglycemia (fasting glucose ≥5.6 mmol/L or the use of glucose lowering drugs). Homeostasis Model Assessment of Insulin Resistance (HOMA-IR) was calculated by:

$$\text{Fasting plasma insulin (mU/L)} \times \text{fasting plasma glucose (mmol/L)}/22.5 \quad (3)$$

2.3. Laboratory Methods

Laboratory methods are reported as described in detail previously [20–24]. Venous blood samples were drawn after an overnight fast while participants had rested for 15 min. Heparinized plasma and serum samples were obtained by centrifugation at $1400 \times g$ for 15 min at 4 °C. Plasma and serum samples were stored at −80 °C until analysis. Glucose was measured directly after blood collection. Plasma total cholesterol, triglycerides and HDL cholesterol were measured as previously described [20,21,24]. Non-HDL cholesterol was calculated as the difference between total cholesterol and HDL cholesterol. LDL cholesterol was calculated by the Friedewald formula if triglycerides were <4.5 mmol/L [30].

Serum ALT and AST were measured using the standardized kinetic method with pyridoxal phosphate activation (Roche Modular P; Roche Diagnostics, Mannheim, Germany). Serum GGT was assayed by an enzymatic colorimetric method (Roche Modular P, Roche Diagnostics, Mannheim, Germany). Standardization of ALT, AST and GGT was performed according to International Federation of Clinical Chemistry guidelines [31–33]. High sensitivity C-reactive protein (hsCRP) was assayed by nephelometry. Serum creatinine was measured by an enzymatic method on a Roche Modular P analyzer (Roche Diagnostics, Mannheim, Germany). Serum cystatin C was measured by Gentian Cystatin C Immunoassay (Gentian AS, Moss, Norway) on a Modular analyzer (Roche Diagnostics, Indianapolis, IN, USA).

Urine samples collected at home were stored cold (4 °C) for a maximum of 4 days. After handing in the urine collections, specimens were stored at −20 °C. Urinary sodium was measured by indirect potentiometry with a MEGA clinical chemistry analyzer (Merck & Co., Inc., Kenilworth, NJ, USA). Urinary albumin excretion (UAE) was measured by nephelometry (Dade Behring Diagnostic, Marburg, Germany).

2.4. Statistical Analysis

IBM SPSS software (IBM Corp, version 23.0, Armonk, NY, USA) was used for data analysis. Results are expressed as mean ± standard deviation (SD), median with interquartile range (IQR) or as numbers (percentages). Normality of distribution was assessed and checked for skewness. HOMA-IR was $\log_e$-transformed for analysis to achieve an approximately normal distribution. Between group differences in variables were determined by unpaired T-tests or ANOVA (Analysis of Variance) test for normally distributed variables, Mann–Whitney U or Kruskal–Wallis test for non-normally distributed variables or by Chi-square tests for categorical variables where appropriate. Multivariable binary regression analyses were carried out to disclose the independent associations of urinary sodium excretion with an elevated FLI and HSI when taking account of clinical covariates and laboratory parameters. In multivariable analyses, urinary sodium excretion was expressed per 1 SD increment. Interactions were tested between 24 h sodium excretion and sex. Two-sided p-values < 0.05 were considered significant.

3. Results

3.1. Patient Characteristics

The study population consisted of 6132 subjects of whom 1936 (31.6%) were categorized with a FLI ≥60, as proxy of NAFLD. Table 1 shows the clinical characteristics and laboratory data of the participants according to FLI categorization. Subjects with a FLI ≥60 were older and more likely to be men (men 66.2% vs. women 33.8%), to be classified with MetS, T2D as well as hypertension and had a positive cardiovascular history more frequently. Consequently, subjects with FLI ≥60 used antihypertensive medication and glucose and lipid lowering drugs more frequently. BMI, BSA, waist circumference, waist/hip ratio, glucose, insulin, HOMA-IR, liver function tests, total cholesterol, non-HDL cholesterol, LDL cholesterol, triglycerides and creatinine were significantly higher in subjects with FLI ≥60, but eGFR and HDL cholesterol were lower. Sodium intake, determined by averaged 24 h urine sodium excretion was higher in participants with an FLI ≥60 (Table 1, 163.63 ± 61.81 mmol/24 h vs. 136.76 ± 50.90 mmol/24 h, $p < 0.001$).

Baseline characteristics categorized according to quartile categories of sodium intake are shown in Table 2. Suspected NAFLD, with either an FLI ≥60 or HSI >36 were both significantly higher in each ascending quartile category of urinary sodium excretion ($p < 0.001$). As the urinary sodium excretion increased, participants were more likely to be male, to be classified with MetS, T2D, had a higher BMI, BSA, waist circumference, waist/hip ratio and higher levels of glucose, insulin and HOMA-IR (all $p < 0.001$). Of note, hypertension and history of CVD were not significantly different between urinary sodium excretion quartiles.

Table 1. Baseline characteristics including averaged 24 h urinary sodium excretion (two collections) in 4196 subjects with a Fatty Liver Index (FLI) < 60 and 1936 subjects with an FLI ≥ 60.

	FLI < 60, n = 4196 (68.4%)	FLI ≥ 60, n = 1936 (31.6%)	p Value
Age (years), mean ± SD	52.2 ± 12.0	57.2 ± 11.4	<0.001
Sex			<0.001
Men, n (%)	1750 (41.7)	1282 (66.2)	
Women, n (%)	2446 (58.3)	654 (33.8)	
MetS, n (%)	372 (8.9)	1193 (61.7)	<0.001
Type 2 diabetes mellitus, n (%)	130 (3.1)	256 (13.2)	<0.001
History of cardiovascular disease, n (%)	192 (4.6)	194 (10.0)	<0.001
Hypertension, n (%)	1014 (24.2)	1050 (54.3)	<0.001
Current smokers, n (%)	1184 (28.2)	513 (26.5)	0.162
Alcohol ≥ 10 g/day, n (%)	139 (3.3)	117 (6.1)	<0.001
Antihypertensive medication, n (%)	613 (14.6)	703 (36.3)	<0.001
Glucose lowering drugs, n (%)	73 (1.7)	152 (7.9)	<0.001
Lipid lowering drugs, n (%)	273 (6.5)	307 (15.9)	<0.001
Systolic blood pressure (mm Hg), mean ± SD	121 ± 17	135 ± 18	<0.001
Diastolic blood pressure (mm Hg), mean ± SD	71 ± 9	77 ± 9	<0.001
BMI (kg/m^2), mean ± SD	24.8 ± 2.9	30.9 ± 4.1	<0.001
BSA (m^2), mean ± SD	1.87 ± 0.17	2.12 ± 0.18	<0.001
Waist circumference (cm), mean ± SD	86.0 + 9.2	105.0 ± 9.2	<0.001
Waist/hip ratio, mean ± SD	0.87 ± 0.07	0.96 ± 0.07	<0.001
Glucose (mmol/L), mean ± SD	4.80 ± 0.86	5.48 ± 1.41	<0.001
Insulin (mU/L), median (IQR)	6.80 (5.1–9.3)	13.00 (9.5–19.0)	<0.001
HOMA-IR (mU mmol/L^2/22.5), median (IQR)	0.36 (0.04–0.70)	1.10 (0.73–1.53)	<0.001
hsCRP (mg/L), median (IQR)	1.02 (0.49–2.32)	2.44 (1.23–4.47)	<0.001
ALT (U/L), median (IQR)	15 (12–20)	23 (17–32)	<0.001
AST (U/L), median (IQR)	22 (19–25)	25 (21–29)	<0.001
ALP (U/L), mean ± SD	65 ± 19	76 ± 23	<0.001
GGT (U/L), median (IQR)	19 (14–27)	40 (28–61)	<0.001
Total cholesterol (mmol/L), mean ± SD	5.32 ± 1.01	5.67 ± 1.06	<0.001
Non-HDL cholesterol (mmol/L), mean ± SD	3.98 ± 0.99	4.57 ± 1.02	<0.001
LDL cholesterol (mmol/L), mean ± SD	3.51 ± 0.92	3.74 ± 0.94	<0.001
HDL cholesterol (mmol/L), mean ± SD	1.34 ± 0.31	1.10 ± 0.24	<0.001
Triglycerides (mmol/L), median (IQR)	0.95 (0.72–1.26)	1.67 (1.28–2.20)	<0.001
Serum creatinine (umol/L), mean ± SD	82.92 ± 22.42	89.56 ± 19.23	<0.001
eGFR (mL/min/1.73 m^2), mean ± SD	93.8 ± 16.4	86.5 ± 17.8	<0.001
Averaged 24 h urine excretion values			
Sodium excretion (mmol/24 h), mean ± SD	136.76 ± 50.90	163.63 ± 61.81	<0.001
UAE (mg/24 h), median (IQR)	7.44 (5.65–11.21)	10.04 (6.69–18.81)	<0.001

Data are given in number with percentages (%), mean ± standard deviation (SD) for normally distributed data or median with interquartile ranges (IQR) for non-normally distributed data. HOMA-IR was log$_e$ transformed for analyses. LDL cholesterol was calculated by the Friedewald formula if triglycerides were <4.5 mmol/L in 6028 subjects. Abbreviations: 24 h, twenty-four hours; ALP, alkaline phosphatase; ALT, aminotransferase; AST, aspartate aminotransferase; BMI, body mass index; BSA, body surface area; FLI, Fatty Liver Index; eGFR, estimated glomerular filtration rate; GGT, gamma-glutamyltransferase; HOMA-IR, Homeostasis Model Assessment of Insulin Resistance; HDL, high density lipoproteins; hsCRP, high sensitivity C-reactive protein; LDL, low density lipoproteins; MetS, metabolic syndrome; T2D, type 2 diabetes mellitus; UAE, urinary albumin excretion.

Table 2. Baseline characteristics of the study population according to quartile categories of averaged 24 h urinary sodium excretion (two collections) in 6132 subjects.

	24 h Urinary Sodium Excretion				p Value
	Quartile 1	Quartile 2	Quartile 3	Quartile 4	
N (%)	1533 (25.0)	1532 (25.0)	1532 (25.0)	1533 (25.0)	
24 h Sodium excretion (mmol/day), mean ± SD	82.14 ± 18.02	122.92 ± 9.79	155.83 ± 10.08	220.06 ± 42.67	<0.001
Suspected NAFLD					
FLI ≥ 60, n (%)	329 (21.5)	386 (25.2)	505 (33.0)	715 (46.6)	<0.001
HSI > 36, n (%)	324 (21.1)	355 (23.2)	445 (29.0)	608 (39.7)	<0.001
Age (years), mean ± SD	54.9 ± 12.6	54.1 ± 12.1	53.0 ± 12.0	52.1 ± 11.3	<0.001
Sex					<0.001
Men, n (%)	466 (30.4)	632 (41.3)	822 (53.7)	1.112 (72.5)	
Women, n (%)	1067 (69.6)	900 (58.7)	710 (46.3)	421 (27.5)	
MetS, n (%)	333 (21.7)	344 (22.5)	375 (24.5)	512 (33.4)	<0.001
Type 2 diabetes mellitus, n (%)	82 (5.3)	89 (5.8)	98 (6.4)	117 (7.6)	0.007
History of cardiovascular disease, n (%)	96 (6.3)	90 (5.9)	100 (6.5)	100 (6.5)	0.605
Hypertension, n (%)	506 (33.0)	491 (32.1)	517 (33.7)	550 (35.9)	0.057
Current smokers, n (%)	493 (32.2)	427 (27.9)	385 (25.1)	391 (25.5)	<0.001
Alcohol ≥ 10 g/day, n (%)	66 (4.3)	65 (4.3)	52 (3.4)	73 (4.8)	0.810
Antihypertensive medication, n (%)	342 (22.3)	317 (20.7)	327 (21.3)	330 (21.5)	0.718
Glucose lowering drugs, n (%)	49 (3.2)	49 (3.2)	56 (3.7)	71 (4.6)	0.027
Lipid lowering drugs, n (%)	152 (9.9)	146 (9.5)	136 (8.9)	146 (9.5)	0.585
Systolic blood pressure (mm Hg), mean ± SD	124 ± 19	125 ± 20	125 ± 18	128 ± 17	<0.001
Diastolic blood pressure (mm Hg), mean ± SD	71 ± 9	73 ± 9	73 ± 9	74 ± 9	<0.001
BMI (kg/m^2), mean ± SD	25.7 ± 4.1	26.1 ± 4.0	26.6 ± 4.3	28.1 ± 4.5	<0.001
BSA (m^2), mean ± SD	1.86 ± 0.19	1.91 ± 0.19	1.97 ± 0.19	2.07 ± 0.20	<0.001
Waist circumference (cm), mean ± SD	87.9 ± 11.8	89.7 ± 11.9	91.8 ± 12.0	97.3 ± 12.9	<0.001
Waist/hip ratio, mean ± SD	0.87 ± 0.08	0.88 ± 0.08	0.90 ± 0.08	0.93 ± 0.08	<0.001
Glucose (mmol/L), mean ± SD	4.94 ± 1.07	4.96 ± 1.09	4.98 ± 1.11	5.12 ± 1.08	<0.001
Insulin (mU/L), median (IQR)	7.50 (5.3–10.9)	7.60 (5.5–11.0)	8.10 (5.8–11.8)	9.50 (6.6–14.3)	<0.001
HOMA-IR (mU mmol/L^2/22.5), median (IQR)	0.49 (0.11–0.91)	0.46 (0.12–0.90)	0.53 (0.15–0.97)	0.72 (0.32–1.20)	<0.001
hsCRP (mg/L), median (IQR)	1.37 (0.65–3.07)	1.35 (0.58–3.09)	1.33 (0.61–2.87)	1.31 (0.63–3.08)	0.036
ALT (U/L), median (IQR)	15 (12–21)	16 (12–22)	17 (13–25)	20 (14–28)	<0.001
AST (U/L), median (IQR)	22 (19–26)	22 (19–26)	23 (20–26)	23 (20–27)	<0.001
ALP (U/L), mean ± SD	69 ± 21	69 ± 23	66 ± 18	69 ± 19	0.089
GGT (U/L), median (IQR)	21 (14–34)	21 (14–33)	23 (16–37)	27 (18–43)	<0.001
Total cholesterol (mmol/L), mean ± SD	5.45 ± 1.05	5.39 ± 1.03	5.37 ± 1.03	5.46 ± 1.04	0.372
Non-HDL cholesterol (mmol/L), mean ± SD	4.14 ± 1.04	4.11 ± 1.02	4.10 ± 1.04	4.26 ± 1.03	<0.001
LDL cholesterol (mmol/L), mean ± SD	3.59 ± 0.94	3.55 ± 0.92	3.54 ± 0.92	3.63 ± 0.94	0.079
HDL cholesterol (mmol/L), mean ± SD	1.31 ± 0.32	1.29 ± 0.32	1.27 ± 0.31	1.20 ± 0.29	<0.001
Triglycerides (mmol/L), median (IQR)	1.08 (0.79–1.45)	1.06 (0.75–1.55)	1.05 (0.80–1.54)	1.22 (0.86–1.75)	<0.001
Serum creatinine (umol/L), mean ± SD	82.77 ± 24.48	83.48 ± 17.41	85.78 ± 26.83	87.67 ± 15.93	<0.001
eGFR (ml/min/1.73 m^2), mean ± SD	89.0 ± 17.1	91.0 ± 17.1	92.7 ± 17.3	94.2 ± 16.7	<0.001
UAE (mg/24 h), median (IQR)	6.90 (5.16–10.76)	7.77 (5.75–12.53)	8.21 (6.0–3.36)	9.24 (6.67–16.45)	<0.001

p-values represent p for trend. Data are given in number with percentages (%), mean ± standard deviation (SD) for normally distributed data or median with interquartile ranges (IQR) for non-normally distributed data. HOMA-IR was log$_e$ transformed for analyses. LDL cholesterol was calculated by the Friedewald formula if triglycerides were <4.5 mmol/L (6028 subjects). Abbreviations: 24 h, twenty-four hours; ALP, alkaline phosphatase; ALT, aminotransferase; AST, aspartate aminotransferase; BMI, body mass index; BSA, body surface area; eGFR, estimated glomerular filtration rate; FLI, Fatty Liver Index; GGT, gamma-glutamyltransferase; HOMA-IR, Homeostasis Model Assessment of Insulin Resistance; HDL, high density lipoproteins; hsCRP, high sensitivity C-reactive protein; HSI, Hepatic Steatosis Index; LDL, low density lipoproteins; MetS, metabolic syndrome; T2D, type 2 diabetes mellitus; UAE, urinary albumin excretion.

3.2. Independent Associations of Suspected NAFLD with Sodium Intake

Multivariable binary regression analyses were subsequently performed in order to establish the independent associations of suspected NAFLD with urinary sodium excretion (Tables 3 and 4). In age- and sex-adjusted analysis, FLI ≥60 was positively associated with an increased sodium excretion (expressed per 1 SD increment) (Table 3, Model 1, odds ratio (OR) 1.54, 95% confidence interval (CI) 1.45–1.64, $p < 0.001$). This positive association was also demonstrated after additional adjustment for presence of T2D, a positive history of CVD, hypertension, alcohol intake (≥10 and <10 g/day) and current smoking (Table 3, Model 2, OR 1.51, 95% CI 1.42–1.61, $p < 0.001$), and for eGFR, UAE, use of antihypertensive medication and glucose and lipid lowering drugs (Table 3, Model 3, OR 1.54, 95% CI 1.44–1.64, $p < 0.001$). Finally, after additional adjustment for HOMA-IR the association of FLI ≥ 60 with urinary sodium excretion was attenuated but remained present (Table 3, Model 4, OR 1.30, 95% CI 1.21–1.41, $p < 0.001$). In alternative analyses with an HSI >36 instead of FLI ≥60 (Table 4), essentially similar positive associations of HSI >36 with urinary sodium excretion were found (Table 4, Model 1–4). When the multivariably adjusted models (Tables 3 and 4) were further adjusted for BSA, as a measure of body size, the positive association of increased sodium excretion remained present both with respect to an FLI ≥60 (OR 1.13, 95% CI 1.05–1.21, $p < 0.001$) and to an HSI >36 (OR 1.21, 95% CI 1.13–1.31, $p < 0.001$). Furthermore, when these models were further adjusted for waist/hip ratio as a measure of body fat distribution, the positive associations of increased sodium excretion remained present both with respect to a FLI ≥60 (OR 1.25, 95% CI 1.15–1.36, $p < 0.001$) and to an HSI > 36 (OR 1.37, 95% CI 1.27–1.48, $p < 0.001$).

The positive association of FLI ≥60 with urinary sodium excretion was present in women and men separately (Table S1). Likewise, the association of HSI >36 with urinary sodium excretion was present in both women and men (Table S2). However, neither the association of FLI ≥60 nor of HSI >36 was different between sexes (fully adjusted model including age; p interaction = 0.396 for FLI ≥ 60 and p interaction = 0.298 for HSI >36, respectively).

Sensitivity analyses in 3221 participants (Tables S3 and S4), after exclusion of subjects with alcohol intake ≥10 g/day, a positive history of CVD, presence of hypertension, impaired eGFR (<60 mL/min/1.73 m^2), elevated UAE (>30 mg/24 h), use of antihypertensive medication, glucose lowering drugs and lipid lowering drugs, also showed positive associations of FLI ≥ 60 with urinary sodium excretion after adjustment for age, sex, T2D and current smoking (Table S3, Models 1 and 2). After additional adjustment for HOMA-IR, the association of FLI ≥ 60 with urinary sodium excretion was again attenuated but remained significant (Table S3, Model 3). Similar sensitivity analyses with HSI > 36 iterated these findings (Table S4).

Table 3. Multivariable regression analysis demonstrating the positive association of an elevated Fatty Liver Index (FLI ≥ 60) with averaged 24 h sodium excretion (two collections) after adjustment for clinical and laboratory covariates in 6132 subjects.

	Model 1		Model 2		Model 3		Model 4	
	OR (95% CI)	p Value	OR (95% CI)	p Value	OR (95% CI)	p Value	OR (95% CI)	p Value
Age (years)	1.04 (1.03–1.04)	<0.001	1.01 (1.01–1.02)	<0.001	1.00 (0.99–1.01)	0.665	0.99 (0.99–1.00)	0.131
Sex (men vs. women)	2.02 (1.79–2.28)	<0.001	2.02 (1.78–2.29)	<0.001	2.07 (1.82–2.37)	<0.001	2.48 (2.13–2.90)	<0.001
Sodium excretion per 24 h (1 SD increment)	1.54 (1.45–1.64)	<0.001	1.51 (1.42–1.61)	<0.001	1.54 (1.44–1.64)	<0.001	1.30 (1.21–1.41)	<0.001
Type 2 diabetes mellitus (yes/no)			3.13 (2.46–3.97)	<0.001	3.30 (2.26–4.82)	<0.001	0.48 (0.30–0.75)	0.001
History of cardiovascular disease (yes/no)			1.00 (0.79–1.27)	0972	0.77 (0.59–1.01)	0.057	0.71 (0.53–0.97)	0.031
Hypertension (yes/no)			2.94 (2.56–3.37)	<0.001	2.39 (1.99–2.88)	<0.001	2.04 (1.64–2.52)	<0.001
Alcohol intake (≥10 g/day vs. <10 g/day)			1.56 (1.18–2.06)	0.002	1.68 (1.26–2.24)	<0.001	2.23 (1.59–3.14)	<0.001
Current smoking (yes/no)			1.09 (0.95–1.25)	0.202	1.05 (0.91–1.21)	0.489	1.24 (1.05–1.46)	0.010
eGFR (mL/min/1.73 m^2)					0.99 (0.98–0.99)	<0.001	0.99 (0.98–0.99)	<0.001
UAE (mg/24 h)					1.00 (1.00–1.00)	0.018	1.00 (1.00–1.00)	0.104
Use of antihypertensive medication (yes/no)					1.22 (0.99–1.50)	0.063	1.01 (0.80–1.29)	0.909
Use of glucose lowering drugs (yes/no)					0.89 (0.55–1.45)	0.647	1.83 (1.05–3.20)	0.033
Use of lipid lowering drugs (yes/no)					1.50 (1.20–1.87)	<0.001	1.25 (0.98–1.61)	0.077
HOMA-IR (mU mmol/L^2/22.5)							9.45 (8.10–11.01)	<0.001

OR, odds ratio; 95% CI, 95% confidence intervals. OR is given per 1 SD increase for urinary sodium excretion. 1 SD change in urinary sodium excretion corresponds to 55.99 mmol sodium per day. HOMA-IR was log$_e$ transformed for analyses. Abbreviations: 24 h, twenty-four hours; eGFR, estimated glomerular filtration rate; FLI, Fatty Liver Index; HOMA-IR, Homeostasis Model Assessment of Insulin Resistance; UAE; urinary albumin excretion. Model 1: adjusted for age, sex, presence of type 2 diabetes, history of cardiovascular disease, presence of hypertension, alcohol intake and current smoking. Model 3: adjusted for age, sex, presence of type 2 diabetes, history of cardiovascular disease, presence of hypertension, alcohol intake, current smoking, estimated glomerular filtration rate, urinary albumin excretion, use of antihypertensive medication, glucose lowering drugs and lipid lowering drugs. Model 4: adjusted for age, sex, presence of type 2 diabetes, history of cardiovascular disease, presence of hypertension, alcohol intake, current smoking, estimated glomerular filtration rate, urinary albumin excretion, use of antihypertensive medication, glucose lowering drugs and lipid lowering drugs and HOMA-IR.

Table 4. Multivariable regression analysis demonstrating the positive association of an elevated Hepatic Steatosis Index (HSI > 36) with averaged 24 h sodium excretion (two collections) after adjustment for clinical and laboratory covariates in 6132 subjects.

	Model 1		Model 2		Model 3		Model 4	
	OR (95% CI)	p Value	OR (95% CI)	p Value	OR (95% CI)	p Value	OR (95% CI)	p Value
Age (years)	1.03 (1.02–1.03)	<0.001	1.00 (1.00–1.01)	0.906	0.99 (0.98–1.00)	0.021	0.99 (0.98–1.00)	0.003
Sex (men vs. women)	0.62 (0.55–0.70)	<0.001	0.59 (0.52–0.67)	<0.001	0.61 (0.53–0.70)	<0.001	0.57 (0.49–0.65)	<0.001
Sodium excretion per 24 h (1 SD increment)	1.63 (1.54–1.74)	<0.001	1.59 (1.49–1.70)	<0.001	1.59 (1.49–1.70)	<0.001	1.40 (1.31–1.51)	<0.001
Type 2 diabetes mellitus (yes/no)			5.01 (3.95–6.35)	<0.001	5.36 (3.69–7.79)	<0.001	1.69 (1.12–2.56)	0013
History of cardiovascular disease (yes/no)			0.82 (0.64–1.05)	0.109	0.66 (0.50–0.88)	0.004	0.62 (0.46–0.84)	0002
Hypertension (yes/no)			2.41 (2.09–2.77)	<0.001	2.07 (1.71–2.50)	<0.001	1.77 (1.44–2.16)	<0.001
Alcohol intake (≥10 g/day vs. <10 g/day)			1.28 (0.95–1.73)	0.106	1.34 (0.99–1.82)	0.060	1.44 (1.04–2.01)	0.029
Current smoking (yes/no)			0.73 (0.63–0.83)	<0.001	0.71 (0.61–0.82)	<0.001	0.74 (0.64–0.87)	<0.001
eGFR (mL/min/1.73 m²)					0.99 (0.99–1.00)	0.001	1.00 (0.99–1.00)	0.198
UAE (mg/24 h)					1.00 (1.00–1.00)	0.800	1.00 (1.00–1.00)	0.730
Use of antihypertensive medication (yes/no)					1.18 (0.96–1.45)	0.120	1.01 (0.81–1.26)	0.942
Use of glucose lowering drugs (yes/no)					0.83 (0.52–1.34)	0.450	1.30 (0.77–2.18)	0.325
Use of lipid lowering drugs (yes/no)					1.42 (1.13–1.77)	0.002	1.25 (0.98–1.58)	0.068
HOMA-IR (mU mmol/L²/22.5)							4.04 (3.56–4.57)	<0.001

OR, odds ratio; 95% CI, 95% confidence intervals. OR is given per 1 SD increase for urinary sodium excretion. 1 SD change in urinary sodium excretion corresponds to 55.99 mmol sodium per day. HOMA-IR was log$_e$ transformed for analyses. Abbreviations: 24 h, twenty-four hours; eGFR, estimated glomerular filtration rate; HOMA-IR, Homeostasis Model Assessment of Insulin Resistance; HSI, Hepatic Steatosis Index; UAE; urinary albumin excretion. Model 1: adjusted for age and sex. Model 2: adjusted for age, sex, presence of type 2 diabetes, history of cardiovascular disease, presence of hypertension, alcohol intake and current smoking. Model 3: adjusted for age, sex, presence of type 2 diabetes, history of cardiovascular disease, presence of hypertension, alcohol intake, current smoking, estimated glomerular filtration rate, urinary albumin excretion, use of antihypertensive medication, glucose lowering drugs and lipid lowering drugs. Model 4: adjusted for age, sex, presence of type 2 diabetes, history of cardiovascular disease, presence of hypertension, alcohol intake, current smoking, estimated glomerular filtration rate, urinary albumin excretion, use of antihypertensive medication, glucose lowering drugs and lipid lowering drugs and HOMA-IR.

4. Discussion

In this large-scale, cross-sectional study in a predominantly Caucasian population we have demonstrated, to the best of our knowledge, for the first time a positive association of higher sodium intake, determined by two consecutive 24 h urinary sodium excretions, with suspected NAFLD. In our study we used an elevated FLI [26], and in alternative analyses an elevated HSI [28] as proxies of NAFLD, in keeping with international recommendations to use biomarkers to categorize subjects with suspected NAFLD in large-scale studies [27]. In multivariable regression analyses, accounting for various clinical variables, including BSA, as a measure of body size, eGFR and UAE, suspected NAFLD remained independently associated with a higher sodium intake. Interestingly, both for FLI and HSI, this association was attenuated after further adjustment for HOMA-IR. Taken together, the present study demonstrates that suspected NAFLD is positively associated with sodium intake. Furthermore, our current findings conceivably suggest that insulin resistance may represent a metabolic intermediate in explaining the relationship of higher sodium intake with NAFLD.

Two earlier cross-sectional studies have investigated the association of sodium intake with NAFLD [14,18]. Huh et al. described that Korean subjects recruited from the general population (Korea National Health and Nutrition Examination Surveys study) with suspected NAFLD had higher urinary sodium spot concentration values [14]. Choi et al. observed that higher sodium intake was associated with a greater prevalence of NAFLD among young and middle-aged asymptomatic Korean adults participating in a hospital-based cohort study. In this report, sodium intake was estimated by dietary recall questionnaires [18]. Very recently, perceived sodium intake determined by questionnaire was found to predict NAFLD development among Chinese people [34]. Our present results corroborate these previous studies. However, these surveys were performed in Asian populations with different dietary habits and sodium intake compared to Europeans [14,18], and used less reliable methods to estimate sodium intake compared to multiple 24 h urinary sodium excretion measurements [19]. The International Consortium for Quality Research on Dietary Sodium/Salt (TRUE) (with representative experts in hypertension, nutrition, statistics and dietary sodium) recommends that for a correct estimation of 24 h dietary sodium consumption, multiple complete 24 h urine samples are preferred to take account of short-term variations in dietary intake [19]. The hospital-based cohort could introduce another potential draw-back of the findings by Choi et al., which may limit extrapolation of their findings to the general population [18]. Hence, to date our study is the first study assessing the association of suspected NAFLD with high sodium intake, as determined by multiple 24 h urinary sodium excretion levels. Second, the current repost is the first to describe the positive association of high sodium intake in NAFLD in a predominantly Caucasian population.

The presently documented sodium intake amounted to 164 mmol/24 h and 137 mmol/24 h in subjects with and without an elevated FLI, respectively. In comparison, the World Health Organization currently recommends a salt intake of 5 grams per day, corresponding to 86 mmol of sodium per day [35], which is evidently lower than that documented in the present report. However, this recommendation was published later than the documentation of urinary sodium excretion used for the present report. In view of higher sodium intake in obese individuals [13,16,17], and the impact of obesity on NAFLD development [1], the here reported association of NAFLD with urinary sodium excretion is not unexpected. Furthermore, it has been suggested that the relationship of sodium intake with obesity is more pronounced in women [36], as presently documented for suspected NAFLD. While excessive sodium intake and its relation with obesity is to an important extent due to consumption of highly processed foods and increased total calorie intake [11,13,15], it should be noted that high sodium intake could directly impact on obesity development even independent from energy intake [37,38]. Chronic salt overload promotes adipocyte hypertrophy in rats [39,40]. Furthermore, increased leptin concentrations have been described in animal studies in response to high sodium intake [39]. Indeed, high sodium induces lipogenesis and inflammatory adipocytokine secretion in adipocytes, which may provide a possible mechanism for inflammatory adipogenesis in sodium-linked obesity [41], and in turn may also play a role in accelerated NAFLD development [42]. On the other hand, in experimental animal models, the inhibition of the renin-angiotensin-aldosterone system reduces the activation of stellate and Kupffer cells and reduces oxidative stress possibly leading to the improvement of NAFLD [43]. Furthermore, higher sodium intake is likely to downregulate the renin-angiotensin-aldosterone system [44], whereas low dietary sodium as well as exogenous angiotensin II suppress plasma adiponectin [45], an adipokine which has been proposed to protect against NAFLD development [46]. Thus, it seems unlikely that renin-angiotensin-aldosterone system-mediated effects are directly involved in the association of NAFLD with higher sodium intake.

Remarkably in our study, the association of a higher sodium intake and NAFLD was attenuated after adjusting for HOMA-IR, raising the possibility that insulin resistance may play a role in explaining the association of NAFLD with higher sodium intake. In rats, a high-salt diet may promote insulin resistance [47]. In humans, insulin resistance was independently and positively associated with a higher sodium intake [48,49]. The results of our cross-sectional study are in agreement with the hypothesis that a higher sodium intake, possibly via an effect on insulin resistance, promotes the development of NAFLD. Of further note, the association of an elevated FLI and HSI with a higher sodium intake remained present after adjustment for the BSA, and alternatively the waist/hip ratio. This would suggest that this association is at least in part independent of body size and body fat distribution, a potentially relevant finding in view of the reported relation of sodium consumption with adiposity [13,16,17]. Obviously, further research is needed to delineate the responsible pathogenic mechanisms more precisely.

The relationship of high sodium intake with cardiometabolic diseases is well established [11,36,50]. It is also evident that a higher sodium intake increases systemic blood pressure [44]. Interestingly, high blood pressure may also independently contribute to the development of NAFLD, even in the absence of obesity and MetS [51]. In turn, a bi-directional effect has been proposed where NAFLD represents an important risk factor for the development of hypertension, possibly involving insulin resistance [51]. From a clinical perspective our study supports the contention that it is important to take sodium intake into account when evaluating the adverse impact of NAFLD on the incidence of cardiovascular disease, hypertension and diabetes [6,9,52].

Our study has several strengths. Our sample size of over 6000 individuals enabled robust calculations on effect sizes and sufficiently powered subgroup and multivariable adjusted analyses. Additionally, the PREVEND population is well characterized, with extensive and standardized measurements [20,21]. On the other hand, a number of limitations and methodological aspects also need to be discerned. First, its cross-sectional design does not allow cause-effect relationships to be established with certainty, nor can we exclude the possibility of reversed causation. Second, an elevated FLI was chosen as a proxy of suspected NAFLD. Notably, the FLI and HSI do not translate in an absolute measure of hepatic fat accumulation. Thus over- and underestimation of suspected NAFLD could have occurred. Nonetheless, the FLI is considered to have sufficient accuracy for NAFLD assessment, and its current use is in line with international guidelines to apply biomarker scores in order to characterize NAFLD in larger-sized cohorts [26,27]. Moreover, the positive association of suspected NAFLD with sodium intake was confirmed using the HSI as an alternative algorithm for NAFLD categorization [28]. Performing liver ultrasound or liver biopsy for the diagnosis of NAFLD, was not feasible in the PREVEND cohort study. Third, we could not differentiate between simple hepatic steatosis and hepatic fibrosis; therefore, no relationship of hepatic fibrosis with sodium intake could be established. Fourth, to preclude collinearity with the FLI and/or HSI in the statistical analyses, variables making part of the equations (i.e., BMI and waist circumference) were not included in multivariable analyses. Instead we used BSA and waist/hip ratio in subsidiary analyses. Fifth, since alcohol intake and medical history were based on self-administered questionnaires, some misreporting by individuals cannot be excluded. However, considering the large number of subjects, this limitation is unlikely to have major effects on the interpretation of our results. Furthermore, the proportion of subjects using alcohol in excess of 30 gram per day in the PREVEND cohort is rather low, i.e., about 5.2% [53]. We adjusted for alcohol consumption and the association of an elevated FLI with sodium intake remained present in sensitivity analyses in which we excluded subjects with alcoholic intake ≥10 g/day. Furthermore, the PREVEND cohort is possibly enriched with people with micro-albuminuria. For this reason, we adjusted for eGFR and UAE in multivariable regression analysis and carried out a sensitivity analysis excluding subjects with impaired eGFR and elevated UAE. Reassuringly these analyses yielded similar positive and independent associations of suspected NAFLD with high sodium intake. Finally, no detailed information of diet composition is available in the PREVEND study. For this reason, we cannot exclude a contribution of unmeasured changes in diet components that are associated with a higher sodium intake in prevalent NAFLD, and the relationship of higher urinary sodium excretion and suspected NAFLD.

In conclusion, this study shows that suspected NAFLD is featured by a higher sodium intake. It seems conceivably that insulin resistance-related processes may explain in part the association of NAFLD with sodium intake.

Supplementary Materials: The following are available online at http://www.mdpi.com/2077-0383/8/12/2157/s1, Table S1: Multivariable regression analysis demonstrating the positive association of an elevated Fatty Liver Index (FLI ≥ 60) with averaged 24 h sodium excretion (two collections) after adjustment for clinical and laboratory covariates in 3100 women and in 3032 men separately., Table S2: Multivariable regression analysis demonstrating the positive association of an elevated Hepatic Steatosis Index (HSI > 36) with averaged 24 h sodium excretion (two collections) after adjustment for clinical and laboratory covariates in 3100 women and in 3032 men separately., Table S3: Sensitivity analyses demonstrating the positive association of an elevated Fatty Liver Index (FLI ≥ 60) with 24 h sodium excretion in 3221 subjects after excluding subjects with alcohol intake ≥10 g/day, a positive cardiovascular history, presence of hypertension, impaired estimated glomerular filtration rate (<60 mL/min/1.73 m^2), elevated urinary albumin excretion (>30 mg/24 h), use of antihypertensive drugs, glucose lowering drugs and lipid lowering drugs., Table S4: Sensitivity analyses demonstrating the positive association of an elevated Hepatic Steatosis Index (HSI > 36) with 24 h sodium excretion in 3221 subjects after excluding subjects with alcohol intake ≥10 g/day, a positive cardiovascular history, presence of hypertension, impaired estimated glomerular filtration rate (<60 mL/min/1.73 m^2), elevated urinary albumin excretion (>30 mg/24 h), use of antihypertensive drugs, glucose lowering drugs and lipid lowering drugs.

Author Contributions: Conceptualization, E.H.v.d.B. and R.P.F.D.; Data curation, E.H.v.d.B., E.G.G., S.J.L.B. and R.P.F.D.; Formal analysis, E.H.v.d.B. and R.P.F.D.; Funding acquisition, S.J.L.B.; Investigation, E.H.v.d.B., E.G.G., H.B., S.J.L.B. and R.P.F.D.; Methodology, E.H.v.d.B., E.G.G. and R.P.F.D.; Resources, S.J.L.B. and R.P.F.D.; Supervision, R.P.F.D.; Validation, E.G.G.; Visualization, E.H. v.d.B., E.G.G., H.B., S.J.L.B. and R.P.F.D.; Writing—original draft, E.H.v.d.B. and R.P.F.D.; Writing—review & editing, E.H.v.d.B., E.G.G., H.B., S.J.L.B. and R.P.F.D.

Funding: The Dutch Kidney Foundation supported the infrastructure of the PREVEND program from 1997 to 2003 (Grant E.033). The University Medical Center Groningen supported the infrastructure from 2003 to 2006. Dade Behring, Ausam, Roche, and Abbott financed laboratory equipment and reagents by which various laboratory determinations could be performed. The Dutch Heart Foundation supported studies on lipid metabolism (Grant 2001–005).

Acknowledgments: J.E. Kootstra-Ros, Laboratory Center, University Medical Center Groningen, supervised performance of the liver function tests.

Conflicts of Interest: The authors declared they do not have anything to disclose regarding conflict of interest with respect to this manuscript.

References

1. Loomba, R.; Sanyal, A.J. The global NAFLD epidemic. *Nat. Rev. Gastroenterol. Hepatol.* **2013**, *10*, 686–690. [CrossRef] [PubMed]
2. Van den Berg, E.H.; Amini, M.; Schreuder, T.C.M.A.; Dullaart, R.P.F.; Faber, K.N.; Alizadeh, B.Z.; Blokzijl, H. Prevalence and determinants of non-alcoholic fatty liver disease in lifelines: A large Dutch population cohort. *PLoS ONE* **2017**, *12*, e0171502. [CrossRef] [PubMed]
3. Puoti, C.; Elmo, M.G.; Ceccarelli, D.; Ditrinco, M. Liver steatosis: The new epidemic of the Third Millennium. Benign liver state or silent killer? *Eur. J. Intern. Med.* **2017**, *46*, 1–5. [CrossRef] [PubMed]
4. Bugianesi, E.; McCullough, A.J.; Marchesini, G. Insulin resistance: A metabolic pathway to chronic liver disease. *Hepatology* **2005**, *42*, 987–1000. [CrossRef] [PubMed]

5. Ballestri, S.; Zona, S.; Targher, G.; Romagnoli, D.; Baldelli, E.; Nascimbeni, F.; Roverato, A.; Guaraldi, G.; Lonardo, A. Nonalcoholic fatty liver disease is associated with an almost twofold increased risk of incident type 2 diabetes and metabolic syndrome. Evidence from a systematic review and meta-analysis. *J. Gastroenterol. Hepatol.* **2016**, *31*, 936–944. [CrossRef] [PubMed]
6. Van den Berg, E.H.; Flores-Guerrero, J.L.; Gruppen, E.G.; de Borst, M.H.; Wolak-Dinsmore, J.; Connelly, M.A.; Bakker, S.J.L.; Dullaart, R.P.F. Non-Alcoholic Fatty Liver Disease and Risk of Incident Type 2 Diabetes: Role of Circulating Branched-Chain Amino Acids. *Nutrients* **2019**, *11*, 705. [CrossRef]
7. Nass, K.J.; van den Berg, E.H.; Faber, K.N.; Schreuder, T.C.M.A.; Blokzijl, H.; Dullaart, R.P.F. High prevalence of apolipoprotein B dyslipoproteinemias in non-alcoholic fatty liver disease: The lifelines cohort study. *Metab. Clin. Exp.* **2017**, *72*, 37–46. [CrossRef]
8. Bril, F.; Sninsky, J.J.; Baca, A.M.; Superko, H.R.; Portillo Sanchez, P.; Biernacki, D.; Maximos, M.; Lomonaco, R.; Orsak, B.; Suman, A.; et al. Hepatic Steatosis and Insulin Resistance, But Not Steatohepatitis, Promote Atherogenic Dyslipidemia in NAFLD. *J. Clin. Endocrinol. Metab.* **2016**, *101*, 644–652. [CrossRef]
9. Targher, G.; Marra, F.; Marchesini, G. Increased risk of cardiovascular disease in non-alcoholic fatty liver disease: Causal effect or epiphenomenon? *Diabetologia* **2008**, *51*, 1947–1953. [CrossRef]
10. Arab, J.P.; Arrese, M.; Trauner, M. Recent Insights into the Pathogenesis of Nonalcoholic Fatty Liver Disease. *Annu. Rev. Pathol.* **2018**, *13*, 321–350. [CrossRef]
11. Bibbins-Domingo, K.; Chertow, G.M.; Coxson, P.G.; Moran, A.; Lightwood, J.M.; Pletcher, M.J.; Goldman, L. Projected effect of dietary salt reductions on future cardiovascular disease. *N. Engl. J. Med.* **2010**, *362*, 590–599. [CrossRef] [PubMed]
12. Vedovato, M.; Lepore, G.; Coracina, A.; Dodesini, A.R.; Jori, E.; Tiengo, A.; Del Prato, S.; Trevisan, R. Effect of sodium intake on blood pressure and albuminuria in Type 2 diabetic patients: The role of insulin resistance. *Diabetologia* **2004**, *47*, 300–303. [CrossRef] [PubMed]
13. Larsen, S.C.; Ängquist, L.; Sørensen, T.I.A.; Heitmann, B.L. 24 h urinary sodium excretion and subsequent change in weight, waist circumference and body composition. *PLoS ONE* **2013**, *8*, e69689. [CrossRef] [PubMed]
14. Huh, J.H.; Lee, K.J.; Lim, J.S.; Lee, M.Y.; Park, H.J.; Kim, M.Y.; Kim, J.W.; Chung, C.H.; Shin, J.Y.; Kim, H.-S.; et al. High Dietary Sodium Intake Assessed by Estimated 24-h Urinary Sodium Excretion Is Associated with NAFLD and Hepatic Fibrosis. *PLoS ONE* **2015**, *10*, e0143222. [CrossRef]
15. Libuda, L.; Kersting, M.; Alexy, U. Consumption of dietary salt measured by urinary sodium excretion and its association with body weight status in healthy children and adolescents. *Public Health Nutr.* **2012**, *15*, 433–441. [CrossRef]
16. Song, H.J.; Cho, Y.G.; Lee, H.-J. Dietary sodium intake and prevalence of overweight in adults. *Metab. Clin. Exp.* **2013**, *62*, 703–708. [CrossRef]
17. Zhao, L.; Cogswell, M.E.; Yang, Q.; Zhang, Z.; Onufrak, S.; Jackson, S.L.; Chen, T.-C.; Loria, C.M.; Wang, C.-Y.; Wright, J.D.; et al. Association of usual 24-h sodium excretion with measures of adiposity among adults in the United States: NHANES, 2014. *Am. J. Clin. Nutr.* **2019**, *109*, 139–147. [CrossRef]
18. Choi, Y.; Lee, J.E.; Chang, Y.; Kim, M.K.; Sung, E.; Shin, H.; Ryu, S. Dietary sodium and potassium intake in relation to non-alcoholic fatty liver disease. *Br. J. Nutr.* **2016**, *116*, 1447–1456. [CrossRef]
19. Campbell, N.R.C.; He, F.J.; Tan, M.; Cappuccio, F.P.; Neal, B.; Woodward, M.; Cogswell, M.E.; McLean, R.; Arcand, J.; MacGregor, G.; et al. The International Consortium for Quality Research on Dietary Sodium/Salt (TRUE) position statement on the use of 24-hour, spot, and short duration (. *J Clin. Hypertens.* **2019**, *18*, 1082. [CrossRef]
20. Borggreve, S.E.; Hillege, H.L.; Wolffenbuttel, B.H.R.; de Jong, P.E.; Bakker, S.J.L.; van der Steege, G.; van Tol, A.; Dullaart, R.P.F.; PREVEND Study Group. The effect of cholesteryl ester transfer protein -629C- >A promoter polymorphism on high-density lipoprotein cholesterol is dependent on serum triglycerides. *J. Clin. Endocrinol. Metab.* **2005**, *90*, 4198–4204. [CrossRef]
21. Kappelle, P.J.W.H.; Gansevoort, R.T.; Hillege, J.L.; Wolffenbuttel, B.H.R.; Dullaart, R.P.F.; PREVEND Study Group. Apolipoprotein B/A-I and total cholesterol/high-density lipoprotein cholesterol ratios both predict cardiovascular events in the general population independently of nonlipid risk factors, albuminuria and C-reactive protein. *J. Intern. Med.* **2011**, *269*, 232–242. [CrossRef] [PubMed]

22. Oterdoom, L.H.; Gansevoort, R.T.; Schouten, J.P.; de Jong, P.E.; Gans, R.O.B.; Bakker, S.J.L. Urinary creatinine excretion, an indirect measure of muscle mass, is an independent predictor of cardiovascular disease and mortality in the general population. *Atherosclerosis* **2009**, *207*, 534–540. [CrossRef] [PubMed]
23. Joosten, M.M.; Gansevoort, R.T.; Mukamal, K.J.; Lambers Heerspink, H.J.; Geleijnse, J.M.; Feskens, E.J.M.; Navis, G.; Bakker, S.J.L.; PREVEND Study Group. Sodium excretion and risk of developing coronary heart disease. *Circulation* **2014**, *129*, 1121–1128. [CrossRef] [PubMed]
24. van den Berg, E.H.; Gruppen, E.G.; James, R.W.; Bakker, S.J.L.; Dullaart, R.P.F. Serum paraoxonase 1 activity is paradoxically maintained in nonalcoholic fatty liver disease despite low HDL cholesterol. *J. Lipid. Res.* **2019**, *60*, 168–175. [CrossRef] [PubMed]
25. Inker, L.A.; Schmid, C.H.; Tighiouart, H.; Eckfeldt, J.H.; Feldman, H.I.; Greene, T.; Kusek, J.W.; Manzi, J.; Van Lente, F.; Zhang, Y.L.; et al. CKD-EPI Investigators Estimating glomerular filtration rate from serum creatinine and cystatin C. *N. Engl. J. Med.* **2012**, *367*, 20–29. [CrossRef] [PubMed]
26. Bedogni, G.; Bellentani, S.; Miglioli, L.; Masutti, F.; Passalacqua, M.; Castiglione, A.; Tiribelli, C. The Fatty Liver Index: A simple and accurate predictor of hepatic steatosis in the general population. *BMC Gastroenterol.* **2006**, *6*, 33. [CrossRef]
27. European Association for the Study of the Liver (EASL). Electronic address: Easloffice@easloffice.eu; European Association for the Study of Diabetes (EASD); European Association for the Study of Obesity (EASO) EASL-EASD-EASO Clinical Practice Guidelines for the management of non-alcoholic fatty liver disease. *J. Hepatol.* **2016**, *64*, 1388–1402. [CrossRef]
28. Lee, J.-H.; Kim, D.; Kim, H.J.; Lee, C.-H.; Yang, J.I.; Kim, W.; Kim, Y.J.; Yoon, J.-H.; Cho, S.-H.; Sung, M.-W.; et al. Hepatic steatosis index: A simple screening tool reflecting nonalcoholic fatty liver disease. *Dig. Liver Dis.* **2010**, *42*, 503–508. [CrossRef]
29. Grundy, S.M.; Cleeman, J.I.; Daniels, S.R.; Donato, K.A.; Eckel, R.H.; Franklin, B.A.; Gordon, D.J.; Krauss, R.M.; Savage, P.J.; Smith, S.C.; et al. American Heart Association; National Heart, Lung, and Blood Institute Diagnosis and management of the metabolic syndrome: An American Heart Association/National Heart, Lung, and Blood Institute Scientific Statement. *Circulation* **2005**, *112*, 2735–2752. [CrossRef]
30. Friedewald, W.T.; Levy, R.I.; Fredrickson, D.S. Estimation of the concentration of low-density lipoprotein cholesterol in plasma, without use of the preparative ultracentrifuge. *Clin. Chem.* **1972**, *18*, 499–502.
31. Schumann, G.; Bonora, R.; Ceriotti, F.; Férard, G.; Ferrero, C.A.; Franck, P.F.H.; Gella, F.J.; Hoelzel, W.; Jørgensen, P.J.; Kanno, T.; et al. International Federation of Clinical Chemistry and Laboratory Medicine IFCC primary reference procedures for the measurement of catalytic activity concentrations of enzymes at 37 degrees C. International Federation of Clinical Chemistry and Laboratory Medicine. Part 4. Reference procedure for the measurement of catalytic concentration of alanine aminotransferase. *Clin. Chem. Lab. Med.* **2002**, *40*, 718–724. [PubMed]
32. Schumann, G.; Bonora, R.; Ceriotti, F.; Férard, G.; Ferrero, C.A.; Franck, P.F.H.; Gella, F.J.; Hoelzel, W.; Jørgensen, P.J.; Kanno, T.; et al. International Federation of Clinical Chemistry and Laboratory Medicine IFCC primary reference procedures for the measurement of catalytic activity concentrations of enzymes at 37 degrees C. International Federation of Clinical Chemistry and Laboratory Medicine. Part 5. Reference procedure for the measurement of catalytic concentration of aspartate aminotransferase. *Clin. Chem. Lab. Med.* **2002**, *40*, 725–733. [PubMed]
33. Schumann, G.; Bonora, R.; Ceriotti, F.; Férard, G.; Ferrero, C.A.; Franck, P.F.H.; Gella, F.J.; Hoelzel, W.; Jørgensen, P.J.; et al. International Federation of Clinical Chemistry and Laboratory Medicine IFCC primary reference procedures for the measurement of catalytic activity concentrations of enzymes at 37 degrees C. International Federation of Clinical Chemistry and Laboratory Medicine. Part 6. Reference procedure for the measurement of catalytic concentration of gamma-glutamyltransferase. *Clin. Chem. Lab. Med.* **2002**, *40*, 734–738. [PubMed]
34. Shen, X.; Jin, C.; Wu, Y.; Zhang, Y.; Wang, X.; Huang, W.; Li, J.; Wu, S.; Gao, X. Prospective study of perceived dietary salt intake and the risk of non-alcoholic fatty liver disease. *J. Hum. Nutr. Diet.* **2019**, *67*, 11. [CrossRef] [PubMed]
35. WHO. *Guideline: Sodium Intake for Adults and Children 2012*; World Health Organization: Geneva, Switzerland, 2012.
36. Yi, S.S.; Kansagra, S.M. Associations of sodium intake with obesity, body mass index, waist circumference, and weight. *Am. J. Prev. Med.* **2014**, *46*, e53–e55. [CrossRef] [PubMed]

37. Ma, Y.; He, F.J.; MacGregor, G.A. High salt intake: Independent risk factor for obesity? *Hypertension* **2015**, *66*, 843–849. [CrossRef] [PubMed]
38. Nam, G.E.; Kim, S.M.; Choi, M.-K.; Heo, Y.-R.; Hyun, T.-S.; Lyu, E.-S.; Oh, S.-Y.; Park, H.-R.; Ro, H.-K.; Han, K.; et al. Association between 24-h urinary sodium excretion and obesity in Korean adults: A multicenter study. *Nutrition* **2017**, *41*, 113–119. [CrossRef]
39. Fonseca-Alaniz, M.H.; Brito, L.C.; Borges-Silva, C.N.; Takada, J.; Andreotti, S.; Lima, F.B. High dietary sodium intake increases white adipose tissue mass and plasma leptin in rats. *Obesity* **2007**, *15*, 2200–2208. [CrossRef]
40. Fonseca-Alaniz, M.H.; Takada, J.; Andreotti, S.; de Campos, T.B.F.; Campaña, A.B.; Borges-Silva, C.N.; Lima, F.B. High sodium intake enhances insulin-stimulated glucose uptake in rat epididymal adipose tissue. *Obesity* **2008**, *16*, 1186–1192. [CrossRef]
41. Lee, M.; Sorn, S.R.; Lee, Y.; Kang, I. Salt Induces Adipogenesis/Lipogenesis and Inflammatory Adipocytokines Secretion in Adipocytes. *Int. J. Mol. Sci.* **2019**, *20*, 160. [CrossRef]
42. Lyon, C.J.; Law, R.E.; Hsueh, W.A. Minireview: Adiposity, inflammation, and atherogenesis. *Endocrinology* **2003**, *144*, 2195–2200. [CrossRef] [PubMed]
43. Kato, J.; Koda, M.; Kishina, M.; Tokunaga, S.; Matono, T.; Sugihara, T.; Ueki, M.; Murawaki, Y. Therapeutic effects of angiotensin II type 1 receptor blocker, irbesartan, on non-alcoholic steatohepatitis using FLS-ob/ob male mice. *Int. J. Mol. Med.* **2012**, *30*, 107–113. [PubMed]
44. Graudal, N.A.; Hubeck-Graudal, T.; Jurgens, G. Effects of low sodium diet versus high sodium diet on blood pressure, renin, aldosterone, catecholamines, cholesterol, and triglyceride. *Cochrane Database Syst. Rev.* **2017**, *4*, CD004022. [CrossRef] [PubMed]
45. Lely, A.T.; Krikken, J.A.; Bakker, S.J.L.; Boomsma, F.; Dullaart, R.P.F.; Wolffenbuttel, B.H.R.; Navis, G. Low dietary sodium and exogenous angiotensin II infusion decrease plasma adiponectin concentrations in healthy men. *J. Clin. Endocrinol. Metab.* **2007**, *92*, 1821–1826. [CrossRef]
46. Finelli, C.; Tarantino, G. What is the role of adiponectin in obesity related non-alcoholic fatty liver disease? *World J. Gastroenterol.* **2013**, *19*, 802–812. [CrossRef]
47. Ogihara, T.; Asano, T.; Ando, K.; Chiba, Y.; Sekine, N.; Sakoda, H.; Anai, M.; Onishi, Y.; Fujishiro, M.; Ono, H.; et al. Insulin resistance with enhanced insulin signaling in high-salt diet-fed rats. *Diabetes* **2001**, *50*, 573–583. [CrossRef]
48. Park, Y.M.; Kwock, C.K.; Park, S.; Eicher-Miller, H.A.; Yang, Y.J. An association of urinary sodium-potassium ratio with insulin resistance among Korean adults. *Nutr. Res. Pract* **2018**, *12*, 443–448. [CrossRef]
49. Kim, Y.M.; Kim, S.H.; Shim, Y.S. Association of sodium intake with insulin resistance in Korean children and adolescents: The Korea National Health and Nutrition Examination Survey 2010. *J. Pediatr. Endocrinol. Metab.* **2018**, *31*, 117–125. [CrossRef]
50. Hoffmann, I.S.; Cubeddu, L.X. Salt and the metabolic syndrome. *Nutr. Metab. Cardiovasc. Dis.* **2009**, *19*, 123–128. [CrossRef]
51. Oikonomou, D.; Georgiopoulos, G.; Katsi, V.; Kourek, C.; Tsioufis, C.; Alexopoulou, A.; Koutli, E.; Tousoulis, D. Non-alcoholic fatty liver disease and hypertension: Coprevalent or correlated? *Eur. J Gastroenterol. Hepatol.* **2018**, *30*, 979–985. [CrossRef]
52. Kunutsor, S.K.; Bakker, S.J.L.; Blokzijl, H.; Dullaart, R.P.F. Associations of the fatty liver and hepatic steatosis indices with risk of cardiovascular disease: Interrelationship with age. *Clin. Chim. Acta* **2017**, *466*, 54–60. [CrossRef]
53. Gruppen, E.G.; Bakker, S.J.L.; James, R.W.; Dullaart, R.P.F. Serum paraoxonase-1 activity is associated with light to moderate alcohol consumption: The PREVEND cohort study. *Am. J. Clin. Nutr.* **2018**, *108*, 1283–1290. [CrossRef]

© 2019 by the authors. Licensee MDPI, Basel, Switzerland. This article is an open access article distributed under the terms and conditions of the Creative Commons Attribution (CC BY) license (http://creativecommons.org/licenses/by/4.0/).

Review

The Role of the Gut Microbiota in Lipid and Lipoprotein Metabolism

Yijing Yu [1], Fitore Raka [1,2] and Khosrow Adeli [1,3,*]

1. Molecular Medicine, Research Institute, The Hospital for Sick Children, Toronto, ON M5G 1X8, Canada; yijing.yu@mail.utoronto.ca (Y.Y.); f.raka@mail.utoronto.ca (F.R.)
2. Department of Physiology, University of Toronto, Toronto, ON M5S 1A8, Canada
3. Departments of Laboratory Medicine & Pathobiology and Biochemistry, University of Toronto, Toronto, ON M5S 1A8, Canada
* Correspondence: khosrow.adeli@sickkids.ca; Tel.: +416-813-8682

Received: 31 October 2019; Accepted: 8 December 2019; Published: 17 December 2019

Abstract: Both environmental and genetic factors contribute to relative species abundance and metabolic characteristics of the intestinal microbiota. The intestinal microbiota and accompanying microbial metabolites differ substantially in those who are obese or have other metabolic disorders. Accumulating evidence from germ-free mice and antibiotic-treated animal models suggests that altered intestinal gut microbiota contributes significantly to metabolic disorders involving impaired glucose and lipid metabolism. This review will summarize recent findings on potential mechanisms by which the microbiota affects intestinal lipid and lipoprotein metabolism including microbiota dependent changes in bile acid metabolism which affects bile acid signaling by bile acid receptors FXR and TGR5. Microbiota changes also involve altered short chain fatty acid signaling and influence enteroendocrine cell function including GLP-1/GLP-2-producing L-cells which regulate postprandial lipid metabolism.

Keywords: gut microbiota; lipoprotein metabolism; metabolic disorder

1. Introduction

The gastrointestinal microbiota represents the largest population of microbial community in the human body. This community contains more than 100 trillion microbial cells inhabiting in the small and large intestine which is estimated to be comparable to the total number of cells in the human body [1]. Additionally, approximately 3.3 million gut microbial genes are detected, which is 150-fold more genes than found in the human genome [2]. Earlier studies have suggested that gut microbiota are involved in several (patho-)physiological processes, including the metabolism of certain nutritional and drug compounds, the development of host immunity, intestinal inflammatory states, and colon cancer development [3–5]. Due to the advance of genomic techniques, gut microbiota studies and findings have dramatically expanded in the past decade [6,7]. The two commonly used approaches, 16S or 18S ribosomal RNA (rRNA) sequencing and metagenomic sequencing, have been frequently employed in the microbiota characterization studies. The former approach can detect phylogenetic profiling of microbiota, while the later analyzes all the community DNA within the samples. Moreover, metatranscriptomics or metaproteomics provide transcriptional or protein information for functional studies [8,9]. Although the gut microbiota is composed of more than 1000 species of bacteria, a set of core microbiota harbor in healthy human individuals [8]. Most of the core bacteria belong to five phyla (*Firmicutes, Proteobacteria, Bacteriodetes, Actinobacteria,* and *Veerrucomicrobia*) in humans [10].

The composition and stability of human microbiota exhibit various changes at different stages of life. The first intestinal bacterial colonization occurs during birth, and the infant gut microbiota could be affected by the mode of newborn delivery, the type of feeding, and other factors (e.g., host genetics, antibiotics, prescribed drugs, probiotic and prebiotics supplementation) [11,12]. At three to five years of age, children start to harbor microbiota that are comparable to the ones in adults. After childhood, the gut microbiota become quite stable, although long-term changes can occur in response to the changes in diet and lifestyle, the usage of antibiotics and probiotics, infection, and surgery [13]. The dysbiosis of microbiota has been linked with several human metabolic diseases, such as obesity, diabetes, and nonalcoholic fatty liver disease (NAFLD) [14–18]. In this review, we will present the available evidence implicating gut microbiota in the regulation of lipid and lipoprotein metabolism and potential mechanisms underlying this regulation (Figure 1).

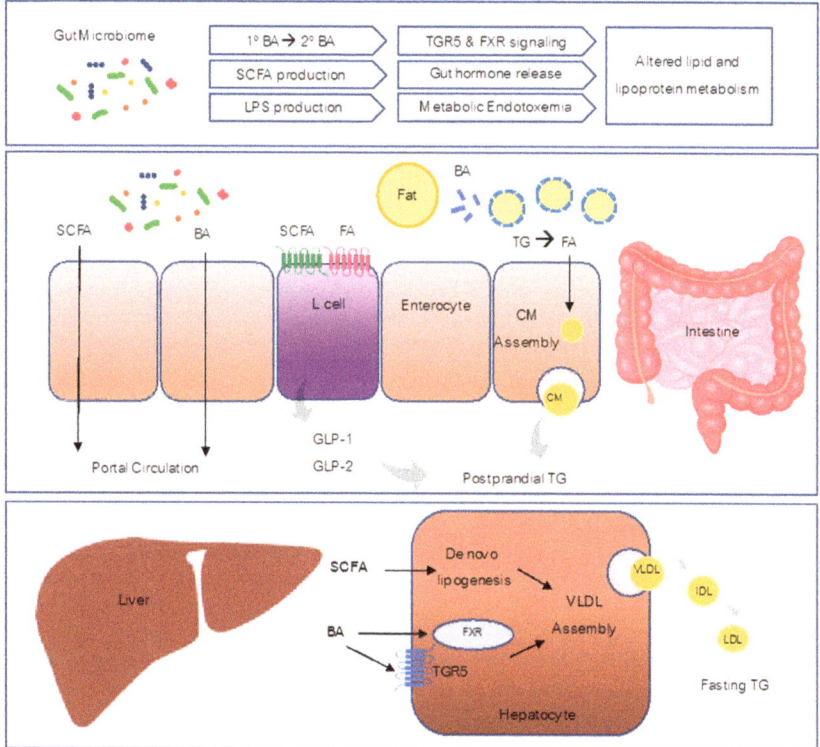

Figure 1. Mechanisms of the role of microbiota in host lipoprotein metabolism. The intestine and liver are two major organs involved in lipid and lipoprotein metabolism. In the intestine, BA emulsify fat into smaller fat particles which allows lipases to breakdown triglycerides into fatty acids. Fatty acids can then be absorbed and be used as substrates for CM assembly. CM production contributes to postprandial TG levels. Microbiota within intestine generates SCFA, secondary BA and LPS. SCFA and FA can activate receptors on enteroendocrine L-cells to release GLP-1/GLP-2 which regulate postprandial CM production. SCFA and BA are absorbed and enter the portal circulation. In the liver, SCFA can act as substrates for de novo lipogenesis and contribute to VLDL production. BA can activate FXR and TGR5 signaling pathways to regulate VLDL assembly. VLDL and its products IDL and LDL contribute to fasting TG levels. BA, bile acids; CM, chylomicron; FXR, farnesoid X receptor; FA, fatty acids; GLP-1/2, Glucagon-like peptide 1/2; IDL, intermediate density lipoprotein; LPS, lipopolysaccharide; LDL, low density lipoprotein; SCFA, short-chain fatty acid; TGR5, G-protein-coupled bile acid receptor; TG, triglyceride; VLDL, very low density lipoproteins.

Role of Microbiota in Metabolic Disorders

Metabolic syndrome is a group of interlinked disorders that includes hypertension, insulin resistance, visceral obesity, low high-density lipoprotein (HDL) cholesterol, and elevated triglycerides [19]. Numerous clinical studies have shown that metabolic syndrome is associated with increased risk of cardiovascular disease and type 2 diabetes [20]. Emerging studies from the last 15 years have implicated gut microbiota as a pathogenic factor affecting many components of the metabolic syndrome. Backhed and colleagues were the first to report a striking difference in body fat content observed between germ-free (GF) mice and conventionally raised mice with the latter having 40% higher body fat content [21]. These observations promoted many future studies exploring the role of gut microbiota in metabolic health. In exploration of the relative abundance of microbial taxa, the gut of obese mice was found to have a 50% reduction in the abundance of *Bacteroidetes* along with a proportional increase in *Firmicutes* [22]. Interestingly, patients with obesity also have a similar alteration in the relative proportion of *Bacteroidetes* and *Firmucutes* and this was shown to improve with a low-calorie diet [23].

Alterations in the gut microbiome from high-fat diet (HFD) feeding result in an increased proportion of lipopolysaccharide (LPS)-containing microbiota in the gut which has been linked to increased glycemia and insulinemia [24]. While earlier studies have determined differences in relative abundance of microbial taxa present in the gut of obese mice versus lean mice, recent studies have focused on particular microbial species and their metabolites that mediate the observed effects. For example, imidazole propionate is identified as a microbially produced histidine-derived metabolite that is higher in subjects with type 2 diabetes and was associated with impaired glucose tolerance and insulin signaling when administered to mice [25]. More recently, the tryptophan-derived microbial metabolite indole was found to upregulate the expression of miR-181 in white adipocytes leading to improvements in body weight gain, glucose tolerance, and insulin sensitivity [26]. While changes in gut microbiota occur due to HFD feeding, it is also important to consider that diets high in fat are typically lower in fiber compared to chow controls, which could also contribute to these phenotypes [27].

There is also evidence of microbiota dependent protection from metabolic disorders. A study reported that transfer of feces from healthy individuals to individuals with metabolic syndrome via a duodenal catheter alters the fecal microbiota of recipients and is associated with improvements in insulin sensitivity [28]. Studies in mice have revealed specific species to be beneficial for improving metabolic disorders: for example, *Akkermansia muciniphila* treatment has the ability to improve diet-induced obesity, fasting glycemia, and adipose tissue metabolism [29]. While these studies provide insight for the development of therapies that target human gut microbiota for treatment of obesity and its associated metabolic disorders, the intricate process of how specific specifies of bacteria and their metabolites regulate energy metabolism remain unclear.

2. Regulation of Lipoprotein Metabolism by the Gut Microbiome

2.1. Introduction of Lipoprotein

Dietary lipid contributes to around 35% of daily energy resource in humans [30,31]. As lipids are insoluble in water, the human digestive organs have developed a complex system of digestive and absorptive processes by transporting lipids in the form of lipoproteins. Lipoprotein particles are synthesized in the liver and intestine and are composed of lipids (such as phospholipids, cholesterol and triglycerides) and apolipoproteins [31]. Based on size and density, lipoproteins are classified into 5 classes: chylomicron (CM), very low density lipoproteins (VLDL), intermediate density lipoprotein (IDL), low density lipoprotein (LDL), and high density lipoprotein (HDL) particles [32]. Each class of lipoprotein has a specific function in lipid metabolism. CM, as the largest lipoprotein particle, transports dietary triglycerides and cholesterol from the intestine to peripheral tissues, while VLDL are synthetized in the liver to export triglycerides. VLDL particles contain apolipoprotein B 100 (ApoB100) as the main structural apolipoprotein and CM contain ApoB48, a truncated form of ApoB100. Both

VLDL and CM are assembled by the microsomal triglyceride transfer protein (MTP) which incorporates lipids into ApoB. LDL particles are the catabolic products of VLDL, while HDL is involved in reverse transport of cholesterol back to the liver [32]. Any deficiency during the lipid digestion process and lipoprotein synthesis cycle could result in dyslipidemia, a key factor in the pathogenesis of metabolic disorders (such as obesity, diabetes, and NAFLD) [33–36].

The intestine is not only the site for lipid digestion and absorption, but also serves as the major home for microbiota. Gut microbiota has the ability to modulate dietary lipid composition, digestion, and absorption, potentially altering intestinal lipoprotein formation. Abnormal levels of lipoproteins have been observed in GF mice and antibiotic-treated mice. GF mice are found to be resistant to develop an obesity phenotype under HFD feeding [37], partly due to the deficiencies in lipid digestion, absorption, and transport [38]. Recent studies have demonstrated that GF mice exhibit decreased plasma triglyceride and LDL compared to control mice, and this alteration was associated with decreased intestinal absorption and hormone peptide secretion [38]. Furthermore, supplementation of specific strain bacteria (e.g., *Lactobacillus rhamnosus gg*) to conventionally reared mice resulted in body weight gain, increased levels of LDL and cholesterol, and the altered expression of several genes involved in lipid transport [38]. Another study has shown that GF mice have significantly lower levels of fasting triglycerides and VLDL production when compared to conventionally reared mice [37]. To date, there are few systematic analyses of the association between the microbiome composition and lipid profiles in human and animal studies, which will be discussed as follows.

2.2. The Association between the Gut Microbiome and Lipid Profile

An early study by Carr et al. showed that supplementation of grain sorghum lipid extract (GSL) significantly improved the HDL/non-HDL cholesterol equilibrium in the Syrian hamster model [39]. A follow-up study by Martinez et al. characterized the interplay among diet, gut microbial ecology, and host metabolism [40]. Using 16S rRNA sequencing, Martinez et al. demonstrated that the abundance of *Bifidobacteria*, which was significantly increased under GSL-supplemented diet feeding, was positively associated with plasma HDL cholesterol levels, while the abundance of *Coriobacteriaceae*, decreased under GSL diet feeding, was positively associated with non-HDL plasma cholesterol levels [40].

In humans, Cotillard et al. analyzed the intestinal bacterial genes in 49 obese or overweight individuals by high-throughput sequencing technology [41]. Two groups of humans with high bacterial gene count (HGC) and low bacterial gene count (LGC) were defined in this study. The LGC group exhibited significantly higher frequency of insulin resistance and higher levels of fasting serum triglycerides, as well as higher LDL cholesterol, when compared to the HGC group [41]. Another human study reported by Le Chatelier et al. performed quantitative metagenomics with fecal DNA extracted from 169 obese individuals and 123 nonobese control individuals. They found that the LGC group was associated with elevated serum leptin, triglycerides and free fatty acids (FFA), decreased serum adiponectin, and decreased levels of HDL-cholesterol [42].

Fu et al. studied the largest correlation cohort to date, for which fecal samples were collected and the lipid profile determined in 893 participants [43]. Based on α diversity analysis, microbiota operational taxonomic unit (OTU) richness was negatively correlated with body mass index (BMI) and triglycerides, but positively correlated with HDL, while there was no significant correlation between microbial richness and LDL or total cholesterol levels. Furthermore, several microbes were correlated with lipid levels or BMI. For instance, the family of *Clostridiaceae/Lachnospiraceae* was specifically associated with LDL but not with BMI and other lipoproteins; the genus of *Eggerthella* was associated with increased triglyceride and decreased HDL levels, while the *Butyricimonas* was strongly associated with decreased triglyceride levels [43].

In addition, a study of 30 hypercholesterolemia individuals and 27 healthy controls reported differences in commensal bacterial profiles between hypercholesterolemic subjects and controls [44]. In this study, Rebolledo et al. analyzed the fasting whole blood lipid profile and bacterial community profiles by denaturing gradient gel electrophoresis. As expected, hypercholesterolemia subjects had higher levels of serum total cholesterol, triglycerides, and LDL-cholesterol than controls, while they harbored a lower richness and diversity of gut microbiota, thus proposing a potential role of the gut microbiota in the development of hypercholesterolemia [44]. Karlsson et al. performed the metagenomic sequencing with fecal samples collected from 145 European women with normal, impaired, or diabetic glucose control, and reported a negative correlation between serum triglycerides and *Clostridum*, and a positive correlation between HDL and *Clostridium* [45].

Recently, a lifestyle intervention study was performed with a low-saturated-fat and reduced-energy intake diet in 1065 metabolic syndrome patients and healthy subjects [46]. The metabolic syndrome patients had at least three of the following factors: central obesity (waist circumference ≥80 cm in women and ≥90 cm in men), fasting blood glucose ≥ 5.50 mmol/L, triglyceride concentration ≥ 1.65 mmol/L, low levels of HDL cholesterol (<1.00 mmol/L in men and <1.25 mmol/L in women), and systolic and/or diastolic blood pressure ≥ 130/85 mm Hg. Positive association between the body mass index and the gut microbiota dysbiosis, and negative association between insulin resistance and HDL-C concentration were found in metabolic syndrome patients. A short period (15 days) of diet intervention decreased the level of serum triglycerides, while the longer-term (75 days) intervention led to a significant reduction in VLDL, CM, LDL, small HDL particle, and improved glucose tolerance in 44.8% of metabolic syndrome patients. Moreover, the lifestyle intervention ameliorated gut dysbiosis, resulting in a reduction in the P/B ratio and an increase in the abundance of *Akkermansia muciniphila* and *Faecalibacterium prausnitzii* [46].

2.3. Evidence from Other Studies

Atherosclerosis-prone apolipoprotein E-deficient (ApoE-/-) mice exhibit delayed lipoprotein clearance and consequently develop dyslipoproteinemia [47]. Antibiotic (i.e., ampicillin) treatment was found to transiently improve glucose tolerance, lower plasma LDL and VLDL cholesterol levels, and reduce aortic atherosclerotic lesion area in Apo E-/- mice, suggesting that alteration of gut microbiota can attenuate dyslipoprotienemia in this model [48]. Additionally, a few studies have reported that gut microbiota could regulate the production of apolipoproteins and MTP [49,50]. Sato et al. found that rats receiving four days of antibiotics (streptomycin and penicillin) had reduced lymphatic transport of triglycerides and phospholipids, and decreased levels of mucosal apolipoproteins B, A-I, and A-IV, suggesting that gut microbiome promotes lipid absorption, probably by regulating intestinal production of apolipoproteins and secretion of CM [51]. Jin et al. also reported that six weeks of antibiotic (penicillin G (Pen G) and erythromycin (Ery)) treatment in mice resulted in increased hepatic lipid accumulation and higher circulating triglyceride levels and expression of key genes involved in FFA synthesis, triglyceride synthesis, and transport including MTP. [49].

A recent study in hamsters suggested that soybean sterols could alleviate HFD-diet-associated liver pathology and abnormal cholesterol metabolism by altering both host gene expression and gut microbiota composition [52]. Specifically, compared to the HFD vehicle group, the supplementation of soybean sterols significantly reduced the levels of triglycerides, total cholesterol, and non-HDL-cholesterol, increased the expression of cholesterol absorption/sterol excretion-related genes (e.g., NPC1L1, ABCG5, and ABCG8), and decreased the expression of cholesterol biosynthesis-related genes (e.g., SREBP2 and LDL-R). The addition of soybean sterols significantly increased the abundance of *Lactobacillus*, *Oscillospira*, and *Akkermansia*, and also the levels of all six short-chain fatty acids in the fecal content. In addition, a few studies have shown plant sterol esters could decrease plasma LDL cholesterol levels in humans [53–55]. A follow-up study revealed the link between gut bacteria and the cholesterol-lowering ability of plant sterol esters [56]. In this study, 13 healthy people with elevated plasma LDL cholesterol were administered with Stearate-enriched plant sterol esters for four weeks. This supplementation resulted in microbiota population change according to 16s rRNA pyrosequencing, and also correlated significantly to changes in serum cholesterol concentrations [56].

3. Mechanisms of the Role of Microbiota in Host Lipid Metabolism

3.1. Bile Acids

Bile acids are the major functional component of bile, which are synthesized in the liver and stored in the gall bladder, and further released in the intestine. One of the major functional roles of bile acids is to emulsify dietary fat [57]. However, bile acids are also known to act as important signaling molecules involved in metabolic pathways (i.e., glucose and lipid metabolism) [58] via interacting with various host bile acid receptors, such as nuclear receptors farnesoid-X-receptor (FXR) and cell membrane receptor Takeda G protein-coupled receptor 5 (TGR5) [58].

Based on recent studies of microbiota depletion or GF animal models, gut microbiota was found to be a key player in the regulation of lipid metabolism via bile acid receptor signaling. Many studies have shown that the remarkable change in gut microbiota induced by antibiotics treatment correlates with decreased intestinal FXR signaling as well as modified bile acid composition [59–62]. When comparing GF and conventionally raised mice, dramatically different composition of the bile acid pool and expression profile of genes involved in bile acid synthesis, conjugation, and reabsorption were observed [62]. Kuno et al. found a five-day antibiotics (vancomycin and polymyxin B) treatment decreased amounts of secondary bile acid-producing bacteria in feces and reduced the levels of secondary bile acid [lithocholic acid (LCA) and deoxycholic acid (DCA)] [63]. Another hamster study reported that four weeks of an antibiotics cocktail alleviated HFD-induced hepatic steatosis and glucose intolerance, which correlated with the modulation of gut microbiota, bile acids, and FXR signaling [64]. However, Reijnders et al. reported that a seven-day antibiotic (amoxicillin or vancomycin) treatment did not have any clinical impact on host metabolism in obese humans, although changes in microbiota, short-chain fatty acids levels, and bile acid levels were observed [65].

Bile acids are originally produced from cholesterol in hepatocytes as conjugated and primary bile acid (cholic acid (CA) and chenodeoxycholic acid (CDCA)). Specific enzymes (i.e., bile salt hydrolase (BSH), bile acid-inducible (BAI) enzymes) deconjugate and dehydroxylate these primary bile acids to generate unconjugated and secondary bile acids (lithocholic acid (LCA), deoxycholic acid (DCA)) [57]. Recent studies have shown that gut microbiota control the homeostasis of bile acids, thus impacting various host pathophysiological processes [58,66]. BSH, which was discovered in 1995, is one of the well-known bacterial bile acid deconjugating enzymes [67,68]. BSH has been detected in various commensal gut microbial species belonging to both gram-positive and gram-negative bacteria, including *Lactobacillus*, *Bifidobacterium*, *Enterococcus*, *Clostridium spp*, and *Bacteroides spp*. A functional metagenomics study revealed that functional BSH is abundantly present in the gut and enriched in the human gut microbiome [69].

In addition, it has been proposed that there is bidirectional interaction between gut microbiota and host bile acid profiles. Changes of the gut microbiota could alter the expression level of BSH and subsequently result in changes of bile acid pool, as evidenced by antibiotics treatment or GF animal studies [62,70]. In addition, bile acids are known to be toxic to bacteria and affect the growth rate of certain bacteria. In vivo studies suggest that bile acids alter microbiota composition and are associated with host metabolic changes. Islam et al. found that supplementation of primary bile acid (i.e., cholic acid) in the diet could alter plylum-level composition of gut microbiota, and similar changes were found in HFD studies [71]. In another study, GF and conventional mice were fed with different diets (control diet, control diet plus primary bile acid CA, and CDCA and HFD) for eight weeks [72]. Enhanced fat mass accumulation and metabolic changes were observed in conventional mice, but not in GF mice. Besides the impaired glucose metabolism, elevated amounts of hepatic triglycerides, cholesteryl esters, and monounsaturated fatty acids were shown in the bile acids treatment group [72]. Additionally, a shift of gut microbiota communities were found in bile acids-fed mice, including increased abundance of *Desulfovibrionaceae*, *Clostridium lactatifermentans*, and *Flintibacter butyricus*, and decreased abundance of *Lachnospiraceae* [72]. Moreover, according to metatranscriptomic analysis, changes of genes involved in lipid and amino acid metabolism were observed in this study [72]. In addition, the bile acid profiles and microbiota were found to be altered in certain disease conditions, such as obesity and type 2 diabetes (T2D), which also indicates a link between bile acid and gut microbiota [66].

Role of Bile Acid Signaling through FXR and TGR5

Recent studies have revealed that alteration of gut microbiota could not only affect the bile acid pool, but also influence the bile acid receptor signaling (i.e., FXR and TGR5). FXR is a nuclear receptor which is expressed in the liver, stomach, duodenum, jejunum, ileum, colon, white adipose tissues, kidney, gall bladder, heart, kidney, and adrenal cortex adrenal gland [73]. Bile acids act as ligands to bind and activate FXR, with CDCA and its conjugated forms being the most potent natural agonists (CDCA >LCA >DCA >CA) [74]. FXR has been reported to be involved in glucose homeostasis, energy expenditure, and lipid metabolism. Watanabe et al. has shown that CA could prevent hepatic triglyceride accumulation, decrease VLDL secretion, and lower serum triglyceride in a type 2 diabetes mouse model of hypertriglyceridemia [75]. Furthermore, the protective effects were due to an increase in short heterodimer partner (SHP) following FXR activation, resulting in a subsequent reduction of hepatic sterol regulatory element binding protein-1c (SREBP-1c) expression and other lipogenic genes [75]. An earlier study has shown that CDCA could activate FXR signaling and lead to decreased expression of MTP and apoB, thereby lowering lipoprotein production in HepG2 cell culture [76]. Using Syrian golden hamsters, Bilz et al. have also shown that activation of FXR by CDCA in a high-fructose diet could reduce VLDL triglycerides, VLDL, and IDL/LDL cholesterol levels, mainly by decreasing de novo lipogenesis and hepatic secretion of triglyceride-rich lipoproteins [77]. FXR stimulatory bacteria strains that were selected based on a luciferase assay with the FXR reporter were able to improve HFD-induced obesity phenotype in mice [78]. Controversial results were, however, reported from the study on human non-alcoholic fatty liver disease, showing that decreased activation of liver FXR was found in these patients and associated with increased expression of liver X receptor, SREBP-1c, and hepatic steatosis [79]. However, another study showed FXR signaling was activated in obese human intestines, and intestine-selective FXR inhibitor improved obesity and metabolic syndrome in HFD-induced obese mice [80].

TGR5 is a transmembrane G-protein-coupled receptor and widely expressed among various tissues, including gallbladder, small intestine, colon, liver, brain, skeletal muscle, heart, and the enteric nervous system. Bile acids can bind to TGR5 and act as potent agonists (with rank of LCA> DCA >CDCA> CA) [81]. Similar to FXR, TGR5 activity can regulate glucose and lipid homeostasis. Several in vivo studies in mice have suggested that TGR5 activation by synthetic agonists could improve insulin and glucose tolerance as well as liver and pancreatic function [82], decrease plasma triglycerides and LDL-cholesterol [83], and ameliorate hepatic steatosis [84]. However, controversial results have been reported with a TGR5 KO mice model, as TGR5 KO female mice gained much more body weight and accumulated more fat mass under HFD feeding, while TGR5 knockout male mice did not show significant difference in body weight gain or fat mass accumulation when compared to the wild-type control [85]. Another study reported that TGR5 KO mice have decreased fasting-induced hepatic lipid accumulation, an increased hepatic fatty acid oxidation rate, and decreased hepatic fatty acid uptake [86].

Additionally, activation of FXR and/or TGR5 was reported to increase glucagon-like peptide-1 (GLP-1) secretion, a hormone peptide which contributes to maintaining the metabolic homeostasis. Pathak et al. has shown that both FXR and TGR5 agonists could increase glucose-induced GLP-1 secretion [87]. In addition, activation of TGR5 increased intracellular cAMP, which further enhanced GLP-1 secretion [82]. Additionally, a TGR5 agonist was able to induce GLP-1 secretion in human studies [88], while glucose-induced GLP-1 secretion was reduced in FXR or TGR5 KO mice compared to control mice [89]. The role of GLP-1 in lipid and lipoprotein metabolism will be discussed below.

3.2. Short Chain Fatty Acids (SCFAs)

SCFAs are the fermentation products of non-digestible carbohydrates by gut microbiota. Gut microbiota have enzymes that the host lacks to break down these carbohydrates. The undigested dietary fibers as well as proteins that "escape" digestion and absorption are sequentially fermented in the small and large intestines by gut bacteria [90]. SCFAs can be divided into five groups: formic acid (C1), acetic acid (C2), propionic acid (C3), butyric acid (C4), and valeric acid (C5). The majority of gut SCFAs are acetate, propionate, and butyrate, which constitute more than 95% of the SCFA content [91]. After production, SCFA can be absorbed in the cecum, colon, and rectum, and then enter the mesenteric vein, drain into the portal vein, and finally into circulation [92]. After entering the circulation, SCFA can affect the metabolism of several peripheral tissues, including liver, adipose tissue, and skeletal muscle [93]. The advent of advanced genomic sequencing approaches has yielded great progress in identifying the bacteria responsible for SCFA production. Among the three major gut SCFAs, propionate and butyrate production are detected within a few conserved and substrate specific bacteria, while acetate production are widely distributed among bacterial groups. For butyrate production, a small number of gut bacteria, including *Faecalibacterium prausnitzii*, *Eubacterium rectale*, *Eubacterium hallii*, and *Ruminococcus bromii* are the major responsible strains [94]. Additionally, a few studies studying specific dietary ingredients have associated SCFA production with certain bacterial species. For instance, resistant starch greatly contributed to butyrate production in the colon predominantly by *Ruminococcus bromii* [95], while mucin was fermented into propionate mainly by *Akkermansia municiphilla* [96].

SCFAs play key roles in various physiological processes, such as regulation of energy intake, energy harvesting, glucose metabolism, lipid metabolism, adipogenesis, and immune responses, as well as pathophysiology of obesity and related metabolic disorders. Both animal and human studies support the relationship between the gut microbiota, SCFA, and metabolic disorders (i.e., obesity, T2D, NAFLD). Obese mice and humans show a changed ratio of *Firmicutes* to *Bacteroidetes* and increased fecal concentration of SCFA [97–100]. The obese phenotype of Toll like receptor 5 KO mice were also associated with gut dysbiosis and enhanced cecal SCFA production [101]. In humans, the butyrate-producing bacteria have been associated with the beneficial metabolic effects of the fecal transplantation from lean donors to obese patients [28], while a lower capacity to ferment complex carbohydrates was found in the microbiota of obesity patients compared to lean controls [102]. In the liver, acetate and butyrate are the major substrates for de novo lipogenesis, as well as substrates in cholesterolgenesis, while propionate appears to be an effective inhibitor of both processes [103]. Under physiological conditions, SCFAs such as acetate and propionate were found to inhibit adipose tissue lipolysis [104], resulting in reduced FFA flux from the adipose tissue to liver, but an opposite phenotype was observed in fatty liver disease. A previous human study also reported that propionate supplementation in the diet was able to prevent body weight gain and liver fat accumulation in NAFLD patients [105].

Many studies have suggested that SCFA are related to the release of gut-derived satiety hormones, in particular GLP-1 and peptide YY (PYY). Based on in vitro studies using intestinal cell lines from both rodents and humans, SCFAs can stimulate the secretion of PYY and GLP-1 from L-cells, which were dependent on the SCFA receptors GPR41 and GPR43 [106–110]. Furthermore, evidence from in vivo studies has shown that carbohydrate feeding (i.e., oligofructose, inulin, and resistant starch) can increase PYY and GLP-1 secretion in rodents [111–113]. Moreover, human studies with either supplementation of SCFA or fermentable polysaccharides (i.e., oligofructose) have shown that weight loss, improved hepatic lipid content, and glucose metabolism were associated with enhanced PYY and GLP-1 release and secretion [105,114–116]. However, it warrants further studies to investigate the effects of physiologically relevant SCFA mixtures on the secretion of GLP-1 and PYY and their metabolic consequences in humans.

3.3. Enteroendocrine Cell Regulation of Hormone Secretion

The intestine harbors specialized cells called enteroendocrine cells that secrete hormones in response to nutrients and microbial metabolites. In turn, these gut hormones can regulate nutrient metabolism and feeding behavior [117]. Classification of enteroendocrine cells was originally based on expression of one or two hormones, however, transcriptional profiling has determined that these cells can express multiple hormones, making them less distinct than previously recognized [118]. There is accumulating evidence that microbiota can regulate enteroendocrine cell development and function. Intestinal enteroendocrine cells like absorptive enterocytes have an apical membrane facing the gut lumen which allows for interaction with microbes and their metabolites.

3.3.1. Microbial Regulation of Enteroendocrine L-Cells

Two gut hormones discovered by our lab as being important regulators of intestinal lipid metabolism are GLP-1 and GLP-2, co-secreted in equimolar amounts by enteroendocrine L-cells. Microbiota-derived products such as SCFAs have been shown to stimulate GLP-1 secretion via activation of the G-protein-coupled receptor FFAR2. Microbiota can also influence the L-cell transcriptome. Using transgenic GLU-Venus mice driving expression of yellow fluorescent protein under the proglucagon promoter, Arora et al. found that genes related to vesicle organization and synaptic vesicle cycle such as synaptophysin, synaptotgamins, Rab, and SNAP proteins were downregulated in L-cells from conventionally raised GLU-Venus mice compared to GF GLU-Venus mice, suggesting GF mice have enhanced hormone secretion [119]. They also observed that L-cells from conventionally raised GLU-Venus mice have upregulated expression of nutrient sensing receptors and downregulated expression of the bile acid receptor TGR5 compared to GF GLU-Venus mice [119]. Studies over the past few years have revealed the link between gut microbiota and GLP-1 signaling. Evidence has shown a higher plasma GLP-1 level observed in GF mice and antibiotic treated animals compared to controls [38,120]. Importantly, activation of TGR5 has been shown to increase GLP-1 secretion, and enhanced TGR5 expression in GF mice could contribute to elevated GLP-1 levels found in GF mice [121]. A few studies have also shown that prebiotics administration results in increased level of GLP-1 in the circulation of rats and humans [111,114,122]. Additionally, the beneficial effects of prebiotics in diabetes and obesity models were dependent on a functional GLP-1, as the protective effects were not observed in GLP-1R KO mice [123]. The fermentation products of gut microbiota, such as SCFAs and bile acids, may underline the mechanisms for the crosstalk between microbiota and GLP-1 since microbiota metabolites are involved in stimulating the secretion of certain gut peptide hormones, including PYY, GLP-1, and GLP-2. SCFAs are able to bind to and activate selected G-protein-coupled receptors in different tissues to regulate peptide hormone secretion (e.g., activation of GPR41 and 43 to release GLP-1 secretion) [106–110]. Thus, microbiota and their metabolites have the capacity to regulate enteroendocrine L-cell function leading to altered hormone release. The roles of GLP-1 and GLP-2 in regulating lipoprotein metabolism will be discussed below.

GLP-1

As an incretin hormone, GLP-1 is well known to induce insulin secretion and subsequently decrease the circulating glucose levels [124,125]. However, GLP-1 also regulates appetite, body weight loss, and lipid metabolism [126–128]. In addition to decreasing adipose lipolysis [129], GLP-1 is involved in regulation of intestinal and liver lipid/lipoprotein metabolism [128]. The role of GLP-1 in modulating intestinal lipoprotein production has been extensively studied in both animal models as well as humans [130]. Administration of GLP-1 receptor agonist exendin-4 or dipeptidyl peptidase 4 (DPP-4; as a GLP-1 degradation enzyme) inhibitor in hamsters both significantly reduce intestinal lipoprotein production and plasma triglycerides and cholesterol within the triglyceride-rich lipoprotein fraction [130]. Conversely, blocking GLP-1 receptor signaling by infusion of the GLP-1 receptor antagonist exendin 9-39 or using GLP-1 receptor knockout mice enhanced apoB48-containing lipoprotein secretion [130]. Similar effects were shown in a rat model, while administration of native GLP-1 decreased intestinal fat absorption and production of apoB and apoAIV [131]. Administration of DDP-4 inhibitor in T2D patients exhibited benefit effects on postprandial levels of triglycerides (TG), apoB48, CM triglycerides [132] and hepatic triglycerides [133,134]. Antidiabetic drug exenatide (53% GLP-1 homology) reduced postprandial triglycerides and apoB48 in T2D patients [135]. Overall, GLP-1 activity has been shown to exert very beneficial effects on lipid and lipoprotein homeostasis.

GLP-2

Alongside GLP-1, its sister peptide GLP-2 is co-secreted from L-cells upon luminal lipid sensing [136]. GLP-2 plays an important role in increasing the absorption of carbohydrates and amino acids from the intestinal lumen by increasing nutrient transporter expression [137–139]. Although secreted in equimolar amounts, these hormones have opposing roles in regulation of postprandial lipids. In a hamster model and in healthy humans, GLP-2 was found to raise postprandial CM production and plasma triglycerides [140,141]. A GLP-2 mediated increase in CM production in hamsters was associated with increased apical expression of the fatty acid translocase CD36 which is involved in fatty acid uptake [142]. Interestingly, coadministration of GLP-1 and GLP-2 for 30 minutes in hamsters results in a GLP-2 dominant effect and an increase in CM secretion [143]. Similarly, in an insulin resistant model, the stimulatory actions of GLP-2 appear to override the inhibitory effects of GLP-1 [143]. Considering that the half-life of GLP-2 is longer than GLP-1, increasing L-cell hormone secretion may result in a GLP-2 dominant effect, leading to excess accumulation of atherogenic CM particles. Thus, microbial regulation of enteroendocrine L-cells can have profound implications for lipoprotein metabolism.

3.3.2. Microbial Regulation of Enterochromaffin Cells

Microbial regulation has also been observed for other enteroendocrine cells. Indigenous spore-forming microbes from colons of specific pathogen free (SPF) mice were found to increase colonic and blood serotonin levels and these spore forming microbes were associated with a particular set of metabolites that positively correlated with serotonin levels [144]. In line with this, mice inoculated with spore-forming *Clostridium ramosum* have an increased number of serotonin-producing cells known as enterochromaffin cells [145]. This was shown to occur through increases in transcription factors regulating enterochromaffin cell development. As a result, mice inoculated with C. *ramosum* had higher levels of serotonin and gained more weight on a HFD. This is consistent with emerging studies proposing a role for gut-derived serotonin in obesity [146–148]. Interestingly, gut-derived serotonin has been shown to positively regulate liver lipid metabolism [149–151]. Using tryptophan hydroxylase 1 knockout mice which lack the enzyme for synthesizing serotonin in the gut, Choi et al. found that these mice are protected from HFD-induced fatty liver [149]. Under HFD feeding, these mice showed decreased expression of genes involved in lipogenesis but no differences in apoB or MTP expression. This would suggest that that serotonin does not affect VLDL secretion, however no experiments besides gene expression were done to directly measure VLDL secretion in this model. Considering the close proximity of enterochromaffin cells to absorptive enterocytes, it is possible that serotonin could modulate aspects of intestinal lipid metabolism. Studies with C. *ramosum* found that colonization by this serotonin-enhancing bacterium increased expression of lipid transporters such as CD36 and FATP4 in the colon [145]. Furthermore, serotonin treatment of Caco-2 cells also increased CD36 and FATP4 protein levels [145]. Given that the colon is generally not a major site of lipid absorption, the significance of these findings in serotonin-mediated lipid absorption are unclear, however the role of serotonin in small intestine lipid absorption and metabolism remains to be elucidated.

3.4. Gut Barrier Function

Barrier function at the intestinal mucosal interface enables absorption of water and essential nutrients, but also protects against ingested or endogenous toxins. The intestinal epithelial layer consists of absorptive enterocytes, goblet cells, enteroendocrine cells, tuft cells, and Paneth cells. Substances can pass through this barrier via a transcellular route involving selective transporters or simple diffusion. Alternatively, they can pass through in the space between epithelial cells, a process known as paracellular diffusion. Some residing bacteria produce LPS which activate Toll-like receptor 4 (TLR4) [152]. LPS is restricted within the gut lumen, however, when intestinal barrier function is weakened, LPS can enter the blood circulation and trigger systemic inflammation [24]. Lebrun et al. have demonstrated that when gut barrier function is compromised as is with dextran sodium sulfate treatment or via ischemia/reperfusion in mice, LPS stimulates a rapid and robust rise in GLP-1 [153]. The authors attributed these findings to mechanisms involving enhanced LPS access to TLR4 receptors located on GLP-1-secreting enteroendocrine cells when gut barrier function was altered. Thus, gut barrier function can have important implications for enteroendocrine cell responses to microbial metabolites.

Additionally, LPS can be internalized by enterocytes through a transcellular route involving a TLR4-dependant mechanism and can be incorporated into CM [154]. LPS absorption from the gut was found to be associated with fat absorption since plasma LPS concentrations increased by 50% after ingestion of a high-fat meal [155]. While increased LPS into the circulation via enhanced gut permeability is known to be associated with systemic inflammation [24], the role of LPS contained within CM is unknown. In states of insulin resistance, postprandial overproduction of CM particles is observed and leads to an increase in atherogenic CM remnant particles [156–159]. Interestingly, long-chain triglyceride ingestion in mice led to an increase in plasma LPS which was mainly associated with the CM remnant fraction [154]. LPS contained within CM remnants could contribute to their proatherogenic nature since elevated concentrations of circulating LPS correlate well with an increased atherosclerosis risk [160]. LPS injection accelerates plaque formation in both mice and rabbits [161,162] and apoE-/- mice with a genetic deletion of the LPS receptor TLR4 have significantly reduced plaque formation [163,164]. Whether LPS alone or LPS associated with CM remnants is responsible for these observations remains unknown.

3.5. Other Microbiota Metabolites

Gut microbiota are known to provide a small proportion of amino acids to host by de novo sythesis. Hoyles et al. reported a positive association of hepatic steatosis with gut microbial amino acid metabolism based on the study of 56 morbidly obese, weight-stable, nondiabetic women [165]. Interestingly, animo acid phenylacetate (PAA) was strongly correlated with steatosis, which was found to be mainly produced by *bacteroides spp* in humans. Both fecal microbiota transplants and chronic phenylacetate treatment revealed that phenylacetate induced liver steatosis. Serum PAA may thus be a useful biomarker and a potential therapeutic target for treatment of hepatic steatosis [166].

4. Concluding Remarks

A key role for gut microbiota in modulating metabolic health is now widely acknowledged, but the underlying mechanisms are not clearly understood. Of particular interest are the links between dietary nutrients, intestinal flora, and the regulation of carbohydrate and lipid metabolism. Although modulation of lipid and lipoprotein homeostasis by gut microbiota is largely unexplored, emerging evidence suggest important roles in regulation of dietary fat absorption and postprandial lipid metabolism. Evidence also suggests a potential role in regulating hepatic lipid accumulation and development of hepatic steatosis. These potential links appear to be mediated by microbiota-mediated regulation of intestinal enteroendocrine system and secretion of specific gut peptides. Complex interactions between bacterial populations in the gut and the intricate network of enteric neurons as well as gut immune system are also likely to play important roles in modulating nutrient metabolism and metabolic health. However, no direct experimental evidence currently exists, and new research efforts are needed to explore these links and elucidate the underlying mechanisms.

Author Contributions: Y.Y., F.R., and K.A. wrote the manuscript, F.R. designed the figure.

Funding: This work was supported by a Foundation Grant to KA by the Canadian Institute of Health Research (CIHR). F.R. is supported by a doctoral scholarship from the Banting and Best Diabetes Centre at the University of Toronto.

Conflicts of Interest: The authors declare no conflict of interest.

References

1. Sender, R.; Fuchs, S.; Milo, R. Are We Really Vastly Outnumbered? Revisiting the Ratio of Bacterial to Host Cells in Humans. *Cell* **2016**, *164*, 337–340. [CrossRef] [PubMed]
2. Qin, J.; Li, R.; Raes, J.; Arumugam, M.; Burgdorf, K.S.; Manichanh, C.; Nielsen, T.; Pons, N.; Levenez, F.; Yamada, T.; et al. A human gut microbial gene catalogue established by metagenomic sequencing. *Nature* **2010**, *464*, 59–65. [CrossRef] [PubMed]
3. Savage, D.C. Microbial ecology of the gastrointestinal tract. *Annu. Rev. Microbiol.* **1977**, *31*, 107–133. [CrossRef] [PubMed]
4. Savage, D.C. Gastrointestinal microflora in mammalian nutrition. *Annu. Rev. Nutr.* **1986**, *6*, 155–178. [CrossRef]
5. Haenel, H.; Bendig, J. Intestinal flora in health and disease. *Prog. Food Nutr. Sci.* **1975**, *1*, 21–64.
6. Human Microbiome Project, C. Structure, function and diversity of the healthy human microbiome. *Nature* **2012**, *486*, 207–214. [CrossRef]
7. Integrative, H.M.P.R.N.C. The Integrative Human Microbiome Project: dynamic analysis of microbiome-host omics profiles during periods of human health and disease. *Cell Host Microbe* **2014**, *16*, 276–289. [CrossRef]
8. Bashiardes, S.; Zilberman-Schapira, G.; Elinav, E. Use of Metatranscriptomics in Microbiome Research. *Bioinform. Biol. Insights* **2016**, *10*, 19–25. [CrossRef]
9. Lai, L.A.; Tong, Z.; Chen, R.; Pan, S. Metaproteomics Study of the Gut Microbiome. *Methods Mol. Biol.* **2019**, *1871*, 123–132. [CrossRef]
10. Arumugam, M.; Raes, J.; Pelletier, E.; Le Paslier, D.; Yamada, T.; Mende, D.R.; Fernandes, G.R.; Tap, J.; Bruls, T.; Batto, J.M.; et al. Enterotypes of the human gut microbiome. *Nature* **2011**, *473*, 174–180. [CrossRef]
11. Collado, M.C.; Cernada, M.; Bauerl, C.; Vento, M.; Perez-Martinez, G. Microbial ecology and host-microbiota interactions during early life stages. *Gut Microbes* **2012**, *3*, 352–365. [CrossRef]
12. Jackson, M.A.; Verdi, S.; Maxan, M.E.; Shin, C.M.; Zierer, J.; Bowyer, R.C.E.; Martin, T.; Williams, F.M.K.; Menni, C.; Bell, J.T.; et al. Gut microbiota associations with common diseases and prescription medications in a population-based cohort. *Nat. Commun.* **2018**, *9*, 2655. [CrossRef] [PubMed]
13. Senghor, B.; Sokhna, C.; Ruimy, R.; Lagier, J.C. Gut microbiota diversity according to dietary habits and geographical provenance. *Hum. Microb. J.* **2018**, 1–9. [CrossRef]
14. Leung, C.; Rivera, L.; Furness, J.B.; Angus, P.W. The role of the gut microbiota in NAFLD. *Nat. Rev. Gastroenterol. Hepatol.* **2016**, *13*, 412–425. [CrossRef] [PubMed]
15. Henao-Mejia, J.; Elinav, E.; Jin, C.; Hao, L.; Mehal, W.Z.; Strowig, T.; Thaiss, C.A.; Kau, A.L.; Eisenbarth, S.C.; Jurczak, M.J.; et al. Inflammasome-mediated dysbiosis regulates progression of NAFLD and obesity. *Nature* **2012**, *482*, 179–185. [CrossRef]
16. Qin, J.; Li, Y.; Cai, Z.; Li, S.; Zhu, J.; Zhang, F.; Liang, S.; Zhang, W.; Guan, Y.; Shen, D.; et al. A metagenome-wide association study of gut microbiota in type 2 diabetes. *Nature* **2012**, *490*, 55–60. [CrossRef]
17. Vatanen, T.; Franzosa, E.A.; Schwager, R.; Tripathi, S.; Arthur, T.D.; Vehik, K.; Lernmark, A.; Hagopian, W.A.; Rewers, M.J.; She, J.X.; et al. The human gut microbiome in early-onset type 1 diabetes from the TEDDY study. *Nature* **2018**, *562*, 589–594. [CrossRef]
18. Turnbaugh, P.J.; Hamady, M.; Yatsunenko, T.; Cantarel, B.L.; Duncan, A.; Ley, R.E.; Sogin, M.L.; Jones, W.J.; Roe, B.A.; Affourtit, J.P.; et al. A core gut microbiome in obese and lean twins. *Nature* **2009**, *457*, 480–484. [CrossRef]
19. Wilson, P.W.; D'Agostino, R.B.; Parise, H.; Sullivan, L.; Meigs, J.B. Metabolic syndrome as a precursor of cardiovascular disease and type 2 diabetes mellitus. *Circulation* **2005**, *112*, 3066–3072. [CrossRef]
20. Stern, M.P.; Williams, K.; Gonzalez-Villalpando, C.; Hunt, K.J.; Haffner, S.M. Does the metabolic syndrome improve identification of individuals at risk of type 2 diabetes and/or cardiovascular disease? *Diabetes Care* **2004**, *27*, 2676–2681. [CrossRef]

21. Backhed, F.; Ding, H.; Wang, T.; Hooper, L.V.; Koh, G.Y.; Nagy, A.; Semenkovich, C.F.; Gordon, J.I. The gut microbiota as an environmental factor that regulates fat storage. *Proc. Natl. Acad. Sci. USA* **2004**, *101*, 15718–15723. [CrossRef] [PubMed]
22. Ley, R.E.; Backhed, F.; Turnbaugh, P.; Lozupone, C.A.; Knight, R.D.; Gordon, J.I. Obesity alters gut microbial ecology. *Proc. Natl. Acad. Sci. USA* **2005**, *102*, 11070–11075. [CrossRef] [PubMed]
23. Ley, R.E.; Turnbaugh, P.J.; Klein, S.; Gordon, J.I. Microbial ecology: human gut microbes associated with obesity. *Nature* **2006**, *444*, 1022–1023. [CrossRef] [PubMed]
24. Cani, P.D.; Amar, J.; Iglesias, M.A.; Poggi, M.; Knauf, C.; Bastelica, D.; Neyrinck, A.M.; Fava, F.; Tuohy, K.M.; Chabo, C.; et al. Metabolic endotoxemia initiates obesity and insulin resistance. *Diabetes* **2007**, *56*, 1761–1772. [CrossRef] [PubMed]
25. Koh, A.; Molinaro, A.; Stahlman, M.; Khan, M.T.; Schmidt, C.; Manneras-Holm, L.; Wu, H.; Carreras, A.; Jeong, H.; Olofsson, L.E.; et al. Microbially Produced Imidazole Propionate Impairs Insulin Signaling through mTORC1. *Cell* **2018**, *175*, 947–961 e917. [CrossRef] [PubMed]
26. Virtue, A.T.; McCright, S.J.; Wright, J.M.; Jimenez, M.T.; Mowel, W.K.; Kotzin, J.J.; Joannas, L.; Basavappa, M.G.; Spencer, S.P.; Clark, M.L.; et al. The gut microbiota regulates white adipose tissue inflammation and obesity via a family of microRNAs. *Sci. Transl. Med.* **2019**, *11*. [CrossRef] [PubMed]
27. Pellizzon, M.A.; Ricci, M.R. The common use of improper control diets in diet-induced metabolic disease research confounds data interpretation: The fiber factor. *Nutr. Metab. (Lond.)* **2018**, *15*, 3. [CrossRef]
28. Vrieze, A.; Van Nood, E.; Holleman, F.; Salojarvi, J.; Kootte, R.S.; Bartelsman, J.F.; Dallinga-Thie, G.M.; Ackermans, M.T.; Serlie, M.J.; Oozeer, R.; et al. Transfer of intestinal microbiota from lean donors increases insulin sensitivity in individuals with metabolic syndrome. *Gastroenterology* **2012**, *143*, 913–916 e917. [CrossRef]
29. Everard, A.; Belzer, C.; Geurts, L.; Ouwerkerk, J.P.; Druart, C.; Bindels, L.B.; Guiot, Y.; Derrien, M.; Muccioli, G.G.; Delzenne, N.M.; et al. Cross-talk between Akkermansia muciniphila and intestinal epithelium controls diet-induced obesity. *Proc. Natl. Acad. Sci. USA* **2013**, *110*, 9066–9071. [CrossRef]
30. Lichtenstein, A.H.; Kennedy, E.; Barrier, P.; Danford, D.; Ernst, N.D.; Grundy, S.M.; Leveille, G.A.; Van Horn, L.; Williams, C.L.; Booth, S.L. Dietary fat consumption and health. *Nutr. Rev.* **1998**, *56*, S19–28. [CrossRef]
31. Hennessy, A.A.; Ross, R.P.; Fitzgerald, G.F.; Caplice, N.; Stanton, C. Role of the gut in modulating lipoprotein metabolism. *Curr. Cardiol. Rep.* **2014**, *16*, 515. [CrossRef] [PubMed]
32. Corvilain, B. Lipoprotein metabolism. *Rev. Med. Brux.* **1997**, *18*, 3–9. [PubMed]
33. Demignot, S.; Beilstein, F.; Morel, E. Triglyceride-rich lipoproteins and cytosolic lipid droplets in enterocytes: Key players in intestinal physiology and metabolic disorders. *Biochimie* **2014**, *96*, 48–55. [CrossRef] [PubMed]
34. Warnakula, S.; Hsieh, J.; Adeli, K.; Hussain, M.M.; Tso, P.; Proctor, S.D. New insights into how the intestine can regulate lipid homeostasis and impact vascular disease: Frontiers for new pharmaceutical therapies to lower cardiovascular disease risk. *Can. J. Cardiol.* **2011**, *27*, 183–191. [CrossRef]
35. Chahil, T.J.; Ginsberg, H.N. Diabetic dyslipidemia. *Endocrinol Metab. Clin. North. Am.* **2006**, *35*, 491–510, vii–viii. [CrossRef]
36. Liu, Q.; Bengmark, S.; Qu, S. The role of hepatic fat accumulation in pathogenesis of non-alcoholic fatty liver disease (NAFLD). *Lipids Health Dis.* **2010**, *9*, 42. [CrossRef]
37. Leone, V.; Gibbons, S.M.; Martinez, K.; Hutchison, A.L.; Huang, E.Y.; Cham, C.M.; Pierre, J.F.; Heneghan, A.F.; Nadimpalli, A.; Hubert, N.; et al. Effects of diurnal variation of gut microbes and high-fat feeding on host circadian clock function and metabolism. *Cell Host Microbe* **2015**, *17*, 681–689. [CrossRef]
38. Martinez-Guryn, K.; Hubert, N.; Frazier, K.; Urlass, S.; Musch, M.W.; Ojeda, P.; Pierre, J.F.; Miyoshi, J.; Sontag, T.J.; Cham, C.M.; et al. Small Intestine Microbiota Regulate Host Digestive and Absorptive Adaptive Responses to Dietary Lipids. *Cell Host Microbe* **2018**, *23*, 458–469. [CrossRef]
39. Carr, T.P.; Weller, C.L.; Schlegel, V.L.; Cuppett, S.L.; Guderian, D.M., Jr.; Johnson, K.R. Grain sorghum lipid extract reduces cholesterol absorption and plasma non-HDL cholesterol concentration in hamsters. *J. Nutr.* **2005**, *135*, 2236–2240. [CrossRef]
40. Martinez, I.; Wallace, G.; Zhang, C.; Legge, R.; Benson, A.K.; Carr, T.P.; Moriyama, E.N.; Walter, J. Diet-induced metabolic improvements in a hamster model of hypercholesterolemia are strongly linked to alterations of the gut microbiota. *Appl. Environ. Microbiol.* **2009**, *75*, 4175–4184. [CrossRef]

41. Cotillard, A.; Kennedy, S.P.; Kong, L.C.; Prifti, E.; Pons, N.; Le Chatelier, E.; Almeida, M.; Quinquis, B.; Levenez, F.; Galleron, N.; et al. Dietary intervention impact on gut microbial gene richness. *Nature* **2013**, *500*, 585–588. [CrossRef] [PubMed]
42. Le Chatelier, E.; Nielsen, T.; Qin, J.; Prifti, E.; Hildebrand, F.; Falony, G.; Almeida, M.; Arumugam, M.; Batto, J.M.; Kennedy, S.; et al. Richness of human gut microbiome correlates with metabolic markers. *Nature* **2013**, *500*, 541–546. [CrossRef] [PubMed]
43. Fu, J.; Bonder, M.J.; Cenit, M.C.; Tigchelaar, E.F.; Maatman, A.; Dekens, J.A.; Brandsma, E.; Marczynska, J.; Imhann, F.; Weersma, R.K.; et al. The Gut Microbiome Contributes to a Substantial Proportion of the Variation in Blood Lipids. *Circ. Res.* **2015**, *117*, 817–824. [CrossRef]
44. Rebolledo, C.; Cuevas, A.; Zambrano, T.; Acuna, J.J.; Jorquera, M.A.; Saavedra, K.; Martinez, C.; Lanas, F.; Seron, P.; Salazar, L.A.; et al. Bacterial Community Profile of the Gut Microbiota Differs between Hypercholesterolemic Subjects and Controls. *Biomed Res. Int.* **2017**, *2017*, 8127814. [CrossRef]
45. Karlsson, F.H.; Tremaroli, V.; Nookaew, I.; Bergstrom, G.; Behre, C.J.; Fagerberg, B.; Nielsen, J.; Backhed, F. Gut metagenome in European women with normal, impaired and diabetic glucose control. *Nature* **2013**, *498*, 99–103. [CrossRef] [PubMed]
46. Guevara-Cruz, M.; Flores-Lopez, A.G.; Aguilar-Lopez, M.; Sanchez-Tapia, M.; Medina-Vera, I.; Diaz, D.; Tovar, A.R.; Torres, N. Improvement of Lipoprotein Profile and Metabolic Endotoxemia by a Lifestyle Intervention That Modifies the Gut Microbiota in Subjects With Metabolic Syndrome. *J. Am. Heart Assoc.* **2019**, *8*, e012401. [CrossRef] [PubMed]
47. Lo Sasso, G.; Schlage, W.K.; Boue, S.; Veljkovic, E.; Peitsch, M.C.; Hoeng, J. The Apoe(-/-) mouse model: A suitable model to study cardiovascular and respiratory diseases in the context of cigarette smoke exposure and harm reduction. *J. Transl. Med.* **2016**, *14*, 146. [CrossRef]
48. Rune, I.; Rolin, B.; Larsen, C.; Nielsen, D.S.; Kanter, J.E.; Bornfeldt, K.E.; Lykkesfeldt, J.; Buschard, K.; Kirk, R.K.; Christoffersen, B.; et al. Modulating the Gut Microbiota Improves Glucose Tolerance, Lipoprotein Profile and Atherosclerotic Plaque Development in ApoE-Deficient Mice. *PLoS ONE* **2016**, *11*, e0146439. [CrossRef]
49. Jin, Y.; Wu, Y.; Zeng, Z.; Jin, C.; Wu, S.; Wang, Y.; Fu, Z. From the Cover: Exposure to Oral Antibiotics Induces Gut Microbiota Dysbiosis Associated with Lipid Metabolism Dysfunction and Low-Grade Inflammation in Mice. *Toxicol. Sci.* **2016**, *154*, 140–152. [CrossRef]
50. Kim, J.; Lee, H.; An, J.; Song, Y.; Lee, C.K.; Kim, K.; Kong, H. Alterations in Gut Microbiota by Statin Therapy and Possible Intermediate Effects on Hyperglycemia and Hyperlipidemia. *Front. Microbiol.* **2019**, *10*, 1947. [CrossRef]
51. Sato, H.; Zhang, L.S.; Martinez, K.; Chang, E.B.; Yang, Q.; Wang, F.; Howles, P.N.; Hokari, R.; Miura, S.; Tso, P. Antibiotics Suppress Activation of Intestinal Mucosal Mast Cells and Reduce Dietary Lipid Absorption in Sprague-Dawley Rats. *Gastroenterology* **2016**, *151*, 923–932. [CrossRef] [PubMed]
52. Li, X.; Zhang, Z.; Cheng, J.; Diao, C.; Yan, Y.; Liu, D.; Wang, H.; Zheng, F. Dietary supplementation of soybean-derived sterols regulates cholesterol metabolism and intestinal microbiota in hamsters. *J. Funct. Foods* **2019**, *59*, 242–250. [CrossRef]
53. Abumweis, S.S.; Barake, R.; Jones, P.J. Plant sterols/stanols as cholesterol lowering agents: A meta-analysis of randomized controlled trials. *Food Nutr. Res.* **2008**, *52*. [CrossRef] [PubMed]
54. Noakes, M.; Clifton, P.M.; Doornbos, A.M.; Trautwein, E.A. Plant sterol ester-enriched milk and yoghurt effectively reduce serum cholesterol in modestly hypercholesterolemic subjects. *Eur. J. Nutr.* **2005**, *44*, 214–222. [CrossRef]
55. Seppo, L.; Jauhiainen, T.; Nevala, R.; Poussa, T.; Korpela, R. Plant stanol esters in low-fat milk products lower serum total and LDL cholesterol. *Eur. J. Nutr.* **2007**, *46*, 111–117. [CrossRef]
56. Dubenetzky, M.C.; Rasmussen, H.; Carr, T.; Walter, J. The role of gut microbiota in the low-density lipoprotein (LDL) cholesterol-lowering effects of plant sterol esters. *FASEB J.* **2011**, *25*.
57. Begley, M.; Gahan, C.G.; Hill, C. The interaction between bacteria and bile. *FEMS Microbiol. Rev.* **2005**, *29*, 625–651. [CrossRef]
58. Li, T.; Chiang, J.Y. Bile acid signaling in metabolic disease and drug therapy. *Pharmacol. Rev.* **2014**, *66*, 948–983. [CrossRef]

59. Jiang, C.; Xie, C.; Li, F.; Zhang, L.; Nichols, R.G.; Krausz, K.W.; Cai, J.; Qi, Y.; Fang, Z.Z.; Takahashi, S.; et al. Intestinal farnesoid X receptor signaling promotes nonalcoholic fatty liver disease. *J. Clin. Investig.* **2015**, *125*, 386–402. [CrossRef]
60. Miyata, M.; Takamatsu, Y.; Kuribayashi, H.; Yamazoe, Y. Administration of ampicillin elevates hepatic primary bile acid synthesis through suppression of ileal fibroblast growth factor 15 expression. *J. Pharmacol. Exp. Ther.* **2009**, *331*, 1079–1085. [CrossRef]
61. Kuribayashi, H.; Miyata, M.; Yamakawa, H.; Yoshinari, K.; Yamazoe, Y. Enterobacteria-mediated deconjugation of taurocholic acid enhances ileal farnesoid X receptor signaling. *Eur. J. Pharmacol.* **2012**, *697*, 132–138. [CrossRef] [PubMed]
62. Sayin, S.I.; Wahlstrom, A.; Felin, J.; Jantti, S.; Marschall, H.U.; Bamberg, K.; Angelin, B.; Hyotylainen, T.; Oresic, M.; Backhed, F. Gut microbiota regulates bile acid metabolism by reducing the levels of tauro-beta-muricholic acid, a naturally occurring FXR antagonist. *Cell Metab.* **2013**, *17*, 225–235. [CrossRef] [PubMed]
63. Kuno, T.; Hirayama-Kurogi, M.; Ito, S.; Ohtsuki, S. Reduction in hepatic secondary bile acids caused by short-term antibiotic-induced dysbiosis decreases mouse serum glucose and triglyceride levels. *Sci. Rep.* **2018**, *8*, 1253. [CrossRef] [PubMed]
64. Sun, L.; Pang, Y.; Wang, X.; Wu, Q.; Liu, H.; Liu, B.; Liu, G.; Ye, M.; Kong, W.; Jiang, C. Ablation of gut microbiota alleviates obesity-induced hepatic steatosis and glucose intolerance by modulating bile acid metabolism in hamsters. *Acta. Pharm. Sin. B* **2019**, *9*, 702–710. [CrossRef] [PubMed]
65. Reijnders, D.; Goossens, G.H.; Hermes, G.D.; Neis, E.P.; van der Beek, C.M.; Most, J.; Holst, J.J.; Lenaerts, K.; Kootte, R.S.; Nieuwdorp, M.; et al. Effects of Gut Microbiota Manipulation by Antibiotics on Host Metabolism in Obese Humans: A Randomized Double-Blind Placebo-Controlled Trial. *Cell Metab.* **2016**, *24*, 341. [CrossRef]
66. Qi, Y.; Jiang, C.; Cheng, J.; Krausz, K.W.; Li, T.; Ferrell, J.M.; Gonzalez, F.J.; Chiang, J.Y. Bile acid signaling in lipid metabolism: metabolomic and lipidomic analysis of lipid and bile acid markers linked to anti-obesity and anti-diabetes in mice. *Biochim. Biophys. Acta* **2015**, *1851*, 19–29. [CrossRef]
67. Brannigan, J.A.; Dodson, G.; Duggleby, H.J.; Moody, P.C.; Smith, J.L.; Tomchick, D.R.; Murzin, A.G. A protein catalytic framework with an N-terminal nucleophile is capable of self-activation. *Nature* **1995**, *378*, 416–419. [CrossRef]
68. Artymiuk, P.J. A sting in the (N-terminal) tail. *Nat. Struct. Biol.* **1995**, *2*, 1035–1037. [CrossRef]
69. Jones, B.V.; Begley, M.; Hill, C.; Gahan, C.G.; Marchesi, J.R. Functional and comparative metagenomic analysis of bile salt hydrolase activity in the human gut microbiome. *Proc. Natl. Acad. Sci. USA* **2008**, *105*, 13580–13585. [CrossRef]
70. Wahlstrom, A.; Kovatcheva-Datchary, P.; Stahlman, M.; Khan, M.T.; Backhed, F.; Marschall, H.U. Induction of farnesoid X receptor signaling in germ-free mice colonized with a human microbiota. *J. Lipid Res.* **2017**, *58*, 412–419. [CrossRef]
71. Islam, K.B.; Fukiya, S.; Hagio, M.; Fujii, N.; Ishizuka, S.; Ooka, T.; Ogura, Y.; Hayashi, T.; Yokota, A. Bile acid is a host factor that regulates the composition of the cecal microbiota in rats. *Gastroenterology* **2011**, *141*, 1773–1781. [CrossRef] [PubMed]
72. Just, S.; Mondot, S.; Ecker, J.; Wegner, K.; Rath, E.; Gau, L.; Streidl, T.; Hery-Arnaud, G.; Schmidt, S.; Lesker, T.R.; et al. The gut microbiota drives the impact of bile acids and fat source in diet on mouse metabolism. *Microbiome* **2018**, *6*, 134. [CrossRef] [PubMed]
73. Forman, B.M.; Goode, E.; Chen, J.; Oro, A.E.; Bradley, D.J.; Perlmann, T.; Noonan, D.J.; Burka, L.T.; McMorris, T.; Lamph, W.W.; et al. Identification of a nuclear receptor that is activated by farnesol metabolites. *Cell* **1995**, *81*, 687–693. [CrossRef]
74. Schneider, K.M.; Albers, S.; Trautwein, C. Role of bile acids in the gut-liver axis. *J. Hepatol.* **2018**, *68*, 1083–1085. [CrossRef] [PubMed]
75. Watanabe, M.; Houten, S.M.; Wang, L.; Moschetta, A.; Mangelsdorf, D.J.; Heyman, R.A.; Moore, D.D.; Auwerx, J. Bile acids lower triglyceride levels via a pathway involving FXR, SHP, and SREBP-1c. *J. Clin. Investig.* **2004**, *113*, 1408–1418. [CrossRef]
76. Hirokane, H.; Nakahara, M.; Tachibana, S.; Shimizu, M.; Sato, R. Bile acid reduces the secretion of very low density lipoprotein by repressing microsomal triglyceride transfer protein gene expression mediated by hepatocyte nuclear factor-4. *J. Biol. Chem.* **2004**, *279*, 45685–45692. [CrossRef]

77. Bilz, S.; Samuel, V.; Morino, K.; Savage, D.; Choi, C.S.; Shulman, G.I. Activation of the farnesoid X receptor improves lipid metabolism in combined hyperlipidemic hamsters. *Am. J. Physiol. Endocrinol. Metab.* **2006**, *290*, E716–E722. [CrossRef]
78. Zhang, X.; Osaka, T.; Tsuneda, S. Bacterial metabolites directly modulate farnesoid X receptor activity. *Nutr. Metab. (Lond.)* **2015**, *12*, 48. [CrossRef]
79. Yang, Z.X.; Shen, W.; Sun, H. Effects of nuclear receptor FXR on the regulation of liver lipid metabolism in patients with non-alcoholic fatty liver disease. *Hepatol. Int.* **2010**, *4*, 741–748. [CrossRef]
80. Jiang, C.; Xie, C.; Lv, Y.; Li, J.; Krausz, K.W.; Shi, J.; Brocker, C.N.; Desai, D.; Amin, S.G.; Bisson, W.H.; et al. Intestine-selective farnesoid X receptor inhibition improves obesity-related metabolic dysfunction. *Nat. Commun.* **2015**, *6*, 10166. [CrossRef]
81. Maruyama, T.; Miyamoto, Y.; Nakamura, T.; Tamai, Y.; Okada, H.; Sugiyama, E.; Nakamura, T.; Itadani, H.; Tanaka, K. Identification of membrane-type receptor for bile acids (M-BAR). *Biochem. Biophys. Res. Commun.* **2002**, *298*, 714–719. [CrossRef]
82. Thomas, C.; Gioiello, A.; Noriega, L.; Strehle, A.; Oury, J.; Rizzo, G.; Macchiarulo, A.; Yamamoto, H.; Mataki, C.; Pruzanski, M.; et al. TGR5-mediated bile acid sensing controls glucose homeostasis. *Cell Metab.* **2009**, *10*, 167–177. [CrossRef] [PubMed]
83. Zambad, S.P.; Tuli, D.; Mathur, A.; Ghalsasi, S.A.; Chaudhary, A.R.; Deshpande, S.; Gupta, R.C.; Chauthaiwale, V.; Dutt, C. TRC210258, a novel TGR5 agonist, reduces glycemic and dyslipidemic cardiovascular risk in animal models of diabesity. *Diabetes Metab. Syndr. Obes.* **2013**, *7*, 1–14. [CrossRef] [PubMed]
84. Finn, P.D.; Rodriguez, D.; Kohler, J.; Jiang, Z.; Wan, S.; Blanco, E.; King, A.J.; Chen, T.; Bell, N.; Dragoli, D.; et al. Intestinal TGR5 agonism improves hepatic steatosis and insulin sensitivity in Western diet-fed mice. *Am. J. Physiol. Gastrointest. Liver Physiol.* **2019**, *316*, G412–G424. [CrossRef] [PubMed]
85. Maruyama, T.; Tanaka, K.; Suzuki, J.; Miyoshi, H.; Harada, N.; Nakamura, T.; Miyamoto, Y.; Kanatani, A.; Tamai, Y. Targeted disruption of G protein-coupled bile acid receptor 1 (Gpbar1/M-Bar) in mice. *J. Endocrinol.* **2006**, *191*, 197–205. [CrossRef] [PubMed]
86. Donepudi, A.C.; Boehme, S.; Li, F.; Chiang, J.Y. G-protein-coupled bile acid receptor plays a key role in bile acid metabolism and fasting-induced hepatic steatosis in mice. *Hepatology* **2017**, *65*, 813–827. [CrossRef] [PubMed]
87. Pathak, P.; Xie, C.; Nichols, R.G.; Ferrell, J.M.; Boehme, S.; Krausz, K.W.; Patterson, A.D.; Gonzalez, F.J.; Chiang, J.Y.L. Intestine farnesoid X receptor agonist and the gut microbiota activate G-protein bile acid receptor-1 signaling to improve metabolism. *Hepatology* **2018**, *68*, 1574–1588. [CrossRef]
88. Wu, T.; Bound, M.J.; Standfield, S.D.; Gedulin, B.; Jones, K.L.; Horowitz, M.; Rayner, C.K. Effects of rectal administration of taurocholic acid on glucagon-like peptide-1 and peptide YY secretion in healthy humans. *Diabetes Obes. Metab.* **2013**, *15*, 474–477. [CrossRef]
89. Pathak, P.; Liu, H.; Boehme, S.; Xie, C.; Krausz, K.W.; Gonzalez, F.; Chiang, J.Y.L. Farnesoid X receptor induces Takeda G-protein receptor 5 cross-talk to regulate bile acid synthesis and hepatic metabolism. *J. Biol. Chem.* **2017**, *292*, 11055–11069. [CrossRef]
90. Flint, H.J.; Duncan, S.H.; Scott, K.P.; Louis, P. Links between diet, gut microbiota composition and gut metabolism. *Proc. Nutr. Soc.* **2015**, *74*, 13–22. [CrossRef]
91. Cummings, J.H.; Pomare, E.W.; Branch, W.J.; Naylor, C.P.; Macfarlane, G.T. Short chain fatty acids in human large intestine, portal, hepatic and venous blood. *Gut* **1987**, *28*, 1221–1227. [CrossRef] [PubMed]
92. Bloemen, J.G.; Venema, K.; van de Poll, M.C.; Olde Damink, S.W.; Buurman, W.A.; Dejong, C.H. Short chain fatty acids exchange across the gut and liver in humans measured at surgery. *Clin. Nutr.* **2009**, *28*, 657–661. [CrossRef] [PubMed]
93. Canfora, E.E.; Jocken, J.W.; Blaak, E.E. Short-chain fatty acids in control of body weight and insulin sensitivity. *Nat. Rev. Endocrinol.* **2015**, *11*, 577–591. [CrossRef] [PubMed]
94. Louis, P.; Young, P.; Holtrop, G.; Flint, H.J. Diversity of human colonic butyrate-producing bacteria revealed by analysis of the butyryl-CoA:acetate CoA-transferase gene. *Environ. Microbiol.* **2010**, *12*, 304–314. [CrossRef] [PubMed]
95. Ze, X.; Duncan, S.H.; Louis, P.; Flint, H.J. Ruminococcus bromii is a keystone species for the degradation of resistant starch in the human colon. *ISME J.* **2012**, *6*, 1535–1543. [CrossRef]

96. Derrien, M.; Vaughan, E.E.; Plugge, C.M.; de Vos, W.M. Akkermansia muciniphila gen. nov., sp. nov., a human intestinal mucin-degrading bacterium. *Int J. Syst. Evol. Microbiol.* **2004**, *54*, 1469–1476. [CrossRef]
97. Turnbaugh, P.J.; Ley, R.E.; Mahowald, M.A.; Magrini, V.; Mardis, E.R.; Gordon, J.I. An obesity-associated gut microbiome with increased capacity for energy harvest. *Nature* **2006**, *444*, 1027–1031. [CrossRef]
98. Murphy, E.F.; Cotter, P.D.; Healy, S.; Marques, T.M.; O'Sullivan, O.; Fouhy, F.; Clarke, S.F.; O'Toole, P.W.; Quigley, E.M.; Stanton, C.; et al. Composition and energy harvesting capacity of the gut microbiota: Relationship to diet, obesity and time in mouse models. *Gut* **2010**, *59*, 1635–1642. [CrossRef]
99. Schwiertz, A.; Taras, D.; Schafer, K.; Beijer, S.; Bos, N.A.; Donus, C.; Hardt, P.D. Microbiota and SCFA in lean and overweight healthy subjects. *Obes. Silver Spring* **2010**, *18*, 190–195. [CrossRef]
100. Fernandes, J.; Su, W.; Rahat-Rozenbloom, S.; Wolever, T.M.; Comelli, E.M. Adiposity, gut microbiota and faecal short chain fatty acids are linked in adult humans. *Nutr. Diabetes* **2014**, *4*, e121. [CrossRef]
101. Singh, V.; Chassaing, B.; Zhang, L.; San Yeoh, B.; Xiao, X.; Kumar, M.; Baker, M.T.; Cai, J.; Walker, R.; Borkowski, K.; et al. Microbiota-Dependent Hepatic Lipogenesis Mediated by Stearoyl CoA Desaturase 1 (SCD1) Promotes Metabolic Syndrome in TLR5-Deficient Mice. *Cell Metab.* **2015**, *22*, 983–996. [CrossRef] [PubMed]
102. Ridaura, V.K.; Faith, J.J.; Rey, F.E.; Cheng, J.; Duncan, A.E.; Kau, A.L.; Griffin, N.W.; Lombard, V.; Henrissat, B.; Bain, J.R.; et al. Gut microbiota from twins discordant for obesity modulate metabolism in mice. *Science* **2013**, *341*, 1241214. [CrossRef] [PubMed]
103. Demigne, C.; Morand, C.; Levrat, M.A.; Besson, C.; Moundras, C.; Remesy, C. Effect of propionate on fatty acid and cholesterol synthesis and on acetate metabolism in isolated rat hepatocytes. *Br. J. Nutr* **1995**, *74*, 209–219. [CrossRef] [PubMed]
104. Hong, Y.H.; Nishimura, Y.; Hishikawa, D.; Tsuzuki, H.; Miyahara, H.; Gotoh, C.; Choi, K.C.; Feng, D.D.; Chen, C.; Lee, H.G.; et al. Acetate and propionate short chain fatty acids stimulate adipogenesis via GPCR43. *Endocrinology* **2005**, *146*, 5092–5099. [CrossRef] [PubMed]
105. Chambers, E.S.; Viardot, A.; Psichas, A.; Morrison, D.J.; Murphy, K.G.; Zac-Varghese, S.E.; MacDougall, K.; Preston, T.; Tedford, C.; Finlayson, G.S.; et al. Effects of targeted delivery of propionate to the human colon on appetite regulation, body weight maintenance and adiposity in overweight adults. *Gut* **2015**, *64*, 1744–1754. [CrossRef]
106. Tolhurst, G.; Heffron, H.; Lam, Y.S.; Parker, H.E.; Habib, A.M.; Diakogiannaki, E.; Cameron, J.; Grosse, J.; Reimann, F.; Gribble, F.M. Short-chain fatty acids stimulate glucagon-like peptide-1 secretion via the G-protein-coupled receptor FFAR2. *Diabetes* **2012**, *61*, 364–371. [CrossRef]
107. Psichas, A.; Sleeth, M.L.; Murphy, K.G.; Brooks, L.; Bewick, G.A.; Hanyaloglu, A.C.; Ghatei, M.A.; Bloom, S.R.; Frost, G. The short chain fatty acid propionate stimulates GLP-1 and PYY secretion via free fatty acid receptor 2 in rodents. *Int. J. Obes. (Lond.)* **2015**, *39*, 424–429. [CrossRef]
108. Reimer, R.A.; Darimont, C.; Gremlich, S.; Nicolas-Metral, V.; Ruegg, U.T.; Mace, K. A human cellular model for studying the regulation of glucagon-like peptide-1 secretion. *Endocrinology* **2001**, *142*, 4522–4528. [CrossRef]
109. Karaki, S.; Mitsui, R.; Hayashi, H.; Kato, I.; Sugiya, H.; Iwanaga, T.; Furness, J.B.; Kuwahara, A. Short-chain fatty acid receptor, GPR43, is expressed by enteroendocrine cells and mucosal mast cells in rat intestine. *Cell Tissue Res.* **2006**, *324*, 353–360. [CrossRef]
110. Samuel, B.S.; Shaito, A.; Motoike, T.; Rey, F.E.; Backhed, F.; Manchester, J.K.; Hammer, R.E.; Williams, S.C.; Crowley, J.; Yanagisawa, M.; et al. Effects of the gut microbiota on host adiposity are modulated by the short-chain fatty-acid binding G protein-coupled receptor, Gpr41. *Proc. Natl. Acad. Sci. USA* **2008**, *105*, 16767–16772. [CrossRef]
111. Cani, P.D.; Dewever, C.; Delzenne, N.M. Inulin-type fructans modulate gastrointestinal peptides involved in appetite regulation (glucagon-like peptide-1 and ghrelin) in rats. *Br. J. Nutr.* **2004**, *92*, 521–526. [CrossRef] [PubMed]
112. Zhou, J.; Martin, R.J.; Tulley, R.T.; Raggio, A.M.; McCutcheon, K.L.; Shen, L.; Danna, S.C.; Tripathy, S.; Hegsted, M.; Keenan, M.J. Dietary resistant starch upregulates total GLP-1 and PYY in a sustained day-long manner through fermentation in rodents. *Am. J. Physiol. Endocrinol. Metab.* **2008**, *295*, E1160–1166. [CrossRef] [PubMed]
113. Cani, P.D.; Neyrinck, A.M.; Maton, N.; Delzenne, N.M. Oligofructose promotes satiety in rats fed a high-fat diet: involvement of glucagon-like Peptide-1. *Obes. Res.* **2005**, *13*, 1000–1007. [CrossRef] [PubMed]

114. Cani, P.D.; Lecourt, E.; Dewulf, E.M.; Sohet, F.M.; Pachikian, B.D.; Naslain, D.; De Backer, F.; Neyrinck, A.M.; Delzenne, N.M. Gut microbiota fermentation of prebiotics increases satietogenic and incretin gut peptide production with consequences for appetite sensation and glucose response after a meal. *Am. J. Clin. Nutr.* **2009**, *90*, 1236–1243. [CrossRef] [PubMed]
115. Parnell, J.A.; Reimer, R.A. Weight loss during oligofructose supplementation is associated with decreased ghrelin and increased peptide YY in overweight and obese adults. *Am. J. Clin. Nutr.* **2009**, *89*, 1751–1759. [CrossRef] [PubMed]
116. Pedersen, C.; Lefevre, S.; Peters, V.; Patterson, M.; Ghatei, M.A.; Morgan, L.M.; Frost, G.S. Gut hormone release and appetite regulation in healthy non-obese participants following oligofructose intake. A dose-escalation study. *Appetite* **2013**, *66*, 44–53. [CrossRef]
117. Gribble, F.M.; Reimann, F. Enteroendocrine Cells: Chemosensors in the Intestinal Epithelium. *Annu. Rev. Physiol.* **2016**, *78*, 277–299. [CrossRef]
118. Habib, A.M.; Richards, P.; Cairns, L.S.; Rogers, G.J.; Bannon, C.A.; Parker, H.E.; Morley, T.C.; Yeo, G.S.; Reimann, F.; Gribble, F.M. Overlap of endocrine hormone expression in the mouse intestine revealed by transcriptional profiling and flow cytometry. *Endocrinology* **2012**, *153*, 3054–3065. [CrossRef]
119. Arora, T.; Akrami, R.; Pais, R.; Bergqvist, L.; Johansson, B.R.; Schwartz, T.W.; Reimann, F.; Gribble, F.M.; Backhed, F. Microbial regulation of the L cell transcriptome. *Sci. Rep.* **2018**, *8*, 1207. [CrossRef]
120. Wichmann, A.; Allahyar, A.; Greiner, T.U.; Plovier, H.; Lunden, G.O.; Larsson, T.; Drucker, D.J.; Delzenne, N.M.; Cani, P.D.; Backhed, F. Microbial modulation of energy availability in the colon regulates intestinal transit. *Cell Host Microbe* **2013**, *14*, 582–590. [CrossRef]
121. Harach, T.; Pols, T.W.; Nomura, M.; Maida, A.; Watanabe, M.; Auwerx, J.; Schoonjans, K. TGR5 potentiates GLP-1 secretion in response to anionic exchange resins. *Sci. Rep.* **2012**, *2*, 430. [CrossRef]
122. Kok, N.N.; Morgan, L.M.; Williams, C.M.; Roberfroid, M.B.; Thissen, J.P.; Delzenne, N.M. Insulin, glucagon-like peptide 1, glucose-dependent insulinotropic polypeptide and insulin-like growth factor I as putative mediators of the hypolipidemic effect of oligofructose in rats. *J. Nutr.* **1998**, *128*, 1099–1103. [CrossRef] [PubMed]
123. Cani, P.D.; Knauf, C.; Iglesias, M.A.; Drucker, D.J.; Delzenne, N.M.; Burcelin, R. Improvement of glucose tolerance and hepatic insulin sensitivity by oligofructose requires a functional glucagon-like peptide 1 receptor. *Diabetes* **2006**, *55*, 1484–1490. [CrossRef] [PubMed]
124. Kreymann, B.; Williams, G.; Ghatei, M.A.; Bloom, S.R. Glucagon-like peptide-1 7-36: A physiological incretin in man. *Lancet* **1987**, *2*, 1300–1304. [CrossRef]
125. Mojsov, S.; Weir, G.C.; Habener, J.F. Insulinotropin: glucagon-like peptide I (7-37) co-encoded in the glucagon gene is a potent stimulator of insulin release in the perfused rat pancreas. *J. Clin. Invest.* **1987**, *79*, 616–619. [CrossRef] [PubMed]
126. Dailey, M.J.; Moran, T.H. Glucagon-like peptide 1 and appetite. *Trends Endocrinol. Metab.* **2013**, *24*, 85–91. [CrossRef]
127. Ottney, A. Glucagon-like peptide-1 receptor agonists for weight loss in adult patients without diabetes. *Am. J. Health Syst. Pharm.* **2013**, *70*, 2097–2103. [CrossRef]
128. Farr, S.; Taher, J.; Adeli, K. Glucagon-like peptide-1 as a key regulator of lipid and lipoprotein metabolism in fasting and postprandial states. *Cardiovasc. Hematol. Disord Drug Targets* **2014**, *14*, 126–136. [CrossRef]
129. Armstrong, M.J.; Hull, D.; Guo, K.; Barton, D.; Hazlehurst, J.M.; Gathercole, L.L.; Nasiri, M.; Yu, J.; Gough, S.C.; Newsome, P.N.; et al. Glucagon-like peptide 1 decreases lipotoxicity in non-alcoholic steatohepatitis. *J. Hepatol.* **2016**, *64*, 399–408. [CrossRef]
130. Hsieh, J.; Longuet, C.; Baker, C.L.; Qin, B.; Federico, L.M.; Drucker, D.J.; Adeli, K. The glucagon-like peptide 1 receptor is essential for postprandial lipoprotein synthesis and secretion in hamsters and mice. *Diabetologia* **2010**, *53*, 552–561. [CrossRef]
131. Qin, X.; Shen, H.; Liu, M.; Yang, Q.; Zheng, S.; Sabo, M.; D'Alessio, D.A.; Tso, P. GLP-1 reduces intestinal lymph flow, triglyceride absorption, and apolipoprotein production in rats. *Am. J. Physiol. Gastrointest. Liver Physiol.* **2005**, *288*, G943–G949. [CrossRef] [PubMed]
132. Matikainen, N.; Manttari, S.; Schweizer, A.; Ulvestad, A.; Mills, D.; Dunning, B.E.; Foley, J.E.; Taskinen, M.R. Vildagliptin therapy reduces postprandial intestinal triglyceride-rich lipoprotein particles in patients with type 2 diabetes. *Diabetologia* **2006**, *49*, 2049–2057. [CrossRef] [PubMed]

133. Tremblay, A.J.; Lamarche, B.; Deacon, C.F.; Weisnagel, S.J.; Couture, P. Effect of sitagliptin therapy on postprandial lipoprotein levels in patients with type 2 diabetes. *Diabetes Obes. Metab.* **2011**, *13*, 366–373. [CrossRef] [PubMed]
134. Masuda, D.; Kobayashi, T.; Sairyou, M.; Hanada, H.; Ohama, T.; Koseki, M.; Nishida, M.; Maeda, N.; Kihara, S.; Minami, T.; et al. Effects of a Dipeptidyl Peptidase 4 Inhibitor Sitagliptin on Glycemic Control and Lipoprotein Metabolism in Patients with Type 2 Diabetes Mellitus (GLORIA Trial). *J. Atheroscler. Thromb.* **2018**, *25*, 512–520. [CrossRef] [PubMed]
135. Schwartz, E.A.; Koska, J.; Mullin, M.P.; Syoufi, I.; Schwenke, D.C.; Reaven, P.D. Exenatide suppresses postprandial elevations in lipids and lipoproteins in individuals with impaired glucose tolerance and recent onset type 2 diabetes mellitus. *Atherosclerosis* **2010**, *212*, 217–222. [CrossRef]
136. Brubaker, P.L.; Crivici, A.; Izzo, A.; Ehrlich, P.; Tsai, C.H.; Drucker, D.J. Circulating and tissue forms of the intestinal growth factor, glucagon-like peptide-2. *Endocrinology* **1997**, *138*, 4837–4843. [CrossRef]
137. Kitchen, P.A.; Fitzgerald, A.J.; Goodlad, R.A.; Barley, N.F.; Ghatei, M.A.; Legon, S.; Bloom, S.R.; Price, A.; Walters, J.R.; Forbes, A. Glucagon-like peptide-2 increases sucrase-isomaltase but not caudal-related homeobox protein-2 gene expression. *Am. J. Physiol. Gastrointest. Liver Physiol.* **2000**, *278*, G425–428. [CrossRef]
138. Kato, Y.; Yu, D.; Schwartz, M.Z. Glucagonlike peptide-2 enhances small intestinal absorptive function and mucosal mass in vivo. *J. Pediatr. Surg.* **1999**, *34*, 18–21. [CrossRef]
139. Cheeseman, C.I. Upregulation of SGLT-1 transport activity in rat jejunum induced by GLP-2 infusion in vivo. *Am. J. Physiol* **1997**, *273*, R1965–R1971. [CrossRef]
140. Hsieh, J.; Trajcevski, K.E.; Farr, S.L.; Baker, C.L.; Lake, E.J.; Taher, J.; Iqbal, J.; Hussain, M.M.; Adeli, K. Glucagon-Like Peptide 2 (GLP-2) Stimulates Postprandial Chylomicron Production and Postabsorptive Release of Intestinal Triglyceride Storage Pools via Induction of Nitric Oxide Signaling in Male Hamsters and Mice. *Endocrinology* **2015**, *156*, 3538–3547. [CrossRef]
141. Dash, S.; Xiao, C.; Morgantini, C.; Connelly, P.W.; Patterson, B.W.; Lewis, G.F. Glucagon-like peptide-2 regulates release of chylomicrons from the intestine. *Gastroenterology* **2014**, *147*, 1275–1284 e1274. [CrossRef] [PubMed]
142. Hsieh, J.; Longuet, C.; Maida, A.; Bahrami, J.; Xu, E.; Baker, C.L.; Brubaker, P.L.; Drucker, D.J.; Adeli, K. Glucagon-like peptide-2 increases intestinal lipid absorption and chylomicron production via CD36. *Gastroenterology* **2009**, *137*, 997–1005, 1005 e1001-1004. [CrossRef] [PubMed]
143. Hein, G.J.; Baker, C.; Hsieh, J.; Farr, S.; Adeli, K. GLP-1 and GLP-2 as yin and yang of intestinal lipoprotein production: evidence for predominance of GLP-2-stimulated postprandial lipemia in normal and insulin-resistant states. *Diabetes* **2013**, *62*, 373–381. [CrossRef] [PubMed]
144. Yano, J.M.; Yu, K.; Donaldson, G.P.; Shastri, G.G.; Ann, P.; Ma, L.; Nagler, C.R.; Ismagilov, R.F.; Mazmanian, S.K.; Hsiao, E.Y. Indigenous bacteria from the gut microbiota regulate host serotonin biosynthesis. *Cell* **2015**, *161*, 264–276. [CrossRef] [PubMed]
145. Mandic, A.D.; Woting, A.; Jaenicke, T.; Sander, A.; Sabrowski, W.; Rolle-Kampcyk, U.; von Bergen, M.; Blaut, M. Clostridium ramosum regulates enterochromaffin cell development and serotonin release. *Sci. Rep.* **2019**, *9*, 1177. [CrossRef]
146. Crane, J.D.; Palanivel, R.; Mottillo, E.P.; Bujak, A.L.; Wang, H.; Ford, R.J.; Collins, A.; Blumer, R.M.; Fullerton, M.D.; Yabut, J.M.; et al. Inhibiting peripheral serotonin synthesis reduces obesity and metabolic dysfunction by promoting brown adipose tissue thermogenesis. *Nat. Med.* **2015**, *21*, 166–172. [CrossRef]
147. Oh, C.M.; Namkung, J.; Go, Y.; Shong, K.E.; Kim, K.; Kim, H.; Park, B.Y.; Lee, H.W.; Jeon, Y.H.; Song, J.; et al. Regulation of systemic energy homeostasis by serotonin in adipose tissues. *Nat. Commun.* **2015**, *6*, 6794. [CrossRef]
148. Sumara, G.; Sumara, O.; Kim, J.K.; Karsenty, G. Gut-derived serotonin is a multifunctional determinant to fasting adaptation. *Cell Metab.* **2012**, *16*, 588–600. [CrossRef]
149. Choi, W.; Namkung, J.; Hwang, I.; Kim, H.; Lim, A.; Park, H.J.; Lee, H.W.; Han, K.H.; Park, S.; Jeong, J.S.; et al. Serotonin signals through a gut-liver axis to regulate hepatic steatosis. *Nat. Commun.* **2018**, *9*, 4824. [CrossRef]
150. Namkung, J.; Shong, K.E.; Kim, H.; Oh, C.M.; Park, S.; Kim, H. Inhibition of Serotonin Synthesis Induces Negative Hepatic Lipid Balance. *Diabetes Metab. J.* **2018**, *42*, 233–243. [CrossRef]
151. Niture, S.; Gyamfi, M.A.; Kedir, H.; Arthur, E.; Ressom, H.; Deep, G.; Kumar, D. Serotonin induced hepatic steatosis is associated with modulation of autophagy and notch signaling pathway. *Cell Commun. Signal.* **2018**, *16*, 78. [CrossRef] [PubMed]

152. Hirschfeld, M.; Ma, Y.; Weis, J.H.; Vogel, S.N.; Weis, J.J. Cutting edge: repurification of lipopolysaccharide eliminates signaling through both human and murine toll-like receptor 2. *J. Immunol.* **2000**, *165*, 618–622. [CrossRef] [PubMed]
153. Lebrun, L.J.; Lenaerts, K.; Kiers, D.; Pais de Barros, J.P.; Le Guern, N.; Plesnik, J.; Thomas, C.; Bourgeois, T.; Dejong, C.H.C.; Kox, M.; et al. Enteroendocrine L Cells Sense LPS after Gut Barrier Injury to Enhance GLP-1 Secretion. *Cell Rep.* **2017**, *21*, 1160–1168. [CrossRef] [PubMed]
154. Ghoshal, S.; Witta, J.; Zhong, J.; de Villiers, W.; Eckhardt, E. Chylomicrons promote intestinal absorption of lipopolysaccharides. *J. Lipid. Res.* **2009**, *50*, 90–97. [CrossRef] [PubMed]
155. Erridge, C.; Attina, T.; Spickett, C.M.; Webb, D.J. A high-fat meal induces low-grade endotoxemia: evidence of a novel mechanism of postprandial inflammation. *Am. J. Clin. Nutr.* **2007**, *86*, 1286–1292. [CrossRef] [PubMed]
156. Haidari, M.; Leung, N.; Mahbub, F.; Uffelman, K.D.; Kohen-Avramoglu, R.; Lewis, G.F.; Adeli, K. Fasting and postprandial overproduction of intestinally derived lipoproteins in an animal model of insulin resistance. Evidence that chronic fructose feeding in the hamster is accompanied by enhanced intestinal de novo lipogenesis and ApoB48-containing lipoprotein overproduction. *J. Biol. Chem.* **2002**, *277*, 31646–31655. [CrossRef] [PubMed]
157. Duez, H.; Lamarche, B.; Uffelman, K.D.; Valero, R.; Cohn, J.S.; Lewis, G.F. Hyperinsulinemia is associated with increased production rate of intestinal apolipoprotein B-48-containing lipoproteins in humans. *Arterioscler. Thromb. Vasc. Biol.* **2006**, *26*, 1357–1363. [CrossRef] [PubMed]
158. Duez, H.; Pavlic, M.; Lewis, G.F. Mechanism of intestinal lipoprotein overproduction in insulin resistant humans. *Atheroscler. Suppl.* **2008**, *9*, 33–38. [CrossRef]
159. Zilversmit, D.B. Atherogenesis: A postprandial phenomenon. *Circulation* **1979**, *60*, 473–485. [CrossRef]
160. Stoll, L.L.; Denning, G.M.; Weintraub, N.L. Potential role of endotoxin as a proinflammatory mediator of atherosclerosis. *Arterioscler. Thromb. Vasc. Biol.* **2004**, *24*, 2227–2236. [CrossRef]
161. Ostos, M.A.; Recalde, D.; Zakin, M.M.; Scott-Algara, D. Implication of natural killer T cells in atherosclerosis development during a LPS-induced chronic inflammation. *FEBS Lett.* **2002**, *519*, 23–29. [CrossRef]
162. Lehr, H.A.; Sagban, T.A.; Ihling, C.; Zahringer, U.; Hungerer, K.D.; Blumrich, M.; Reifenberg, K.; Bhakdi, S. Immunopathogenesis of atherosclerosis: endotoxin accelerates atherosclerosis in rabbits on hypercholesterolemic diet. *Circulation* **2001**, *104*, 914–920. [CrossRef] [PubMed]
163. Doddapattar, P.; Gandhi, C.; Prakash, P.; Dhanesha, N.; Grumbach, I.M.; Dailey, M.E.; Lentz, S.R.; Chauhan, A.K. Fibronectin Splicing Variants Containing Extra Domain A Promote Atherosclerosis in Mice Through Toll-Like Receptor 4. *Arterioscler. Thromb. Vasc. Biol.* **2015**, *35*, 2391–2400. [CrossRef] [PubMed]
164. Michelsen, K.S.; Wong, M.H.; Shah, P.K.; Zhang, W.; Yano, J.; Doherty, T.M.; Akira, S.; Rajavashisth, T.B.; Arditi, M. Lack of Toll-like receptor 4 or myeloid differentiation factor 88 reduces atherosclerosis and alters plaque phenotype in mice deficient in apolipoprotein E. *Proc. Natl. Acad. Sci. USA* **2004**, *101*, 10679–10684. [CrossRef]
165. Hoyles, L.; Fernandez-Real, J.M.; Federici, M.; Serino, M.; Abbott, J.; Charpentier, J.; Heymes, C.; Luque, J.L.; Anthony, E.; Barton, R.H.; et al. Molecular phenomics and metagenomics of hepatic steatosis in non-diabetic obese women. *Nat. Med.* **2018**, *24*, 1070–1080. [CrossRef]
166. Delzenne, N.M.; Bindels, L.B. Microbiome metabolomics reveals new drivers of human liver steatosis. *Nat. Med.* **2018**, *24*, 906–907. [CrossRef]

© 2019 by the authors. Licensee MDPI, Basel, Switzerland. This article is an open access article distributed under the terms and conditions of the Creative Commons Attribution (CC BY) license (http://creativecommons.org/licenses/by/4.0/).

Article

The Systemic Redox Status Is Maintained in Non-Smoking Type 2 Diabetic Subjects Without Cardiovascular Disease: Association with Elevated Triglycerides and Large VLDL

Peter R. van Dijk [1],*, Amaal Eman Abdulle [2], Marian L.C. Bulthuis [3], Frank G. Perton [4], Margery A. Connelly [5], Harry van Goor [3],† and Robin P.F. Dullaart [1],†

[1] Department of Endocrinology, University Medical Center Groningen, University of Groningen, 9700 RB Groningen, The Netherlands; r.p.f.dullaart@umcg.nl
[2] Department of Internal Medicine, division vascular medicine, University Medical Center Groningen, University of Groningen, 9700 RB Groningen, The Netherlands; a.eman.abdulle@umcg.nl
[3] Department of Pathology and Medical, Biology, Section Pathology, University Medical Center Groningen, University of Groningen, 9700 RB Groningen, The Netherlands; m.l.c.bulthuis@umcg.nl (M.L.C.B.); h.van.goor@umcg.nl (H.v.G.)
[4] Laboratory Center, University Medical Center Groningen, University of Groningen, 9700 RB Groningen, The Netherlands; F.G.Perton@umcg.nl
[5] Laboratory Corporation of America®Holdings (LabCorp), Morrisville, NC 27560, USA; connem5@labcorp.com
* Correspondence: p.r.van.dijk@umcg.nl
† These authors contributed equally to the manuscript.

Received: 1 November 2019; Accepted: 22 December 2019; Published: 24 December 2019

Abstract: Decreased circulating levels of free thiols (R-SH, sulfhydryl groups) reflect enhanced oxidative stress, which plays an important role in the pathogenesis of cardiometabolic diseases. Since hyperglycemia causes oxidative stress, we questioned whether plasma free thiols are altered in patients with type 2 diabetes mellitus (T2DM) without cardiovascular disease or renal function impairment. We also determined their relationship with elevated triglycerides and very low density lipoproteins (VLDL), a central feature of diabetic dyslipidemia. Fasting plasma free thiols (colorimetric method), lipoproteins, VLDL (nuclear magnetic resonance spectrometry), free fatty acids (FFA), phospholipid transfer protein (PLTP) activity and adiponectin were measured in 79 adult non-smoking T2DM subjects (HbA1c 51 ± 8 mmol/mol, no use of insulin or lipid lowering drugs), and in 89 non-smoking subjects without T2DM. Plasma free thiols were univariately correlated with glucose (r = 0.196, $p < 0.05$), but were not decreased in T2DM subjects versus non-diabetic subjects ($p = 0.31$). Free thiols were higher in subjects with (663 ± 84 µmol/L) versus subjects without elevated triglycerides (619 ± 91 µmol/L; $p = 0.002$). Age- and sex-adjusted multivariable linear regression analysis demonstrated that plasma triglycerides were positively and independently associated with free thiols ($\beta = 0.215$, $p = 0.004$), FFA ($\beta = 0.168$, $p = 0.029$) and PLTP activity ($\beta = 0.228$, $p = 0.002$), inversely with adiponectin ($\beta = -0.308$, $p < 0.001$) but not with glucose ($\beta = 0.052$, $p = 0.51$). Notably, the positive association of free thiols with (elevated) triglycerides appeared to be particularly evident in men. Additionally, large VLDL were independently associated with free thiols ($\beta = 0.188$, $p = 0.029$). In conclusion, circulating free thiols are not decreased in this cohort of non-smoking and generally well-controlled T2DM subjects. Paradoxically, higher triglycerides and more large VLDL particles are likely associated with higher plasma levels of thiols, reflecting lower systemic oxidative stress.

Keywords: adiponectin; free thiols; nuclear magnetic resonance spectroscopy; phospholipid transfer protein activity; triglycerides; type 2 diabetes mellitus; large very low density lipoproteins

1. Introduction

Reactive oxygen species (ROS) play essential roles in cell signalling and homeostasis [1,2]. Enhanced oxidative stress results from an imbalance between ROS production and antioxidant defence, and is prominently involved in the pathogenesis of cardiometabolic disorders including diabetes mellitus and atherosclerotic cardiovascular disease (CVD) [3]. Yet, current information with respect to the extent to which systemic oxidative stress is enhanced in the diabetic state is still limited. It has been well established that systemic ROS are increased under chronic hyperglycemic circumstances leading to progression of beta-cell deterioration, insulin resistance and atherosclerosis [4]. Remarkably however, it is not well known whether systemic oxidative stress is already enhanced in patients with type 2 diabetes mellitus (T2DM) free of clinically manifest CVD, chronic kidney disease (CKD) and albuminuria.

Thiols are organic compounds that are characterised by the presence of a sulfhydryl (R-SH) moiety [5]. R-SH groups are readily oxidised by ROS and other reactive species. The plasma concentration of free thiols has been proposed to mirror systemic or local ROS production in such a way that a decline in circulating free thiols indicates an enhanced oxidative tone [6,7]. In line with this supposition, decreased plasma levels of free thiols have been observed among persons with CVD as compared to healthy subjects [8,9]. In stable heart failure patients, higher levels of serum free thiols were associated with favourable disease outcomes [7]. In hemodialyis patients, free thiols tend to be decreased in the context of coinciding non-alcoholic fatty liver disease [10]. Furthermore, among renal transplant recipients higher free thiols concentrations were associated with a beneficial cardiovascular risk profile and a better patient and graft survival [11]. Recently, we demonstrated that in T2DM patients treated in primary care with high circulating free thiols concentrations were associated with lower HbA1c levels and less microvascular complications [12].

Elevated triglycerides, which are mainly carried by plasma very low density lipoproteins (VLDL), are a central hallmark of diabetic dyslipidemia [13–15]. Despite intense study, the pathogenesis of altered (hepatic) VLDL metabolism in the diabetic state has still not been fully elucidated. Among other mechanisms, enhanced free fatty acid (FFA) and glucose availability contribute to increased hepatic VLDL production in the context of T2DM [13,16,17]. Notably, the secretory control of VLDL particles by the liver is a complex process, governed by endoplasmic reticulum (ER)/proteasome-associated degradation, post ER pre-secretory proteolysis and receptor-mediated degradation, i.e., re-uptake mediated via low density lipoprotein (LDL) receptors and proteoglycan abundancy [18–21]. Of particular note, in vitro studies have revealed that intra-hepatic degradation of VLDL particles via post ER pre-secretory proteolysis is triggered by intracellular ROS, coinciding with an enhanced oxidant tone [18,21,22]. Given the role of the liver in the metabolism of fatty acids, glucose and other substrates it is not surprising that the liver is crucially involved in generating ROS [23]. Additionally, since the plasma pool of free thiols is mainly associated with albumin, it is conceivably that the liver contributes at least in part to systemic ROS status. Combined, these findings make it plausible to hypothesise that systemic ROS status is linked to triglyceride metabolism.

The present study was initiated, firstly, to discern the extent to which plasma levels of free thiols, as a proxy of systemic ROS status, are altered in T2DM subjects without clinically manifest CVD disease and preserved renal function. Secondly, we aimed to determine associations of plasma triglycerides and VLDL particle characteristics with free thiols, thereby testing a hitherto unexplored relationship of elevated triglycerides and systemic ROS.

2. Experimental Section

2.1. Subjects and Clinical Measurements

The study was performed at the University Medical Center Groningen (Groningen, the Netherlands) and was approved by the local medical ethics committee (METC2002/174c). Caucasian participants of European descent (aged >18 years) were recruited by advertisement. Written informed consent was obtained from all participants. T2DM had been diagnosed by primary care physicians using guidelines from the Dutch College of General Practitioners (fasting plasma glucose ≥7.0 mmol/L and/or non-fasting plasma glucose ≥11.1 mmol/L). Only T2DM patients who were treated with diet alone or in combination with metformin and/or sulfonylurea were allowed to participate. T2DM subjects using insulin or other glucose lowering medication were excluded. Further exclusion criteria were clinically manifest CVD, renal insufficiency (estimated glomerular filtration rate (e-GFR) <60 mL/min/1.73 m^2 and/or urinary albumin >20 mg/L). e-GFR was calculated based on the serum creatinine-based Chronic Kidney Disease Epidemiology Collaboration equation [24]. Subjects with liver disease (serum transaminase levels >2 times above the upper reference limit), pregnant women and subjects who were using lipid lowering drugs were also excluded. Subjects who used other medications (except for oral contraceptives), current smokers and subjects who used >3 alcoholic drinks daily were additionally excluded. Subjects using anti-hypertensive medication were allowed to participate. Physical examination did not reveal pulmonary or cardiac abnormalities. BMI (kg/m^2) was calculated as weight (kg) divided by height (m) squared. Waist circumference was measured as the smallest circumference between rib cage and iliac crest. Blood pressure was measured after 15 min of rest at the left arm using a sphygmomanometer. The participants were evaluated between 08.00 and 10.00 h after an overnight fast.

The metabolic syndrome (MetS) was defined according to NCEP-ATP III criteria [25]. Three or more of the following criteria were required for categorization of subjects with MetS: waist circumference > 102 cm for men and > 88 cm for women; hypertension (blood pressure ≥130/85 mmHg or the use of antihypertensive drugs); fasting plasma triglycerides ≥1.70 mmol/L; HDL cholesterol <1.00 mmol/L for men and <1.30 mmol/L for women; fasting glucose ≥5.60 mmol/L. Insulin sensitivity was estimated by homeostasis model assessment of insulin resistance (HOMA-IR) applying the following equation: fasting plasma insulin (mU/L) × glucose (mmol/L)/22.5 [26].

2.2. Laboratory Measurements

EDTA-anticoagulated plasma samples were stored at −80 °C until analysis, except for plasma glucose and glycated hemoglobin (HbA1c) levels which were measured shortly after blood collection. Glucose was measured on an APEC glucose analyser (APEC Inc., Danvers, MA, USA).

High sensitivity C-reactive protein (hs-CRP) was measured by nephelometry with a threshold of 0.18 mg/L (BNII, Dade Behring). Glycated hemoglobin (HbA1c) was measured by high performance liquid chromatography (Bio-Rad, Veenendaal, The Netherlands). Plasma total cholesterol and triglycerides were assayed by routine enzymatic methods (Roche/Hitachi cat nos 11,876,023 and 11,875,540, respectively; Roche Diagnostics GmBH, Mannheim, Germany). HDL cholesterol was measured with a homogeneous enzymatic colorimetric test (Roche/Hitachi, cat no 04,713,214; Roche Diagnostics GmBH, Mannheim, Germany). LDL cholesterol was calculated by the Friedewald formula if plasma triglycerides were <4.5 mmol/L. Apolipoprotein B (apo B) was measured by immunoturbidimetry (Roche/Cobas Integra Tina-quant, cat. Number 03032574, respectively; Roche Diagnostics, Roche Diagnostics GmBH, Mannheim, Germany). Plasma free fatty acids (FFA) were measured with a kit from Wako Diagnostics (HR Series NEFA-HR (2), Wako Chemicals GmbH, Neuss, Germany).

VLDL particle concentrations and VLDL subfractions were measured by nuclear magnetic resonance (NMR) spectroscopy with the LP3 algorithm (LabCorp, Morrisville, North Carolina, USA) [27,28]. VLDL particle classes and subfractions were quantified from the amplitudes of their spectroscopically distinct lipid methyl group NMR signals. The diameter range estimate for VLDL (including chylomicrons if present) was >60 nm to 29 nm, The VLDL particle concentration (VLDL-P) was calculated as the weighted average of the respective lipoprotein subclasses. The intra-assay coefficient of variation (CV) for VLDL-P amounts to 11.0%.

Plasma PLTP activity was assayed with a phospholipid vesicles-HDL system, using (^{14}C)-labelled dipalmitoyl phosphatidylcholine as described [29,30]. Briefly, plasma samples (1 µL) were incubated with (^{14}C)-phosphatidylcholine-labelled phosphatidylcholine vesicles and excess pooled normal HDL for 45 min. at 37 °C. The method is specific for PLTP activity. Plasma PLTP activity levels vary linearly with the amount of plasma added to the incubation system. PLTP activity was related to the activity in human reference pool plasma and was expressed in arbitrary units (AU; 100 AU corresponds to 13.6 µmol phosphatidylcholine transferred per mL per h). The intra-assay CV of PLTP activity was 5%.

Insulin was assayed by microparticle enzyme immunoassay (AxSYM insulin assay: Abbott Laboratories, Abbott Park, IL, USA). Total adiponectin was measured by enzyme-linked immunosorbent assay (Linco Research Inc., St. Charles, MO, USA, cat no EZHADP-61k). The provided by this assay are strongly correlated with results obtained by enzyme-linked immunoassay measurement [31].

Plasma free thiols were assayed as described with minor modifications [32,33]. After thawing, samples were four-fold diluted using 0.1 M Tris buffer (pH 8.2). Background absorption was measured at 412 nm using the Varioskan microplate reader (ThermoScientific, Breda, the Netherlands), together with a reference measurement at 630 nm. Subsequently, 20 µL 1.9 mM 5,5'-dithio-bis (2-nitrobenzoic acid) (DTNB, Ellman's Reagent, CAS no. 69-78-3, Sigma Aldrich Corporation, St. Louis, MO, USA) was added to the samples in 0.1 M phosphate buffer (pH 7.0). After an incubation time of 20 min at room temperature, absorbance was measured again. Final serum free thiol concentrations were established by parallel measurement of an L-cysteine (CAS no. 52-90-4, Fluka Biochemika, Buchs, Switzerland) calibration curve (range from 15.6 µM to 1000 µM) in 0.1 M Tris/10 mM EDTA (pH 8.2). The intra-assay CV of free thiols amount to 6%.

2.3. Statistical Analysis

SPSS (IBM SPSS Statistics for Windows, Version 23.0. IBM Corp. Armonk, NY, USA) was used for data analysis. The distributions of all variables were examined using histograms and Q-Q plots. Data are expressed as mean (with standard deviation (SD)) or median (with interquartile range (IQR)) for normally distributed and non-normally distributed data, respectively. Nominal data are presented as n (with percentage (%)). Non-parametrically distributed data were $\log_e$ transformed for statistical analysis. Between-group differences in continuous variables were determined by unpaired T-tests. Between-group differences in dichotomous variables were determined by χ^2-analysis. Univariate relationships were assessed using Pearson correlation coefficients. Multivariable linear regression analysis was applied to disclose the independent contribution of variables. A p-value <0.05 (two-sided) was used to indicate statistical significance.

3. Results

3.1. Study Population

The study population comprised 79 T2DM subjects and 89 non-diabetic subjects (Table 1). Median diabetes duration was 5 (interquartile range (IQR) 4–6.5) years. Fourteen T2DM subjects used metformin and 11 used sulfonylurea alone, whereas these drugs were used combined in 24 patients. No other glucose lowering drugs were taken. The other T2DM subjects were given lifestyle advice only. Anti-hypertensive medication (mostly angiotensin-converting enzyme inhibitors, angiotensin II receptor antagonists and diuretics, alone or in combination) were taken by 33 T2DM subjects and by none of the non-diabetic subjects ($p < 0.001$). Estrogens were used by two pre- and two postmenopausal non-diabetic women. Other medications were not used.

As shown in Table 1, T2DM subjects were on average 4 years older than the non-diabetic subjects, were more obese, had higher blood pressure, fasting plasma glucose and HbA1c levels, were more insulin resistant and categorised with MetS more frequently. Sex distribution was not different between the groups. e-GFR was similar in the groups.

Table 1. Clinical and laboratory characteristics in 79 subjects with Type 2 diabetes mellitus (T2DM) and 89 subjects without T2DM.

	T2DM Subjects (n = 79)	Non-Diabetic Subjects (n = 89)	p-Value
Age (years)	59 ± 9	55 ± 9	0.011
Sex (men/women)	49/30	49/40	0.45
Metabolic syndrome (yes/no)	54/25	21/68	<0.001
Systolic blood pressure (mm Hg)	144 ± 20	131 ± 19	<0.001
Diastolic blood pressure (mm Hg)	87 ± 9	83 ± 11	0.007
BMI (kg/m^2)	28.6 ± 4.8	25.9 ± 3.9	<0.001
Waist (cm)	100 ±13	89 ± 9	<0.001
Glucose (mmol/L)	8.9 ± 2.3	5.6 ± 0.7	<0.001
HbA1c (mmol/mol)	51 ± 8	39 ± 5	<0.001
Insulin (mU/L)	10.8 (7.2,15.3)	6.5 (4.7,8.6)	<0.001
HOMA-IR (mU mmol/L^2/22.5)	3.96 (2.37,6.64)	1.57 (1.09,2.29)	<0.001
e-GFR (mL./min/1.73 m^2)	91 ± 13	87 ± 14	0.12
Total cholesterol (mmol/L)	5.38 ± 0.96	5.74 ± 0.99	0.022
Non-HDL cholesterol (mmol/L)	4.12 ± 1.04	4.25 ± 1.06	0.44
LDL cholesterol (mmol/L)	3.27 ± 0.85	3.53 ± 0.89	0.06
HDL cholesterol (mmol/L)	1.26 ± 0.38	1.49 ± 0.40	<0.001
Triglycerides (mmol/L)	1.70 (1.17,2.30)	1.31 (0.89,1.95)	0.034
ApoB (g/L)	0.95 ± 0.23	0.96 ± 0.23	0.55
VLDL-P (nmol/L)	69.5 (46.2,88.7)	61.8 (51.6,101.6)	0.83
Large VLDL (nmol/L)	7.1 (2.7,11.1)	3.5 (2.0,9.4)	0.019
Medium VLDL (nmol/L)	28.9 (14.9,41.3)	25.3 (13.2,44.8)	0.86
Small VLDL (nmol/L)	27.7 (16.8,43.9)	33.9 (21.2,44.5)	0.076
FFA (mmol/L))	0.35 ± 0.12	0.30 ± 0.12	0.001
PLTP activity (AU)	104 ± 11	94 ± 10	<0.001
Adiponectin (mg/L)	14.3 (10.7,26.3)	20.1 (14.9,43.0)	<0.001
Free Thiols (μmol/L)	645 ± 102	630 ± 78	0.31

Data are expressed in means ± SD, medians (interquartile range) or numbers. Abbreviations: ApoB: apolipoprotein b; BMI: body mass index; e-GFR: estimated glomerular filtration rate; FFA: free fatty acids; HbA1c: glycated hemoglobin; HDL: high density lipoproteins; LDL: low density lipoproteins; PLTP: phospholipid transfer protein; VLDL: very low density lipoproteins: VLDL-P: very low density particle concentration. Differences between subjects with and without T2DM were determined by T-tests (using log$_e$ transformed values in case of not-normally distributed data) and by χ^2 tests, where appropriate. VLDL particle concentrations and VLDL subfractions were measured in 66 subjects with diabetes and in 56 subjects without diabetes. LDL cholesterol was calculated by the Friedewald formula in 75 T2DM subjects and in 87 non-diabetic subjects.

3.2. Comparison of Laboratory Variables between Subject Categories

Plasma total cholesterol was lower in T2DM subjects, which was due to lower HDL cholesterol. Triglycerides were higher in T2DM subjects but non-HDL cholesterol, LDL cholesterol and apolipoprotein b (ApoB) levels were not different between the groups. The total VLDL particle concentration was not different between the groups, but large VLDL particles were increased, whereas small VLDL particles tended to be decreased in T2DM subjects. Plasma FFA and PLTP activity were elevated, whereas adiponectin was decreased in T2DM subjects (Table 1).

In all subjects combined, plasma free thiol concentrations were higher in men (n = 98, 655 ± 87 μmol/L) than in women (n = 70, 614 ± 91 μmol/L, p = 0.003). Free thiols were not different between T2DM and non-diabetic subjects (Table 1), and were not significantly related to diabetes duration (r = −0.136, p = 0.23). Free thiols were also not different in T2DM subjects compared to non-diabetic subjects after exclusion of T2DM subjects using glucose lowering medication (n = 20, 600 ± 109 μmol/L, p = 0.14), antihypertensive medication (n = 44, 652 ± 110 μmol/L, p = 0.21) or both classes of drugs (n = 15, 598 ± 117μmol/L, p = 0.17). Likewise, free thiols were also not different between subjects with MetS (n = 75, 643 ± 89 μmol/L) and without MetS (n = 93, 633 ± 92 μmol/L, p = 0.49). Free thiols were not different in subjects with elevated glucose (p = 0.39), enlarged waist circumference (p = 0.44) or elevated blood pressure (p = 0.29) compared to subjects who did not fulfil these MetS criteria (data not shown). Notably, free thiols were higher in subjects with elevated triglycerides (n = 71, 663 ± 84 μmol/L, p = 0.002) or with low HDL cholesterol (n = 45, 661 ± 84 μmol/L, p = 0.042) compared to subjects who did not fulfil these MetS criteria (triglyceride criterion: 619 ± 91 μmol/L; HDL cholesterol criterion (629 ± 92 μmol/L).

3.3. Univariate Correlations

In all subjects combined, univariate analysis showed that plasma free thiols were inversely correlated with age, HDL cholesterol, FFA and adiponectin, and positively with glucose, e-GFR, non-HDL cholesterol, triglycerides, VLDL particle concentration and large-and medium-sized VLDL particles (Table 2, panel A). Comparable correlations were observed in T2DM subjects separately (Table 2, panel B and panel C). The univariate correlations of plasma free thiols with triglycerides and with large VLDL particles are shown in Figure 1, panel A and panel B, respectively. In non-diabetic subjects, free thiols were correlated positively with total cholesterol, non-HDL cholesterol, triglycerides and ApoB, and inversely with FFA and adiponectin (Table 2, panel C). Neither in all subjects combined nor in T2DM subjects and non-diabetic subjects separately were free thiols correlated with blood pressure, BMI, waist circumference and PLTP activity (Table 2, panel A, B and C). In addition, in all subjects combined, PLTP activity was correlated positively with triglycerides (r = 0.325, p < 0.001) and FFA (r = 0.162, p = 0.036; data not shown). Similar correlations of PLTP activity with triglycerides were found in T2DM subjects and non-diabetic subjects separately (data not shown).

Table 2. Univariate relationships of plasma free thiols with clinical and laboratory variables in all subjects combined (n = 168, panel **A**), Type 2 diabetic (T2DM) subjects (n = 79, panel **B**) and in non-diabetic subjects (n = 89, panel **C**).

	(A) All Subjects (n = 168) Free Thiols	(B) T2DM Subjects (n = 79) Free Thiols	(C) Non-Diabetic Subjects (n = 89) Free Thiols
Age	−0.183 *	−0.297 **	−0.098
Systolic blood pressure	−0.053	−0.074	−0.093
Diastolic blood pressure	0.031	0.145	−0.119
BMI	0.059	−0.012	0.108
Waist	0.131	0.057	0.172
Glucose	0.196 *	0.262 *	0.012
HbA1c	0.151	0.186	−0.013
Insulin	0.095	0.022	0.131
HOMA-IR	0.141	0.106	0.132
e-GFR	0.213 **	0.271 *	0.144
Total cholesterol	0.084	−0.057	0.281 **
Non-HDL cholesterol	0.168 *	0.056	0.314 **
LDL cholesterol	0.034	−0.148	0.247 *
HDL cholesterol	−0.232 **	−0.302 **	−0.137
Triglycerides	0.285 ***	0.290 **	0.260 *
ApoB	0.129	0.063	0.349 ***
VLDL-P	0.204 *	0.247 *	0.138
Large VLDL	0.266 **	0.333 **	0.129
Medium VLDL	0.234 **	0.333 **	0.081
Small VLDL	−0.013	0.007	−0.020
FFA	−0.187 *	−0.210	−0.224 *
PLTP activity	0.034	−0.096	0.133
Adiponectin	−0.261 ***	−0.259 *	−0.247 *

Pearson correlation coefficients are shown. Non-parametrically distributed data are $\log_e$ transformed. * p < 0.05; ** p ≤ 0.01; *** p ≤ 0.001. Abbreviations: ApoB: apolipoprotein B; BMI: body mass index; e-GFR: estimated glomerular filtration rate; FFA: free fatty acids: HbA1c: glycated hemoglobin; HDL: high density lipoproteins; LDL: low density lipoproteins; PLTP: phospholipid transfer protein; VLDL: very low density lipoproteins; VLDL-P: very low density particle concentration. VLDL particle concentrations and VLDL subfractions were measured in 66 subjects with diabetes and in 56 subjects without diabetes. LDL cholesterol was calculated by the Friedewald formula in 75 T2DM subjects and in 87 non-diabetic subjects.

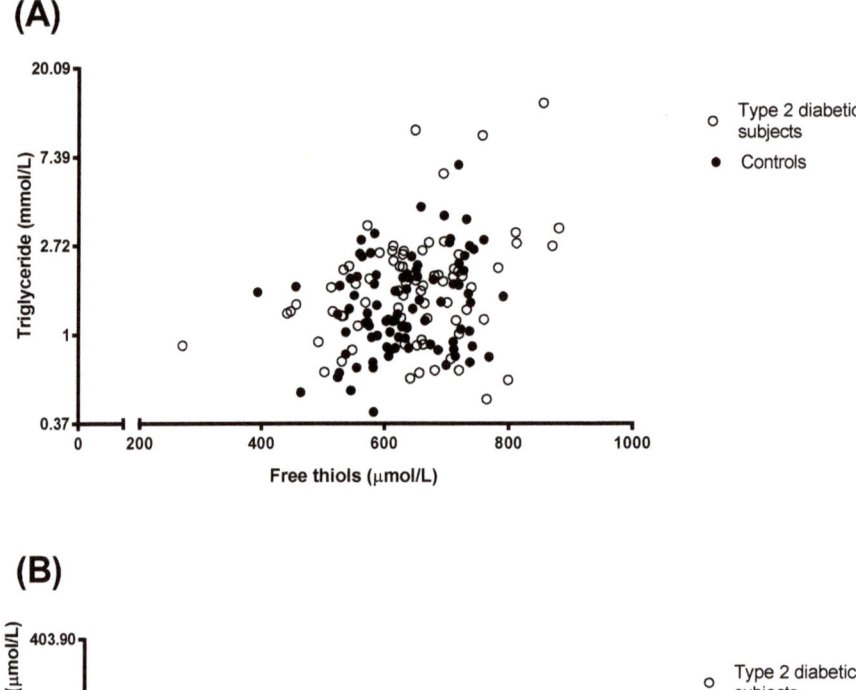

Figure 1. Univariate correlations of plasma free thiols with plasma triglycerides (panel **A**; 79 type 2 diabetic subjects and 89 subjects without diabetes) and with very large very low density lipoprotein (VLDL) (panel **B**; 66 type 2 diabetic subjects and 58 subjects without diabetes). Legend: closed symbols represent subjects without diabetes; open symbols represent subjects with type 2 diabetes. The log transformed data (i.e., triglyceride and large very low density lipoproteins) were used. The back transformed measures are shown on the Y-axis. Pearson correlation coefficients: free thiols with triglycerides; all subjects, $r = 0.285$, $p < 0.001$; Type 2 diabetic subjects, $r = 0.290$, $p = 0.01$; subjects without diabetes, $r = 0.260$, $p = 0.014$. Free thiols with large VLDL: all subjects, $r = 0.266$, $p = 0.003$; Type 2 diabetic subjects, $r = 0.333$, $p = 0.006$; subjects without diabetes, $r = 0.138$, $p = 0.31$.

In all men combined (n = 98), plasma free thiols were inversely correlated with age and adiponectin, and positively with HOMA-IR, e-GFR, triglycerides, VLDL particle concentration and large VLDL particles (Table S1). In all women combined (n = 70), plasma free thiols were not significantly correlated with any of the variables listed in Table 2 (Table S1).

3.4. Multivariable Linear Regression Analyses

Multivariable linear regression analyses were first carried out to disclose the association of free thiols with the diabetic state when taking account of age and sex. In this analysis, free thiols were not significantly related to the presence of T2DM after adjustment for age and sex ($p = 0.159$; data not shown). Next, we examined the extent to which plasma free thiol concentrations were independently associated with the individual MetS components. This analysis showed that free thiols were positively associated with elevated triglycerides but not with other MetS components (Table 3). Furthermore, free thiols were positively associated with male sex (Table 3). In an additional analysis, now also including e-GFR, there was no independent association of free thiols with e-GFR ($\beta = 0.122$, $p = 0.20$), while the relationship with elevated triglycerides remained significant ($\beta = 0.209$, $p = 0.011$). In a sex-stratified analysis, plasma free thiols were positively associated with elevated triglycerides in men (Table S2). In women, plasma free thiols were not significantly related to any of the individual MetS components (Table S2).

Table 3. Multivariable linear regression analysis demonstrating independent associations of plasma free thiols with age, sex and individual metabolic syndrome components in 168 subjects (79 Type 2 diabetic subjects and 89 non-diabetic subjects).

	β	p-Value
Age	−0.205	0.011
Sex (men versus women)	0.213	0.005
Elevated glucose	0.074	0.35
Elevated blood pressure	−0.069	0.40
Enlarged waist	0.021	0.80
Elevated triglycerides	0.200	0.015
Low HDL cholesterol	0.060	0.47

Abbreviations: β: standardised regression coefficient; HDL: high density lipoproteins.

Given the association between triglycerides and free thiols, we next determined whether this relationship remained when taking account of the diabetic state instead of the MetS criterion of elevated plasma glucose. In this analysis, plasma triglycerides were positively associated with free thiols ($\beta = 0.271$, $p < 0.001$) after adjustment for age ($\beta = 0.01$, $p = 0.89$), sex ($\beta = 0.001$, $p = 0.97$) and the presence of diabetes ($\beta = 0.245$, $p = 0.059$). Subsequently, we tested whether other variables that were univariately related to plasma triglycerides or free thiols including PLTP activity, FFA and adiponectin (Table 2). Plasma glucose was also included in these models in view of its univariate correlation with free thiols. In this multivariable model, with plasma triglycerides as dependent variable, plasma triglycerides were positively and independently associated with free thiols, FFA and PLTP activity, and inversely with adiponectin but not with glucose (Table 4). These associations remained significant after further adjustment for the use of sulfonylurea, metformin and antihypertensive drugs (data not shown). Similar associations of plasma triglycerides with free thiols, FFA, PLTP activity and adiponectin were found with HbA1c or the presence of T2DM instead of plasma glucose (data not shown).

In sex-stratified analysis, it was found that plasma triglycerides were positively associated with free thiols, FFA, PLTP activity and inversely with adiponectin in men (Table S3). In women, plasma triglycerides were positively associated with PLTP activity and tended to be inversely associated with adiponectin, but the association with free thiols was not significant (Table S3).

Table 4. Multivariable linear regression analysis demonstrating independent associations of plasma triglycerides with age, sex, free thiols, free fatty acids, phospholipid transfer protein activity and adiponectin and glucose in 168 subjects (79 Type 2 diabetic subjects and 89 non-diabetic subjects).

	β	p-Value
Age	−0.044	0.55
Sex (men versus women)	0.016	0.84
Free thiols	0.215	0.004
FFA	0.168	0.029
PLTP activity	0.228	0.002
Adiponectin	−0.308	<0.001
Glucose	0.052	0.51

Triglycerides and adiponectin are $\log_e$ transformed. Abbreviations: β: standardised regression coefficient; FFA: free fatty acids: PLTP phospholipid transfer protein.

Finally, we determined the association of free thiols with VLDL subfractions. Large-sized VLDL particles were positively and independently associated with free thiols and FFA, and inversely with adiponectin (Table 5, panel A). Medium-sized VLDL particles tended to be associated positively with free thiols and were inversely associated with adiponectin (Table 5, panel B). Small-sized VLDL particles were not independently associated with any of these variables (Table 5, panel C). In sex-stratified analyses, the independent association of large VLDL particles with plasma free thiols did not reach statistical significance in men (β = 0.184, p = 0.14) or in women (β = 0.188, p = 0.14; data not shown).

Table 5. Multivariable linear regression analysis demonstrating independent associations of very low density lipoprotein (VLDL) subfractions with age, sex, free thiols, free fatty acids, PLTP activity, adiponectin and glucose in 122 subjects (66 Type 2 diabetic subjects and 56 non-diabetic subjects). Panel **A**, Large VLDL; panel **B**, medium VLDL, panel **C**, small VLDL.

	A Large VLDL		B Medium VLDL		C Small VLDL	
	β	p-Value	β	p-Value	β	p-Value
Age	−0.042	0.630	0.036	0.71	0.021	0.82
Sex (men versus women)	0.074	0.440	0.049	0.65	−0.156	0.17
Free thiols	0.188	0.029	0.176	0.067	−0.023	0.82
FFA	0.242	0.008	−0.014	0.91	−0.157	0.41
PLTP activity	0.147	0.094	0.075	0.45	−0.066	0.52
Adiponectin	−0.294	0.002	−0.243	0.021	−0.047	0.67
Glucose	0.097	0.320	−0.052	0.630	0.029	0.80

VLDL subfractions and adiponectin are $\log_e$ transformed. Abbreviations: β: standardised regression coefficient; FFA: free fatty acids; PLTP phospholipid transfer protein; VLDL: very low density lipoproteins.

4. Discussion

In this cohort of non-smoking and in general appropriately controlled T2DM subjects, we report that circulating levels of free thiols are still maintained. In the current study, we excluded T2DM subjects who smoked, those with clinically manifest CVD or impaired renal function as well as subjects using insulin or lipid lowering medication. Therefore, the present findings suggest that decreased free thiol concentrations, as a proxy of enhanced systemic oxidative stress, are unlikely to represent an early feature of relatively moderate chronic hyperglycemia. Second, we report here for the first time that plasma triglycerides and in particular large VLDL particles are positively and independently related to free thiols in multivariable linear regression analysis in which we accounted for age, sex, glucose (or alternatively the presence of diabetes), FFA, PLTP activity and adiponectin. Notably, the association of free thiols with triglycerides appeared to be particularly evident in men. We consider this novel and seemingly surprising finding to agree with the hypothesis that systemic ROS status may impact on triglyceride metabolism. As a mechanism, we postulate that this association may at least in part be attributable to an effect of hepatic ROS to inhibit VLDL release in the bloodstream.

To underscore the lack of effect of the presence of T2DM on plasma free thiol concentrations as observed in the present study, we also carried out sensitivity analyses after exclusion of T2DM subjects using any glucose lowering medication, antihypertensive medication or both classes of drugs. Again, it was found that plasma free thiols were not decreased in the cohort of T2DM subjects studied. Our results can be compared with another study on thiols in T2DM by Pasaoglu et al. [34]. In a small study among 40 T2DM subjects (20 newly diagnosed and 20 treated using oral blood glucose lowering therapy only) and 20 non-diabetic subjects they demonstrated lower plasma total thiol levels among persons with T2DM as compared to control subjects (605 ± 36 versus 665 ± 60 µmol/L) [34]. Since a negative correlation between HbA1c and thiols has been reported [12], it should be taken into account that T2DM subjects in their study were poorly regulated with an average HbA1c of 74 ± 30 mmol/mol and of 78 mmol/mol, in newly diagnosed and medically treated T2DM subjects, respectively. Recently, we measured free thiols in a population of 929 T2DM outpatients treated in primary care [12]. In this cohort, we found significant multivariable associations between free thiols and the presence of macrovascular complications, age, BMI, diastolic blood pressure, HbA1c (all inverse associations), male sex, diabetes duration and use of platelet aggregation inhibitors (all positive associations). In addition, there was (modest) additional value for free thiols for risk prediction of long-term complications. In agreement with these and previous results, we observed in the current study that free thiols declined with ageing and were higher in men than in women [11,12], whereas we also confirmed a positive relationship with e-GFR at least in univariate regression analysis [11].

Data concerning the relationship of ROS status and circulating free thiols in particular on fasting plasma triglycerides in humans is scarce. In the aforementioned study amongst primary care treated T2DM patients, there was no association between plasma free thiols and triglyceride concentrations in multivariable analysis [12]. It should be noted, however, that in this population, samples were non-fasting and 80% of the patients used cholesterol-lowering medication. Amongst persons with type 1 diabetes, no association between plasma free thiols and triglycerides was present [35]. Interestingly, in mice receiving a high-saturated fat diet, the intake of the anti-oxidative 'thiol donor' n-acetyl cysteine decreased plasma triglyceride concentrations possibly by lowering the activity and mRNA of malic enzyme, fatty acid synthase and 3-hydroxy-3-methylglutaryl coenzyme A reductase [36,37].

In the present study, we assessed the relationship of plasma free thiols with triglycerides not only when taking account of FFA and glucose, which are well-known substrates of hepatic triglyceride synthesis [13,16], but also of PLTP activity and adiponectin. We found that plasma triglycerides as well as large VLDL particles were associated with FFA levels in agreement with the contention that FFA provide an important source for hepatic VLDL synthesis [13,17,38]. Although hepatic overproduction of large VLDL particles, a central feature of diabetic dyslipidemia, has been proposed to be to an important extent driven by hyperglycemia [16], plasma triglycerides were not independently associated with glucose, which we ascribe to the modestly elevated fasting glucose levels in the current T2DM cohort.

Importantly, studies in mice with genetic *Pltp* deficiency have revealed that PLTP is able to promote the hepatic secretion of apoB-containing lipoproteins [39], effect which was subsequently found to be attributable to the ability of PLTP to regulate ROS-dependent intracellular degradation of newly assembled VLDL [40,41]. In humans, plasma PLTP activity has repeatedly been shown to be associated with plasma triglyceride metabolism [42–45]. Moreover, genetic variation in *Pltp* associates with lipid traits including triglycerides [46]. Plasma triglycerides were indeed positively associated with PLTP activity in the current study. Of note, this relationship was independent from its positive association with plasma free thiols. In establishing the association of plasma triglycerides with free thiols, we also accounted for plasma adiponectin in multivariable regression analysis. Adiponectin may stimulate hepatic fatty acid oxidation and ameliorate the development of hepatic steatosis [47]. In line, we previously reported an inverse relationship of plasma adiponectin with alanine aminotransferase [48]. Importantly, adiponectin has been implicated in preventing mitochondrial dysfunction, and hence, may inhibit ROS production in vitro, possibly by promoting uncoupling protein 2 expression [47]. Of note, we observed an inverse univariate relationship of plasma free thiols with adiponectin, thereby providing a rationale to adjust for plasma levels of this adipokine when evaluating the association of plasma triglycerides with free thiols. Still, plasma triglycerides as well as large VLDL particles remained associated with plasma free thiols when considering adiponectin in multivariable linear regression analysis. Taken together, our current study provides support for the hypothesis that ROS status may influence triglyceride metabolism even independent of plasma glucose, FFA, PLTP activity and adiponectin. However, since plasma free thiols were not different between the currently studied T2DM and non-diabetic subjects, it seems unlikely that alterations in ROS status explain to a considerable extent why plasma triglycerides and large VLDL particle concentrations are increased in T2DM. Additionally, the mechanisms responsible for apparent sex-differences in the association of free thiols with triglycerides are not yet clear. Plasma free thiols are higher in men than in women as confirmed in the current study [11,12]. Additionally, fasting plasma triglycerides may also be higher in men (as derived from [49]). Among other explanations it seems, therefore, plausible that there is less between subject variation in free thiols and triglycerides in women compared to men, which could to some extent blunt the strength of the correlation between free thiols and triglycerides in women.

Our findings are relevant from a diagnostic perspective. In transplant patients and heart failure patients the systemic redox status may have predictive power for outcome [7,11]. In recent work we also showed that the systemic level of free thiols is a better reflection of disease activity in inflammatory bowel disease than faecal calprotectin, the classical marker [50]. However, free thiols are also receptive to therapeutic modulation, and can therefore be considered a potential therapeutic target for therapy [51,52]. In clinical trials, a cysteine derivative N-Acetylcysteine directly reduces disulfide bonds and act thereby as a glutathione (GSH) precursor [52]. Although promising, therapeutic results are inconsistent and these compounds may be effective in selected patients with low free thiol concentrations.

Several other methodological aspects of our study need to be discussed. First, we carried out a cross-sectional study. For this reason, cause-effect relationships cannot be established with certainty, nor can we exclude the possibility of reversed causation. Hence, we cannot exclude that alterations in triglyceride metabolism could affect the balance between ROS production and antioxidant defence. In this vein, a scenario is also possible whereby increased hepatic substrate availability leads to enhanced β-oxidation which in turn affects hepatic ROS production. Second, the sample size of the current study was rather limited, possibly masking relatively modest associations between variables of interest. Third, only diabetic subjects who did not use lipid lowering medication were allowed to participate, thereby avoiding confounding due to effects of statins on triglycerides and PLTP activity [53,54]. As a consequence, it is likely that T2DM subjects with mild dyslipidemia were preferentially included. Fourth, only non-smokers participated. This exclusion criterion was implemented in order to avoid effects of smoking on systemic ROS [55]. Moreover, cigarette smoking affects lipid metabolism, including elevations in PLTP activity [56].

5. Conclusions

Plasma free thiols, which mirror the redox status, are still maintained in generally adequately controlled non-smoking T2DM subjects, without clinically manifest cardiovascular disease or chronic kidney disease indicating limited organ and cell damage. Higher fasting plasma triglycerides and increased large VLDL particles concentrations are inversely associated with free thiols, raising the possibility that enhanced oxidative stress could be paradoxically implicated in less severe dyslipidemia.

Supplementary Materials: The following are available online at http://www.mdpi.com/2077-0383/9/1/49/s1, Table S1: Univariate relationships of plasma free thiols with clinical and laboratory variables in males (n = 98) and females (n = 70). Table S2: Multivariable linear regression analysis demonstrating independent associations of plasma free thiols with age and individual metabolic syndrome components in males (n = 98) and females (n = 70). Table S3: Multivariable linear regression analysis demonstrating independent associations of plasma triglycerides with age, sex, free thiols, free fatty acids, PLTP activity and adiponectin and glucose in males (n = 98) and females (n = 70). Table S4: Multivariable linear regression analysis demonstrating independent associations of large very low density lipoprotein subfractions with age, sex, free thiols, free fatty acids, PLTP activity, adiponectin and glucose in males (n = 98) and females (n = 70).

Author Contributions: P.R.v.D., A.E.A. and R.P.F.D. analysed the data. P.R.v.D., H.v.G. and R.P.F.D. wrote the manuscript. M.L.C.B., F.G.P. and M.A.C. were essential in providing measurement of free thiols, lipids and lipoproteins. All authors discussed the results and contributed to the final manuscript. All authors approved the final version of this article. All authors have read and agreed to the published version of the manuscript.

Funding: This research received no external funding.

Acknowledgments: Plasma PLTP activity was measured in the laboratory of GM. Dallinga-Thie, Department of Experimental Vascular Medicine, Academic Medical Center, Amsterdam, The Netherlands. Lipoprotein subfractions were measured at LabCorp (Raleigh, North Carolina, USA) at no cost.

Conflicts of Interest: M.A.C. is an employee of LabCorp. The rest of the authors declare that they have no competing interests.

References

1. Ray, P.D.; Huang, B.-W.; Tsuji, Y. Reactive oxygen species (ROS) homeostasis and redox regulation in cellular signaling. *Cell Signal.* **2012**, *24*, 981–990. [CrossRef] [PubMed]
2. Jones, D.P.; Sies, H. The Redox Code. *Antioxid. Redox Signal.* **2015**, *23*, 734–746. [CrossRef] [PubMed]
3. Sies, H. Oxidative stress: A concept in redox biology and medicine. *Redox Biol.* **2015**, *4*, 180–183. [CrossRef] [PubMed]
4. Kaneto, H.; Katakami, N.; Matsuhisa, M.; Matsuoka, T. Role of reactive oxygen species in the progression of type 2 diabetes and atherosclerosis. *Mediat. Inflamm.* **2010**, *2010*, 453892. [CrossRef] [PubMed]
5. Sen, C.K.; Packer, L. Thiol homeostasis and supplements in physical exercise. *Am. J. Clin. Nutr.* **2000**, *72*, 653S–669S. [CrossRef]
6. Gupta, S.; Gambhir, J.K.; Kalra, O.; Gautam, A.; Shukla, K.; Mehndiratta, M.; Agarwal, S.; Shukla, R. Association of biomarkers of inflammation and oxidative stress with the risk of chronic kidney disease in Type 2 diabetes mellitus in North Indian population. *J. Diabetes Complicat.* **2013**, *27*, 548–552. [CrossRef]
7. Koning, A.M.; Meijers, W.C.; Pasch, A.; Leuvenink, H.G.D.; Frenay, A.-R.S.; Dekker, M.M.; Feelisch, M.; de Boer, R.A.; van Goor, H. Serum free thiols in chronic heart failure. *Pharmacol. Res.* **2016**, *111*, 452–458. [CrossRef]
8. Banne, A.F.; Amiri, A.; Pero, R.W. Reduced level of serum thiols in patients with a diagnosis of active disease. *J. Anti-Aging Med.* **2003**, *6*, 327–334. [CrossRef]
9. Kundi, H.; Ates, I.; Kiziltunc, E.; Cetin, M.; Cicekcioglu, H.; Neselioglu, S.; Erel, O.; Ornek, E. A novel oxidative stress marker in acute myocardial infarction; thiol/disulphide homeostasis. *Am. J. Emerg. Med.* **2015**, *33*, 1567–1571. [CrossRef]
10. Wu, P.-J.; Chen, J.-B.; Lee, W.-C.; Ng, H.-Y.; Lien, S.-C.; Tsai, P.-Y.; Wu, C.-H.; Lee, C.-T.; Chiou, T.T.-Y. Oxidative Stress and Nonalcoholic Fatty Liver Disease in Hemodialysis Patients. *BioMed Res. Int.* **2018**, *2018*, 3961748. [CrossRef]

11. Frenay, A.-R.S.; de Borst, M.H.; Bachtler, M.; Tschopp, N.; Keyzer, C.A.; van den Berg, E.; Bakker, S.J.L.; Feelisch, M.; Pasch, A.; van Goor, H. Serum free sulfhydryl status is associated with patient and graft survival in renal transplant recipients. *Free Radic. Biol. Med.* **2016**, *99*, 345–351. [CrossRef] [PubMed]
12. Schillern, E.E.M.; Pasch, A.; Feelisch, M.; Waanders, F.; Hendriks, S.H.; Mencke, R.; Harms, G.; Groenier, K.H.; Bilo, H.J.G.; Hillebrands, J.-L.; et al. Serum free thiols in type 2 diabetes mellitus: A prospective study. *J. Clin. Transl. Endocrinol.* **2019**, *16*, 100182. [CrossRef] [PubMed]
13. Taskinen, M.-R. Diabetic dyslipidaemia: From basic research to clinical practice. *Diabetologia* **2003**, *46*, 733–749. [CrossRef] [PubMed]
14. Borggreve, S.E.; De Vries, R.; Dullaart, R.P.F. Alterations in high-density lipoprotein metabolism and reverse cholesterol transport in insulin resistance and type 2 diabetes mellitus: Role of lipolytic enzymes, lecithin:cholesterol acyltransferase and lipid transfer proteins. *Eur. J. Clin. Investig.* **2003**, *33*, 1051–1069. [CrossRef] [PubMed]
15. Soran, H.; Schofield, J.D.; Adam, S.; Durrington, P.N. Diabetic dyslipidaemia. *Curr. Opin. Lipidol.* **2016**, *27*, 313–322. [CrossRef]
16. Adiels, M.; Borén, J.; Caslake, M.J.; Stewart, P.; Soro, A.; Westerbacka, J.; Wennberg, B.; Olofsson, S.-O.; Packard, C.; Taskinen, M.-R. Overproduction of VLDL1 driven by hyperglycemia is a dominant feature of diabetic dyslipidemia. *Arterioscler. Thromb. Vasc. Biol.* **2005**, *25*, 1697–1703. [CrossRef]
17. Adiels, M.; Westerbacka, J.; Soro-Paavonen, A.; Häkkinen, A.M.; Vehkavaara, S.; Caslake, M.J.; Packard, C.; Olofsson, S.O.; Yki-Järvinen, H.; Taskinen, M.R.; et al. Acute suppression of VLDL1 secretion rate by insulin is associated with hepatic fat content and insulin resistance. *Diabetologia* **2007**, *50*, 2356–2365. [CrossRef]
18. Fisher, E.A.; Pan, M.; Chen, X.; Wu, X.; Wang, H.; Jamil, H.; Sparks, J.D.; Williams, K.J. The triple threat to nascent apolipoprotein B. Evidence for multiple, distinct degradative pathways. *J. Biol. Chem.* **2001**, *276*, 27855–27863. [CrossRef]
19. Williams, K.J.; Fisher, E.A. Atherosclerosis: Cell biology and lipoproteins–three distinct processes that control apolipoprotein-B secretion. *Curr. Opin. Lipidol.* **2001**, *12*, 235–237. [CrossRef]
20. Shrestha, P.; van de Sluis, B.; Dullaart, R.P.F.; van den Born, J. Novel aspects of PCSK9 and lipoprotein receptors in renal disease-related dyslipidemia. *Cell Signal.* **2019**, *55*, 53–64. [CrossRef]
21. Lagrost, L. Plasma phospholipid transfer protein: A multifaceted protein with a key role in the assembly and secretion of apolipoprotein B-containing lipoproteins by the liver. *Hepatol. Baltim. Md* **2012**, *56*, 415–418. [CrossRef] [PubMed]
22. Pan, M.; Cederbaum, A.I.; Zhang, Y.-L.; Ginsberg, H.N.; Williams, K.J.; Fisher, E.A. Lipid peroxidation and oxidant stress regulate hepatic apolipoprotein B degradation and VLDL production. *J. Clin. Investig.* **2004**, *113*, 1277–1287. [CrossRef] [PubMed]
23. Klieser, E.; Mayr, C.; Kiesslich, T.; Wissniowski, T.; Fazio, P.D.; Neureiter, D.; Ocker, M. The Crosstalk of miRNA and Oxidative Stress in the Liver: From Physiology to Pathology and Clinical Implications. *Int. J. Mol. Sci.* **2019**, *20*, 5266. [CrossRef] [PubMed]
24. Levey, A.S.; Stevens, L.A.; Schmid, C.H.; Zhang, Y.L.; Castro, A.F.; Feldman, H.I.; Kusek, J.W.; Eggers, P.; Van Lente, F.; Greene, T.; et al. A new equation to estimate glomerular filtration rate. *Ann. Intern. Med.* **2009**, *150*, 604–612. [CrossRef]
25. Grundy, S.M.; Cleeman, J.I.; Daniels, S.R.; Donato, K.A.; Eckel, R.H.; Franklin, B.A.; Gordon, D.J.; Krauss, R.M.; Savage, P.J.; Smith, S.C.; et al. Diagnosis and management of the metabolic syndrome: An American Heart Association/National Heart, Lung, and Blood Institute Scientific Statement. *Circulation* **2005**, *112*, 2735–2752. [CrossRef]
26. Matthews, D.R.; Hosker, J.P.; Rudenski, A.S.; Naylor, B.A.; Treacher, D.F.; Turner, R.C. Homeostasis model assessment: Insulin resistance and beta-cell function from fasting plasma glucose and insulin concentrations in man. *Diabetologia* **1985**, *28*, 412–419. [CrossRef]
27. Jeyarajah, E.J.; Cromwell, W.C.; Otvos, J.D. Lipoprotein particle analysis by nuclear magnetic resonance spectroscopy. *Clin. Lab. Med.* **2006**, *26*, 847–870. [CrossRef]
28. Dullaart, R.P.F.; de Vries, R.; Kwakernaak, A.J.; Perton, F.; Dallinga-Thie, G.M. Increased large VLDL particles confer elevated cholesteryl ester transfer in diabetes. *Eur. J. Clin. Investig.* **2015**, *45*, 36–44. [CrossRef]
29. Riemens, S.; van Tol, A.; Sluiter, W.; Dullaart, R. Elevated plasma cholesteryl ester transfer in NIDDM: Relationships with apolipoprotein B-containing lipoproteins and phospholipid transfer protein. *Atherosclerosis* **1998**, *140*, 71–79. [CrossRef]

30. Abbasi, A.; Dallinga-Thie, G.M.; Dullaart, R.P.F. Phospholipid transfer protein activity and incident type 2 diabetes mellitus. *Clin. Chim. Acta Int. J. Clin. Chem.* **2015**, *439*, 38–41. [CrossRef]
31. Liu, M.Y.; Xydakis, A.M.; Hoogeveen, R.C.; Jones, P.H.; Smith, E.O.; Nelson, K.W.; Ballantyne, C.M. Multiplexed analysis of biomarkers related to obesity and the metabolic syndrome in human plasma, using the Luminex-100 system. *Clin. Chem.* **2005**, *51*, 1102–1109. [CrossRef] [PubMed]
32. Ellman, G.L. Tissue sulfhydryl groups. *Arch. Biochem. Biophys.* **1959**, *82*, 70–77. [CrossRef]
33. Hu, M.L.; Louie, S.; Cross, C.E.; Motchnik, P.; Halliwell, B. Antioxidant protection against hypochlorous acid in human plasma. *J. Lab. Clin. Med.* **1993**, *121*, 257–262. [PubMed]
34. Pasaoglu, H.; Sancak, B.; Bukan, N. Lipid peroxidation and resistance to oxidation in patients with type 2 diabetes mellitus. *Tohoku J. Exp. Med.* **2004**, *203*, 211–218. [CrossRef] [PubMed]
35. Dijk, P.R.V.; Waanders, F.; Logtenberg, S.J.J.; Groenier, K.H.; Vriesendorp, T.M.; Kleefstra, N.; Goor, H.V.; Bilo, H.J.G. Different routes of insulin administration do not influence serum free thiols in type 1 diabetes mellitus. *Endocrinol. Diabetes Metab.* **2019**, *2*, e00088. [CrossRef]
36. Lin, C.; Yin, M. Effects of cysteine-containing compounds on biosynthesis of triacylglycerol and cholesterol and anti-oxidative protection in liver from mice consuming a high-fat diet. *Br. J. Nutr.* **2008**, *99*, 37–43. [CrossRef]
37. Lin, C.; Yin, M.; Hsu, C.; Lin, M. Effect of five cysteine-containing compounds on three lipogenic enzymes in Balb/cA mice consuming a high saturated fat diet. *Lipids* **2004**, *39*, 843–848. [CrossRef]
38. Lewis, G.F.; Uffelman, K.D.; Szeto, L.W.; Weller, B.; Steiner, G. Interaction between free fatty acids and insulin in the acute control of very low density lipoprotein production in humans. *J. Clin. Investig.* **1995**, *95*, 158–166. [CrossRef]
39. Jiang, X.C.; Qin, S.; Qiao, C.; Kawano, K.; Lin, M.; Skold, A.; Xiao, X.; Tall, A.R. Apolipoprotein B secretion and atherosclerosis are decreased in mice with phospholipid-transfer protein deficiency. *Nat. Med.* **2001**, *7*, 847–852. [CrossRef]
40. Jiang, X.-C.; Li, Z.; Liu, R.; Yang, X.P.; Pan, M.; Lagrost, L.; Fisher, E.A.; Williams, K.J. Phospholipid transfer protein deficiency impairs apolipoprotein-B secretion from hepatocytes by stimulating a proteolytic pathway through a relative deficiency of vitamin E and an increase in intracellular oxidants. *J. Biol. Chem.* **2005**, *280*, 18336–18340. [CrossRef]
41. Yazdanyar, A.; Jiang, X.-C. Liver phospholipid transfer protein (PLTP) expression with a PLTP-null background promotes very low-density lipoprotein production in mice. *Hepatol. Baltim. Md.* **2012**, *56*, 576–584. [CrossRef] [PubMed]
42. de Vries, R.; Dallinga-Thie, G.M.; Smit, A.J.; Wolffenbuttel, B.H.R.; van Tol, A.; Dullaart, R.P.F. Elevated plasma phospholipid transfer protein activity is a determinant of carotid intima-media thickness in type 2 diabetes mellitus. *Diabetologia* **2006**, *49*, 398–404. [CrossRef] [PubMed]
43. Riemens, S.C.; van Tol, A.; Sluiter, W.J.; Dullaart, R.P. Plasma phospholipid transfer protein activity is lowered by 24-h insulin and acipimox administration: Blunted response to insulin in type 2 diabetic patients. *Diabetes* **1999**, *48*, 1631–1637. [CrossRef] [PubMed]
44. Tzotzas, T.; Desrumaux, C.; Lagrost, L. Plasma phospholipid transfer protein (PLTP): Review of an emerging cardiometabolic risk factor. *Obes. Rev. Off. J. Int. Assoc. Study Obes.* **2009**, *10*, 403–411. [CrossRef] [PubMed]
45. Nass, K.J.; van den Berg, E.H.; Gruppen, E.G.; Dullaart, R.P.F. Plasma lecithin: cholesterol acyltransferase and phospholipid transfer protein activity independently associate with nonalcoholic fatty liver disease. *Eur. J. Clin. Investig.* **2018**, *48*, e12988. [CrossRef]
46. Li, N.; van der Sijde, M.R.; LifeLines Cohort Study Group; Bakker, S.J.L.; Dullaart, R.P.F.; van der Harst, P.; Gansevoort, R.T.; Elbers, C.C.; Wijmenga, C.; Snieder, H.; et al. Pleiotropic effects of lipid genes on plasma glucose, HbA1c, and HOMA-IR levels. *Diabetes* **2014**, *63*, 3149–3158. [CrossRef]
47. Finelli, C.; Tarantino, G. What is the role of adiponectin in obesity related non-alcoholic fatty liver disease? *World J. Gastroenterol.* **2013**, *19*, 802–812. [CrossRef]
48. Dullaart, R.P.F.; van den Berg, E.H.; van der Klauw, M.M.; Blokzijl, H. Low normal thyroid function attenuates serum alanine aminotransferase elevations in the context of metabolic syndrome and insulin resistance in white people. *Clin. Biochem.* **2014**, *47*, 1028–1032. [CrossRef]

49. Borggreve, S.E.; Hillege, H.L.; Wolffenbuttel, B.H.R.; de Jong, P.E.; Bakker, S.J.L.; van der Steege, G.; van Tol, A.; Dullaart, R.P.F. PREVEND Study Group The effect of cholesteryl ester transfer protein -629C->A promoter polymorphism on high-density lipoprotein cholesterol is dependent on serum triglycerides. *J. Clin. Endocrinol. Metab.* **2005**, *90*, 4198–4204. [CrossRef]
50. Bourgonje, A.R.; Gabriëls, R.Y.; de Borst, M.H.; Bulthuis, M.L.C.; Faber, K.N.; van Goor, H.; Dijkstra, G. Serum Free Thiols Are Superior to Fecal Calprotectin in Reflecting Endoscopic Disease Activity in Inflammatory Bowel Disease. *Antioxidants* **2019**, *8*, 351. [CrossRef]
51. Turell, L.; Radi, R.; Alvarez, B. The thiol pool in human plasma: The central contribution of albumin to redox processes. *Free Radic. Biol. Med.* **2013**, *65*, 244–253. [CrossRef] [PubMed]
52. Atkuri, K.R.; Mantovani, J.J.; Herzenberg, L.A.; Herzenberg, L.A. N-Acetylcysteine–a safe antidote for cysteine/glutathione deficiency. *Curr. Opin. Pharmacol.* **2007**, *7*, 355–359. [CrossRef] [PubMed]
53. Dallinga-Thie, G.M.; van Tol, A.; Hattori, H.; Rensen, P.C.N.; Sijbrands, E.J.G. Plasma phospholipid transfer protein activity is decreased in type 2 diabetes during treatment with atorvastatin: A role for apolipoprotein E? *Diabetes* **2006**, *55*, 1491–1496. [CrossRef] [PubMed]
54. Cholesterol Treatment Trialists' (CTT) Collaborators; Kearney, P.M.; Blackwell, L.; Collins, R.; Keech, A.; Simes, J.; Peto, R.; Armitage, J.; Baigent, C. Efficacy of cholesterol-lowering therapy in 18,686 people with diabetes in 14 randomised trials of statins: A meta-analysis. *Lancet Lond. Engl.* **2008**, *371*, 117–125.
55. Polidori, M.C.; Mecocci, P.; Stahl, W.; Sies, H. Cigarette smoking cessation increases plasma levels of several antioxidant micronutrients and improves resistance towards oxidative challenge. *Br. J. Nutr.* **2003**, *90*, 147–150. [CrossRef]
56. Dullaart, R.P.; Hoogenberg, K.; Dikkeschei, B.D.; van Tol, A. Higher plasma lipid transfer protein activities and unfavorable lipoprotein changes in cigarette-smoking men. *Arterioscler. Thromb. J. Vasc. Biol.* **1994**, *14*, 1581–1585. [CrossRef]

© 2019 by the authors. Licensee MDPI, Basel, Switzerland. This article is an open access article distributed under the terms and conditions of the Creative Commons Attribution (CC BY) license (http://creativecommons.org/licenses/by/4.0/).

MDPI
St. Alban-Anlage 66
4052 Basel
Switzerland
Tel. +41 61 683 77 34
Fax +41 61 302 89 18
www.mdpi.com

Journal of Clinical Medicine Editorial Office
E-mail: jcm@mdpi.com
www.mdpi.com/journal/jcm

www.ingramcontent.com/pod-product-compliance
Lightning Source LLC
LaVergne TN
LVHW071942080526
838202LV00064B/6656